Application of Group Theory in Chemistry

for **BSc (Hons), MSc and MPhil (Chemistry)**

Applications to

- Atomic Orbitals
- Valence Bond Theory
- Crystal Field Theory
- Inorganic Chemistry
- Organic Chemistry
- Electronic Spectroscopy—UV and Visible
- Vibrational Spectroscopy

Application of Group Theory in Chemistry

for **BSc (Hons), MSc and MPhil (Chemistry)**

Ranajit Kumar Roy

Former Associate Professor
Department of Chemistry
Ramakrishna Mission Vivekananda Centenary College
Rahara, Kolkata

CBS

CBS Publishers & Distributors Pvt Ltd

New Delhi • Bengaluru • Chennai • Kochi • Kolkata • Mumbai
Bhopal • Bhubaneswar • Hyderabad • Jharkhand • Nagpur • Patna • Pune
Uttarakhand • Dhaka (Bangladesh) • Kathmandu (Nepal)

Application of Group Theory in Chemistry

ISBN: 978-93-88725-65-1

Copyright © Author and Publisher

First Edition: 2020

Published by Satish Kumar Jain and produced by Varun Jain for

CBS Publishers & Distributors Pvt Ltd
4819/XI Prahlad Street, 24 Ansari Road, Daryaganj, New Delhi 110 002, India.
Ph: 23289259, 23266861, 23266867   Fax: 011-23243014   Website: www.cbspd.com
e-mail: delhi@cbspd.com; cbspubs@airtelmail.in.
Corporate Office: 204 FIE, Industrial Area, Patparganj, Delhi 110 092
Ph: 4934 4934   Fax: 4934 4935   e-mail: publishing@cbspd.com; publicity@cbspd.com

Branches

- **Bengaluru:** Seema House 2975, 17th Cross, K.R. Road,
  Banasankari 2nd Stage, Bengaluru 560 070, Karnataka
  Ph: +91-80-26771678/79   Fax: +91-80-26771680   e-mail: bangalore@cbspd.com
- **Chennai:** 7, Subbaraya Street, Shenoy Nagar, Chennai 600 030, Tamil Nadu
  Ph: +91-44-26680620, 26681266   Fax: +91-44-42032115   e-mail: chennai@cbspd.com
- **Kochi:** 42/1325, 1326, Power House Road, Opp KSEB Power House,
  Ernakulam 682 018, Kochi, Kerala
  Ph: +91-484-4059061-65   Fax: +91-484-4059065   e-mail: kochi@cbspd.com
- **Kolkata:** 6/B, Ground Floor, Rameswar Shaw Road, Kolkata-700 014, West Bengal
  Ph: +91-33-22891126, 22891127, 22891128   e-mail: kolkata@cbspd.com
- **Mumbai:** 83-C, Dr E Moses Road, Worli, Mumbai-400018, Maharashtra
  Ph: +91-22-24902340/41   Fax: +91-22-24902342   e-mail: mumbai@cbspd.com

Representatives

• Bhopal	0-8319310552	• Bhubaneswar	0-9911037372	• Hyderabad	0-9885175004
• Jharkhand	0-9811541605	• Nagpur	0-9421945513	• Patna	0-9334159340
• Pune	0-9623451994	• Uttarakhand	0-9716462459	• Dhaka (Bangladesh)	01912-003485
• Kathmandu (Nepal)	977-9818742655				

Printed at: Glorious Printers, Daryaganj, Delhi, India

to
_____

my mother Late Lila Ray

my father Late Kshitish Chandra Ray

my eldest brother Late Dilip Kumar Ray

my teacher Shri Gyan Chandra Das

my philosopher and guide Prof BN Mukherjee

Preface

For understanding chemistry, it is absolutely necessary to have a clear picture on the structure and configuration of the molecule. The use of spectroscopic techniques for structural elucidation has now become very common in chemistry and it is well known that group theory is closely related to spectroscopy. Thus, the modern chemist must have a clear knowledge of group theory and its applications in chemistry. Unfortunately, group theory is dreaded by students as a complicated mathematical concept. Teaching this subject in postgraduate classes for several years, I have gained the impression that the actual coverage of group theory prescribed by different Indian universities is very scattered.

Chemical Applications of Group Theory has been established from my lectures that I have given for several years to students of chemistry as well as teachers at different institutes in West Bengal. There are many books on group theory written by eminent authors are available in the market, however, hardly any book is designed for examination purposes.

In this book, the first four chapters embody the basic concept of group theory. Only the minimum mathematical part, necessary to use as a tool for understanding chemical problems, has been discussed. The latter chapters are devoted to the applications of group theory to atomic and molecular structures and spectroscopy. Each concept has been explained in detail with numerous illustrations, elaborate explanations, problems and their solutions, study questions, etc. for better understanding.

The book will serve as a core text for BSc (Hons), MSc, MPhil students in chemistry and researchers as well.

It is inevitable in a book of this size and complexity that there will be occasional errors, these are mine alone, and I will endeavor to correct them in future editions. Suggestions and criticism from learned professors and inquisitive students towards the improvement of the book will be thankfully received and gratefully acknowledged. I will find my attempt amply rewarded, if the book meets the recommendations of the teachers and the expectations of the taught. I wish the readers a pleasant journey through this book and believe that it will inspire them to explore deeper into the exciting field of group theory.

Ranajit Kumar Roy

Acknowledgement

For the preparation of this book, I have received commendable support and co-operation from my teacher Prof RL Dutta, Department of Chemistry, Burdwan University; Prof GB Kauffman, California State University, USA; Prof GS Sanyal, Kalyani University; Prof A Mishra, Vidyasagar University; Prof R Banerjee, Jadavpur University and Dr Debasis Banerjee, Sri Chaitanya College, Habra. I have frequently discussed various questions starting from elementary to advanced level with my friend, Dr Asoke Kumar Banerjee, Ex-Professor, Department of Chemistry, Surendranath College, Kolkata, I tender by best regards to him. My sincere thanks are also due to my students, past and present, who offered much valuable criticism and advice. I heartily thank them all.

I like to thank my friend, Dr Durgadas Mukherjee, Mahadevananda Mahavidyalaya, Barrackpore; Dr Gurucharan Mukhopadhyay, Bidhannagar Govt College, and one of my former students, Shri Kushal Chakraborty, MSc, for inspiring me to do this write-up for the benefit of young learners.

I am thankful to CBS Publishers & Distributors. I would like to put on record the sincere efforts of Mr Satish K Jain, CMD, CBS Publishers & Distributors, Mr YN Arjuna, Senior Vice President Publishing, Editorial and Publicity and Mr CB Bhattacharjee, for bringing out the book in the present form.

My family has contributed much to this book. They have tolerated too much over the years when I was totally aloof from their social works. Their continuous support, inspiration, devotion, encouragement, sacrifice and understanding, inspired me very much. I am really grateful to them all.

Ranajit Kumar Roy

Contents

Bibliography

Appendices

Molecular Symmetry and the Symmetry Groups

1

1.1 SYMMETRY—AN INTRODUCTION

The term symmetry implies a structure in which the parts of the structure are in harmony with each other, as well as to the whole structure, i.e. the structure is proportional and well balanced. For example, a flower, a butterfly or a human body looks symmetrical around a point on an imaginary axis on a plane and this particular symmetry gives us a sense of beauty (Fig. 1.1).

(a)

(b)

(c)

Fig. 1.1 The shapes and patterns of some pleasing designs found in nature (a) the flower of the black-eyed Susan, *Rudbeckia hirta*; (b) a monarch butterfly, *Danaus plexippus*; (c) a red eft, *Notophthalmus viridescens*

The term symmetry has become synonymous with beauty because nature has made most of its creations symmetrical. In nature, many types of flowers and plants, snowflakes, insects, certain fruits and vegetables and animals exhibit characteristic symmetry. Many engineering achievements such as Lotus Temple of Delhi, the Taj Mahal at Agra, the Pyramids of ancient Egypt, and the Eiffel Tower of Paris, have a degree of symmetry that contributes to their esthetic appeal (Fig. 1.2 a–j).

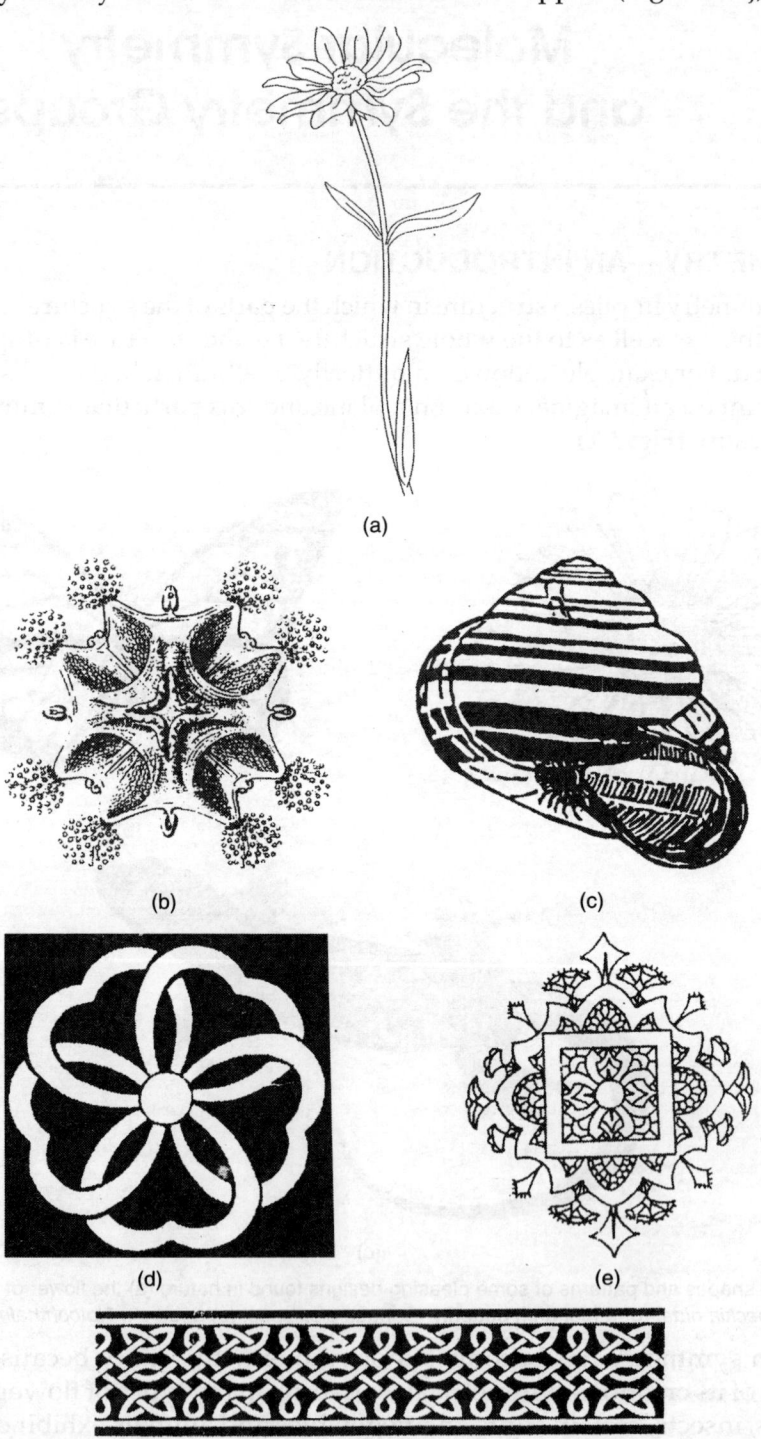

(a)

(b)

(c)

(d)

(e)

(f)

Fig. 1.2 Symmetry in nature (a, b, c), art (d, e, f) and architecture (g, h, i, j)

Scientifically, an object is said to be in symmetry, if it can take up more than one equivalent (indistinguishable) orientations in space. Two possible orientations for H_2 molecule is illustrated in Fig. 1.3, only by labeling the two equivalent hydrogen atoms with prime and double prime marks. Here, the two hydrogen atoms are indistinguishable, the two orientations are equivalent, and the molecule has symmetry.

The two orientations I and II in Fig. 1.3 can be obtained by rotation of the molecule through 180° about an axis through the centre of and perpendicular to the H—H bond axis. Orientation II can not be distinguished from orientation I, i.e. I and II are equivalent orientations.

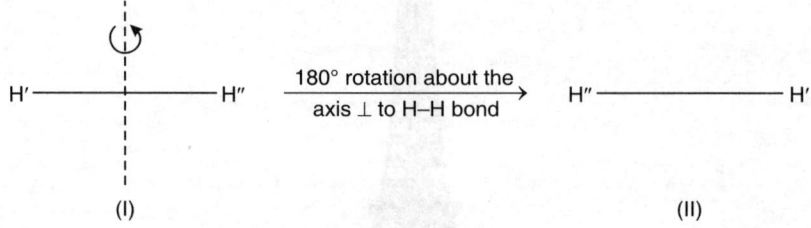

Fig. 1.3 Equivalent orientations for H_2

Symmetry is present in geometrical figures, crystalline solids and molecules. Let us consider two geometrical figures, a rectangle and a square. The square is said to be more symmetrical than the rectangle due to the fact that all the sides of a square are equal whereas in a rectangle opposite sides are equal.

Symmetry in language is known. Language is a medium of communication. It is either prose or poetry. Each of the poetic line ends with a word of repeating phonetic sounds, implying a particular pattern or symmetry. The symmetry in prose is reflected in the usage of palindromes*, which exists in most of the languages. Palindromes in languages occur in the form of either words or sentences. Some examples are as follows.

Palindromes in words: Level, tool, refer, madam, etc.

Palindromes in sentences: Pull up; No melon; no lemon; live not to Nevil; Lewd did I live, evil I did devil.

Symmetry in music is also known. Whether it is vocal, verbal or instrumental, it involves some notes which bring in harmony on symmetry of sounds that sensitize the emotions.

Symmetry concept is extremely useful in chemistry. By analysing the symmetry of molecules, we can predict infrared spectra, describe the type of orbitals used in bending, predict optical activity, interpret electronic spectra, and study a number of additional molecular properties. Theoretically, chemists generally deal with bending of the molecule on spectra of the molecule, by quantum mechanics in which Schrödinger equation is framed and than it is solved. This needs advanced knowledge on physics and mathematics. In such cases, symmetry properties are very much helpful for solving chemical, physical as well as spectral properties of the studied compound. With the application of symmetry, the problem of quantum chemistry can be solved with less difficulty.

1.2 SYMMETRY ELEMENTS AND SYMMETRY OPERATIONS

Symmetry of a molecule can be defined in terms of *symmetry elements* and *symmetry operations. A symmetry element is an imaginary line, plane or point in a molecule (or in any object) about which one or more symmetry operations are carried out.* Operations here mean movement (rotation, reflection, inversion, etc.) about a symmetry element.

---

* A palindrome is a word on a sentence which is invariant to being read backwards as forwards.

A symmetry operation moves a molecule about an axis, a point, or a plane (the symmetry element) into a position indistinguishable from the original position. For example, let us consider H_2 molecule. If it is rotated about an imaginary line passing through the centre of the covalent bond (as shown by dotted line in Fig. 1.4) by an angle of 180°, we will get equivalent configuration, but if it is further rotated about the same imaginary line by 180°, and the identical configuration. Here H' and H" are indistinguishable hydrogen atoms. For convenience, we have written them as H' and H" in order to understand the movement of H-atoms on rotation.

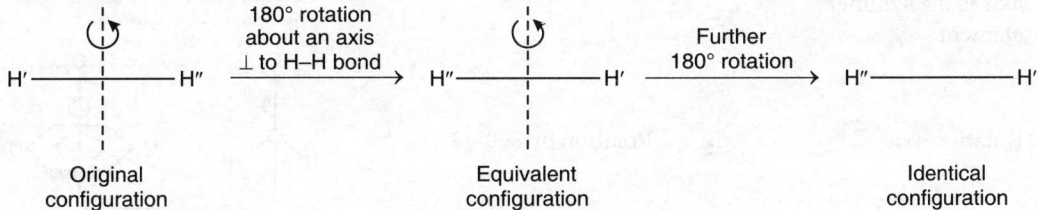

Original configuration   Equivalent configuration   Identical configuration

Fig. 1.4 Equivalent and identical orientations for H_2

Symmetry operation is thus the process carried out on the molecule, which brings it from the original orientation to another equivalent orientation. In Fig. 1.4, the rotation of H_2 molecule by 180° is a symmetry operation. A rotation by an angle less then 180° will leave the molecule in an orientation different (distinguishable) from the original one. Hence, it is not a symmetry operation. Thus, a symmetry operation is genuine only if the two orientations before and after the operation look exactly alike.

The symmetry operations should be carried out within the molecule and not out side the molecule, since operation out side the molecule will lead to translation motion of the molecule, which is not a part of our discussion. In symmetry operations, if there is a point in the molecule that remains unchanged under all of the symmetry operations, the resultant symmetry is referred to as *point symmetry*.

In all, there are five different symmetry elements.

   i. The axis of symmetry (also called proper or principal axis of symmetry or *n* fold axis of symmetry) C_n.

  ii. Plane of symmetry (or mirror plane) σ

 iii. Centre of symmetry (or inversion centre) i

 iv. The rotation–reflection axis (also called improper rotation axis) S_n

  v. The identity E

The symmetry operations based on the above symmetry elements are:

   i. Rotation operation $\hat{C}_n$ about an axis (a line) is also called proper rotation operation.

  ii. Reflection operation $\hat{\sigma}$: reflection of all atoms through the inversion centre.

 iii. Rotation–reflection operation $\hat{S}_n$: Rotation about an axis followed by reflection of all atoms through a plane which is perpendicular to the rotation axis.

To distinguish a symmetry operation from symmetry element, a circumflex (cap) is written over the symbol of the symmetry element. The symmetry elements and their corresponding symmetry operations are summarised in Table 1.1.

Table 1.1 A summary of symmetry elements and symmetry operations

Symmetry element		Symmetry operation		Example
Descriptions	**Symbol**	**Descriptions**	**Symbol**	
1. Proper axis of C_n symmetry or *n*-fold axis (*n* order) of symmetry or rotational axis of symmetry. So rotational axis is the symmetry element.	C_n	Rotational once or several times about an axis by an angle $\left(\theta = \dfrac{360°}{n} \right)$	$\hat{C}_n$	
Rotation axis	C_2	Rotation by 360°/2	$\hat{C}_2$	C_2 H_2O C_2 *p*-dichlorobenzene
	C_3	Rotation by 360°/3	$\hat{C}_3$	C_3 NH_3
	C_4	Rotation by 360°/4	$\hat{C}_4$	C_4 $[PCl_4]^{2-}$
	C_5	Rotation by 360°/5	$\hat{C}_5$	C_5 Cyclopentadienyl group
	C_6	Rotation by 360°/6	$\hat{C}_6$	C_6 Benzene
2. Plane of symmetry (mirror plane is the symmetry element)	σ	Reflection through a mirror plane	$\hat{\sigma}$	C_2, $\sigma_{vertical}$ H_2O C_2, $\sigma_{vertical}$ H_2O C_2, $\sigma_{vertical}$, $\sigma_{vertical}$ H_2O

Contd...

Symmetry element		Symmetry Operation		Example
Descriptions	**Symbol**	**Descriptions**	**Symbol**	
3. Centre of symmetry So the inversion centre (point) is the symmetry.	i	Inversion of all atoms through the centre.	$\hat{i}$	Ethane Ferrocene (staggered) The center of symmetry 1, 2-dimethyl 1, 2-di-phenyldipho disulfide
4. Improper axis of symmetry. So the symmetry element is rotation reflection axis.	S_n (n order)	Rotation about an axis followed by reflection in a plane perpendicular to the rotation axis.	$\hat{S}_n$	
Rotation-reflection	S_4	Rotation by $360°/4$, followed by reflection in the plane perpendicular to the rotation axis.	$\hat{S}_4$	CH_4
	S_6		$\hat{S}_6$	Ethane (staggered)
	S_{10}		$\hat{S}_{10}$	Ferrocene (staggered)
5. Identity element, i.e. symmetry element is none.	E	Identity operation leaves the molecule unchanged, i.e. all atoms unshifted.	$\hat{E}$	CHFClBr All molecules

The following are the symmetry elements and associated symmetry operations possible.

1.2.1 The Axis of Symmetry (or Proper/Principal Axis of Symmetry or *n*-fold Axis of Symmetry) C_n

It is an imaginary line passing through the molecule, around which the molecule can be rotated clockwise to take the nuclear framework from our position to another indistinguishable position, i.e. the rotation around this axis takes the molecule from one orientation to another equivalent orientation. This is represented by C_n, n being the order of the axis, i.e. the number of terms the equivalent orientation arises on rotation through 360° on the axis. For C_n, successive n rotations (each time by 360°/n) give n equivalent orientations for the molecule. After the end of n terms operations, it will generate the original orientation of the molecule.

In H_2O molecule, the axis C_2 passing through **O** atom and in between two **H** atoms has order 2. If we now make a C_2 rotation (180°) we get an indistinguishable orientation where the two H atoms are interchanged. A second C_2 rotation brings back all atoms in their initial positions (Fig. 1.5). Rotation by 180° is thus a symmetry operation. The axis about which the rotation takes place is the symmetry element.

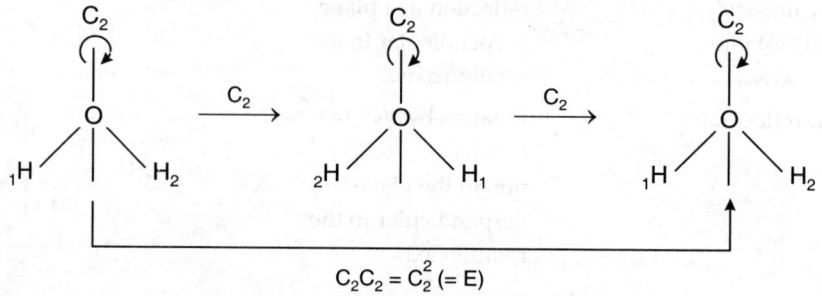

$$C_2 C_2 = C_2^2 \ (= E)$$

Fig. 1.5 The operations C_2 and C_2^2 on H_2O molecule

Successive operations are denoted by C_n (i.e. rotation by 360°), C_n^2 (i.e. rotation by $2 \times 360°/n$), C_n^n (i.e. rotation by $n \times 360°/n = 360°$). For $n = 1, 2, 3, ...$ indicates rotation by 360° (360°/1), 180° (360°/2), 120° (360°/3), etc. respectively around the axis in a symmetrical operation and the corresponding symmetry elements are C_1, C_2, C_3, ... respectively.

Symmetry element	Symmetry Operation
C_1	rotation by 360°
C_2	rotation by 180°
C_3	rotation by 120°
C_n	rotation by 360°/n

Note that $C_n^n = n \times \dfrac{360°}{n}$ *. So $C_4^2 = 2 \times \dfrac{360°}{4} = 180°$

---

* In C_n^n, the subscript n represents order of axis of rotation and the superscript n represents number of times the rotation operation is carried out.

In general, the notation C_n^m represents n-fold axis of rotation and the angle through which the molecule is rotated clockwise is given by $m \times \dfrac{360°}{n}$.

But $C_2 = \dfrac{360°}{2} = 180°$. Thus $C_4^2 \equiv C_2$

For C_6 axis, the symmetry operations may be expressed as

$$C_6^1, C_6^2 (\equiv C_3), C_6^3 (\equiv C_2), C_6^4 (\equiv C_3^2), C_6^5, C_6^6 (\equiv E)$$

Trigonal planer BF_3 molecule possesses the C_3 axis of rotation. The plane of the molecule is the plane of the paper. The rotation axis C_3 passes through B and perpendicular to the plane of the molecule, i.e. the paper. The three fluorine atoms have been marked F_1, F_2 and F_3 respectively, just for our examination though such marking has no physical reality. The first C_3 rotation will give equivalent orientation marked II. The second C_3 rotation followed by third C_3 rotation will give the equivalent orientations marked III and IV (Fig. 1.6). It is to be noted that orientation (IV) is the original starting orientation. Thus rotation of BF_3 successively thrice about the C_3 axis bring the molecule to its original position symbolically, it is written as $C_3^3 = E$ [$\because$ $C_n^n = E$]. C_3 rotation axis has the order 3 as three rotations provide the starting orientation.

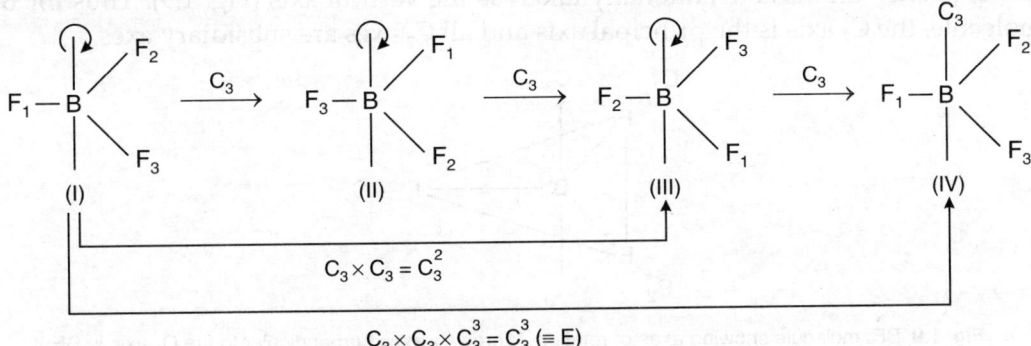

Fig. 1.6 The three C_3 operations in succession on BF_3

Ammonia (NH_3), a trigonal pyramidal molecule, possesses the C_3 axis of rotation which passes through N atom and the centroid of the triangle formed by the three H atoms. C_3 axis is perpendicular to the plane of the paper. The molecule has to be rotated by 120° to get equivalent orientation (Fig. 1.7).

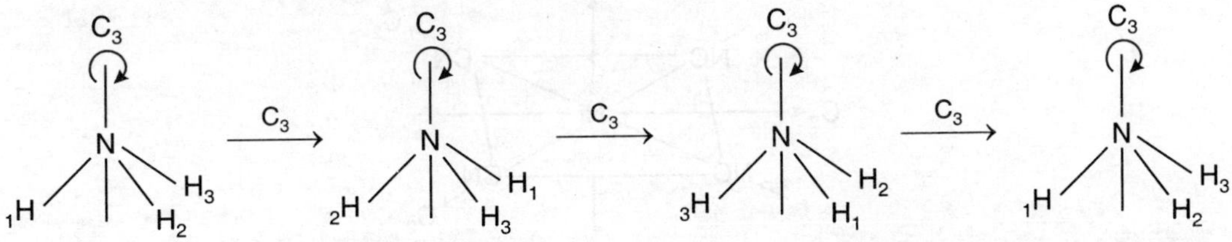

Fig. 1.7 C_3 axis and its symmetry operations in NH_3 molecule

Thus the order of the axis (n) = 360°/angle of rotation, necessary for getting equivalent orientation.

In square planar $[Ni (CN)_4]^{2-}$, the axis passing through Ni atom and perpendicular to the plane is an axis of four fold symmetry (C_4). The angle of rotation is 90°. In benzene molecule, the axis passing through the centre of the hexagonal ring and perpendicular to the plane of the ring is an axis of six fold symmetry (C_6). The angle of rotation is 60° (Fig. 1.8).

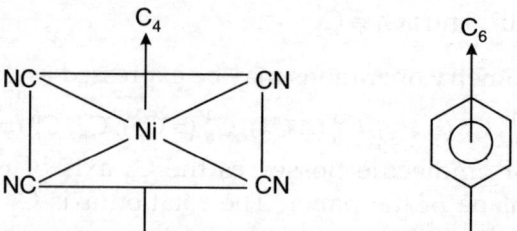

Fig. 1.8. C_4 axis for [Ni $(CN)_4]^{2-}$ and C_6 axis for C_6H_6

A molecule may possess more than one rotation axis through H_2O and NH_3 molecules have only one axis of symmetry. In case of BF_3, $[Ni(CN)_4]^{2-}$ and C_6H_6 there are more than one axis. In BF_3 molecule besides the three fold axis of symmetry (C_3), there are 3 two fold axes (C_2) each containing a B–F bond. These three C_2 axes are in the plane of the molecule. The highest fold axis of rotation is usually referred to as the *principal axis of rotation* and is conventionally taken as the vertical axis (Fig. 1.9). Thus for BF_3 molecule, the C_3 axis is the principal axis and all C_2 axes are subsidiary axes.

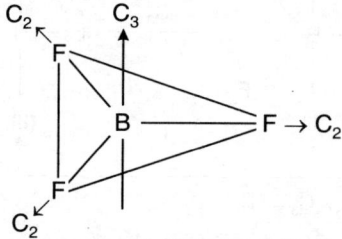

Fig. 1.9 BF_3 molecule showing axes of rotation. [Three C_2 axes perpendicular to the C_3 axis in BF_3]

In $[Ni (CN)_4]^{2-}$, besides the four fold symmetry axis (C_4), there are four two fold symmetry axes (C_2), two C_2 axes passing through Ni atom and our pair of the trans-cyanides, and other two C_2 axes passing Ni and centre of two opposite edges of the square plane (Fig. 1.10). These are in the plane of the molecule.

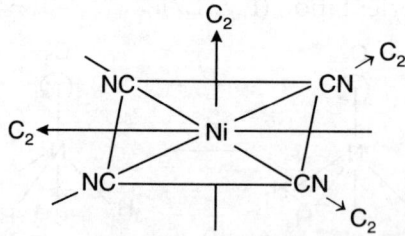

Fig. 1.10 [Ni $(CN)_4]^{2-}$ anion showing axes of rotation, four C_2 axes perpendicular to the C_4 axis in [Ni $(CN)_4]^{2-}$

In benzene, besides the six fold axis of symmetry, there are six two fold axes of symmetry in the plane of the molecule (Fig. 1.11).

Methane (CH_4) has four three fold rotation axes (four C_3) and three two fold rotation axes (three C_2). Each C_3 passing through one of the hydrogen atoms and carbon atom. The four C_3 axes coincide with the C—H bonds (Fig. 1.12(a)). On the other hand, each C_2 bisecting the bond angles between the two pairs of bonds pointing opposite to each other. These axes also pass through the centre of two opposite faces of the cube in which tetrahedron structure of CH_4 is drawn (Fig. 1.12(b)).

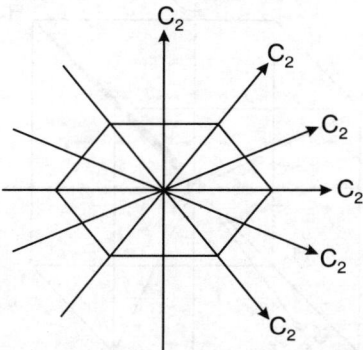

Fig. 1.11 Six C_2 axes (at the molecular plane) of C_6H_6 C_6 axis perpendicular to the molecular plane passes through the centre of the molecule (C_6 not shown)

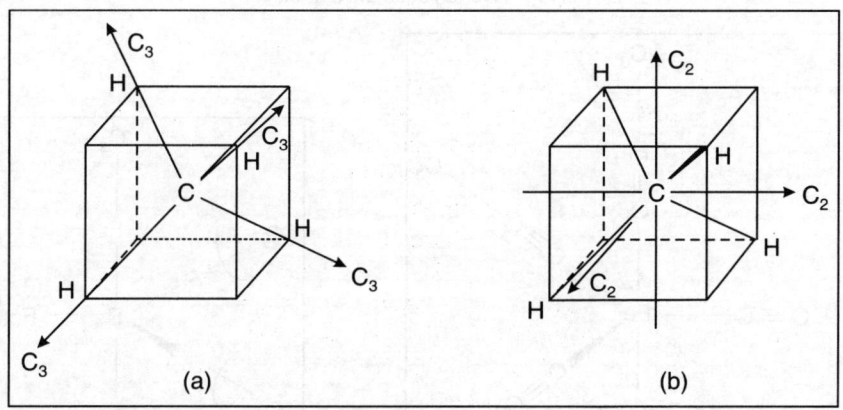

Fig. 1.12 A tetrahedron drawn within a cube and depiction of (a) C_3 axes and (b) C_2 axes in methane

Like CH_4, a tetrahedral NH_4^+ ion has four C_3 rotation axes, each one passing through the N atom and one H atom. Rotation by 120° (360°/3) about the C_3 axes will give as many as three equivalent orientations. This ion has a total of three C_2 axes perpendicular to each other, one C_2 axes bisects the line between H_1 and H_2 and pases through N. The C_2 rotation (by 180°) will give an equivalent orientation with H_1 and H_2 interchanging their positions as also the H_3 and H_4 atoms. The second C_2 rotation axis will bisect the line H_1 and H_3 and pass through the N atom. The third C_2 rotation axis will bisect the line joining H_1 and H_4 and will pass through the N-atom (Fig. 1.13).

Iron pentacarbonyl $Fe(CO)_5$, a trigonal bipyramidal compound, has 1 three fold axis (C_3) and 3 two fold axes (three C_2). Each C_2 axis passes through each of the carbonyl groups in the triangular plane that is perpendicular to the C_3 axis (Fig. 1.14).

Another trigonal bipyramidal compound PF_5, has one C_3 axis passing through the F_{axial}—P—F_{axial}, linear segment and three C_2 axes each passing through one P—$F_{equatorial}$. These C_2 axes are perpendicular to the C_3 axis (Fig. 1.15).

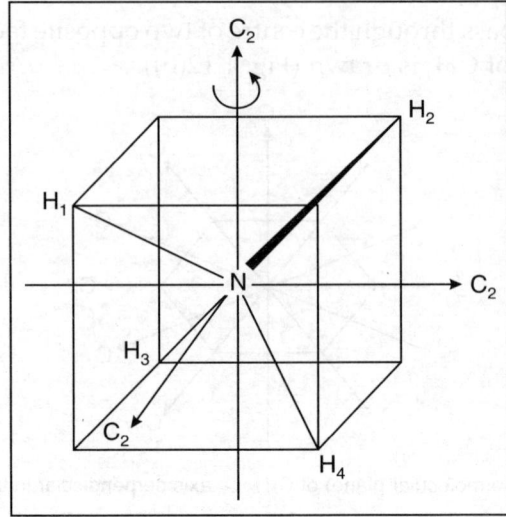

Fig. 1.13 C_2 axes of NH_4^+ ion

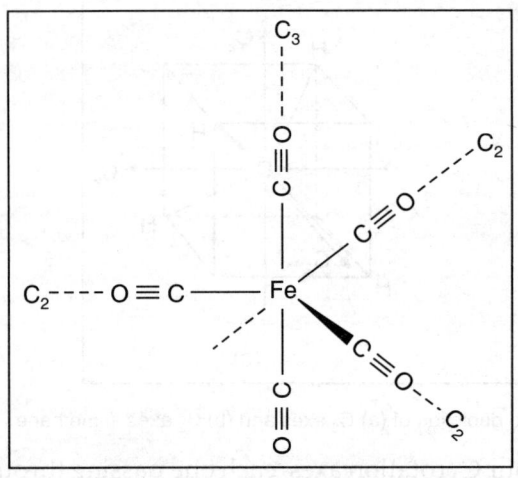

Fig. 1.14

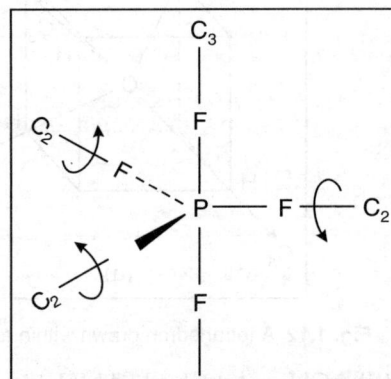

Fig. 1.15

Tungsten hexacarbonyl, $W(CO)_6$ has three C_4 axes one running from top to bottom, the second one running from left to right and the remaining one running from front to back (Fig. 1.16).

Continuing with examples of proper axes in typical molecules, we may cite the planar $[PtCl_4]^{2-}$ ion, which has a C_4 axis perpendicular to the plane of the ion and four C_2 axes in the plane of the ion (Fig. 1.17).

Cyclopentadienyl anion, $C_5H_5^-$ possesses a C_5 axis perpendicular to the molecular plane and five C_2 axes in the molecular plane (Fig. 1.18). Benzene possesses a C_6 axis perpendicular to the plane of the molecule and it passes through the centre of the molecule. There are six C_2 axes in the plane of the molecule (Fig. 1.18).

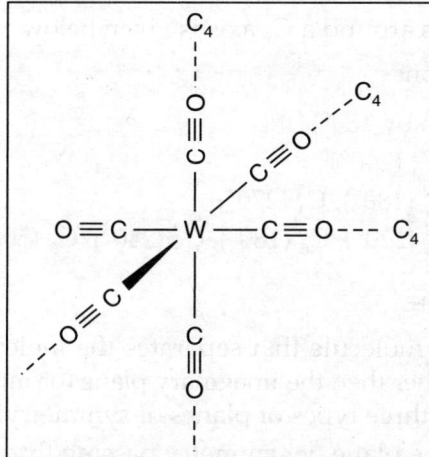

Fig. 1.16 Tungsten hexacarbonyl molecule

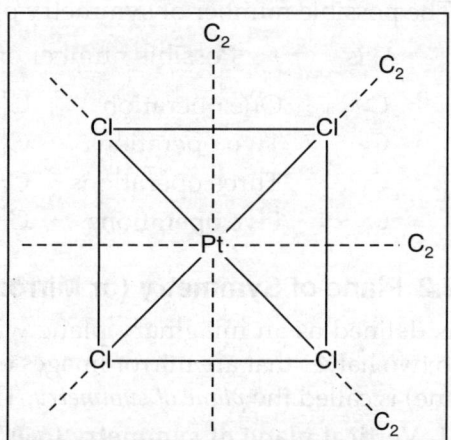

Fig. 1.17 C_4 and C_2 axes in square planar $[PtCl_4]^{2-}$ ion. Four C_2 axes shown in the molecular plane and one C_4 axis (not shown) perpendicular to the molecular plane and passing through Pt

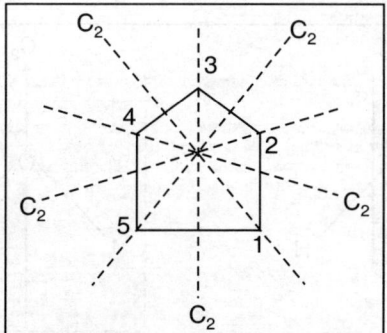

Fig. 1.18 Five C_2 axes shown in the molecular plane ($C_5H_5^-$) and one C_5 axis (not shown) perpendicular to the molecular plane

In case, where there are more than one axis of same order, the axis passing through maximum number of atoms is called the *principal axis*. In allene molecule ($CH_2=C=CH_2$) (Fig. 1.19), the molecular axis passing through the three carbon atoms is an axis of two fold symmetry (C_2). Two other axes, (C_2' and C_2'') perpendicular to the molecular axis and passing through the centre carbon atom, are also axes of two fold symmetry. The molecular axis is considered to be the principal axis.

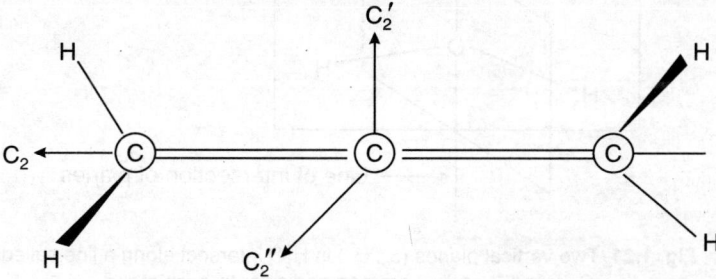

Fig. 1.19 Allene molecule with three C_2 axes (C_2, C_2', C_2'')

The possible number of symmetry operations around a C_n axis is given below.

Axis	Possible number of operations	
C_2	One operation	C_2^1 (rotation by 180°)
C_3	Two operations	C_3^1 (120°), C_3^2 (240°)
C_4	Three operations	C_4^1 (90°), C_4^2 (180°), C_4^3 (270°)
C_6	Five operations	C_6^1 (60°), C_6^2 (120°), C_6^3 (180°), C_6^4 (240°), C_6^5 (300°)

1.2.2 Plane of Symmetry (or Mirror Plane) σ

It is defined as an imaginary plane within the molecule that separates the molecule into two halves that are mirror images of each other then the imaginary plane (or mirror plane) is called the *plane of symmetry*. There are three types of planes of symmetry:

1. Vertical plane of symmetry (σ_v): This is the plane of symmetry passing through the principal axis (C_n) and one of the subsidiary axes (if present). In other words, it is a plane containing the principal axis of symmetry. This is represented as σ_v (Fig. 1.20). Two vertical planes σ_v and σ_v' in H_2O molecule intersect along a line called C_2 axis. σ_v and σ_v' are perpendicular to each other (Fig. 1.21).

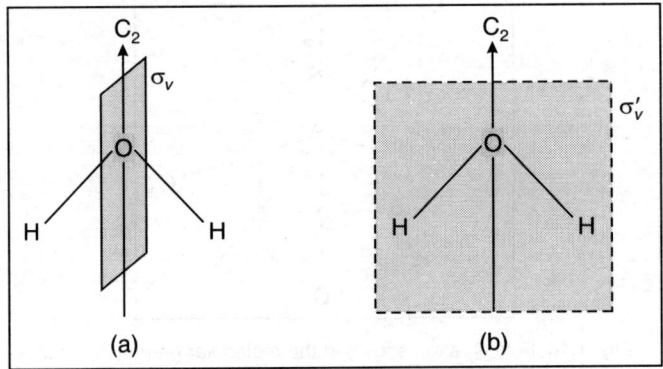

(a) (b)

Fig. 1.20 Reflection planes in water molecule

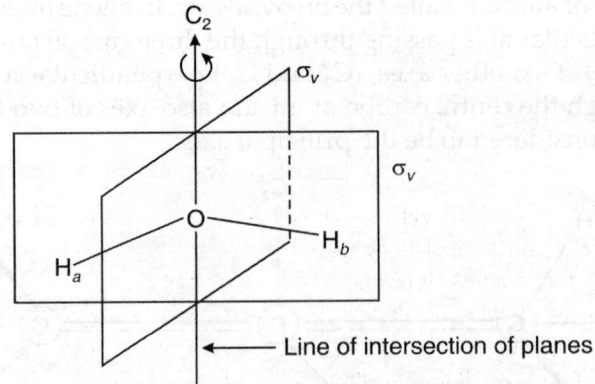

← Line of intersection of planes

Fig. 1.21 Two vertical planes (σ_v, σ_v') in H_2O intersect along a line called C_2 axis. [σ_v and σ_v' are perpendicular to each other]

2. Horizontal plane of symmetry (σ_h): Any plane perpendicular to the principal axis (C_n) is called *horizontal plane*. This is represented as σ_h (Fig. 1.22). Nonplanar molecule PCl_5 contains both σ_h and $3\sigma_h$ planes (Fig. 1.23). Only one of the three vertical planes (σ_v) is shown in Fig. 1.23.

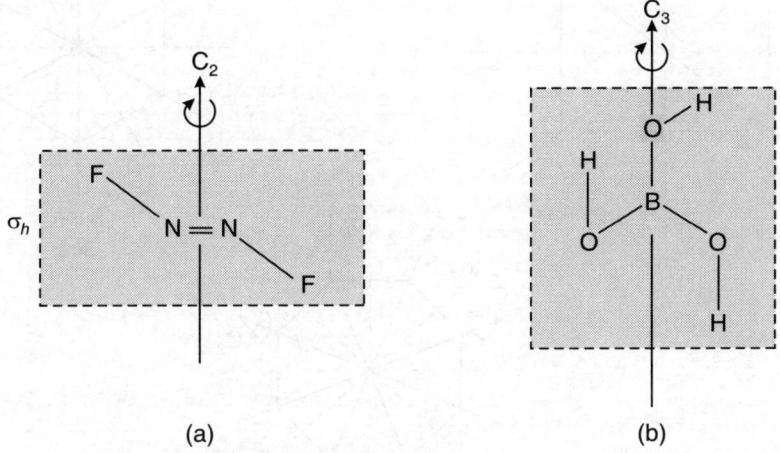

(a)   (b)

Fig. 1.22 (a) C_2 is perpendicular to σ_h plane (b) C_3 is perpendicular to σ_h plane

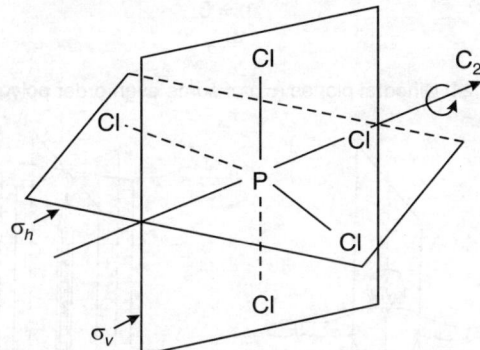

Fig. 1.23 σ_h and one σ_v planes in PCl_5 molecule

3. Dihedral plane of symmetry (σ_d): This is the plane passing through the principal axis (C_n) and also bisecting the angle between two successive subsidiary axes (C_2 axes). It is represented as σ_d (Fig. 1.24).

The human body has a mirror plane that runs through the skull, nose, spinal cord and between the legs (Fig. 1.25(a)). This plane divides the body into right-half and the left-half. Linear objects such as round wood pencil (Fig. 1.25(b)) or linear molecules such as acetylene (H–C≡C–H) or hydrogen cyanide (H–C≡N) (Fig. 1.25(c)) or carbon-dioxide (O=C=O) have infinite number of mirror planes that the center line of the object.

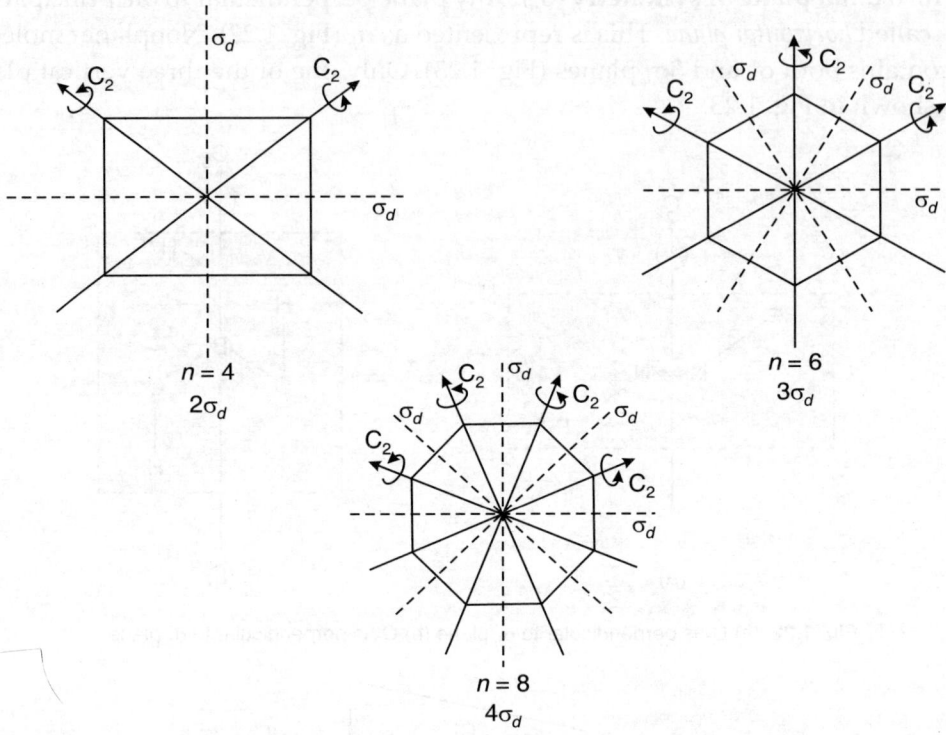

Fig. 1.24 Dihedral planes (σ_d) of some even order polygons

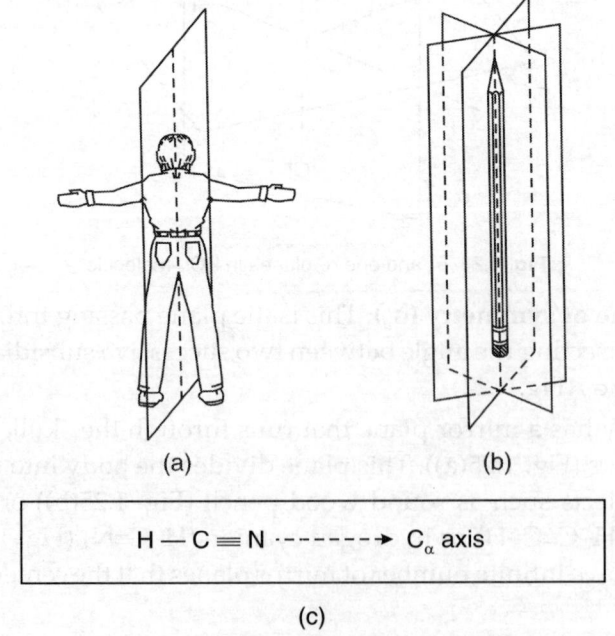

(a)   (b)

H — C ≡ N ------→ C_α axis

(c)

Fig. 1.25 C_α axis of symmetry in HCN molecule

BF$_3$, a trigonal planar molecule, has three vertical planes σ_v, each passing through B atom and one of the F atoms and principal axis C$_3$. Besides three vertical planes (σ_v) there is another plane (molecular plane) σ_h. σ_h plane contains all the four atoms. Reflection of the molecule through this plane provides an indistinguishable orientation. This σ_h plane is perpendicular to the highest fold rotation axis (C$_3$) (Fig. 1.26).

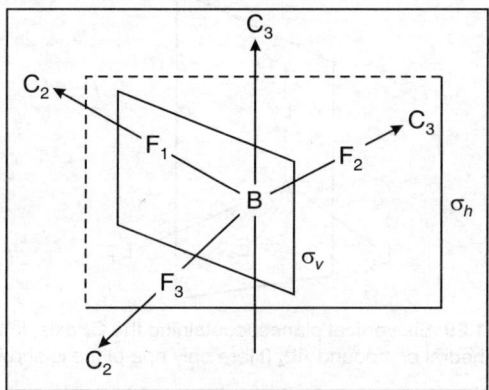

Fig. 1.26 Mirror planes in BF$_3$. [Only one σ_v and one σ_h shown]

Borazine (BoN$_3$H$_6$) has four mirror planes: three are vertical types (σ_v) and the fourth is the plane of the ring (Fig. 1.27).

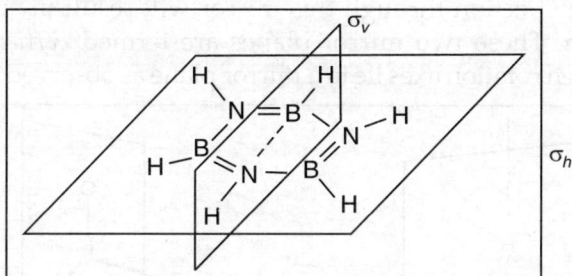

Fig. 1.27 One mirror plane σ_v shown in borazene and the other is the plane (σ_h) of the ring

Square planar [Ni (CN)$_4$]$^{2-}$ ion contains a molecular plane and four vertical planes. The molecular plane passing through nickel and four cyanides is σ_h. Each vertical plane (σ_v) passes through the C$_4$ axis and hence through the central Ni(II) ion and one of the subsidiary axes (C$_2$). Two σ_v planes pass through the two cyanides (CN) at opposite corners and the other two σ_v pass through the centres of opposite edges, i.e. between two CN (Fig. 1.28).

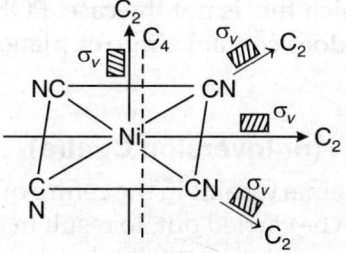

Fig. 1.28 Reflection planes in [Ni (CN)$_4$]$^{2-}$

Tetrahedral compound AL_4 (e.g. CH_4, Si F_4, NH_4^+, PH_4^+, etc). Contains six planes of symmetry. In each plane A atom and two L atoms lie. Other two L atoms lie in different planes. If L atoms are numbered as 1, 2, 3 and 4 the six planes containing one A and two L atoms are as follows (Fig. 1.29).

$$AL_1L_2, AL_1L_3, AL_1L_4, AL_2L_3, AL_2L_4, AL_3L_4$$

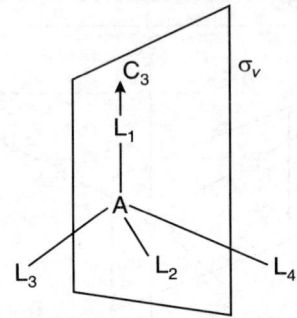

Fig. 1.29 Six vertical planes, containing the C_3 axis, in the tetrahedral compound AL_4 [Here only one plane is shown]

Water molecule contains two symmetry planes: One plane passing through **O** and bisecting the $\widehat{HOH}$ angle [Fig. 1.20(a)]. This plane (σ_v) is perpendicular to the molecular plane (σ_h). The two **H** atoms are interchanged on reflection through this vertical mirror plane. The other plane passing through all the three atoms, i.e. through the molecular plane (Fig. 1.20(b)). Reflection through this mirror will result in an equivalent (rather identical) orientation. These two mirror planes are termed *vertical mirror planes* and labelled σ_v and σ'_v, often rotation axes lie in a mirror plane as observed in BCl_3 [Fig. 1.30(a)].

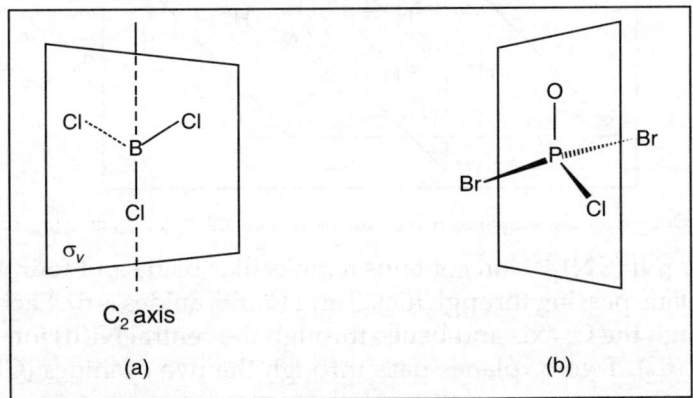

Fig. 1.30 (a) A mirror plane in BCl_3 (b) The mirror plane in $POBr_2Cl$

But there are examples in which this is not the case. $POBr_2Cl$, a tetrahedral molecule, contains no rotation axis but does contain a mirror plane. P, Cl and O atoms lie in the mirror plane (σ) [Fig. 1.30(b)].

1.2.3 Centre of Symmetry *i* (or Inversion Centre)

Centre of symmetry is an imaginary point in the centre of the molecule, through which the reflection of each atom can be carried out, to result in its coincidence with an equivalent atom. In other words, if any atom in the molecule is connected with the centre of symmetry and extended equally on the other side, it meets another equivalent atom. For example: C_2H_4, N_2O_2, CO_2, H_2, square planar $[Ni(CN)_4]^{2-}$, octahedral $[CO(NH_3)_6]^{3+}$, 1, 2-dimethyl 1, 2-diphenyl phosphine disulphide (Fig. 1.31).

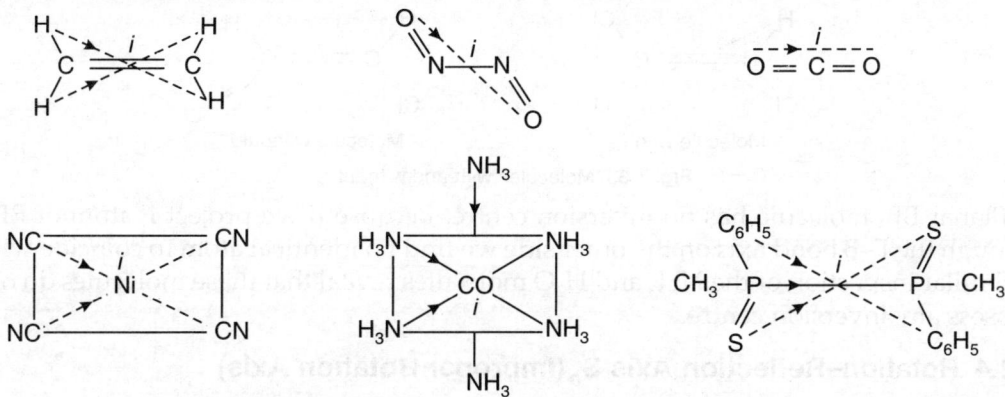

Fig. 1.31 The centre of symmetry of C_2H_4, N_2O_2, CO_2, $[Ni(CN)_4]^{2-}$, $[CO(NH_3)_6]^{3+}$, 1, 2-dimethyl 1, 2-diphenyl phosphine disulphide

The process of reflection through the centre of symmetry leading to an equivalent orientation is a symmetry operation. As a result of this operation the molecule gets completely inverted to an equivalent orientation. Hence, the operation is termed inversion (i) and the centre of symmetry is called inversion centre. Only one inversion operation is possible. Second inversion gives back the original orientation. Therefore, we may write $i.i = i_2 = E$ (identity).

In case of H_2 molecule, we see that there exists a point exactly at the mid point of two H–atoms about which inversion operation is carried out and an indistinguishable (equivalent) configuration is obtained. This point is the *centre of symmetry* (i).

$$H' - H'' \xrightarrow{\text{Inversion}} H'' - H'$$

But if we take HCl then it is easy to point out that there exists no centre of symmetry because after inversion we get distinguishable configuration of HCl molecule (Fig. 1.32).

Many molecules that seem at first glance to have an inversion centre but they do not; for example, methane and other tetrahedral molecules lack inversion symmetry. In methane the two H atoms are in the vertical plane on the right and two H atoms in the horizontal plane on the left as in Fig. 1.32. Inversion results in H atoms in the horizontal plane on the right and two H atoms in the vertical plane on the left. Inversion is therefore not a symmetry operation of methane, since the orientation of the molecule following i operation differs from the original orientation (Fig. 1.32).

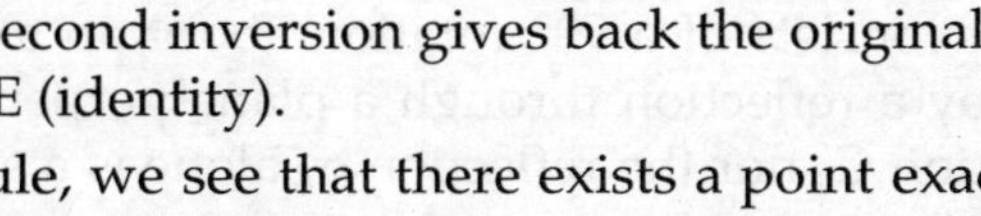

$$H - Cl \xrightarrow{\text{Inversion}} Cl - H$$

Fig. 1.32 Molecule without i

Presence of atoms in pairs does not definitely ensure the presence of centre of symmetry. For example: trans-dichloroethylene has a centre of symmetry, but its dichloroethylene has no centre of symmetry though the atoms are in pairs (Fig. 1.33).

Fig. 1.33 Molecules with and without *i*

Planar BF_3 molecule has no inversion centre. Because if we project F atom of BF_3^- through the F–B bond axis on the other side we find no identical atom to coincide with it. Similar inspection of the NH_3 and H_2O molecules reveal that these molecules do not possess any inversion centre.

1.2.4 Rotation–Reflection Axis S_n (Improper Rotation Axis)

A rotation–reflection operation (sometimes called *improper rotation*) requires rotation of $360°/n$, followed by reflection through a plane perpendicular to the axis of rotation $(\sigma\, C_n)^*$ on the two actions performed is the reverse order, i.e. first reflection and then rotation $(C_n\, \sigma)^{**}$. The rotation–reflection is equivalent to reflection–rotation. The symbol is S_n, where the subscript n indicates rotation (clockwise by our convention) through $2\pi/n$.

$$S_n = C_n\, \sigma$$

Figure 1.34 shows a fourfold improper rotation of a tetrahedral CH_4 molecule, in this case, the operation consists of a 90° rotation about an axis bisecting two HCH bond angles followed by a reflection through a plane perpendicular to the rotation axis. Neither the operation C_4 nor the reflection σ_h alone is a symmetry operation for CH_4, but their product $C_4 \times \sigma_h$ is a symmetry operation, the improper rotation S_4. The symmetry element, the improper rotation axis S_n (S_4 in the example), is the corresponding combination of an n-fold rotational axis and a perpendicular mirror plane.

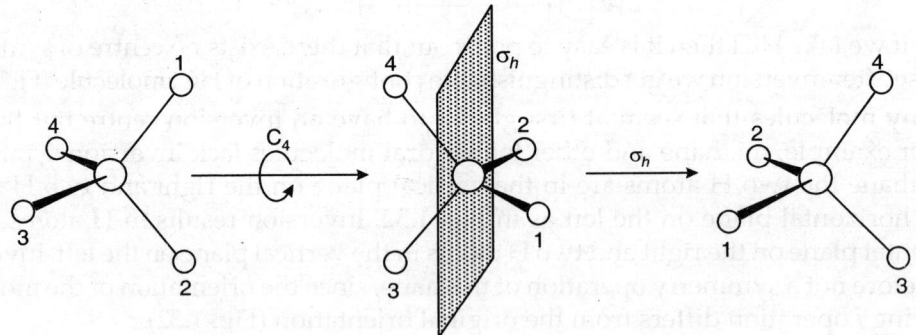

Fig. 1.34 A fourfold axis of improper rotation S_4 in the CH_4 molecule

An S_1 operation means rotation through 360° (360°/1) followed by reflection in a horizontal plane. This amounts to reflection in a horizontal plane, i.e. S_1 and σ mean the same action (Fig. 1.35(a)) ($S_1 \equiv \sigma$). Similarly, S_2 means rotation by 180° (360°/2) being followed by reflection in the horizontal plane. Therefore, S_2 is simply as inversion operation i (Fig. 1.35(b)).

---

* In case of $\sigma\, C_n$, first rotation and their reflection operations are to be performed.

** In carrying out $C_n\sigma$, first reflection and their rotation operations are done.

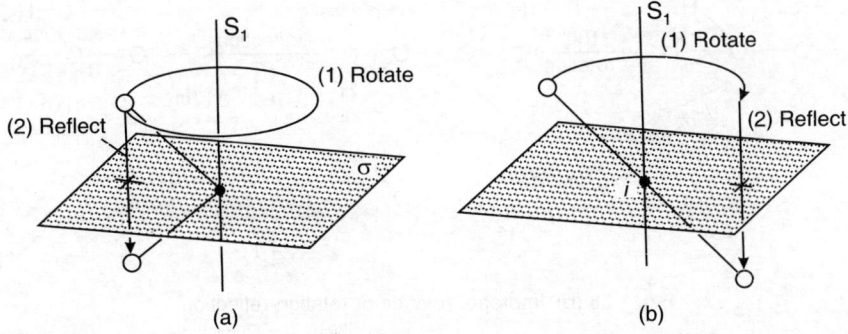

Fig. 1.35 (a) An S_1 axis is equivalent to a mirror plane and (b) an S_2 axis is equivalent to a center of inversion

Thus we represent the operation $S_2 \equiv i$ and $S_1 \equiv \sigma_v$ by using matrices shown below:

For a $C_2(z)$ operation (C_2 rotation about z), the transformation matrix is

$$\begin{pmatrix} -1 & 0 & 0 \\ 0 & -1 & 0 \\ 0 & 0 & 1 \end{pmatrix}$$

For $\sigma_v\,(xy)$, it is

$$\begin{pmatrix} 1 & 0 & 0 \\ 0 & 1 & 0 \\ 0 & 0 & -1 \end{pmatrix}$$

and for inversion (i) it is

$$\begin{pmatrix} -1 & 0 & 0 \\ 0 & -1 & 0 \\ 0 & 0 & -1 \end{pmatrix}$$

$$S_2 = C_2\sigma = \underbrace{\begin{pmatrix} -1 & 0 & 0 \\ 0 & -1 & 0 \\ 0 & 0 & 1 \end{pmatrix}}_{C_2} \underbrace{\begin{pmatrix} 1 & 0 & 0 \\ 0 & 1 & 0 \\ 0 & 0 & -1 \end{pmatrix}}_{\sigma(xy)} = \begin{pmatrix} -1 & 0 & 0 \\ 0 & -1 & 0 \\ 0 & 0 & -1 \end{pmatrix} = i$$

$$S_1 = C_1\sigma = \underbrace{\begin{pmatrix} 1 & 0 & 0 \\ 0 & 1 & 0 \\ 0 & 0 & 1 \end{pmatrix}}_{C_2} \underbrace{\begin{pmatrix} 1 & 0 & 0 \\ 0 & 1 & 0 \\ 0 & 0 & -1 \end{pmatrix}}_{\sigma_v(xy)} = \begin{pmatrix} 1 & 0 & 0 \\ 0 & 1 & 0 \\ 0 & 0 & -1 \end{pmatrix} = \sigma_v(xy)$$

In the transform of H_2O_2, the molecular axis is S_2 and $S_2 = C_2 + \sigma_h$ (Fig. 1.36(a))

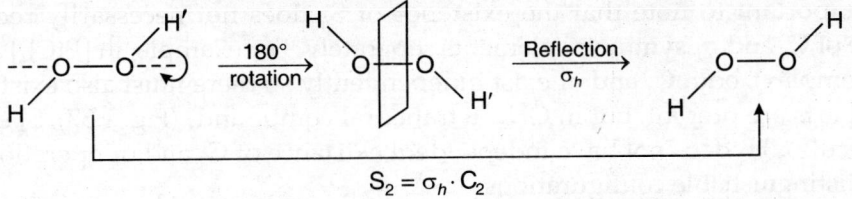

$$S_2 = \sigma_h \cdot C_2$$

Fig. 1.36(a) Improper rotation or rotation reflection

Note that S_2^1 operation is equivalent to inversion operation, i.e. $S_2 \equiv i$ and hence S_2^1 can not be considered as a new operation. S_2^2 gives the original orientation (Fig. 1.36(b)).

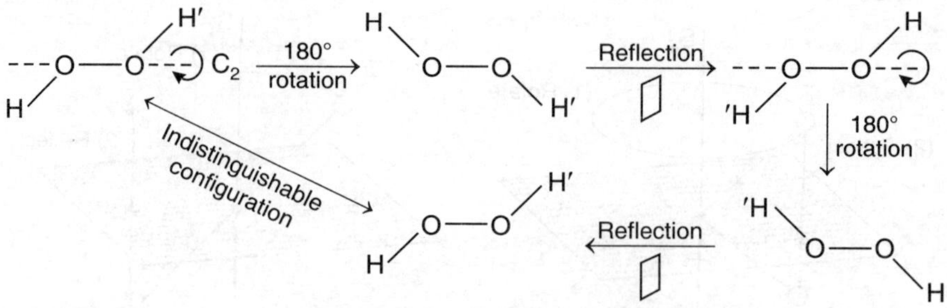

Fig. 1.36 (b) Improper rotation or rotation-reflection

In CH_4 molecule a line through the carbon and bisecting the angle between two hydrogen atoms on each side is an S_4 axis. 90° rotation of CH_4 molecule followed by reflection through a plane perpendicular to the axis of rotation is shown in Fig. 1.37. Two S_4 operations is succession generate a C_2 operation.

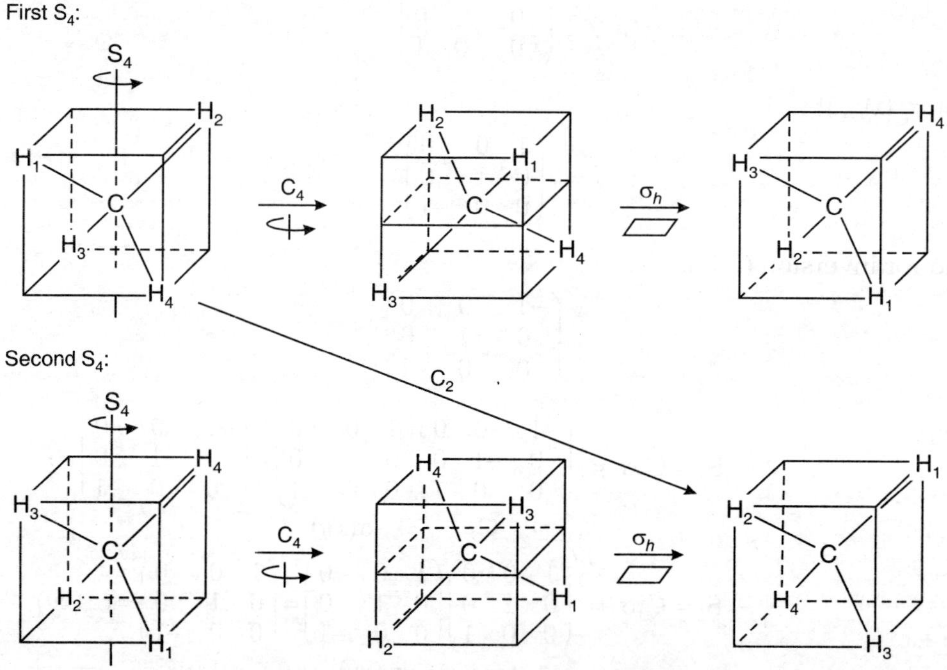

First S_4:

Second S_4:

Fig. 1.37 Improper rotation of rotation-reflection

It is important to note that the existence of S_n does not necessarily require the presence of C_n and σ_h symmetry elements separately. For example, in $[PtCl_4]^{2-}$ (square planar complex), both C_4 and σ_h exist independently so there must also exist S_4, since its both parts are present, but in CH_4 (tetrahedral compound) (Fig. 1.37), S_4 exists but the molecule CH_4 does not have independent existence of C_4 and σ_h operations result in an indistinguishable configurations.

We shall consider another example, in which improper rotation axis S_n exists when neither C_n nor the mirror plane of perpendicular to it exists separately. In the staggered form of ethane (Fig. 1.38(a)) the C–C bond defines a C_3 axis, but there is no perpendicular mirror plane. But if we rotate the molecule 60° $(2\pi/6)$ and then it reflects through a plane perpendicular to the C–C bond, we get an equivalent configuration. Consequently, an S_6 exists and, clearly there is no C_6 (Fig. 1.38(b)).

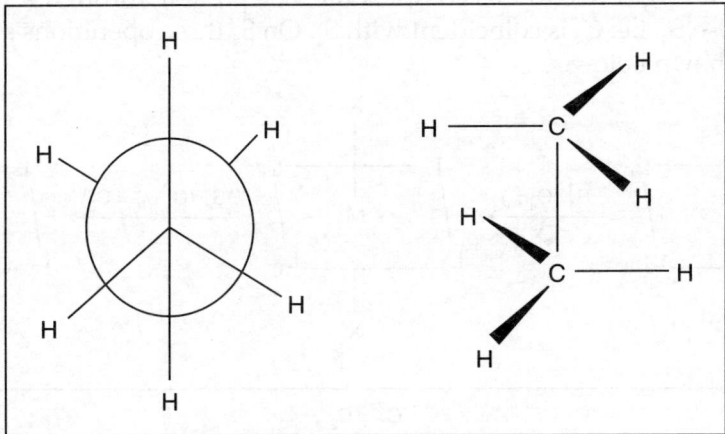

Fig. 1.38 (a) Staggered form of ethane

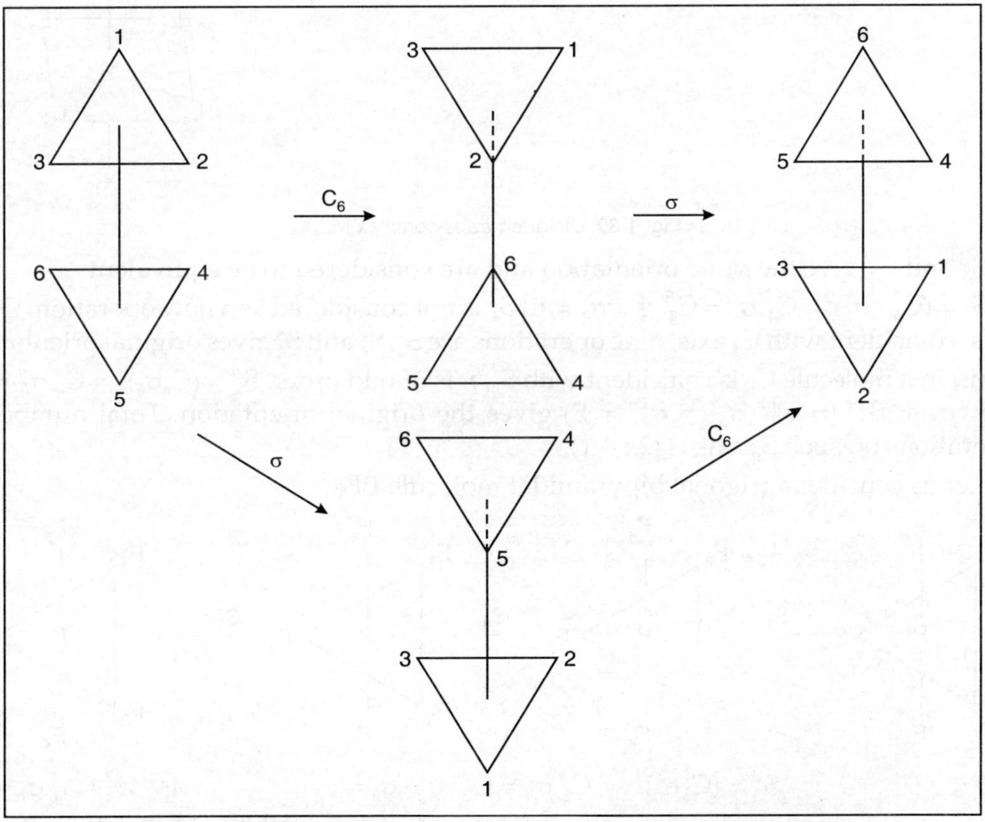

Fig. 1.38 (b) Improper rotation axis in the staggered form of ethane

In a molecule, an axis of improper rotation may be coincident with an axis of proper rotation. Following cases may arise:

(i) In a molecule a C_n (axis of symmetry) can be coincident with S_n (rotation-reflection axis) of the same order (n, being even). Let us consider a distorted octahedral *trans*-complex ML_4X_2 (Fig. 1.39). In this complex the axis passing through X-M-X is C_4. The same axis is also S_4, i.e. C_4 is coincident with S_4. On S_4 three operations are possible S_4^1, S_4^2 and S_4^3 as shown below.

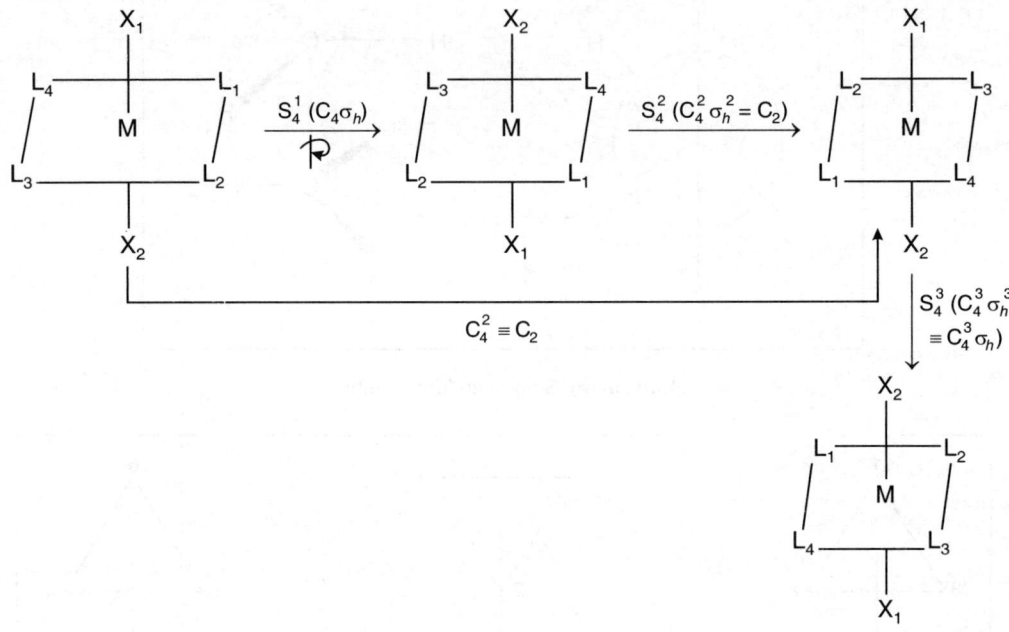

Fig. 1.39 Distorted *trans*-complex ML_4X_2

S_4^2 and C_4^2 give the same orientation and are considered to be equivalent.

$S_4^2 = (C_4^1 \cdot \sigma_h)^2 = C_4^2 \sigma_h^2 = C_4^2$ [$\because \sigma_h^2 = E$] S_4^2 is not considered as a new operation. On S_4 axis, coincident with C_4 axis, new operations are S_4^1, S_4^3 and S_4^4 gives original orientation.

(ii) In a molecule C_n is coincident with S_n, n is of odd order. $S_n^n = (C_n \sigma_h)^n = C_n^n \sigma_h^n = \sigma_h$. However, S_n^{2n} ($= \sigma_h^{2n} \cdot \sigma_h^{2n} = \sigma_h^{2n} = E$) gives the original orientation. Total number of operations on such S_n axis is $(2n - 1)$.

Let us consider a trigonal bipyramidal molecule PF_5.

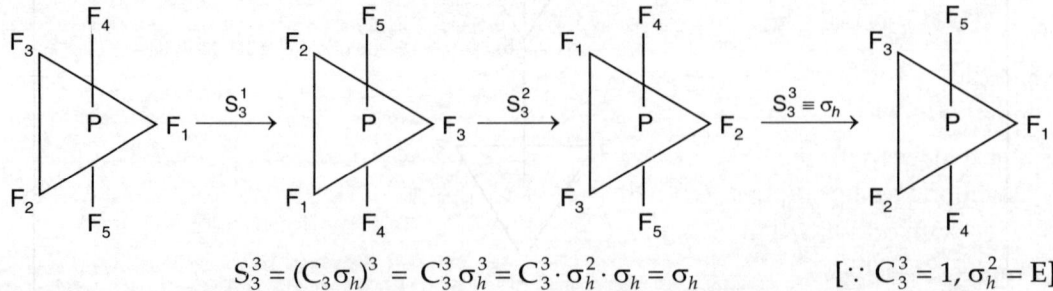

$$S_3^3 = (C_3 \sigma_h)^3 = C_3^3 \sigma_h^3 = C_3^3 \cdot \sigma_h^2 \cdot \sigma_h = \sigma_h \qquad [\because C_3^3 = 1, \sigma_h^2 = E]$$

Original orientation can be achieved on performing $(S_3^3)^v$, i.e. S_3^6. Total number of possible operations on S_3 are $S_3^1, S_3^2, S_3^3, S_3^4, S_3^5$, i.e. $2n - 1 = 2 \cdot 3 - 1 = 5$

$$S_3^1 = C_3^1 \sigma_h = 120° + \sigma_h, \text{new operation}$$

$$S_3^2 = (C_3^1 \sigma_h)^2 = C_3^2 \sigma_h^2 = C_3^2 = 240°$$

$$S_3^3 = (C_3^1 \sigma_h)^3 = C_3^3 = \sigma_h^3 = C_3^3 \sigma_h = 360° + \sigma_h = \sigma_h$$

$$S_3^4 = (C_3^1 \sigma_h)^4 = 480°, \text{i.e.} 120° (= C_3^1)$$

$$S_3^5 = (C_3^1 \sigma_h)^5 = 240° + \sigma_h, \text{new operation}$$

Thus there are only two new operations S_3^1 and S_3^5.

Figure 1.37 shows that each S_4 axis is coincident with a C_2 axis, and that there is no C_4 axis. In fact, when n is even, an S_n axis always has a $C_n/2$ axis coincident with it.

We shall consider some differences in improper rotation axes of even and odd order. With n even, S_n^m generates the set $S_n^1, S_n^2, S_n^3, ..., S_n^n$.

$$S_n^1 = C_n^1 \sigma$$

$$S_n^2 = C_n^2 \sigma^2$$

$$S_n^3 = C_n^3 \sigma^3$$

$$S_n^n = C_n^n \sigma^n$$

We have the relation $\quad \sigma_m^m = \sigma$ when m is odd

and $\quad\quad\quad\quad\quad\quad \sigma^m = E$ when m is even

Thus $\quad\quad S_n^m = C_n^m \sigma^m = C_n^m E \ (m \text{ is even}) = C_n^m$

It is to be noted that the existence of an S_n axis of even order always requires a $C_{n/2}$, since

$$S_n^2 = C_n^2 \sigma^2 = C_n^2 E = C_n^2 = C_{n/2}$$

Let us consider the set $S_6, S_6^2, S_6^3, S_6^4, S_6^5, S_6^6$ by carrying out the operations described below on the staggered ethane example.

S_6 can not be written any other way.

$$S_6^2 = C_6^2 \sigma^2 = C_6^2 E = C_6^2 = C_3$$

$$S_6^3 = S_2 = C_2 \sigma = i \quad \text{(Fig. 1.34(b))}$$

$$S_6^4 = C_6^4 \sigma^4 = C_3^2$$

S_6^5 can not be written any other way

$$S_6^6 = E$$

The complete set would normally be written

$$S_6 \quad C_3 \quad i \quad C_3^2 \quad S_6^5 \quad E$$

When n is odd, S_n^m generates the set $S_n, S_n^2, S_n^3, S_n^4, ..., S_n^{2n}$. An odd-order S_n requires that C_n and a σ perpendicular to it exist. The operation S_n^n (n is odd) is equivalent to $C_n^n \sigma^n = C_n^n \sigma = \sigma$

For example, consider S_3. First one would reflect, producing a configuration that is equivalent to the starting configuration (because of the existence of σ in the molecule). Then one rotates by C_3 to give the configuration corresponding to S_3^1.

S_n with odd n will generate $2n$ operations. By carrying out the operations with a BF_3 molecule (Fig. 1.40) we can have $S_3^1, S_3^2, S_3^3, S_3^4, S_3^5, S_3^6$ operations possible on S_3 which is collinear with C_3 and passing through the centre of boron atom and perpendicular to the molecular plane (plane of the paper).

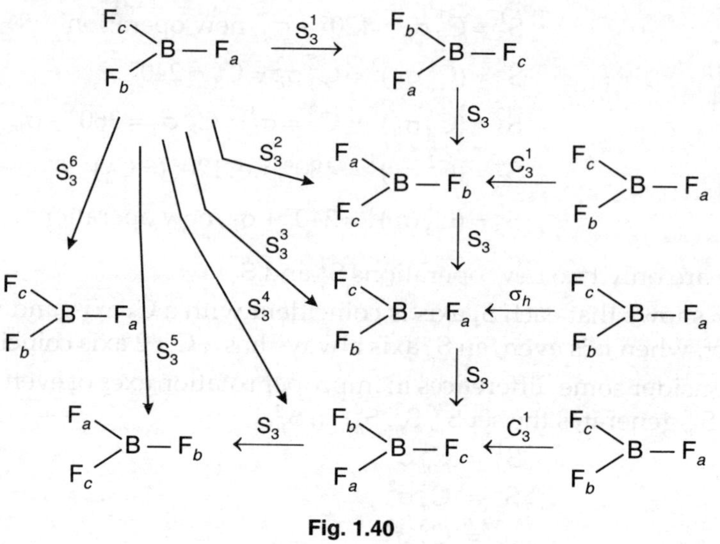

$$S_3^1 \xrightarrow{\quad} $$

Fig. 1.40

It is to be noted that only S_3^1 and S_3^5 are the distinct operations generated by S_3 axis since Fig. 1.39 shows that

$$S_3^2 = C_3^2, S_3^3 = \sigma_h, S_3^4 = C_3^1 \text{ and } S_3^6 = E$$

These results can be shown mathematically (without taking models) as follows.

$$S_3^2 = (C_3 \,\sigma)^2 = C_3^2 \sigma^2 = C_3^2 \, E = C_3^2$$
$$S_3^4 = (C_3 \,\sigma)^4 = C_3^3 C_3^1 \, (\sigma^2)^2 = E \cdot C_3^1 \cdot E < C_3^1$$

Any molecule that lacks in improper rotation axis (S_n) is optically active. In 1, 3, 5, 7-tetramethylcyclo–octatetraene molecule S_4 axis is present. So the component is optically inactive. The compound trans-1, 2-dichlorocyclopropane is dis-symmetric or chiral. No S_n axis is present in the molecule. Hence the compound is optically active (Figs 1.41 and 1.42). Thus if a molecule possesses S_n with any n, it can not be optically active.

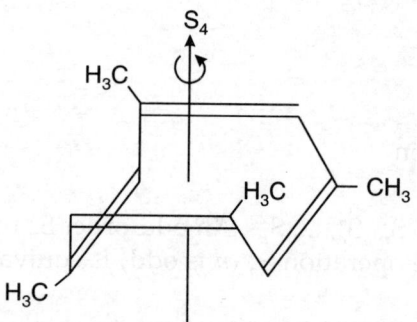

Fig. 1.41 Structure of 1, 3, 5, 7-tetramethylcyclo–octatetraene. S_4-axis is present. Not dissymmetric–optically inactive compound

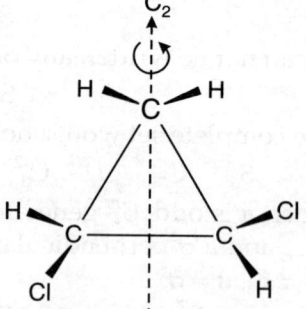

Fig. 1.42 Trans-1, 2-dichlorocyclopropane contains a C_2-axis. Dissymmetric–optically active compound

1.2.5 Identity (E)

The operation which brings back the molecule to the original orientation is called *identity operation*. It is represented as E, from the German word Einhert, meaning unity. Thus when we do not do anything to the molecule (leave it above) we get indistinguishable configuration of the molecule. Therefore, no operation is the operation for the identity element of symmetry. So this, element is present in all the molecules. If a molecule is rotated by 360° or reflected twice or inverted twice we get identical configuration to the original one (Fig. 1.43).

$C_1 = E$, $\sigma^2 = E$, $i^2 = E$

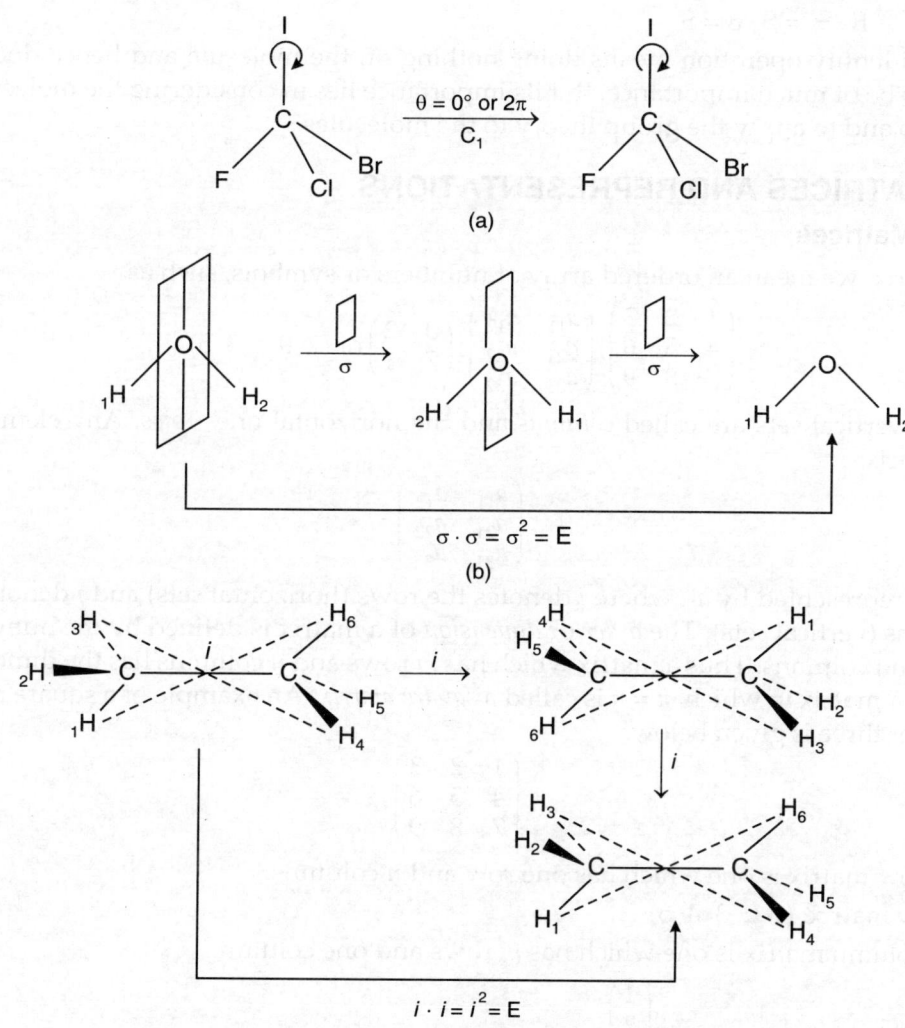

Fig. 1.43

Some other symmetry operations performed on a molecule lead to identity. Thus

$$2\,C_2 \text{ operations in succession} = C_2^2 = E$$
$$3\,C_3 \text{ operations in succession} = C_3^3 = E$$
$$n\,C_n \text{ operations in succession} = C_n^n = E$$
$$n\,S_n \text{ operations in succession} = C_n^n = E$$

The product of an element and its inverse gives identity.

$$(C_n)\,(C_n)^{-1} = E$$
$$C_3^1\,C_3^{-1} = E$$

C_3^1 is the inverse of C_3^2 and C_3^2 is the inverse of C_3^1. Since $\sigma \cdot \sigma = E$, it implies that σ is inverse of itself. So inverse operation is that operation which when combined with the operation, result into identity. Secondly if Identity element is combined with any operation it always result the same operation,

i.e. $E \cdot C_n = C_n$

$E \cdot \sigma = \sigma E = \sigma$

$E \cdot S_n = S_n E = S_n$

The identity operation means doing nothing on the molecule and hence does not seem to be of much importance. But its importance lies in considering the molecule as a group and to apply the group theory to the molecules.

1.3 MATRICES AND REPRESENTATIONS

1.3.1 Matrices

By matrix, we mean an ordered array of numbers or symbols, such as

$$\begin{pmatrix} 1 & 2 & 3 \\ 4 & 5 & 6 \\ 7 & 8 & 9 \end{pmatrix}, \begin{pmatrix} a_{11} & a_{12} \\ a_{21} & a_{22} \\ a_{31} & a_{32} \end{pmatrix}, \begin{pmatrix} 3 & 2 \\ 7 & 1 \end{pmatrix} \text{or} (2\ 0\ 1\ 3\ 5)$$

The vertical sets are called *columns* and the horizontal ones *rows*. Any element of the matrix

$$\begin{pmatrix} a_{11} & a_{12} \\ a_{21} & a_{22} \\ a_{31} & a_{32} \end{pmatrix}$$

can be represented by a_{ij}, where i denotes the rows (horizontal sets) and j denotes the columns (vertical sets). The *order* or *dimension* of a matrix is defined by the number of rows and columns. Thus a matrix which has m rows and n columns has the dimension $m \times n$. A matrix in which $m = n$ is called a *square matrix*. An example of a square matrix of order three is given below.

$$\begin{pmatrix} 1 & 2 & 3 \\ 4 & 5 & 6 \\ 7 & 8 & 9 \end{pmatrix}$$

A row matrix is one which has one row and n columns.

Row matrix: (1 2 3 4 5)

A column matrix is one which has m rows and one column.

Column matrix: $\begin{pmatrix} 1 \\ 2 \\ 3 \\ 4 \end{pmatrix}$

Let us consider a square matrix of the type:

$$\begin{pmatrix} a_{11} & a_{12} & a_{13} \\ a_{21} & a_{22} & a_{23} \\ a_{31} & a_{32} & a_{33} \end{pmatrix}$$

The elements a_{11}, a_{22}, a_{37} are called diagonal elements. The matrix is called *diagonal matrix* if all elements other than a_{11}, a_{22}, a_{33} are zero.

Example of a diagonal matrix is shown below.

Diagonal matrix:
$$\begin{pmatrix} 4 & 0 & 0 \\ 0 & 5 & 0 \\ 0 & 0 & 6 \end{pmatrix}$$

A square matrix is called a *unit matrix* if all the diagonal elements are equal to 1 and all other elements are equal to zero. The unit matrix is an example of a diagonal matrix.

Unit matrix:
$$\begin{pmatrix} 1 & 0 & 0 \\ 0 & 1 & 0 \\ 0 & 0 & 1 \end{pmatrix}$$

A square or rectangular matrix is called a *zero matrix* or *null matrix* when all the elements are equal to zero.

$\begin{pmatrix} 0 & 0 \\ 0 & 0 \end{pmatrix}$ is a null matrix of order 2

A *singular matrix* is one for which the determinant is zero. The matrix

$$\begin{pmatrix} 2 & 4 \\ 4 & 8 \end{pmatrix}$$

has the value zero for its determinant and is therefore called a singular matrix.

Formation of Submatrices:

Any square matrix of order n can be divided into submatrices as shown below

$$\begin{pmatrix} 1 & 0 & 0 & 0 & 0 & 0 \\ 1 & 2 & 0 & 0 & 0 & 0 \\ 0 & 0 & 3 & 0 & 0 & 0 \\ 0 & 0 & 0 & 1 & 3 & 2 \\ 0 & 0 & 0 & 1 & 2 & 2 \\ 0 & 0 & 0 & 4 & 0 & 1 \end{pmatrix}$$

is a matrix of order 6. This matrix can be blocked into there submatrices:

$$\begin{pmatrix} 1 & 0 \\ 1 & 2 \end{pmatrix} \quad (3) \quad \begin{pmatrix} 1 & 3 & 2 \\ 1 & 2 & 2 \\ 4 & 0 & 1 \end{pmatrix}$$

$$(a) \qquad (b) \qquad (c)$$

Inverse of a Matrix:

A square matrix B is called the inverse of A if

$$A = BA = 1$$

The inverse of a matrix may be obtained by using following rules.

1. Compute the determinant of the matrix A = $\begin{pmatrix} a_{11} & a_{12} \\ a_{21} & a_{22} \end{pmatrix}$
2. Interchange the elements a_{11} and a_{22}
3. Change the signs of a_{12} and a_{21}
4. Divide each element of the matrix thus formed by the determinant of A.

To determine the inverse of a matrix B = $\begin{pmatrix} 3 & -2 \\ 1 & 4 \end{pmatrix}$ first of all we have to determine the determinant of the matrix B which is equal to 14 $(12 - (-2) = 14)$. Interchange of the elements 3 and 4 leads to the following matrix.

$$\begin{pmatrix} 4 & -2 \\ 1 & 3 \end{pmatrix}$$

where the signs of -2 and 1 are changed we have another new matrix:

$$\begin{pmatrix} 4 & 2 \\ -1 & 3 \end{pmatrix}$$

Now each element of the row matrix is divided by 14 to obtain the inverse of B. Thus

$$B^{-1} = \left(\frac{1}{14}\right) \times \begin{pmatrix} 4 & 2 \\ -1 & 3 \end{pmatrix}$$

Matrix Addition/Subtraction/Multiplication:

Matrices can be added or subtracted only if they have the same number of rows and columns. The element of the resulting matrix are given by $a_{ij} + b_{ij}$. Thus if

$$A = \begin{pmatrix} -3 & 6 & 4 \\ 1 & 0 & 2 \end{pmatrix} \quad \text{and} \quad B = \begin{pmatrix} 2 & 1 & 1 \\ -6 & 4 & 3 \end{pmatrix}$$

then

$$A + B = C = \begin{pmatrix} -3 & 6 & 4 \\ 1 & 0 & 2 \end{pmatrix} + \begin{pmatrix} 2 & 1 & 1 \\ -6 & 4 & 3 \end{pmatrix}$$

$$= \begin{pmatrix} -3+2 & 6+1 & 4+1 \\ 1+(-6) & 0+4 & 2+3 \end{pmatrix}$$

$$= \begin{pmatrix} -1 & 7 & 5 \\ -5 & 4 & 5 \end{pmatrix}$$

$$3A - 2B = D = 3\begin{pmatrix} -3 & 6 & 4 \\ 1 & 0 & 2 \end{pmatrix} - 2\begin{pmatrix} 2 & 1 & 1 \\ -6 & 4 & 3 \end{pmatrix}$$

$$= \begin{pmatrix} -9 & 18 & 12 \\ 3 & 0 & 6 \end{pmatrix} - \begin{pmatrix} 4 & 2 & 2 \\ -12 & 8 & 6 \end{pmatrix}$$

$$= \begin{pmatrix} -9-4 & 18-2 & 12-2 \\ 3+12 & 0-8 & 6-6 \end{pmatrix} = \begin{pmatrix} -13 & 16 & 10 \\ 15 & -8 & 0 \end{pmatrix}$$

Let us consider two matrices

$$A = \begin{pmatrix} a_{11} & a_{12} \\ a_{21} & a_{22} \end{pmatrix} \quad \text{and} \quad B = \begin{pmatrix} b_{11} & b_{12} \\ b_{21} & b_{22} \end{pmatrix}$$

Their product $(AB = C)$ $C = \begin{pmatrix} c_{11} & c_{12} \\ c_{21} & c_{22} \end{pmatrix}$

The element C_{11} is obtained by multiplying the elements of row 1 of B with the elments of column 1 of A, on by the scheme

$$\begin{pmatrix} b_{11} & b_{12} \\ b_{21} & b_{22} \end{pmatrix}\begin{pmatrix} a_{11} & a_{12} \\ a_{21} & a_{22} \end{pmatrix} = \begin{pmatrix} b_{11}\,a_{11} + b_{12}\,a_{21} & \cdots \\ \vdots & \end{pmatrix}$$

and C_{12} by $\begin{pmatrix} b_{11} & b_{12} \\ b_{21} & b_{22} \end{pmatrix} \begin{pmatrix} a_{11} & a_{12} \\ a_{21} & a_{22} \end{pmatrix} = \begin{pmatrix} \cdots & b_{11}\,a_{12} + b_{12}\,a_{22} \\ \cdots & \cdots \end{pmatrix}$

If $B = \begin{pmatrix} 1 & 2 & 1 \\ 3 & 0 & -1 \\ -1 & -1 & 2 \end{pmatrix}$ and $A = \begin{pmatrix} -3 & 0 & -1 \\ 1 & 4 & 0 \\ 1 & 1 & 1 \end{pmatrix}$ their product

$$BA = C = \begin{pmatrix} 1 & 2 & 1 \\ 3 & 0 & -1 \\ -1 & -1 & 2 \end{pmatrix} \times \begin{pmatrix} -3 & 0 & -1 \\ 1 & 4 & 0 \\ 1 & 1 & 1 \end{pmatrix}$$

$$= \begin{pmatrix} (1)(-3)+(2)(1)+(1)(1) & (1)(0)+(2)(4)+(1)(1) & (1)(-1)+(2)(0)+(1)(1) \\ (3)(-3)+(0)(1)+(-1)(1) & (3)(0)+(0)(4)+(-1)(1) & (3)(-1)+(0)(0)+(-1)(1) \\ (-1)(-3)+(-1)(1)+(2)(1) & (-1)(0)+(-1)(4)+(2)(1) & (-1)(-1)+(-1)(0)+(2)(1) \end{pmatrix}$$

$$= \begin{pmatrix} 0 & 9 & 0 \\ -10 & -1 & -4 \\ 4 & 2 & 3 \end{pmatrix}$$

Similarly, $\begin{pmatrix} 5 & 1 & 3 \\ 4 & 2 & 2 \\ 1 & 2 & 3 \end{pmatrix} \times \begin{pmatrix} 2 & 1 & 1 \\ 1 & 2 & 3 \\ 5 & 4 & 3 \end{pmatrix}$

$$= \begin{pmatrix} (5)(2)+(1)(1)+(3)(5) & (5)(1)+(1)(2)+(3)(4) & (5)(1)+(1)(3)+(3)(3) \\ (4)(2)+(2)(1)+(2)(5) & (4)(1)+(2)(2)+(2)(4) & (4)(1)+(2)(3)+(2)(3) \\ (1)(2)+(2)(1)+(3)(5) & (1)(1)+(2)(2)+(3)(4) & (1)(1)+(2)(3)+(3)(3) \end{pmatrix}$$

$$= \begin{pmatrix} 26 & 19 & 17 \\ 20 & 16 & 16 \\ 19 & 17 & 16 \end{pmatrix}$$

Example 1: $\begin{pmatrix} 1 & -1 & -2 \\ 0 & 1 & -1 \\ 1 & 0 & 0 \end{pmatrix} \times \begin{pmatrix} 2 \\ 1 \\ 3 \end{pmatrix}$

Solution: $\begin{pmatrix} (1)(2)+(-1)(1)+(-2)(3) \\ (0)(2)+(1)(1)+(-1)(3) \\ (1)(2)+(0)(1)+(0)(3) \end{pmatrix} = \begin{pmatrix} -5 \\ -2 \\ 2 \end{pmatrix}$

Example 2: $\begin{pmatrix} 1 & 2 & 3 \end{pmatrix} \times \begin{pmatrix} 1 & -1 & -2 \\ 2 & 1 & -1 \\ 3 & 2 & 1 \end{pmatrix}$

Solution: $\begin{pmatrix} (1)(1)+(2)(2)+(3)(3) & (1)(-1)+(2)(1)+(3)(2) & (1)(-2)+(2)(-1)+(3)(1) \end{pmatrix}$

$= \begin{pmatrix} 14 & 7 & -1 \end{pmatrix}$

Diagonalisation of a Matrix

The process of reducing a matrix to the diagonal matrix is referred to as *diagonalisation*. Let A is a square matrix of order n. If P is similarly the transformation matrix which reduces A to the diagonal matrix D then

$$P^{-1} AP = D$$

Let us take an example. The matrix $A = \begin{pmatrix} -7 & 6 \\ -18 & 14 \end{pmatrix}$

The similarity transformation matrix $P = \begin{pmatrix} 2 & 1 \\ 3 & 2 \end{pmatrix}$

$$\therefore \qquad P^{-1} = \begin{pmatrix} 2 & -1 \\ -3 & 2 \end{pmatrix}$$

Now $\qquad P^{-1} AP = \begin{pmatrix} 3 & -1 \\ -3 & 2 \end{pmatrix} \times \begin{pmatrix} -7 & 6 \\ -18 & 14 \end{pmatrix} \times \begin{pmatrix} 2 & -1 \\ -3 & 2 \end{pmatrix}$

$$= \begin{pmatrix} 2 & 0 \\ 0 & 5 \end{pmatrix}$$

1.3.2 Symmetry Operations as Matrices

We have employed previously five types of operations E, σ, i, C_n, S_n in describing the symmetry of a molecule or other object. Each of these types of operations can be described by a matrix. Matrix notations will now be worked out for all the symmetry elements.

Let P (x, y, z) be the position of a point. Let this point P be shifted to the point Q (x', y', z') after carrying out one of the symmetry operations (Fig. 1.44). The coordinates (x', y', z') may be correlated with (x, y, z) in the form of matrix notations.

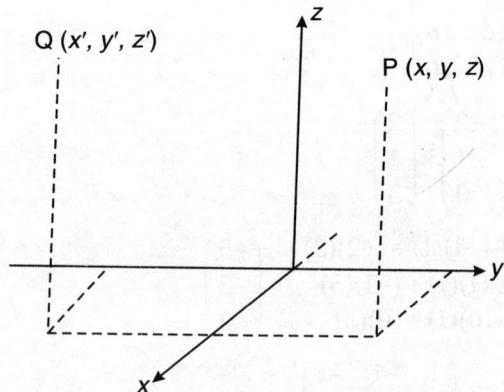

Fig. 1.44 Cartesian axes

1. Identity Operation and E Matrix

This operation causes no change in the molecule. In this operation new coordinates (x', y', z') are the same as the initial one (x, y, z).

The transformation equations are

$$x' = x = x + 0 + 0 = 1 \cdot x + 0 \cdot y + 0 \cdot z$$
$$y' = y = 0 + y + 0 = 0 \cdot x + 1 \cdot y + 0 \cdot z$$
$$z' = z = 0 + 0 + z = 0 \cdot x + 0 \cdot y + 1 \cdot z$$

Each symmetry operation may be expressed as a transformation matrix. The above three equations may be written more compactly in matrix notation.

$$\begin{pmatrix} x' \\ y' \\ z' \end{pmatrix} = \begin{pmatrix} 1 & 0 & 0 \\ 0 & 1 & 0 \\ 0 & 0 & 1 \end{pmatrix} \begin{pmatrix} x \\ y \\ z \end{pmatrix}$$

(New coordinates) = (transformation matrix for E) (old coordinates)

In case of identity operation, transformation matrix

$$\begin{pmatrix} 1 & 0 & 0 \\ 0 & 1 & 0 \\ 0 & 0 & 1 \end{pmatrix} = \text{unit matrix} = E. \text{ Thus}$$

$$\begin{pmatrix} x' \\ y' \\ z' \end{pmatrix} = \begin{pmatrix} 1 & 0 & 0 \\ 0 & 1 & 0 \\ 0 & 0 & 1 \end{pmatrix} \begin{pmatrix} x \\ y \\ z \end{pmatrix} = 1 \cdot \begin{pmatrix} x \\ y \\ z \end{pmatrix} = \begin{pmatrix} x \\ y \\ z \end{pmatrix}$$

Therefore
$$\begin{pmatrix} x' \\ y' \\ z' \end{pmatrix} = \begin{pmatrix} x \\ y \\ z \end{pmatrix}$$

2. Rotation Operation and C_n Matrix

Defining the rotation axis as the z axis, the z coordinate will be unchanged by any rotation about the z axis. Thus the matrix in part will be

$$\begin{pmatrix} & & 0 \\ & & 0 \\ 0 & 0 & 1 \end{pmatrix}$$

The four missing elements can be solved as a two-dimensional problem in the xy plane. Let us consider a point P in the xy plane with coordinates x_1 and y_1 (Fig. 1.45).

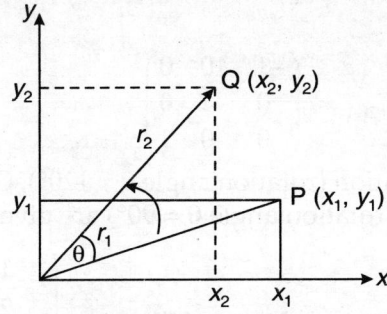

Fig. 1.45 A two-dimensional coordinate used to demonstrate the rotation of a vector through an angle θ in the xy plane

---

* A line joining the point P (x_1, y_1) and the origin is called r-vector.

This point defines a vector r_1 between itself and the origin and can be expressed as a column matrix r_1.

$$r_1 = \begin{pmatrix} x_1 \\ y_1 \end{pmatrix} \qquad \qquad ...(1.1)$$

x_1 and y_1 are the components of the vector r_1

Let the vector be rotated (in anticlockwise directions) through an angle θ such that components of the vector become x_2, y_2. The resulting vector r_2 is expressed as another column matrix.

$$r_2 = \begin{pmatrix} x_2 \\ y_2 \end{pmatrix} \qquad \qquad ...(1.2)$$

x_2 and y_2 are related to x_1 and y_1 by the follow relations.

$$x_2 = x_1 \cos\theta + y_1 \sin\theta \qquad \qquad ...(1.3)$$
$$y_2 = -x_1 \sin\theta + y_1 \cos\theta \qquad \qquad ...(1.4)$$

Thus Eqs (1.3) and (1.4) are represented in the matrix form as follows.

$$\begin{pmatrix} x_2 \\ y_2 \end{pmatrix} = \begin{pmatrix} \cos\theta & \sin\theta \\ -\sin\theta & \cos\theta \end{pmatrix} \begin{pmatrix} x_1 \\ y_1 \end{pmatrix} \qquad \qquad ...(1.5)$$

Using Eqs (1.1), (1.2) and (1.5), we get

$$r_2 = \begin{pmatrix} \cos\theta & \sin\theta \\ -\sin\theta & \cos\theta \end{pmatrix} \times r_1 \qquad ...(1.6) \qquad \left[\because r_1 = \begin{pmatrix} x_1 \\ y_1 \end{pmatrix} \right]$$

If C_n represents rotation about the axis by angle θ, their the rotation operation can be expressed symbolically as

$$r_2 = C_n \times r_1 \qquad \qquad ...(1.7)$$

From Eqs (1.6) and (1.7), we get

$$C_n = \frac{r_2}{r_1} = \begin{pmatrix} \cos\theta & \sin\theta \\ -\sin\theta & \cos\theta \end{pmatrix} \qquad \qquad ...(1.8)$$

C_n represents the matrix for rotation operation. In three dimension, the matrix $C_{n(z)}$ which represents rotation about z axis, is written as

$$C_{n(z)} = \begin{pmatrix} \cos\theta & \sin\theta & 0 \\ -\sin\theta & \cos\theta & 0 \\ 0 & 0 & 1 \end{pmatrix} \qquad \qquad ...(1.9)$$

During C_2 operation

$$C_{2(z)} = \begin{pmatrix} -1 & 0 & 0 \\ 0 & -1 & 0 \\ 0 & 0 & 1 \end{pmatrix}$$

The matrices for C_3^1 operation (rotation angle $\theta = 120°$), C_3^2 operation (rotation angle $\theta = 240°$) and C_4^1 operation (rotation angle $\theta = 90°$) are given below.

$$C_3^1 = \begin{pmatrix} \cos 120° & \sin 120° & 0 \\ -\sin 120° & \cos 120° & 0 \\ 0 & 0 & 1 \end{pmatrix} = \begin{pmatrix} -\dfrac{1}{2} & -\dfrac{\sqrt{3}}{2} & 0 \\ \dfrac{\sqrt{3}}{2} & -\dfrac{1}{2} & 0 \\ 0 & 0 & 1 \end{pmatrix} \qquad ...(1.10)$$

$$C_3^2 = \begin{pmatrix} \cos 240° & \sin 240° & 0 \\ -\sin 240° & \cos 240° & 0 \\ 0 & 0 & 1 \end{pmatrix} = \begin{pmatrix} -\dfrac{1}{2} & -\dfrac{\sqrt{3}}{2} & 0 \\ \dfrac{\sqrt{3}}{2} & -\dfrac{1}{2} & 0 \\ 0 & 0 & 1 \end{pmatrix} \qquad \text{...(1.11)}$$

$$C_4^1 = \begin{pmatrix} \cos 90° & \sin 90° & 0 \\ -\sin 90° & \cos 90° & 0 \\ 0 & 0 & 1 \end{pmatrix} = \begin{pmatrix} 0 & 1 & 0 \\ -1 & 0 & 0 \\ 0 & 0 & 1 \end{pmatrix}$$

3. Reflection Operation and σ Matrix

Let the reflection planes (Fig. 1.46) be represented as (xy), (xz) and (yz), respectively.

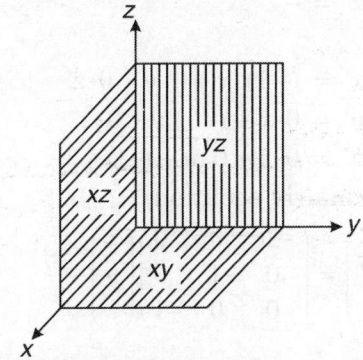

Fig. 1.46 Reflection represented by *xy*, *xz* and *yz*

When rotation operations are carried out in such planes, only the coordinates perpendicular to the plane change sign while the other two coordinates remain the same. Hence, we have

$$P\,(x, y, z) \xrightarrow[\sigma_{xy}\text{ plane}]{\text{reflection}} Q\,(x', y', z')$$

Here
$$x' = x$$
$$y' = y$$
$$z = -z$$

On expanding the above relations, we get
$$x' = 1\cdot x + 0\cdot y + 0\cdot z$$
$$y' = 0\cdot x + 1\cdot y + 0\cdot z$$
$$z' = 0\cdot x + 0\cdot y + (-1)z$$

and the matrix equation is
$$\begin{pmatrix} x' \\ y' \\ z' \end{pmatrix} = \begin{pmatrix} 1 & 0 & 0 \\ 0 & 1 & 0 \\ 0 & 0 & 1 \end{pmatrix} \begin{pmatrix} x \\ y \\ z \end{pmatrix}$$

Therefore, the σ_{xy} matrix is
$$\sigma_{xy} : \begin{pmatrix} 1 & 0 & 0 \\ 0 & 1 & 0 \\ 0 & 0 & -1 \end{pmatrix}$$

Similarly, the matrix for σ_{xz} and σ_{yz} can be obtained as

$$\sigma_{xz} : \begin{pmatrix} 1 & 0 & 0 \\ 0 & -1 & 0 \\ 0 & 0 & 1 \end{pmatrix} \qquad \sigma_{yz} : \begin{pmatrix} -1 & 0 & 0 \\ 0 & 1 & 0 \\ 0 & 0 & 1 \end{pmatrix}$$

4. Inversion Operation and i Matrix

$$P\,(x, y, z) \xrightarrow{\;i,\ \text{inversion}\;} Q\,(x', y', z')$$

The changes can be represented as

$$x' = -x$$
$$y' = -y$$
$$z' = -z$$

On expansion we get

$$x' = (-1)\cdot x + 0\cdot y + 0\cdot z$$
$$y' = 0\cdot x + (-1)\,y + 0\cdot z$$
$$z' = 0\cdot x + 0\cdot y + (-1)\,z$$

which finally gives rise to the matrix equation,

$$\begin{pmatrix} x' \\ y' \\ z' \end{pmatrix} = \begin{pmatrix} -1 & 0 & 0 \\ 0 & -1 & 0 \\ 0 & 0 & -1 \end{pmatrix} \begin{pmatrix} x \\ y \\ z \end{pmatrix}$$

i matrix is of the form,

$$i : \begin{pmatrix} -1 & 0 & 0 \\ 0 & -1 & 0 \\ 0 & 0 & -1 \end{pmatrix}$$

5. Improper rotation Operation and S_n Matrix

The matrix for improper rotation is given by the product of $C_{n(z)}$ matrix and σ_{xy} matrix

$$S_{n(z)} = C_{n(z)} \times \sigma_{xy}$$

$C_{n(z)}$ and σ_{xy} are perpendicular to each other

$$S_{n(z)} = \begin{pmatrix} \cos\theta & \sin\theta & 0 \\ -\sin\theta & \cos\theta & 0 \\ 0 & 0 & 1 \end{pmatrix} \begin{pmatrix} 1 & 0 & 0 \\ 0 & 1 & 0 \\ 0 & 0 & -1 \end{pmatrix}$$

$$= \begin{pmatrix} \cos\theta & \sin\theta & 0 \\ -\sin\theta & \cos\theta & 0 \\ 0 & 0 & -1 \end{pmatrix}$$

Similarly,

$$S_{n(x)} = C_{n(x)} \cdot \sigma_{yz}$$

$$= \begin{pmatrix} 1 & 0 & 0 \\ 0 & \cos\theta & \sin\theta \\ 0 & -\sin\theta & \cos\theta \end{pmatrix} \begin{pmatrix} -1 & 0 & 0 \\ 0 & 1 & 0 \\ 0 & 0 & 1 \end{pmatrix}$$

$$= \begin{pmatrix} -1 & 0 & 0 \\ 0 & \cos\theta & \sin\theta \\ 0 & -\sin\theta & \cos\theta \end{pmatrix}$$

$$S_{n(y)} = C_{n(y)} \cdot \sigma_{xz}$$

$$= \begin{pmatrix} \cos\theta & 0 & \sin\theta \\ 0 & 1 & 0 \\ -\sin\theta & 0 & \cos\theta \end{pmatrix} \begin{pmatrix} 1 & 0 & 0 \\ 0 & -1 & 0 \\ 0 & 0 & 1 \end{pmatrix}$$

$$= \begin{pmatrix} \cos\theta & 0 & \sin\theta \\ 0 & -1 & 0 \\ -\sin\theta & 0 & \cos\theta \end{pmatrix}$$

1.3.3 Multiplication of Symmetry Operations

The transformation matrices for the symmetry operations of water molecule are:

$$E = \begin{pmatrix} 1 & 0 & 0 \\ 0 & 1 & 0 \\ 0 & 0 & 1 \end{pmatrix}, C_2 = \begin{pmatrix} -1 & 0 & 0 \\ 0 & -1 & 0 \\ 0 & 0 & 1 \end{pmatrix}, \sigma_v(xz) = \begin{pmatrix} 1 & 0 & 0 \\ 0 & -1 & 0 \\ 0 & 0 & 1 \end{pmatrix}$$

$$\sigma_v(yz) = \begin{pmatrix} -1 & 0 & 0 \\ 0 & 1 & 0 \\ 0 & 0 & 1 \end{pmatrix}, \sigma_v(xy) = \begin{pmatrix} 1 & 0 & 0 \\ 0 & 1 & 0 \\ 0 & 0 & -1 \end{pmatrix}$$

The combination of two operations can be shown as the combination of their matrices

$$EE = \begin{pmatrix} 1 & 0 & 0 \\ 0 & 1 & 0 \\ 0 & 0 & 1 \end{pmatrix} \begin{pmatrix} 1 & 0 & 0 \\ 0 & 1 & 0 \\ 0 & 0 & 1 \end{pmatrix}$$

$$= \begin{pmatrix} 1+0+0 & 0+0+0 & 0+0+0 \\ 0+0+0 & 0+1+0 & 0+0+0 \\ 0+0+0 & 0+0+0 & 0+0+1 \end{pmatrix} = \begin{pmatrix} 1 & 0 & 0 \\ 0 & 1 & 0 \\ 0 & 0 & 1 \end{pmatrix} = E$$

$$C_2\,\sigma_v(xz) = \begin{pmatrix} -1 & 0 & 0 \\ 0 & -1 & 0 \\ 0 & 0 & 1 \end{pmatrix} \begin{pmatrix} 1 & 0 & 0 \\ 0 & -1 & 0 \\ 0 & 0 & 1 \end{pmatrix}$$

$$= \begin{pmatrix} -1+0+0 & 0+0+0 & 0+0+0 \\ 0+0+0 & 0+1+0 & 0+0+0 \\ 0+0+0 & 0+0+0 & 0+0+1 \end{pmatrix}$$

$$= \begin{pmatrix} -1 & 0 & 0 \\ 0 & 1 & 0 \\ 0 & 0 & -1 \end{pmatrix}$$

$$= \sigma_v(yz)$$

$$EC_2 = \begin{pmatrix} 1 & 0 & 0 \\ 0 & 1 & 0 \\ 0 & 0 & 1 \end{pmatrix} \begin{pmatrix} -1 & 0 & 0 \\ 0 & -1 & 0 \\ 0 & 0 & 1 \end{pmatrix} \begin{pmatrix} -1 & 0 & 0 \\ 0 & 1 & 0 \\ 0 & 0 & 1 \end{pmatrix}$$

$$= \begin{pmatrix} -1+0+0 & 0+0+0 & 0+0+0 \\ 0+0+0 & 0-1+0 & 0+0+0 \\ 0+0+0 & 0+0+0 & 0+0+1 \end{pmatrix}$$

$$= \begin{pmatrix} -1 & 0 & 0 \\ 0 & -1 & 0 \\ 0 & 0 & 1 \end{pmatrix} = C_2$$

$$\sigma_v(xz)\,\sigma_v(yz) = \begin{pmatrix} 1 & 0 & 0 \\ 0 & -1 & 0 \\ 0 & 0 & 1 \end{pmatrix} \begin{pmatrix} -1 & 0 & 0 \\ 0 & 1 & 0 \\ 0 & 0 & 1 \end{pmatrix}$$

$$= \begin{pmatrix} -1+0+0 & 0+0+0 & 0+0+0 \\ 0+0+0 & 0-1+0 & 0+0+0 \\ 0+0+0 & 0+0+0 & 0+0+1 \end{pmatrix}$$

$$= \begin{pmatrix} -1 & 0 & 0 \\ 0 & -1 & 0 \\ 0 & 0 & 1 \end{pmatrix} = C_2$$

$$C_2\,C_2 = \begin{pmatrix} -1 & 0 & 0 \\ 0 & -1 & 0 \\ 0 & 0 & 1 \end{pmatrix} \begin{pmatrix} -1 & 0 & 0 \\ 0 & -1 & 0 \\ 0 & 0 & 1 \end{pmatrix}$$

$$= \begin{pmatrix} 1+0+0 & 0+0+0 & 0+0+0 \\ 0+0+0 & 0+1+0 & 0+0+0 \\ 0+0+0 & 0+0+0 & 0+0+1 \end{pmatrix}$$

$$= \begin{pmatrix} 1 & 0 & 0 \\ 0 & 1 & 0 \\ 0 & 0 & 1 \end{pmatrix} = E$$

$$E\,\sigma_v(xz) = \begin{pmatrix} 1 & 0 & 0 \\ 0 & 1 & 0 \\ 0 & 0 & 1 \end{pmatrix} \begin{pmatrix} 1 & 0 & 0 \\ 0 & -1 & 0 \\ 0 & 0 & 1 \end{pmatrix}$$

$$= \begin{pmatrix} 1+0+0 & 0+0+0 & 0+0+0 \\ 0+0+0 & 0-1+0 & 0+0+0 \\ 0+0+0 & 0+0+0 & 0+0+1 \end{pmatrix}$$

$$= \begin{pmatrix} 1 & 0 & 0 \\ 0 & -1 & 0 \\ 0 & 0 & 1 \end{pmatrix} = \sigma_v(xz)$$

The combination of different operations of H_2O molecule is shown in the Table 1.2.

Table 1.2 Different operations of H_2O molecule

	E	C_2	$\sigma_v(xz)$	$\sigma_v(yz)$
E	E	C_2	$\sigma_v(xz)$	$\sigma_v(yz)$
C_2	C_2	E	$\sigma_v(yz)$	$\sigma_v(xz)$
$\sigma_v(xz)$	$\sigma_v(xz)$	$\sigma_v(yz)$	E	C_2
$\sigma_v(yz)$	$\sigma_v(yz)$	$\sigma_v(xz)$	C_2	E

1.4 SYMMETRY AND OPTICAL ACTIVITY

A complex which is not superimposable on its mirror image is called a *chiral* (or asymmetric) complex. A chiral complex does not have a

(a) mirror plane (σ) through the central atom or

(b) a centre of inversion (**i**)

An example of a chiral organic molecule is C Br Cl F I

Chiral objects are termed dis-symmetric*. This term does not mean that these objects necessarily have no symmetry. In general, we can say that a molecule is chiral if it has

(a) no symmetry operations (other than identity operation, E) or

(b) it has only proper rotation axis (C_n)

Cis-[Co(en)$_2$ Cl$_2$]$^+$ cation exists in two distinct mirror image forms which are not superimposable (Fig. 1.47). On the contrary, the transform (Fig. 1.48) being symmetrical provides only superimposable mirror images, i.e. only one distinct form. It is an optically inactive compound.

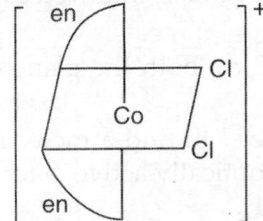 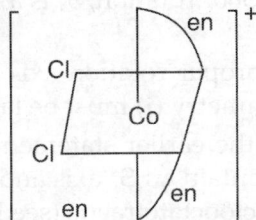 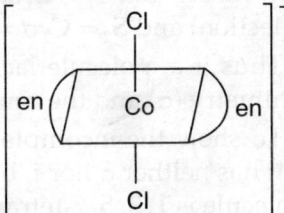

Fig. 1.47 Optically active *cis*-[Co(en)$_2$ Cl$_2$]$^+$ cation *cis*-isomers (*d* and *l*)   **Fig. 1.48** Optically inactive *trans*-[Co(en)$_2$Cl$_2$]$^+$ cation

This (didentate) metal complexes such as [Co(en)$_3$]$^{3+}$ (Fig. 1.49), [Ru(bipy)$_3$]$^{2+}$ have no centre of inversion (*i*) nor a mirror plane (σ) passing through the central metal atom. Hence they are chiral and have been resolved into optical isomers (*d* and *l*).

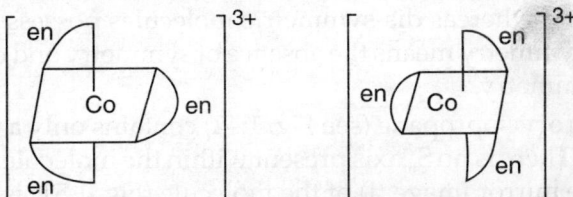

Fig. 1.49 Optical isomers of [Co(en)$_3$]$^{3+}$

---

* Molecule with C_n axis is said to be asymmetric when $n = 1$ and the molecule is called dis-symmetric when $n > 1$.

Ethylene diaminetetraacetatocobaltate (III) or ethylenediaminetetracetatochromate (III) have no centre of inversion (*i*) and no mirror plane (σ). So [Co(EDTA)]⁻ (Fig. 1.50) and [Cr(EDTA)]⁻ are chiral and give nonsuperimposable mirror images. Such complexes have been resolved into optical *d*- and *l*- isomers.

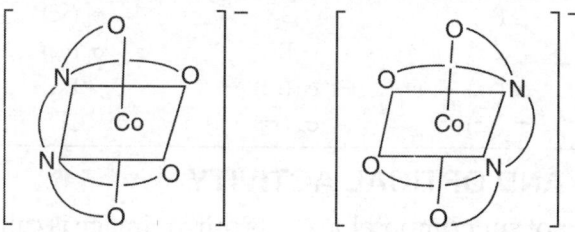

Fig. 1.50 Optical isomers of [Co(EDTA)]⁻

We see that a molecule is optically active if the mirror image of the molecule cannot be superimposed on the original. But if it can be superimposed, the molecule is optically inactive. In using this criterion, the mirror is understood to be external to the whole molecule, and reflection through the mirror gives an image of the whole molecule.

With complicated molecules, the visualisation of superimposability is difficult. Accordingly, it is our advantage to have a symmetry basis for establishing the existence of optically active isomers.

Any molecule that has no improper rotation axis (S_n) is said to be dis-symmetric, and optically active molecules must be dis-symmetric.

One often hears the incomplete statement that in order for optical isomerism to exist, the molecule must lack a plane of symmetry (σ) or centre of symmetry (**i**). We have seen earlier that $S_1 = C_1 \sigma = \sigma$ ($C_1 = 360°$ rotation; S_1 is a rotation by 360° followed by a reflection) and $S_2 = C_2 \sigma = \mathbf{i}$.

Thus if a molecule lacks an improper rotation axis ($S_n = C_n \sigma$), both the plane of symmetry (σ) and the centre of symmetry (**i**) must be lacking.

To show the incompleteness of the earlier statement, we need to find a molecule that has neither σ nor **i**, but does contain an S_n axis and is not optically active. Such a molecule is 1, 3, 5, 7-tetramethyl cyclooctatetraene (see Fig. 1.41).

This molecule does not have a plane or centre of symmetry. However, since it has an S_4 axis, it is optically inactive. Therefore any molecule that lacks in S_n-axes is optically active.

Optically active compounds must be dis-symmetric and need not necessarily be asymmetric. These two terms must be distinguished carefully since they often lead to confusion. A simple differentiation is that asymmetric molecules lack every type of symmetry element, whereas dis-symmetric molecules possess only C_n axes and no other element. So asymmetry means the absence of symmetry and dis-symmetry means absence of some symmetry.

Trans-1, 2-dichlorocyclopropane (see Fig. 1.42) contains only a C_2-axis and no other symmetry element. There is no S_n axis present within the molecule. So the compound is optically active. The mirror image (i) of the molecule (Fig. 1.51) is not superimposable on (a) and hence (a) and (b) are enantiomorphs.

Molecules such as *trans*-1, 2-dichlorocyclopropane (see Fig. 1.42) can not correctly be said to be asymmetric because asymmetric means without symmetry. Such molecules are described as dis-symmetric. An optically active compound need not be asymmetric, but must be dis-symmetric. Thus dis-symmetric compounds need not be asymmetric, but all asymmetric compounds are dis-symmetric.

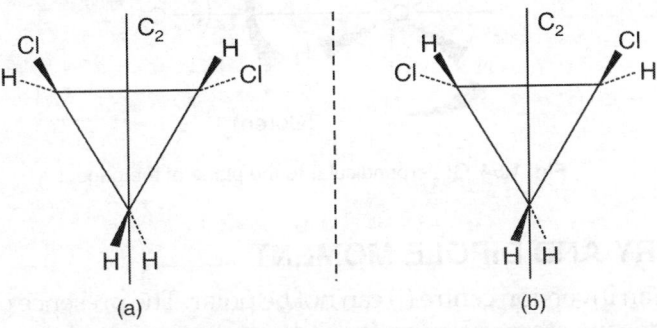

Fig. 1.51 (a) An optically active compound and (b) its mirror image

Spiro shown in Fig. 1.52 has neither a plane nor a centre of symmetry, but the vertical axis that bisects both rings and goes through the spiro nitrogen atom is an S_4 axis. This molecule also has a C_2 axis coincident with the S_4 axis; but the presence of S_4 and not of the C_2 axis is responsible for lack of optical activity.

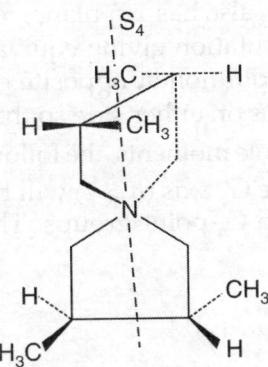

Fig. 1.52 An optically inactive molecule [because of the presence of S_4 axis]

Lactic acid (Fig. 1.53) is an asymmetric molecule. When $n = 1$, the molecule with C_n axis is said to be asymmetric, whereas $n > 1$ the molecule is called dis-symmetric. Thus, from the symmetry point of view, molecules belonging to C_n ($n = 1, 2, 3, ...$) and D_n ($n = 1, 2, 3, ...$) groups are optically active, since all of these groups lack S_n axis.

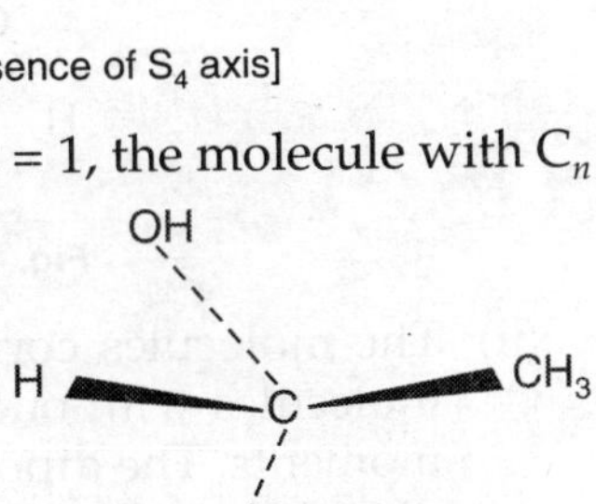

[Co(en)$_3$]$^{3+}$ ion belongs to D_3 point group (D_n contains C_n principal axis along with nC_2 axis). D_3 point group contains E_1C_3 and $3C_2$ elements but does not contain improper rotation axis (S_n). The compound is chiral and hence may be optically active.

Fig. 1.53

In summary, we can state that if a molecule possesses only C_n, it is in dis-symmetry and optically active. If $n = 1$, the molecules is asymmetric as well as dis-symmetric. If $n > 1$, the molecule is dis-symmetric (Fig. 1.54). If a molecule possesses S_n with any n, it can not be optically active.

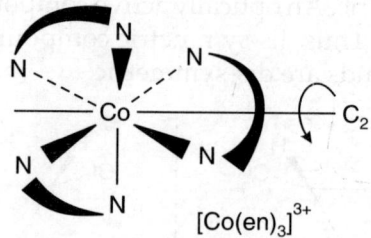

Fig. 1.54 C_3 perpendicular to the plane of the page

1.5 SYMMETRY AND DIPOLE MOMENT

A molecule with an inversion centre (i) can not be polar. The presence of inversion centre in the molecule means atoms of opposite ends can be projected through the inversion centre to get to an equivalent orientation. The molecule therefore has the same charge distributions at opposite ends. Examples: linear CO_2, BeF_2, etc. Linear HCl molecule without inversion centre is polar.

A molecule having (1) a C_n axis and a C_2 axis $\perp$ to the C_n axis or (2) a C_n axis and a σ_n plane $\perp$ to the C_n axis is nonpolar. The planar BF_3 molecule has a C_3 rotation axes and three C_2 axes $\perp$ to the C_3 axis. BF_3 also has a σ_h plane, containing the B atom and the three F atoms $\perp$ to the C_3 axis. A C_2 rotation giving equivalent orientation for the molecule will ensure similar charge distribution at opposite ends (Fig. 1.55). Horizontal plane σ_h means having identical atoms on either side so that there can not be any polarity.

For molecules possessing dipole moments, the following symmetry restrictions apply.

(i) Those molecules with one C_n-axis ($n > 1$) will have permanent dipole moments. Such molecules belong to C_n-point groups. The dipole moment vector will be along the C_n-axis.

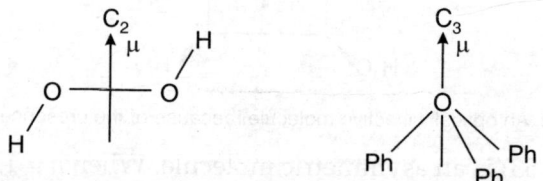

Fig. 1.55 C_2 and C_3 point groups with dipole moment (μ)

(ii) The molecules corresponding to C_s-point groups (e.g. those molecules with one σ-plane and no C_n-axis) will have dipole moments. The dipole moment vector lies on this plane and the direction of the vector depends on the nature of the groups/atoms present in the molecule. Examples: $C_3H_3Cl_3$, SO_2 BrF, $POBr_2Cl$, CH_2BrCl (Fig. 1.56).

(iii) The molecules which have a C_n-axis along with $n\sigma_v$-planes (C_{nv}) have permanent dipole moments. The dipole moment will be along the C_n axis in the C_{nv} group (Fig. 1.57).

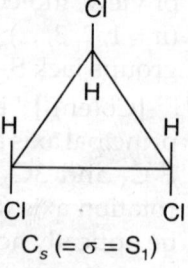

$C_s (= \sigma = S_1)$

Fig. 1.56

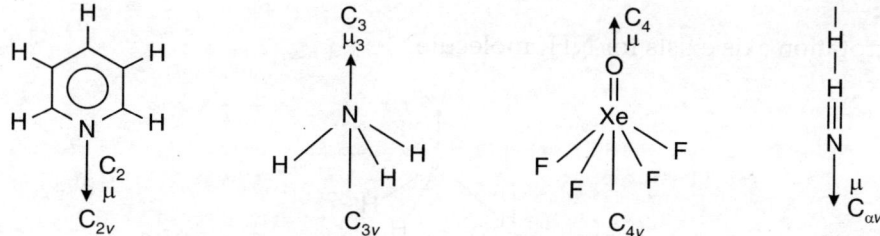

Fig. 1.57 Examples of C_{2v}, C_{3v}, C_{4v} and $C_{\alpha v}$ with dipole moment (μ)

(iv) The molecules which have no symmetry element (μ) are dipolar. They belong to C_1 point group. Common examples are HNClF, CHFClBr, SiFClBrI, etc. The direction of dipole moment vector in their molecules can lie in any direction.

PROBLEMS AND SOLUTIONS

Problem 1. Find out the principal axis of symmetry in C_6H_6, $PtCl_4^{2-}$, NH_3, BF_3, CH_4 and NH_4^+ ion.

Solution

C_6H_6:

Benzene (C_6H_6) is a planar molecule and has a hexagonal structure. It has six C_2 axes and one C_6 axes, which passes through the centre of the molecule and perpendicular to the plane of the molecule (not shown in figure). When a molecule possesses more than one rotation axis, the highest fold axis of rotation is referred to as the principal axis of rotation and is conventionally taken as the vertical axis. Thus the principal axis of symmetry in benzene C_6 axis.

$PtCl_4^{2-}$:

Planar $[PtCl_4]^{2-}$ ion has a C_4 axis perpendicular to the plane of the ion and four C_2 axes in the plane of the ion. One C_4 axis not shown in figure is $\perp$ to the molecular plane and passing through Pt. So C_4 is the principal axes.

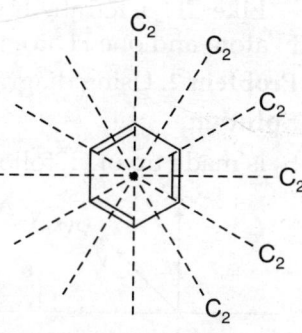

NH₃:

A C_3 rotation axis exists for NH_3 molecule

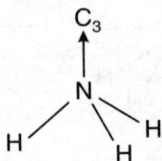

BF₃:

Planar BF_3 molecule possesses one three fold (C_3) axis and three two fold axes (C_2). C_3 is the principal axis.

CH₄:

Tetrahedral CH_4 molecule has four three fold rotation axes ($4C_3$) and three two fold rotation axes ($3C_2$). Each C_3 passing through C—H bond and each C_2 bisecting the bond angle $H\hat{C}H$. The principal axis is C_3

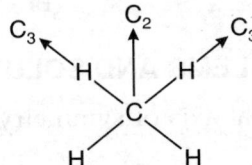

NH₄⁺:

Like CH_4 a tetrahedral NH_4^+ ion has four C_3 rotation axes, each passing through the N atom and one H atom and three C_2 axis. The principal axis is C_3.

Problem 2. Using diagrams as necessary, show that $S_2 \equiv i$ and $S_1 \equiv \sigma$.

Solution

S_2 is made up of C_2 followed by $\sigma_\perp$ which is shown in the figure.

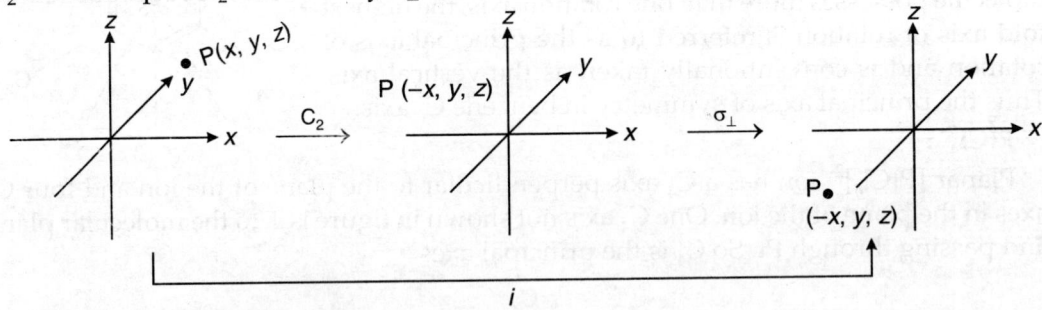

S_1 is made up of C_1 followed by $\sigma_\perp$

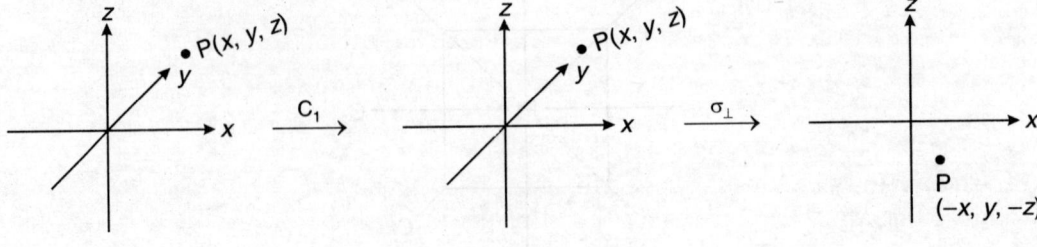

Problem 3. Using transformation matrices, show that S_1 operation is identical to the σ operation, and that the S_2 operation is identical to the i operation.

Solution

$$S_1 = \sigma_h \, C_1 \, \sigma_{xy} \, C_1 = \begin{bmatrix} 1 & 0 & 0 \\ 0 & 1 & 0 \\ 0 & 0 & -1 \end{bmatrix} \begin{bmatrix} 1 & 0 & 0 \\ 0 & 1 & 0 \\ 0 & 0 & 1 \end{bmatrix} = \begin{bmatrix} 1 & 0 & 0 \\ 0 & 1 & 0 \\ 0 & 0 & -1 \end{bmatrix} = \sigma_{xy}$$

$$S_2 = \sigma_{xy} \, C_2 = \begin{bmatrix} 1 & 0 & 0 \\ 0 & 1 & 0 \\ 0 & 0 & -1 \end{bmatrix} \begin{bmatrix} -1 & 0 & 0 \\ 0 & -1 & 0 \\ 0 & 0 & 1 \end{bmatrix} = \begin{bmatrix} -1 & 0 & 0 \\ 0 & -1 & 0 \\ 0 & 0 & -1 \end{bmatrix} = i$$

Problem 4. Show that the product of the identity transformation matrix and any other transformation matrix leaves the other matrix unchanged.

Solution

$$\begin{bmatrix} 1 & 0 & 0 \\ 0 & 1 & 0 \\ 0 & 0 & 1 \end{bmatrix} \begin{bmatrix} a & b & c \\ e & f & g \\ h & i & j \end{bmatrix} = \begin{bmatrix} (a+0+0) & (b+0+0) & (c+0+0) \\ (0+e+0) & (0+f+0) & (0+g+0) \\ -(0+0+h) & (0+0+i) & (0+0+f) \end{bmatrix}$$

$$= \begin{bmatrix} a & b & c \\ e & f & g \\ h & i & j \end{bmatrix}$$

Problem 5. Show that C_6 axis generates a total number of six operations in benzene.

Solution

The set of operations generated by C_6 will be C_6, C_6^2 ($= C_3$), C_6^4 ($= C_3^2$), C_6^3 ($= C_2$), C_6^5 and C_6^6 ($= E$), i.e. C_6, C_3, C_3^2, C_2, C_6^5, E.

It is to be noted that the possible number of symmetry operations of C_n axes is $n - 1$, since the nth operation is nothing but identity operation (E). Therefore, strictly speaking the possible number of operations generated by C_6 is five, since in C_6^6, the identity operation is (E).

Problem 6. For rotation–reflection axis, show that $S_n^{-1} \cdot S_n = E$

Solution
$$\begin{aligned} S_n^{-1} \cdot S_n &= (C_n^{-1} \, \sigma_h)(C_n \, \sigma_h) \\ &= (C_n^{-1} \cdot C_n)(\sigma_h \cdot \sigma_h) \\ &= E \cdot E \\ &= E \end{aligned}$$

Problem 7. Show that $S_n^n = E$, when n is even

Solution
$$\begin{aligned} S_n^n &= C_n^n \cdot \sigma_h^n \\ &= E \cdot E = E \end{aligned}$$

Problem 8. Show that $S_n^{2n} = E$ when n is odd.

Solution
$$\begin{aligned} S_n^{2n} &= (C_n \cdot \sigma_h)^{2n} \\ &= C_n^{2n} \cdot \sigma_h^{2n} \\ &= E \cdot E = E \end{aligned}$$

Problem 9. Find all the symmetry elements in the following molecules: H_2O, p-dichlorobenzene, NH_3, cyclohexane (boat conformation) XeF_2.

Solution

H₂O :

H_2O has two planes of symmetry. One in the plane of the molecule other $\perp$ to the molecular plane.

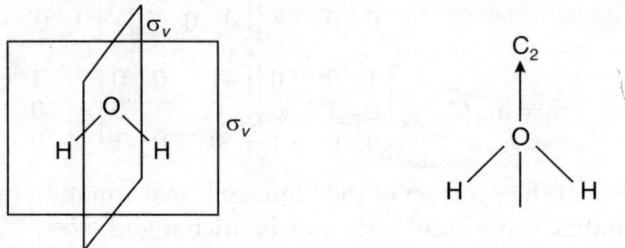

It also has a C_2 axis. Therefore the total symmetry elements $= E + C_2 + 2\sigma_v$

p-dichlorobenzene :

p-dichlorobenzene has their mirror planes:

(i) molecular plane,

(ii) a plane $\perp$ to the molecule, passing through both chlorines

(iii) a plane $\perp$ to the above two planes, bisecting the molecule between the chlorines

It has three C_2 axes, one perpendicular to the molecular plane and two within the plane. One passing through both chlorines and one perpendicular to the axis passing through the chlorines. The molecule has an inversion centre (i).

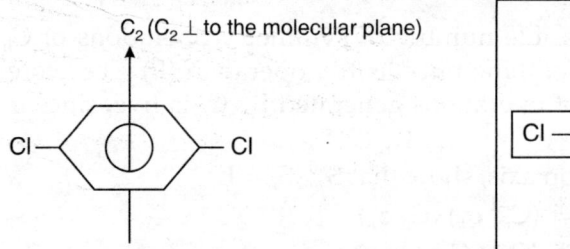

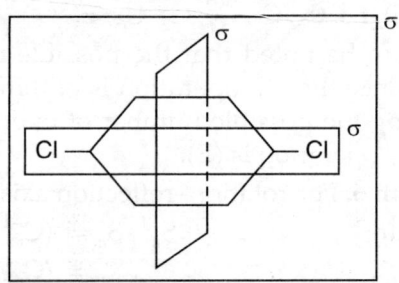

Therefore, the total number of symmetry elements $= E + 3C_2 + 3\sigma + i$

NH₃ :

NH_3 has a three fold axis through the N atom $\perp$ to the plane of the three H-atoms. There are also three mirror planes, each including the N-atom and one H-atom. Therefore, the total number of symmetry elements $= E + C_3 + 3\,\sigma_v$

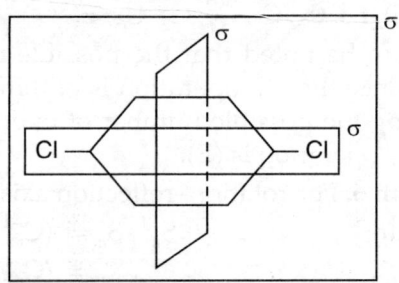

Cyclohexane (boat conformation):

Cyclohexane in the boat conformation has a C_2 axis perpendicular to the plane of the lower four carbon atoms and two mirror planes $(2\sigma_v)$ that include C_2 axis and the two mirror planes are $\perp$ to each other.

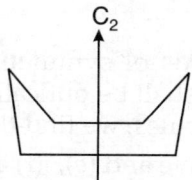

XeF$_2$:

XeF_2 is a linear molecule with a C_α axis through F, Xe, F nuclei and an infinite number of perpendicular C_2 axes, a horizontal mirror plane (which is also an inversion centre), and an infinite number of mirror planes that include C_α axis.

$$F \longrightarrow Xe \longrightarrow F \longrightarrow C_\alpha \text{ axis}$$

Thus we have $E + C_\alpha + \alpha\, C_2 + \sigma_h + \alpha\, \sigma_v + i$

STUDY QUESTIONS

1. What do you mean by symmetry elements and symmetry operations? Distinguish between the terms.
2. Explain with suitable examples the following symmetry elements and the associated symmetry operations.
   (a) Axis of rotation
   (b) Symmetry plane
   (c) Centre of inversion
   (d) Improper axis of rotation
3. Identify the axis of symmetry and plane of symmetry present in the following objects:
   (a) a spoon   (b) butterfly   (c) a ceramic coffee mug with handle (no decoration)
   (d) a dumbbell   (e) a tennis ball   (f) a pencil (no nob).
4. Distinguish amongst the symmetry elements σ_v, σ_h and σ_d.
5. Show that S_n^n (for n = odd) = σ_h and $S_n^n = i$ (m is odd and equal to $n/2$).
6. Show that for H$_2$O molecule, the following relations of operations are correct.
   (i) $\hat{\sigma}_v(xz) \cdot \hat{C}_2 = \hat{\sigma}_v(yz)$
   (ii) $\hat{C}_2 \cdot \hat{\sigma}_v(xz) = \hat{\sigma}_v(yz)$
   (iii) $\hat{\sigma}_v(xz) \cdot \hat{\sigma}_v(yz) = \hat{C}_2$
7. Find the symmetry elements in the following molecule: H$_2$O, p-dichlorobenzene, ethane (staggered conformation), NH$_3$, XeF$_2$.
8. Using necessary diagrams, show that $S_2 \equiv i$ and $S_1 = \sigma$.
9. How does symmetry and optical actively of a compound is related? What is the symmetry criterion of a molecule to be optically active? Which of the following molecules are optically active?

(a) Spirane (b) Allene (c) $[Co(en)_3]^{3+}$ (d) H_2O_2 (e) CF_2Br_2

(f) 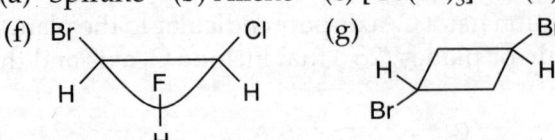 (g)

Hint: Molecules with a S_n-axis of symmetry are optically inactive whereas molecules not having S_n-axis shall be optically active. With this criterion, if we analyse the above given molecules, we find that:

(a), (b), (d), (e), (g) are inactive and (c), (f) are active molecules.

10. What is the relation between symmetry and dipole movement of a molecule? Explain the zero dipole moment of the molecules like CH_4, CO_2, BF_3, ethane (staggered), BeF_2 and definite dipole moment of HCl, H_2O.

11. $[Co(en)_3]^{3+}$ belongs to D_3 point group. Is it optically active? Why?

12. Does $POCl_2Br$ possess a rotation–reflection axis coincident with the P—O bond axis?

13. Is S_2, a symmetry operation for the molecule.

14. What important symmetry element is absent in PF_3 molecule that is present in the D_{3h} point group?

15. How does a D_{2d} complex of formula MCl_4^{2-} differ from a T_d complex (i.e. which symmetry elements differ)?

16. Is the following compound optically active?

17. Locate the σ_d planes in C_6H_6. Are these equivalent to σ_v? Why?

2

Groups, Subgroups, Classes, Generators and Molecular Point Groups

A group is a collection of elements*. All the symmetry operations of a molecule taken together constitute a group, known as *point group*. In water molecule there are four symmetry operation such as E, C_2 and two σ_v planes. This set of four symmetry operations generated by these elements is said to form a symmetry group or point group.

2.1 MATHEMATICAL REQUIREMENTS FOR A POINT GROUP

P, Q, R, S, ... elements form a group provided the following rules are obeyed.

2.1.1 Combination** of Elements

If P, Q and R are the three elements of a group, the combination (product) of two elements on square of an element should produce an other element which is a member of the group.

$$PQ = R,$$
$$P^2 = Q \text{ on } R$$

PQ does not mean P multiplied by Q but P combined with Q according to combining rule. If PQ = QP, in that case the elements are said to be commutable. If all the elements of a group are commutable, the group is said to be Abelian. However, the elements of a group need not necessarily be commutative.

Let us take infinite number of integers, both positive, negative and even zero, ..., –4, –3, –2, –1, 0, 1, 2, 3, 4, ... these elements constitute a group law, the combination law is one of arithmetic addition. The "product" (here arithmetic addition only) of any low elements, say 16 and – 4, which is 12 is already present as a number in the collection, satisfying condition 1.

2.1.2 Associative Law of Combination on Multiplication***

If P, Q, R are the elements of a group, then

$$P(QR) = (PQ)R,$$

i.e. combine P with the combination of QR = Combine PQ with R.

The associative law II is also obeyed for the product (here arithmetic addition only) of any three elements, e.g. 4 1/2 and –6 is {4 + 12 } + (–6) = 4 + {12 + (–6)} = 10

---

* The group elements could be numbers, matrices, symmetry or geometrical operations on a body.
** In group theory the term "combination" is used instead of the term "product" and the term "combine" instead of multiply.
*** Multiplication means simple addition

2.1.3 Commutation of Elements (Identity Rule)

One element (E) in the group must commute with all other members and leave changed. This element E is called the identity element. If P and Q are other elements, we have

$$PE = EP = P \text{ and } QE = EQ = Q$$

E commutes with P, Q in the group.

2.1.4 Reciprocal of Elements (Inverse Rule)

Every member of a group has a reciprocal, which is also an element of the group. For example, the inverse of P is P^{-1}, the inverse of Q is Q^{-1}. If P is an element of a group, P^{-1} should also be another element of the group. An inverse has the property that

$$PP^{-1} = P^{-1}P = E$$
$$QQ^{-1} = Q^{-1}Q = E \text{ and so on.}$$

C_3 has reciprocal element C_3^2 such that $C_3 \cdot C_3^2 = C_3^3 = E$

C_2 has its own inverse $C_2 \cdot C_2 = C_3^3 = E$

i has its own inverse $i \cdot i = i^3 = E$

σ has its own inverse $\sigma \cdot \sigma = \sigma^3 = E$

The inverse of any number is simply the negative of that number and that is also an element of the group. Inverse of 2 is – 2.

$$2 + (-2) = (-2) + 2 = 0$$

Here 2, –2 and 0 are the members of a group.

The above four rules are put into a compact form in Table 2.1.

Table 2.1 Definition of groups

1.	PQ = R	R in the group
2.	P(QR) = (PQ)R	For all elements
3.	PE = EP = P	E in the group
4.	$PP^{-1} = P^{-1}P = E$	P^{-1} in the group

Groups are of the following types:

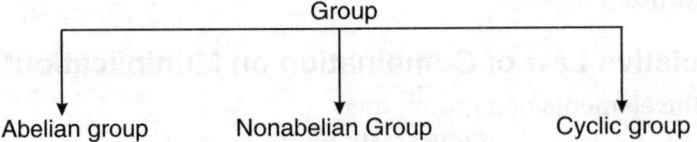

Group

Abelian group   Nonabelian Group   Cyclic group

Abelian group: Any group in which all combination of elements commute is called an abelian group. Two elements A and B commute if AB = BA. For example, in case of H_2O molecule, which belongs to C_{2v} point group possess E, C_2 and two σ_v symmetry operations, i.e four symmetry operations. Combination of the elements do commute

$$\hat{C}_2 \cdot \hat{\sigma}_v(xz) = \hat{\sigma}_v(xz) \cdot \hat{C}_2$$

Nonabelian group: A group is said to be nonabelian if all the elements do not commute with one another.

Cyclic Group: A group is said to be cyclic if all the elements of a group can be generated from one element. For example, X and $X^2, X^3, X^3, X^4, \ldots, X^n$, are the elements of the group with $X^n = E$ (identity), such group is called a *cyclic group*.

G_1^3	E	X	X^2
E	E	E	X^2
X	X	X^2	E
X^2	X^2	E	X

G_1^3 means it is group of order 3 of one type.

Let us consider the set of symmetry operations of molecules, and see pointwise, if they satisfy the condition, are called a group.

   i. We shall consider first H_2O molecule (C_{2v} point group). In C_{2v} point group, the operation elements are E, C_2 σ_v (xz) and σ_v (yz).

      Figure 2.1 illustrates that the symmetry operations of H_2O molecule (C_{2v} point group) obey the commutative law. C_2 operation followed by σ_v operation leads to the same configuration (Fig. 2.1).

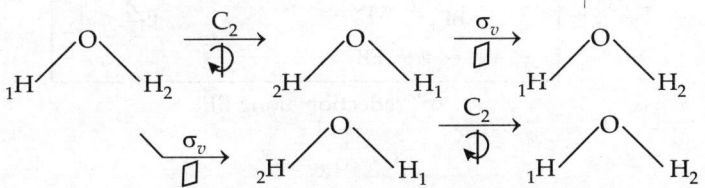

Fig. 2.1 Diagram illustrating the idea that symmetry operations of water molecule obey the commutative law of a group

Thus $\sigma_v C_2 = C_2 \sigma_v$. It can be seen from the multiplication table (Table 2.2) that the combination of low symmetry operation or square of an operation is equal to another symmetry operation

$$C_2 \sigma_v \ (xz) = \sigma_v \ (yz)$$

Rule I: Combination of two elements C_2 and $\sigma_v(xz)$ produce a third element $\sigma_v \ (xz)$ which is a member of C_{2v} point group.

Rule II: $C_2 \cdot C_2 = C_2^2 = E$ (Combination of two elements) $\sigma_v(xz) \ \sigma_v(xz) = \sigma_v(xz) \ C_2$ and C_2 or $\sigma_v(xz)$ and $\sigma_v(xz)$ produce a third element E. E is a member of C_{2v} group.

All symmetry operations of H_2O molecule are commutable hence it is an abelian group.

$C_2 \sigma_v(xz) \ \sigma_v(xz) = (C_2 \ \sigma_v(xz)) \ \sigma_v(yz) = E$ (by rule II). It can be shown by considering the change in the direction of vectors on performing C_2 and $\sigma_v(xz)$ operations. On performing operation C_2, the coordinate X, Y change but Z remains unchanged. On performing $\sigma_v(xz)$ operation, the following effects will arise.

$$\begin{bmatrix} X \\ Y \\ Z \end{bmatrix} \xrightarrow{C_2} \begin{bmatrix} -X \\ -Y \\ Z \end{bmatrix} \xrightarrow{\sigma_v(xz)} \begin{bmatrix} -X \\ Y \\ Z \end{bmatrix} \xrightarrow{\sigma_v(yz)} \begin{bmatrix} X \\ -Y \\ Z \end{bmatrix} \xrightarrow{C_2} \begin{bmatrix} -X \\ Y \\ Z \end{bmatrix}$$

In case of BF_3 molecule, with symmetry operation E, $2C_3$, $3\sigma_v$, $3C_2$, $2S_3$, σ_h, the combination of two symmetry operations results in another symmetry operation but they are not commutable.

$$\sigma_v' \cdot C_v' = \sigma_v''$$

but

$$\sigma_3' \cdot C_v' = \sigma_v'''$$

This relation is shown below (Fig. 2.2)

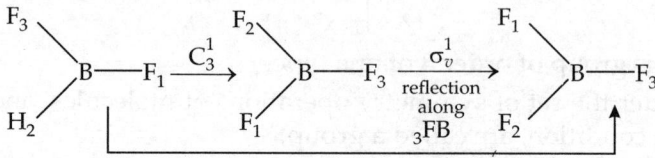

σ_v'' reflection along BF_2

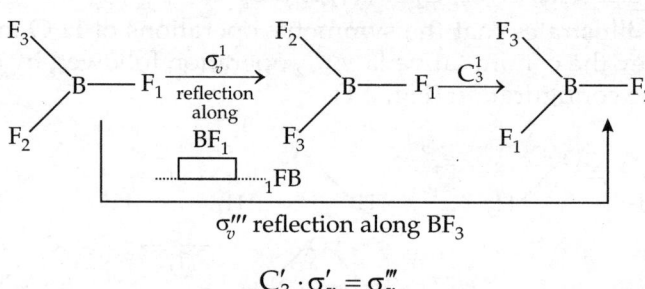

σ_v''' reflection along BF_3

$$C_3' \cdot \sigma_v' = \sigma_v'''$$

Fig. 2.2. Symmetry operation of BF_3 molecule does not obey the commutative law of a group

ii. In sets of symmetry operations of all molecules, the identity operation (E) is present. This is commutable with all other operations and leaves them unchanged.

In H_2O, $C_2 E = E.C_2 = C_2$

In BF_3, $C_3^1 \cdot E = E \cdot E \cdot C_3^1 = C_3^1$

Thus, through E operation means doing nothing, but is important in qualifying the molecule to be called a group.

iii. Associative law of combination holds good in all the molecules.

For example in H_2O molecule,

$$C_2 \sigma_v(xz)\, \sigma_v(yz) = (C_2 \cdot \sigma_v(xz)) \cdot \sigma_v(yz)$$

as

$$C_2\,(\sigma_v(xz)\, \sigma_v(xz)) = C_2 \cdot C_2 = C_2^2 = E$$

and

$$C_2 \cdot \sigma_v(xz) \cdot \sigma_v(yz) = \sigma_v(yz) \cdot \sigma_v(yz) = E$$

iv. The symmetry operations in molecule have their inverse which are also symmetry operations of that molecule.

In H_2O molecule,

$$C_2 \cdot C_2 = C_2^2 = E$$

$$\sigma_v(xz)\, \sigma_v(xz) = E$$

$$\sigma_v(yz)\, \sigma_v(yz) = E$$

In NH_3 molecule, however, a rotation of 120° (C_3^1), if followed by rotation of 240° (C_3^2), will give back the original orientation (E).

Thus $C_3^1 \cdot C_3^2 = C_3^2 = E$. Hence C_3^2 is the reciprocal of C_3^1.

In general for a rotation C^{n-m}.

Hence a molecule can be called a group if the symmetry operations of the molecule satisfy all the conditions of a group.

2.2 GROUP MULTIPLICATION TABLE AND MULTIPLICATION OF SYMMETRY OPERATIONS

2.2.1 The Group Multiplication Table

The symmetry operations of a molecule constitute a group. Each group is characterised by its multiplication table. In multiplication table, the operation indicated in the top row is corrected out first, followed by the operation indicated in the first column. The total number of elements (symmetry operations) present in a group is known as *order of the group*. In a multiplication table each symmetry operation occurs once and only once in each column and row. Each matrix element is the result of the product of the column element (C) times the row element (R). In the product (RC), we take column element (C) first followed by row (R). The result of the binary combination is shown in the body of the table. Thus the relationship between elements of a group for the binary combinations is reflected in the multiplication table (Table 2.2).

Every operation (A) has a reciprocal (A^{-1}) such that

$$A^{-1} \cdot A = A \cdot A^{-1} = B$$
$$EE = B; C_2C_2 = E; \sigma_v \sigma_v = E \text{ and } \sigma_v \sigma_v' = E.$$

Table 2.2 Group multiplication table of the point group C_{2v}

C_{2v}	E	C_2	$\sigma_v(xz)$	$\sigma_v(yz)$
E	E	C_2	$\sigma_v(xz)$	$\sigma_v(yz)$
C_2	C_2	E	$\sigma_v(yz)$	$\sigma_v(xz)$
$\sigma_v(xz)$	$\sigma_v(xz)$	$\sigma_v(yz)$	E	C_2
$\sigma_v(yz)$	$\sigma_v(yz)$	$\sigma_v(xz)$	C_2	E

Thus each operation in this case happens to be its own inverse. For a proper rotation C_n^m, the inverse is

$$C_n^m \times C_n^{n-m} = E$$

For example, C_4 is the reciprocal of C_4^3 (and *vice versa*) since

$$C_4 \cdot C_4^3 = C_4^3 \cdot C_4 = E$$

Similarly, C_3 is the reciprocal of C_3^2

$$C_3 \cdot C_3^2 = C_3^2 \cdot C_3 = E$$

C_2 has its own reciprocal, $C_2 \cdot C_2 = E$.

For any mirror plane the inverse is the identical mirror plane, e.g.

$$\sigma_v \times \sigma_v = E, \ \sigma_v' \times \sigma_v' = E$$

The four symmetry operations (E, C_2, σ_v, σ_v') of H_2O constitute a point group of order 4 and the multiplication table for the group is presented in Table 2.3.

Table 2.3 Group multiplication table of point group C_{2v}

Second operation	First operation			
	E	C_2	σ_v	σ_v'
E	$\boxed{E}$	C_2	σ_v	σ_v'
C_2	C_2	$\boxed{E}$	σ_v'	σ_v
σ_v	σ_v	σ_v'	$\boxed{E}$	C_2
σ_v'	σ_v'	σ_v	C_2	$\boxed{E}$

2.2.2 Groups of Order 1, 2 and 3

We shall now systematically examine the possible abstract groups of low order using their multiplication tables to define them. A group of order 1 consists of identity element alone. It is designated a G_1^1.

The symbol G_1^1 means that it is a group of order 1 of one type with one element E.

$$\begin{array}{c|c} G_1^1 & E \\ \hline E & E \end{array}$$

Group of order 2 contains two elements E and A. It appears as G_1^2 with elements E and A.

$$\begin{array}{c|cc} G_1^2 & E & A \\ \hline E & E & A \\ A & A & E \end{array} \qquad \begin{aligned} E \times E &= E \\ E \times A &= A \times E = A \\ A \times A &= E \end{aligned}$$

For a group of order 3, the number of elements are 3 (E, A and B). The symbol G_1^3 means that it is a group of order three of one type. Part of the multiplication table is given below.

$$\begin{array}{c|ccc} G_1^3 & E & A & B \\ \hline E & E & A & B \\ A & A & & \\ B & B & & \end{array} \qquad \begin{aligned} E \times E &= E \\ A \times E &= E \times A = A \\ B \times E &= E \times B = B \end{aligned}$$

There are two ways to complete the above table. It we take AA = E then BB will also be E(BB = E) and we may augment the above incomplete table as below.

$$\begin{array}{c|ccc} & E & A & B \\ \hline E & E & A & B \\ A & A & \text{\textcircled{E}} & \\ B & B & & \text{\textcircled{E}} \end{array} \qquad \begin{aligned} A \times A &= B \\ B \times B &= E \end{aligned}$$

In order to fill up the remaining gap present in last column last row we assume

$$BA = A \text{ or } BA = B$$

when BA = A, makes A element repeat in a column. So this cannot be accepted

	E	A	B
E	E	A	B
A	A	E	
B	B	A	E

when BA = B, makes B element repeat in a row. This also cannot be accepted.

	E	A	B
E	E	A	B
A	A	E	
B	B	B	E

In order to make the table proper, we can consider this entire group to be generated by taking an element and all its powers, e.g. A, A^2 (= B) and A^3 (= E). Such a group is a *cyclic* group. For such a group, all multiplications must commute. Thus the complete table may be written as below.

$B = A^2, C = A^3$

$$\therefore \quad AB = A \cdot A^2 = A^3 = E$$
$$BA = A^2 \cdot A = A^3 = E$$
$$B^2 = (A^2)^2 = A^3 \cdot A = E \cdot A = A$$

G_1^3	E	A	B
E	E	A	B
A	A	B	E
B	B	E	A

An important property of the *cyclic group** is that they are *abelian*, a group for which all the elements do commute.

2.2.3 Groups of Order 4

There will be a cyclic group of order 4.

Following usual format we may first write the the partial table as below.

G_4^4	E	A	B	C
E	E	A	B	C
A	A			
B	B			
C	C			

---

* Cyclic group: If all the elements of a group can be generated from one element, say A, and A, A^2, A^3, ..., A^n are the elements of the cyclic group of order n.

We shall now employ the following relationship among the elements E, A, B, C.

$$A = X; B = X^2; C = X^3; E = X^4$$

Thus $A \cdot A = X = X \cdot X = X^2 = B$ $\quad$ $B \cdot A = X^2 \cdot X = X^3 = C$
$\quad\quad$ $A \cdot B = X \cdot X^2 = X^3 = C$ $\quad\quad$ $B \cdot B = X^2 \cdot X^2 = X^4 = E$
$\quad\quad$ $A \cdot C = X \cdot X^3 = X^4 = E$ $\quad\quad$ $B \cdot C = X^2 \cdot X^3 = X^4 \cdot X = E \cdot A = A$

$$C \cdot A = X^3 \cdot X = X^4 = E$$
$$C \cdot B = X^3 \cdot X^2 = X^4 \cdot X = E \cdot A = A$$
$$C \cdot C = X^3 \cdot X^3 = X^4 \cdot X^2 = E \cdot B = B$$

The above relations suggest the following table.

G_1^4	E	A	B	C
E	E	A	B	C
A	A	B	C	E
B	B	C	Ⓔ	A
C	C	Ⓔ	A	B

We note that for G_1^4 only one element, namely B, is its own inverse. This cannot be accepted. On the other hand, if we accept all elements have their own inverse then we shall get the part of the table as

	E	A	B	C
E	E	A	B	C
A	A	Ⓔ		
B	B		Ⓔ	
C	C			Ⓔ

and the complete multiplication table will be

G_2^4	E	A	B	C
E	E	A	B	C
A	A	E	C	Ⓑ
B	B	C	E	A
C	C	Ⓑ	A	E

It is to be noted that $A \cdot C = C \cdot A = E$. But in second row and last column E, A, C elements are already occupied in their respective places. There E element should not be put in the second row and last column instead we should place B element at the place. This is equally true for 2nd column and last row.

The four symmetry operations (E, C_2, σ_v, σ_v') of water molecule constitute a point group of order 4 and the multiplication table for the group is presented in Table 2.3.

We shall now consider group multiplication table of the C_{3v} point group. Example of this point group belongs to PCl_3. This molecule is characterised by the existence of a C_3 axis and three σ_v planes (Fig. 2.3).

The six symmetry operations E, C^3, C_3^2, $\sigma_v(1)$, $\sigma_v(2)$, $\sigma_v(3)$ of PCl_3 constitute a point group of order 6 and the multiplication table for the group is presented in Table 2.4.

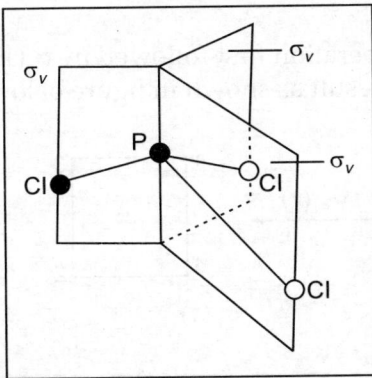

Fig. 2.3

2.3 THE MULTIPLICATION OF SYMMETRY OPERATIONS

Let us consider a square planar molecule AB_4. This molecule has E, C_4, C_4^2, (= C_2), C_4^3, C_2' (1), C_2' (2), $C_2'' = (2)$, i, S_4, S_4^3, (σ_h) $\sigma_v(1)$, $\sigma_v(2)$, $\sigma_d(1)$, $\sigma_d(2)$. We shall carry out two operations, for example $\sigma_v(1)$ followed by $C_2''(1)$. This is expressed as $C_2'' \cdot \sigma_v(1)$, the illustration that follows depicts these two successive operations (Fig. 2.4).

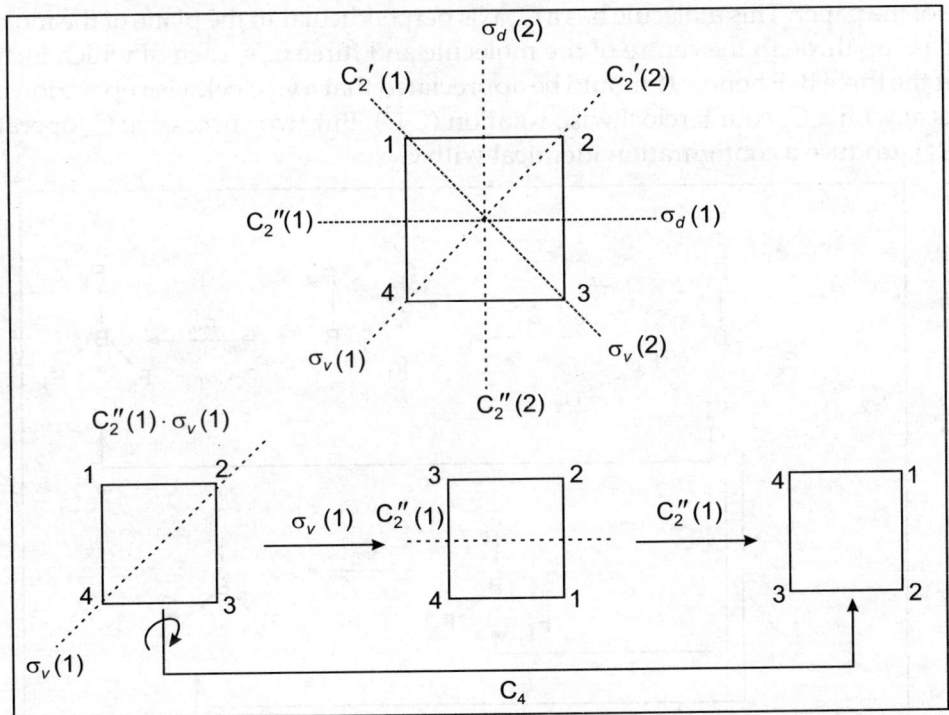

Fig. 2.4 C_2'' (1) . $\sigma_v(1)$ = C_4 in D_{4h}, two successive operations on a square are equal to the other operation in the group

Hence C_2'' (1) $\cdot$ $\sigma_v(1) = C_4$.

But if we carry C_2'' (1) operation first followed by $\sigma_v(1)$ in a square planar molecule AB_4 we shall get different result as shown in figure below.

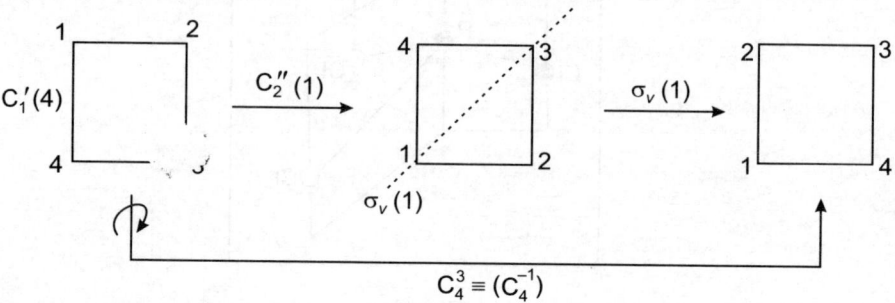

$$C_4^3 \equiv (C_4^{-1})$$

For $\sigma_v(1), C_2'' = (1): C_2'' = (1).$

Here we get $\sigma_v(1) \cdot C_2''$ (1), $C_4^3 \equiv C_4^{-2}$ (counter-clockwise rotation).

Hence

$$C_2'' (1) \cdot \sigma_v(1) = C_4 \neq C_4^{-1} = \sigma_v(1), C_2'' (1).$$

So the operations do not commute.

A planar BF_3 molecule to be oriented with the three fluorines labelled 1, 2 and 3 in the plane of the paper. This molecule has a C_3 axis perpendicular to the plane of the molecule and passing through the centre of the molecule and three σ_v's, each of which includes one of the three B–F bonds. It should be appreciated that a C_3 clockwise operation is not identical with a C_3 counterclockwise rotation (C_3^{-1}). But two successive C_3 operations (i.e. C_3^2) produce a configuration identical with C_3^{-1}.

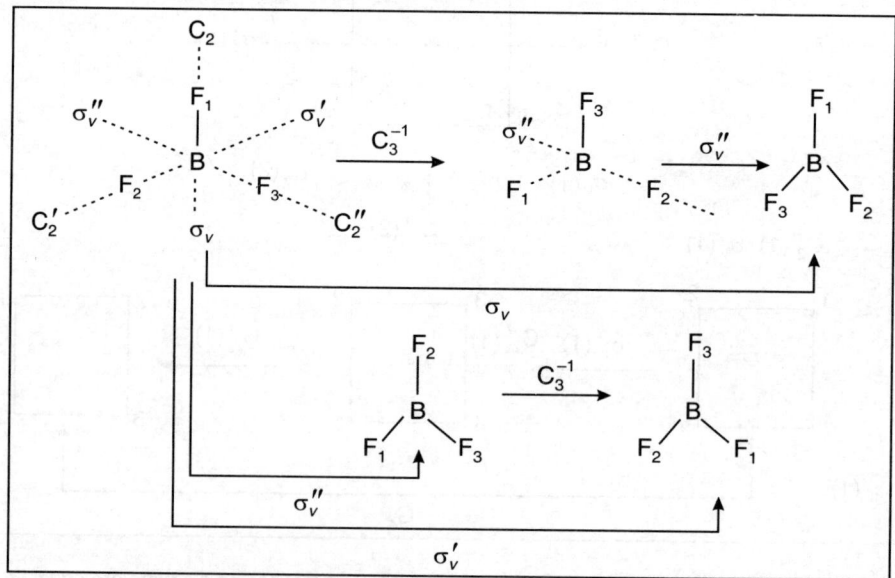

Fig. 2.5 Noncommutative operations σ_v'', $C_3^{-1} = \sigma_v \neq \sigma_v' = C_3^{-1} \cdot \sigma_v''$

Figure 2.5 indicates that $\sigma_v'' \cdot C_3^{-1} = \sigma_v \neq \sigma_v' = C_3^{-1} \cdot \sigma_v''$ and the operations do not commute.

Cis-dichloroethylen $\overset{Cl}{\underset{H}{\diagdown}}C=C\overset{Cl}{\underset{H}{\diagup}}$ belongs to C_{2v} group. The C_2 axis is collinear with the z axis in the plane of the molecule. There are two symmetry planes $\sigma_v(xz)$ and $\sigma_v(yz)$ · $\sigma_v(xz)$ is coplanar with xz plane and $\sigma_v(yz)$ with the molecular plane yz.

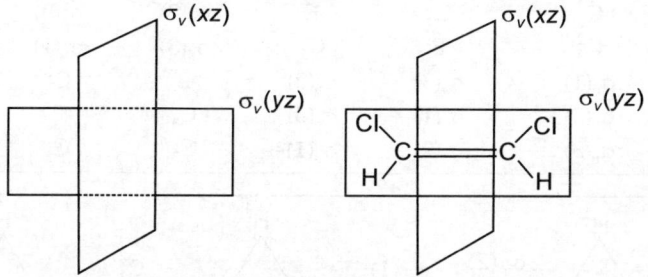

The symmetry operation of this molecule are E, C_2, $\sigma_v(xz)$ and $\sigma_v(yz)$. To follow the specific entries in the multiplication table (Table 2.3), one may refer to Fig. 2.6, where the steps leading to the final results for $\sigma_v(xz)$ C_2 and $\sigma_v(xz)$ $\sigma_v(yz)$.

Fig. 2.6 Products of $\sigma_v(xz)$, C_2(A) and $\sigma_v(xz)$, $\sigma_v(yz)$ (B) in *cis*-dichloroethylene

Ammonia molecule belongs to C_{3v} group. The six symmetry operations $\hat{E}, \hat{C}_3, \hat{C}_3^2$, $\sigma_v(1)$, $\sigma_v(2)$, $\sigma_v(3)$ constitute C_{3v} group. Ammonia molecule is not planar but trigonal pyramidal. The three H's occupy the three vertices of an equilateral triangle. The N atom is raised above the centre of the basal plane defined by the three H's. Rotation axis C_3 passes through the N atom and the centroid of the equilateral triangle. The entries in Table 2.4 can be ascertained by checking the effects of the successive symmetry operations on the equilateral triangle in the manner as shown for the sample cases $C_3^2\sigma_v(1)$, $C_3\sigma_v(1)$, $\sigma_v(2)$, $\sigma_v(1)$, $\sigma_v(2)$, $\sigma_v(3)$ and $\sigma_v(3) \cdot \sigma_v(2)$ (Fig. 2.7).

Table 2.4 Group multiplication table for the point group C_{3v}

Second Operation	First Operation					
	E	C_3	C_3^2	$\sigma_v(1)$	$\sigma_v(2)$	$\sigma_v(3)$
E	E	C_3	C_3^2	$\sigma_v(1)$	$\sigma_v(2)$	$\sigma_v(3)$
C_3	C_3^2	C_3^2	E	$\sigma_v(2)$	$\sigma_v(3)$	$\sigma_v(1)$
C_3^2	C_3^2	E	C_3	$\sigma_v(3)$	$\sigma_v(1)$	$\sigma_v(2)$
$\sigma_v(1)$	$\sigma_v(1)$	$\sigma_v(3)$	$\sigma_v(2)$	E	C_3^2	C_3
$\sigma_v(2)$	$\sigma_v(2)$	$\sigma_v(1)$	$\sigma_v(3)$	C_3	E	C_3^2
$\sigma_v(3)$	$\sigma_v(3)$	$\sigma_v(2)$	$\sigma_v(1)$	C_3^2	C_3	E

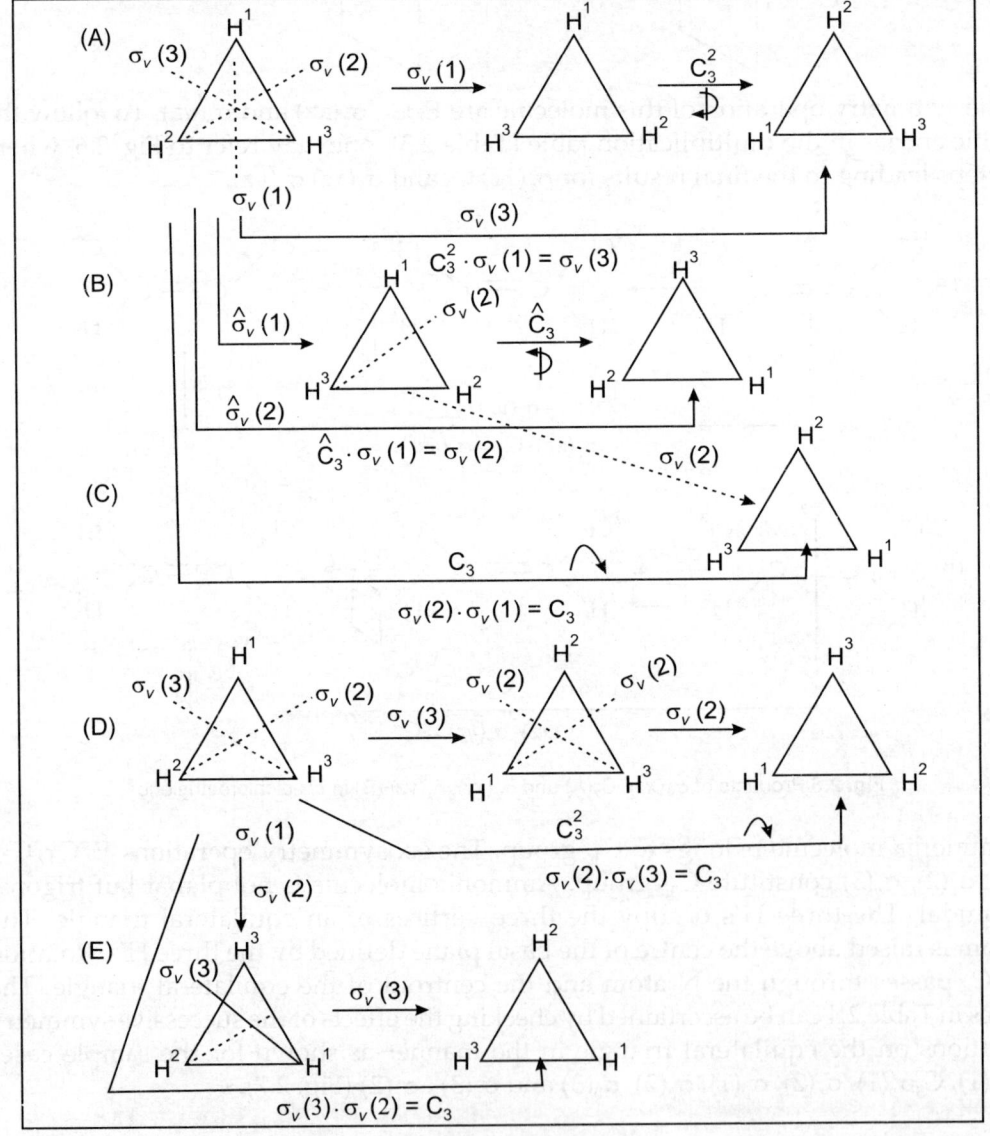

Fig. 2.7 Products $C_3^2 \cdot \sigma_v(1)$ (A), $C_3 \cdot \sigma_v(1)$ (B), $\sigma_v(2)$, $\sigma_v(1)$ (C), $\sigma_v(2)$, $\sigma_v(3)$ (D), $\sigma_v(3)$, $\sigma_v(2)$ (E) in ammonia molecule

2.4 GENERATOR OR GROUP GENERATING ELEMENTS

Generator can be defined as the minimum number of symmetry operation of a molecule, which of the self product on or inter product gives all the symmetry operations of the point group. For example, a cyclic group of order n has only one generator say X and its elements are X, X^2, X^3, X^4, ..., X^n (= E). C_3 group needs only a single element C_3' as the group generation. The complete group has the following elements:

$$C^3 \rightarrow E, C_3^1, C_3^2$$

C_3^2 (C_3^1) and E ($C_3^1 \cdot C_3^1 \cdot C_3^1$) are generated with the help of C_3^1 ($C_3^1 \cdot {}^1_3$) are generated with the help of C_3' element above. In general, for all types of group C_n' is the only generator

D_{3h} point group contains 12 elements

$$D_{3h} \rightarrow E, C_3^1, C_3^2, C_2, C_2^1, C_2', C_2'', \sigma_v, \sigma_v', \sigma_v'', \sigma_h, S_3^1, S_3^5$$

The group generating elements are only three for this group. They are:

C_3^1, C_2 perpendicular to C_3^1 and σ_h. C_3^1, generate E, C_3^2

C_3^1, C_2 combination gives other C_{2s} and σ_v S

C_3^1, and σ_h generates S_3 type of elements, Table 2.5 lists some of the common point groups along with their group generating elements.

Table. 2.5 Point groups along with their group generating elements

Point group	Group generating elements
C_1	C_1
C_s	σ
C_i	i
C_n	C_n^1
C_{nv}	C_n^1, σ_v
C_{nh}	C_n^1, σ_h
D_n	$C_n^1, \sigma_2{}^*$
D_{nd}	$C_n^1, C_2{}^*, \sigma_d$
D_{nh}	$C_n^1, C_2{}^*, \sigma_h$
S_n (n = even)	S_n^1
$C_{\infty v}$	C_∞, σ_v
$D_{\infty h}$	$C_\infty, C_2{}^*, \sigma_h$

*C_2 is perpendicular to C_n or C_∞, or as the case may be.

2.5 CLASSES

A set of elements which are conjugate to one another is called a *class* of the group, C_{3v} point group contains six elements E, $C_3^1, C_3^2, \sigma_v, \sigma_v', \sigma_v''$. But *there* will be only *three* classes. These are E class (h = 1), C_3 class (h = 2) and σ_v class (h = 3).

$\therefore C_{3v}$: E, $2C_3$, $3\sigma_v$. It is written as

C_{3v}	E	$2C_3$	$3\sigma_v$
Γ_1			
Γ_2			
Γ_3			

Note: $2C_3$ is C_3^1 and C_3^2, $3\sigma_v$ is σ_v, σ_v', σ_v''

C_{2v} contains 4 symmetry elements E, C_2, σ_v and σ_v' and four classes C_{2v} : E, C_2, $\sigma_v(xz)$, $\sigma_v'(yz)$

C_{2v}	E	C_2	σ_v	σ_v^1
Γ_1				
Γ_2				
Γ_3				

Note: $\sigma_v(xz)$, $\sigma_v^1(yz)$ are not the same class.

In order to understand classes in a group, similarity transformation of elements has to be appreciated.

2.5.1 Classes of Symmetry Operations

Similarity Transformation

If A and B are the elements of a group and X is another element of the same group such that

$$X^{-1} A X = B \qquad \qquad ...(2.1)$$

the point element B is called the similarity transform of A by X. Here A and B are conjugate elements.

$\sigma_v C_3^1 \sigma_v = C_3^2$ (here C_3^1 and C_3^2 are conjugate elements)

$\sigma_v C_3^2 \sigma_v = C_3^1$

In a group, elements which are conjugate to each other are said to belong to the *same class*. All σ_v planes are in one class. Similarly C_3 and C_3^2 belong to the same class. In a molecular point group, the operations which are conjugate to each other are consider to belong to the same class.

Properties of Conjugate Elements

1. Every element is conjugate to itself (self conjugation), i.e $X^{-1} A X = A$, here X is identity operation

   $E C_3^1 E = C_3^1$, $E \sigma_v E = \sigma_v$; $E C_3^2 E = C_3^2$

2. If A is conjugate with B, B must be conjugate with A (mutual conjugation), i.e. if $X^{-1} A X = B$, then there must be an element Y present such that

   $$Y^{-1} B Y = A$$
   $$C_3^2 \sigma_v C_3' = \sigma_v'', \sigma_v'' \sigma' = \sigma_v;$$
   $$\sigma_v C_3^1 \sigma_v = C_3^2; \sigma_v', C_3^2 \sigma_v' = C_3^1$$

3. If A is conjugate with B and C, then B and C are conjugate with each other (associative conjugation).

$$X^{-1} A X = B \text{ and } Y^{-1} A Y = C \text{ then}$$
$$X^{-1} B X = C \text{ and } Y^{-1} C^{-1} Y = B.$$
$$C_3^2 \sigma_v C_3^1 = \sigma_v''; C_3^1 \sigma_v C_3^2 = \sigma_v'$$

then
$$C_3^2 \sigma_v'' C_3^1 = \sigma_v'; C_3^1 \sigma_v' C_3^2 = \sigma_v''$$

It is clear that elements which are conjugate to each other belong to C_3^1 and C_3^2 are mutually conjugate. So they can be in one class. All planes σ_v, σ_v', σ_v'' are in one class.

Let us see whether two σ_v planes in C_{2v} point group form a class or not ?

$$C_{2v} \text{ has } E + 1C^2 + 2\sigma_v.$$

Identity (E) forms its order class of order 1.

C_2 operation also forms a class of its order of the order 1 we apply similarity transformation on one σ_v and if we get other σ_v, that the two σ_v will form a class because they will be conjugate to each other.

Consider H_2O molecule. It has C_2 as an element, we consider C_2^{-1}, $\sigma_v(yz)$. C_2

We perform symmetry operations on H_2O molecule one by one from right to left.

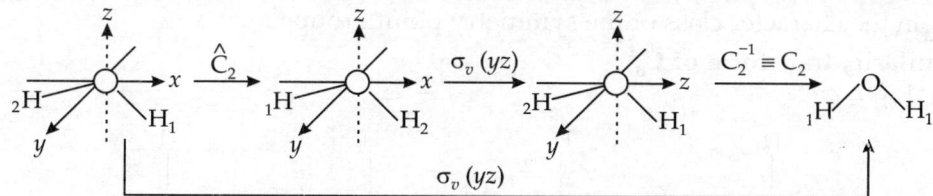

Thus 3 consecutive operation C_2, $\sigma_v(yz)$ and C_2^{-1} ($\equiv C_2$) can be obtained by a simple operation $\sigma_v(xz)$

$\therefore$
$$C_2^{-1}, \sigma_v(yz) . C_2 = \sigma_v(yz) \neq \sigma_v(xz)$$

Similarly we take $\sigma_v(xz)$ and apply similarity transformation by C_2, we get

$$C_2^{-1}, \sigma_v(xz) . C_2 = \sigma_v(xz) \neq \sigma_v(xz)$$

Thus we see that $\sigma_v(xz)$ and $\sigma_v(yz)$ are not related by similarity transformation relation. So $\sigma_v(xz)$ and $\sigma_v(yz)$ will not form a class of the C_{2v} point group. We have to put them individually and C_{2v} will have **four** classes E, C_2, $\sigma_v(xz)$, $\sigma_v(yz)$.

In case of C_{3v} point group, we have three vertical places σ_v, σ_v', σ_v''. They form a class and will be put as $3\sigma_v$ in the character table of C_{3v} point group.

Verify σ_v, σ_v', σ_v'' form a class.

We take NH_3 molecule. We assume NH_3 as a planar molecule. It belong to C_{3v} point group.

We apply similarity transformation as $C_3 \sigma_v C_3^1$

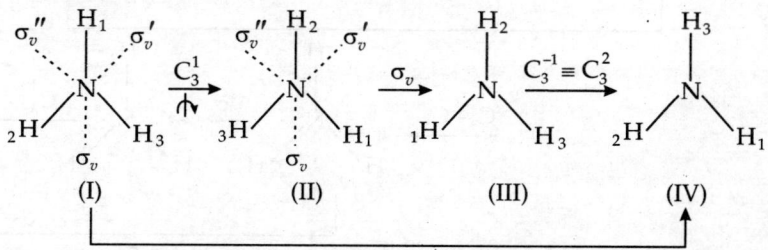

Note that planes always retain σ_v' their orientation
wrt the organic configuration of the molecule (I)

Thus $$C_3^{-1}\,\sigma_v\,C_3^{1} = \sigma_v{}' \neq \sigma_v{}''$$

Hence σ_v and $\sigma_v{}'$ are conjugate to each other. So they will form a class.

Now take $\sigma_v{}'$ and apply similarity transformation as $C_3^{1}\,\sigma_v{}'\,C_3^{-1}$.

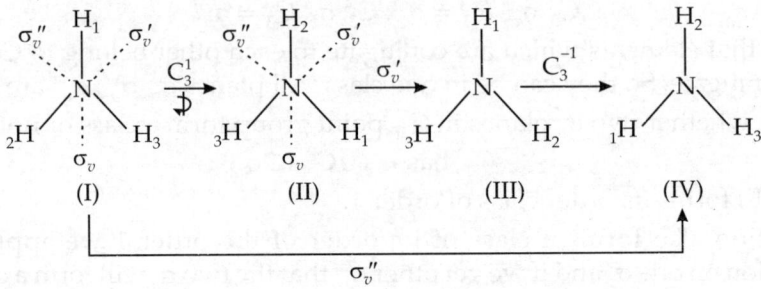

(I) (II) (III) (IV)

Thus $$C_3^{-1}\,\sigma_v{}'\,C_3^{1} = \sigma_v{}'' \neq \sigma_v{}'$$

Hence $\sigma_v{}'$ and $\sigma_v{}''$ are conjugate to each other and will form a class.

Thus σ_v, $\sigma_v{}'$, $\sigma_v{}''$ are conjugate to each other and form a class. We put them together as $3\sigma_v$ in the character class of the symmetry point group C_{3v}.

Similarity transform of C_3^{1}

$$\sigma_v\,C_3^{1}\,\sigma_v = C_3^{2}$$

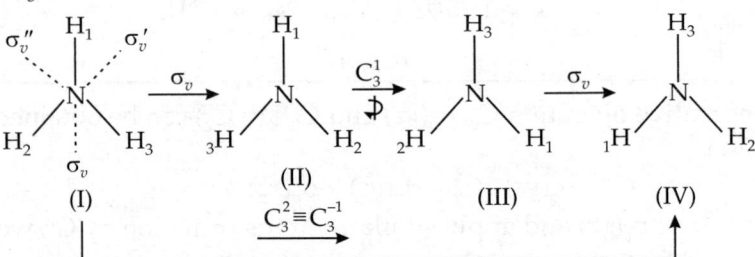

(I) (II) (III) (IV)

Note that planes always retain their orientation with respect to the original configuration of the molecule (I).

Similarity transform of σ_v

$$C_3^{2}\,\sigma_v\,C_3^{1} = \sigma'' \text{ and } C_3^{1}\,\sigma_v\,C_3^{2}\,\sigma_v{}'$$

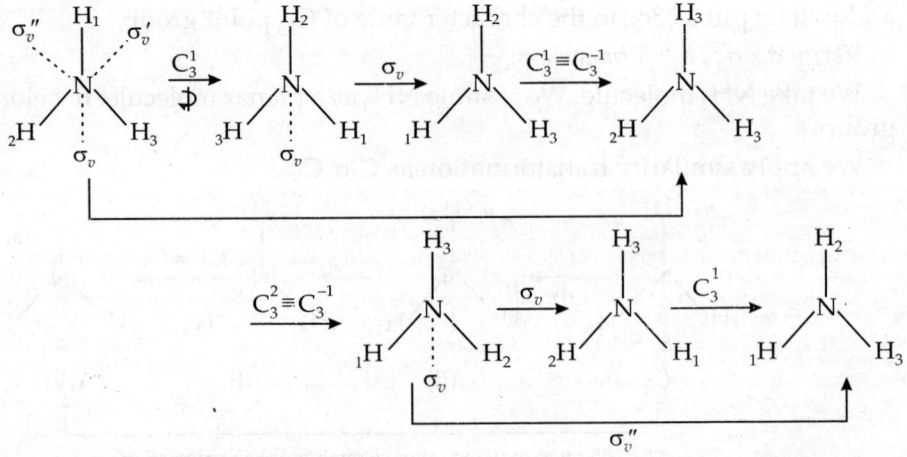

Thus $$C_3^{1}\,\sigma_v\,C_3^{2} = \sigma_v{}''$$

2.6 MOLECULAR POINT GROUPS

In performing all the symmetry operations, the physical properties of the molecule do not change.

The centre of symmetry, or in its absence the centre of gravity of the molecule remains unstable during the symmetry operation. This is because all the symmetry operations pass through the centre of symmetry or centre of gravity. Hence the molecular group is termed as point group.

2.6.1 Notation of Point Groups

There are two types of symbolism in this context, they are Schoenflies notation and Hermann–Mauguin notation. We shall discuss, here in this book, only the molecular symmetry point groups using Schoenflies notations.

Schoenflies notation

 (i) The main symbol (alphabetic) used refers to the axis of highest symmetry in the molecule thus

C stands for highest-fold proper axis

S stands for highest-fold improper axis and

D stands for highest-fold proper axis in combination with nC_2 perpendicular to it
T, O and I are the specially chosen symbols to represent the highly symmetric tetrahedral, octahedral (cube) and icosahedral (dodecahedral) groups.

 (ii) The numerical subscripts indicate the order of the highest order rotational axis. For example, C_n point group contains only C_n axis and the order of the group $h = n$, C_2 group has order = 2

(iii) Further labeling with alphabetical subscripts indicates the presence of certain type of planes of symmetry. Thus V is used for the presence of vertical planes h is used for a place perpendicular to a rotational axis (σ_h) d is used for the presence of dihedral planes (σ_d). In C_{nv} point group, order of the group of the group $h = 2n$ and contains only $C_2 + 2\sigma_v$.

(iv) The subscript i alone (without numerical subscripts) is used where the molecule contains only i element. The point group is C_i.

 (v) The subscript S alone is used when the molecule contains only a plane of symmetry (σ). The point group is C_s.

(vi) The linear molecules are given the symbols $C \propto h$ and $D \propto h$ depending on the absence or the presence of i, in centre of inversion in linear molecules.

Thus we see that point groups have levels such as C_2, C_3, C_{2v}, C_{3v}, D_{2h}, D_{3h}, T_d, O_h etc. A molecule in the point group C_n has only one symmetry element, an n-fold rotation axis (of course, are molecules posses the identity element E). If a molecule contains a horizontal mirror plane perpendicular to its C_n rotation axis then that molecule will belong to the point group C_{nh}. The point group C_{nv} includes those molecules which have n vertical planes containing the rotation axis C_n. Water molecule has a C_2 axis and two vertical planes ($\sigma_v(xz)$, $\sigma_v(yz)$).

Most of the molecules and all crystals belong to one of following four broad divisions into which the point symmetry groups can be classified.

They are

 i. Molecules of low symmetry or nonaxial point groups
 ii. Molecules of high symmetry or axial point groups and

iii. Cubic point groups and
iv. Special point groups of linear molecules.

2.6.2 Molecules of Low symmetry (Nonaxial Groups)

A point group which does not have an axis of proper rotation except C_1, is called a nonaxial point group.

They can have following generations:

(a) **Only identity, (i.e. C_1) is present as generator**

Only possible operation is C_1 (or E) and the point group symbol is C_1.

Such molecules can be said to be completely unsymmetrical (Fig. 2.8). Examples are given below.

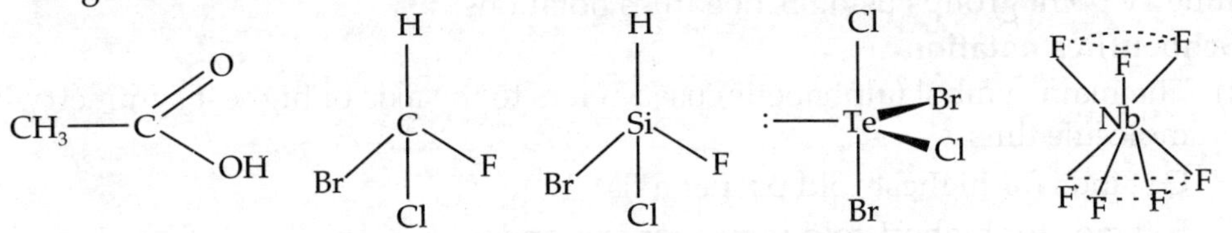

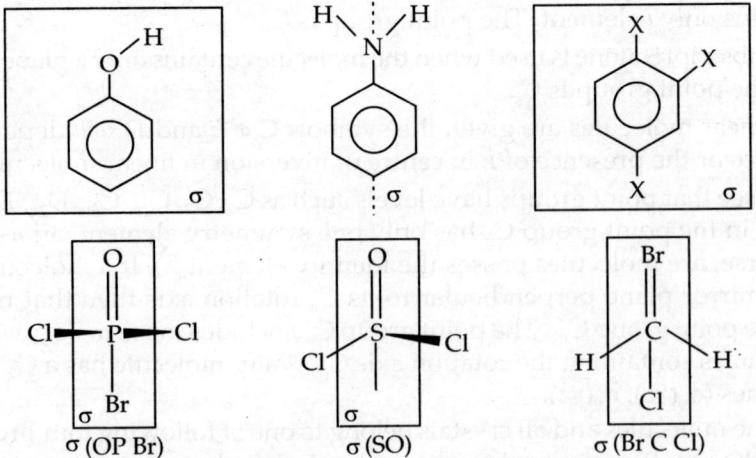

Fig. 2.8 Molecules belonging to C_1 point group (order $h = 1$; E)

(b) **Only σ is the generator**

The possible symmetry operations are σ and E, note that E can be generated by self product of, i.e. $\sigma \cdot \sigma = E$. For such molecules, the point group symbol is C_s. The subscript s stands for the presence of lone mirror plane. The order (h) of this point group is two ($h = 2$; E, σ) (Fig. 2.9).

Example: Hypochlorous acid (HOCl) has only one plane of symmetry, i.e. molecular plane.

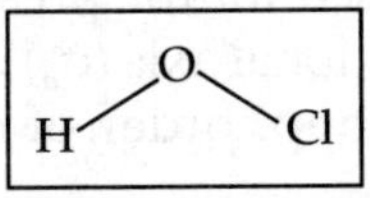

$\leftarrow$ Molecular plane (σ_h)

Other examples of C_s with group element E and σ_h ($h = 2$) are given below.

Fig. 2.9 Molecules belonging to C_s point group which contain only a σ-plane (h-2; E and σ)

(c) **Only *i* is the generator**

If *i* is present as generator then the symmetry operations are *i* and E (E = *i.i*). For such molecules the point group is C_i. The subscript *i* stands for the inversion centre. For example 1,2-dichloro,1,2-difluoro ethane in the staggered form (Fig. 2.10), the order of this point group is two ($h = 2$; E_1 *i*).

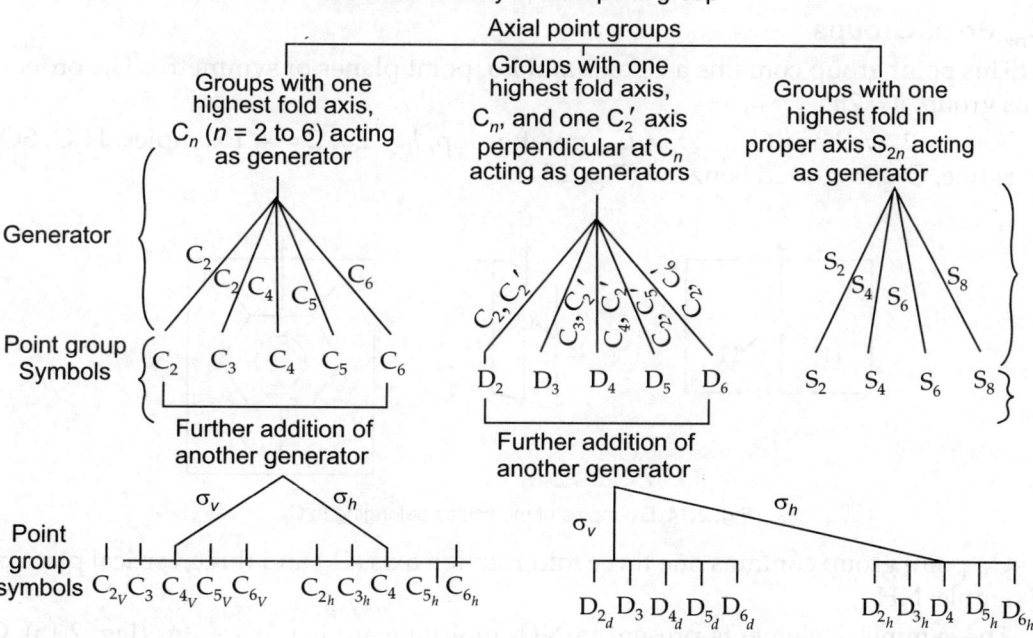

Fig. 2.10 Staggered conformation of 1,2-dichloro,1,2-difluoroethane

2.6.3 Molecules of High Symmetry (Axial Point Groups)

They can have following basic rotational generators.

(a) an axis C_n is present

(b) C_n is present with *n* number of C_2

(c) One improper axis is present

Besides basic rotational generators, there can be additional generators, which together form the family of axial point groups (Table 2.6).

The family tree for the axial group helps explain the involvement of generators in the branching of the axial groups.

Table 2.6 Family of axial point group

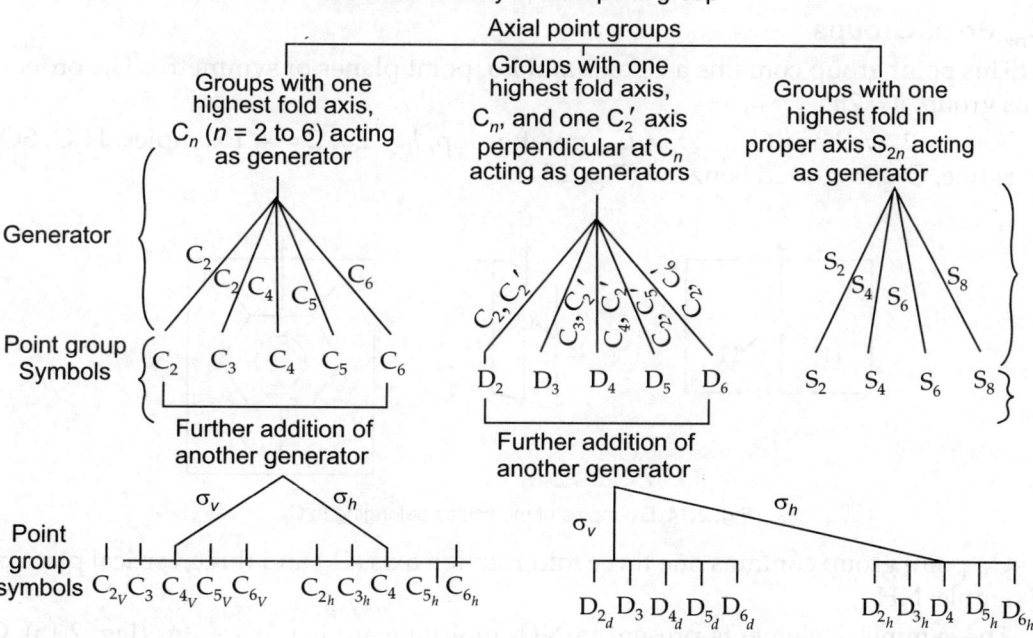

C_n Point Groups

Let us consider those molecules which contain only one C_n proper axis. The presence of C_n implies the presence of ($n - 1$) distinct symmetry elements whether *n* is even or odd. Since C_n generates a set of *n* elements including E, the order (h) of this group is *n*, i.e. $h = n$. The molecules belonging to this are designated as C_n point groups.

C_n Type Point Groups

For $n = 1$, the point group is C_1, For example, CH_3COOH, $CHClBrF$, etc. (see Fig. 2.8). Symmetry element is C_1 or E (Figs 2.11 and 2.12).

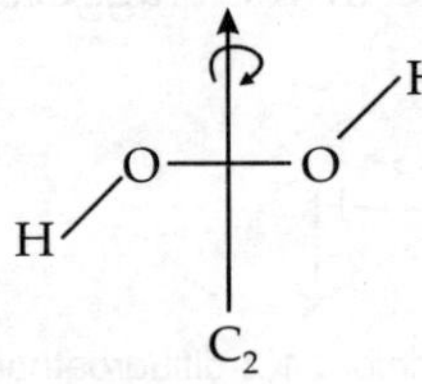

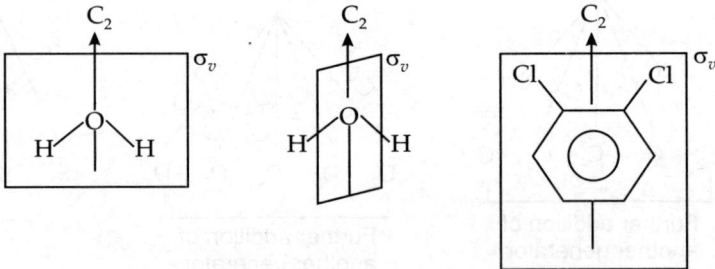

Fig. 2.11	**Fig. 2.12**

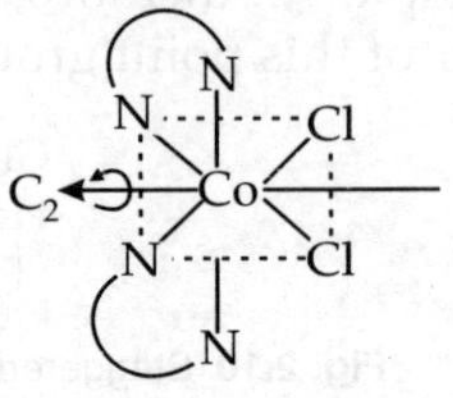

For $n = 2$, the point group is C_2, For example: H_2O_2, *cis*-$[CO(en)_2 Cl_2]^+$. The order of this group is two point group, i.e. C_3, For example NPh_3, PPh_3, $AsPh_3$, etc. order of this group = 3 ($h = n$) (Fig. 2.13).

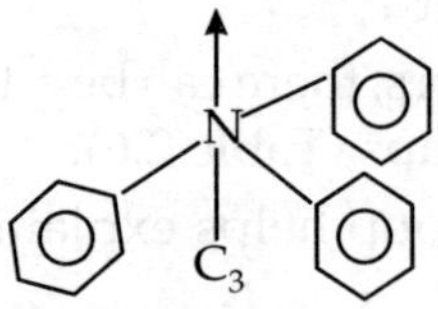

Fig. 2.13

C_{nv} Point Groups

This point group contains a C_n axis and $n\sigma_v$ point planes of symmetry. The order of this group, $h = 2n$.

For $n = 2$, we have $C_2 + 2\sigma_v = C_{2v}$ point group, $h = 2 \times 2 = 4$. Examples: H_2O, SO_2, pyridine, O-substituted benzene (Fig. 2.14).

Fig. 2.14 Examples of molecules belonging to C_{2v}

C_{3v} point group contains one three fold rotation axis (C_3) and three vertical place σ_v. Example: NH_3.

The symmetry elements present in NH_3 molecule are $E + 1C_3 + 3\sigma_v$ (Fig. 2.15). C_3 axis of symmetry generates two distinct symmetry operations, i.e. C_3^1 and C_3^2 because $C_3^3 = E$ and one operation is generated by each vertical planes of symmetry (σ_v).

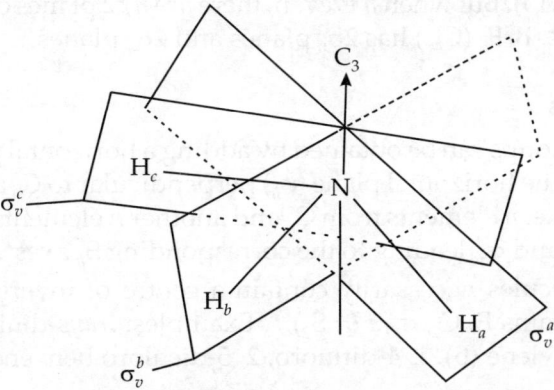

Fig. 2.15 Ammonia molecule contains $3\sigma_v$ and C_3 rotation axis

Thus the total district operations are sin $(E, C_3^1, C_3^2, \sigma_v^{\ a}, \sigma_N^{\ b}, \sigma_v^{\ c})$ and the order of this C_{3v} group = 6. Point group C_{4v} has C_4 axis and four vertical planes. Example: BrF_5. The C_4 axis passes through the apex F and the central Br atoms of the four vertical planes, two are labeled as σ_v and the other two are labeled as σ_d (Fig. 2.16).

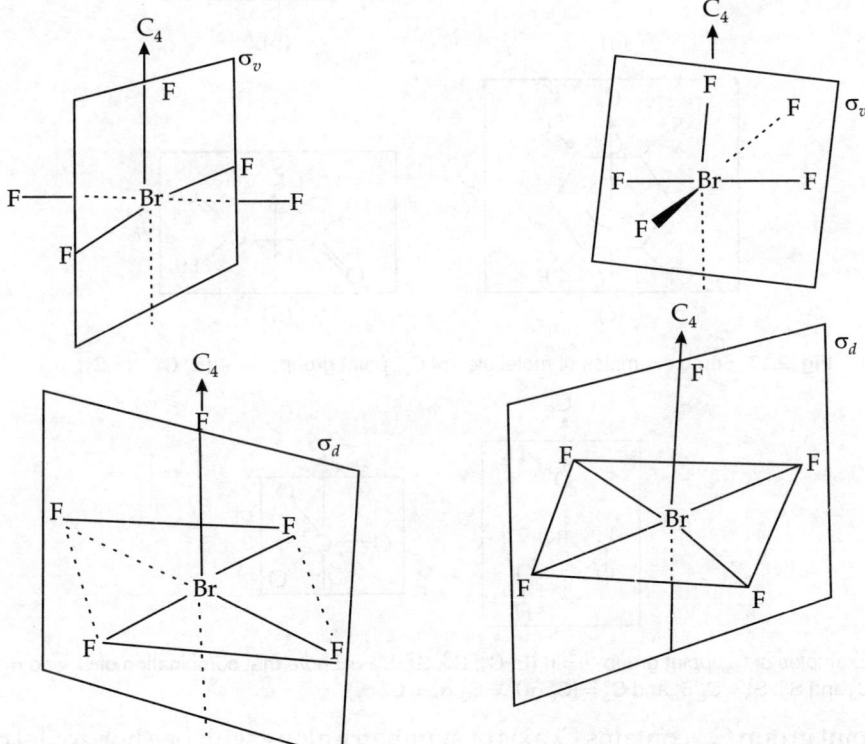

Fig. 2.16 An example of the point group C_{4v}

The set of n elements generated by C_n and $n\sigma_v$ planes constitute this point group and hence order of this group is $2n$ elements. Thus order h of C_{4v} group is 8. An important point has to be made here that when n is odd, all the planes are σ_v type only. NH_3 (C_{3v})

has $3\sigma_v$ planes (Fig. 2.15). But when n is even, there are $n/2$ planes of σ_v type and another $n/2$ planes of σ_d type. BrF_5 (C_{4v}) has $2\sigma_v$ planes and $2\sigma_d$ planes.

C_{nh} Point Groups

This set of point groups can be obtained by adding a horizontal plane (σ_n) to a proper rotational axis (C_n). The horizontal plane (σ_h) perpendicular to C_n axis. This group has a total of $2n$ elements, i.e. n elements from C_n and another n elements can be generated by a combination of C_n and σ_h, leading to the corresponding S_n axes*. When n is even, C_{nh}.

Point group molecules necessarily contain a centre of inversion (i). For example C_{2h} point group contains E, C_2, σ_h, i (= S_2)**. Examples: *trans*-dinitrogen difluoride(a), *trans*-1,2 dichloroethylene (b), 1, 4-difluoro, 2, 5-dichloro benzene (c), glyoxal (d), etc. (Fig. 2.17).

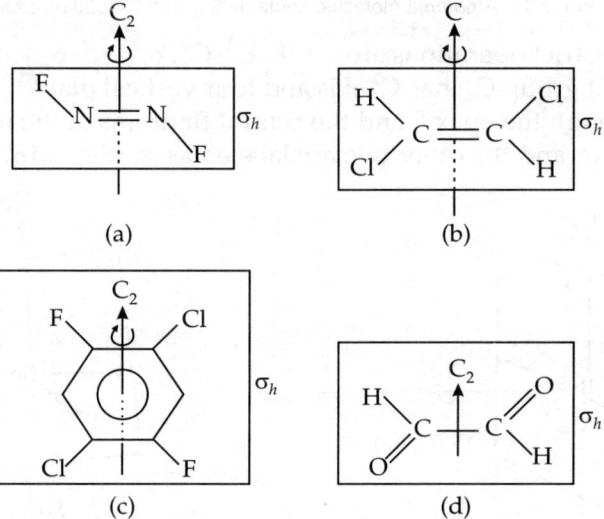

(a)   (b)

(c)   (d)

Fig. 2.17 Some examples of molecules of C_{2h} point group; $h = 4$ (E, C_2, i (= S_2), σ_h)

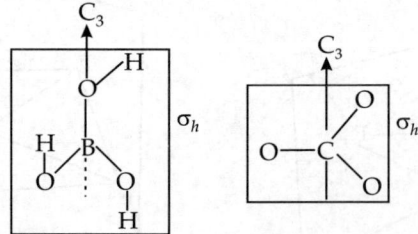

Fig. 2.18 Examples of C_{3h} point group; $h = 6$ (E, C_3^1, C_3^2, S_3^1, S_3^5, σ_h; note that combination of C_3 and σ_h leading to S_3^1 and S_3^5. $S_3^1 = C_3^1 \sigma_h$ and $S_3^5 = (C_3^1 \sigma_h)^5 = C_3^5 \sigma_h^5 = C_3^2 \sigma_h$)

The point group C_{3h} contains C_3 axis of symmetry along with one horizontal σ_h plane. Examples: H_3BO_3 (boric acid), CO_3^{2-} ion (Fig. 2.18).

---

* If a C_n and a σ_h plane $\perp$ to C_n exists separately then S_n is necessarily present in the molecule

** When $n = 2$, instead of being called S_2, is called C_i, $S_2 = i$

D$_n$ Type Point Groups (Dihedral Groups)

D$_n$ Point Groups:

D$_n$ point groups are purely rotational groups. They contain only rotational axes of symmetry. When the molecule contains only one type of C$_n$ axis, it is classified as C$_n$ point group. If, in addition to C$_n$ axis, a set of nC$_2$ axes perpendicular to C$_n$ are added, it results D$_n$ point group D$_n$ = C$_n$ + nC$_2$, thus molecules possessing nC$_2$ axes perpendicular to the principal axis (C$_n$) belong to the dihedral groups. Some of the available examples are presented in Fig. 2.19.

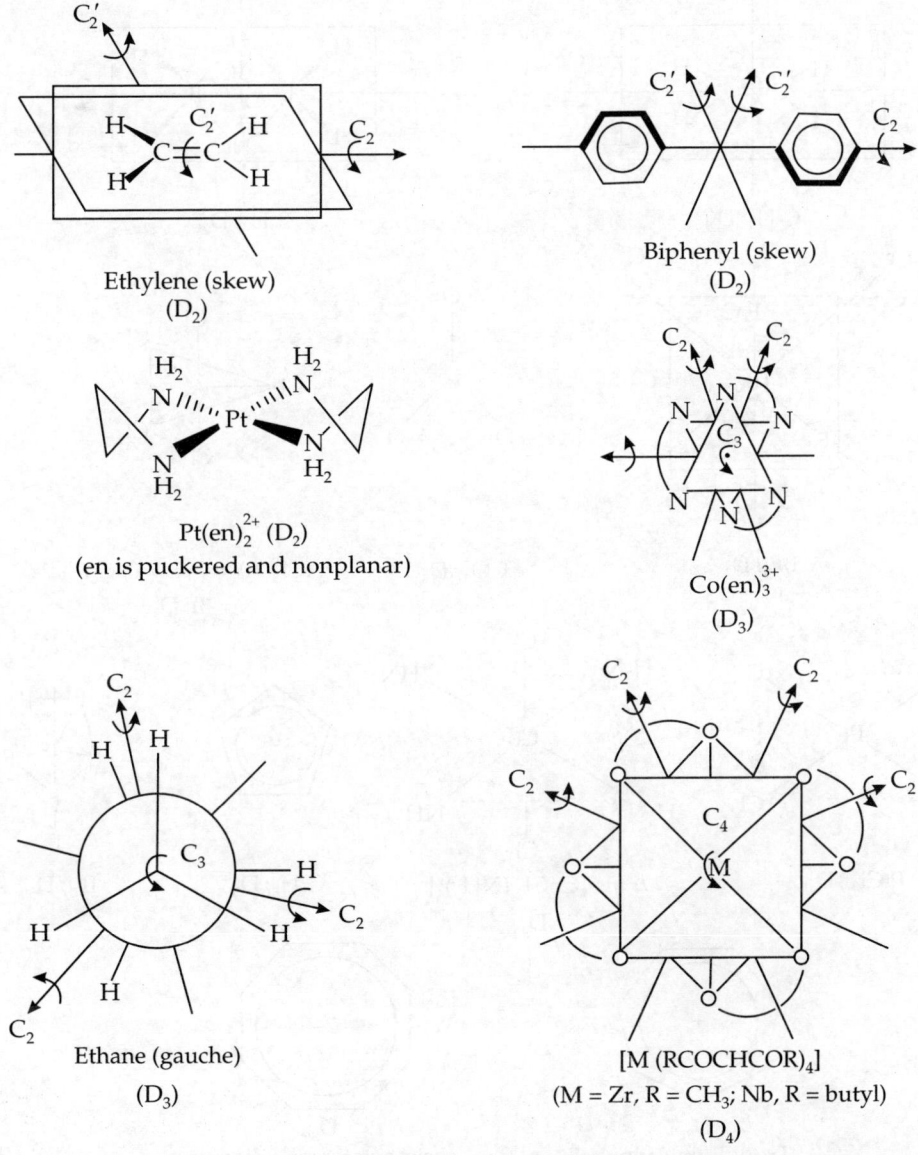

Ethylene (skew)
(D$_2$)

Biphenyl (skew)
(D$_2$)

Pt(en)$_2^{2+}$ (D$_2$)
(en is puckered and nonplanar)

Co(en)$_3^{3+}$
(D$_3$)

Ethane (gauche)
(D$_3$)

[M (RCOCHCOR)$_4$]
(M = Zr, R = CH$_3$; Nb, R = butyl)
(D$_4$)

Fig. 2.19 Examples of molecules belonging to D$_n$ point group (n = 2, 3, 4). Defining rotational axes, C$_n$ + nC$_2$ s perpendicular to C$_n$ are shown

The order (h) of this rotational group is 2n. The order (h) of D$_2$ group 4 (E, C$_2$, 2C$_2'$) that of D$_3$ is 6 (E, C$_3^1$, C$_3^2$, 3C$_2$ and that of D$_4$ is 8 (E, 2C$_4$, C$_2$, 4C$_2$).

D$_{nh}$ Point Groups

Addition of a horizontal plane (σ_h) perpendicular to the principal axis (C$_n$) results in the D$_{nh}$ group, D$_{nh}$ = C$_n$ + nC$_2$ + σ_h. The order of this D$_{nh}$ group is 4n. The order of D$_{2h}$ is 8 (E, C$_2$, 2C$_2'$, i (= S$_2$), σ_h, 2σ_v), that of D$_{3h}$ is 12 (E, 2C$_3$, 3C$_2$, σ_h, 3σ_v, 2S$_3$ (S$_3^1$, S$_3^5$)) and in case of D$_{4h}$, it is 16 (E, 2C$_4$ (C$_4^1$, C$_4^3$), C$_2$ [(= C$_4^2$, 2C$_2'$, 2C$_2''$, σ_h, 2σ_v, 2σ_d, i, 2S$_4$ (S$_4^1$, S$_4^3$)] (Fig. 2.20).

C$_{nh}$ Point Groups

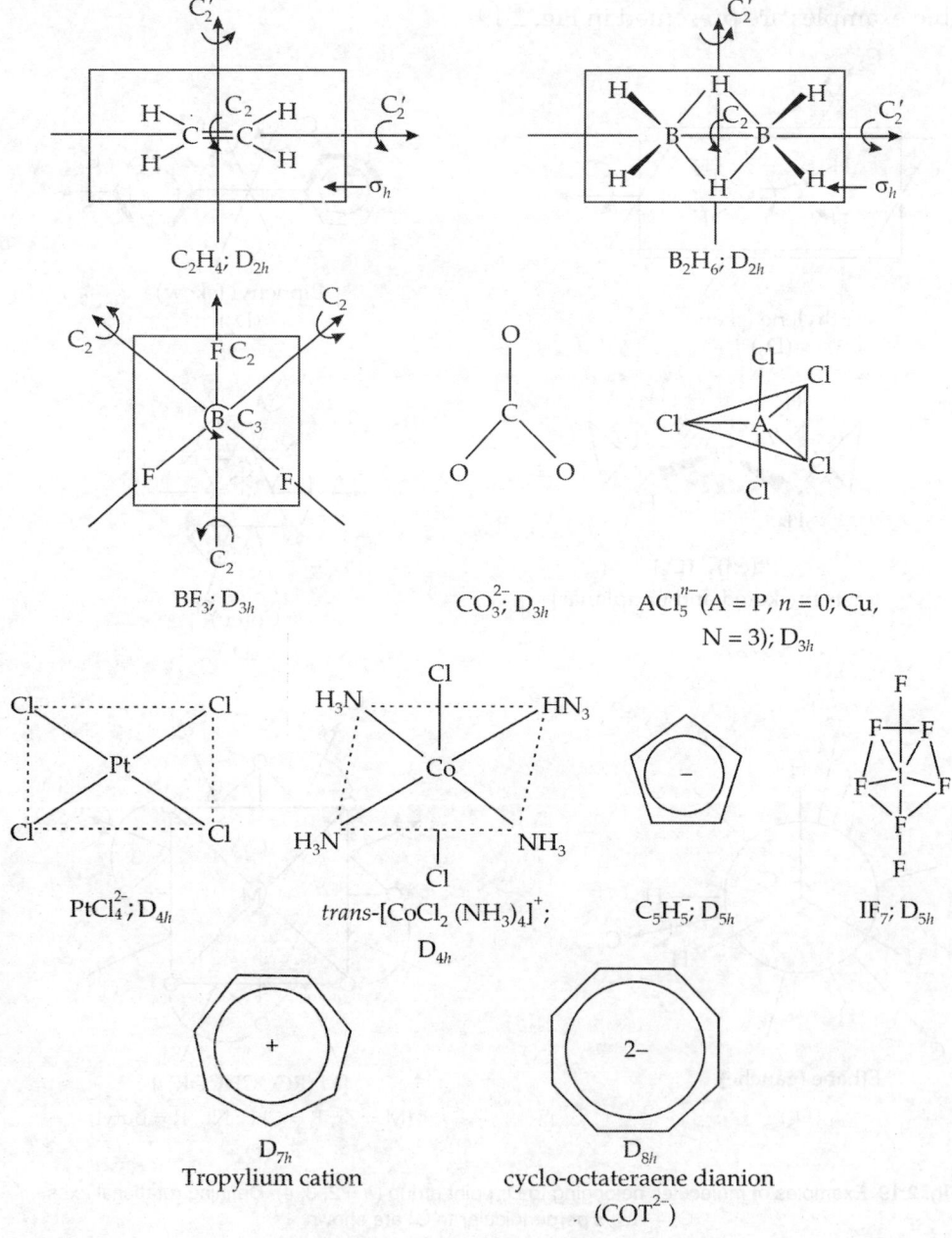

Fig. 2.20 Examples of D$_{nh}$ point groups (n = 2, 3, 4, 5). σ_h and rotational axes are shown for only a few molecules

D_{2h} Point Groups

Molecules belonging to the D_{2h} point group have three mutually perpendicular C_2 axes and a σ_h plane. $D_{2h} = C_2 + 2C_2 + \sigma_h$. Examples of D_{2h} point group are N_2O_4, C_2H_4, B_2H_6 (Fig. 2.20). Planar N_2O_4 molecule has three C_2 axes: (i) one passing through two N along (say along x axis) and (ii) the other two in between the two nitrogens and perpendicular to each other (say along y and z axes).

It has two σ_v mirror planes, one plane passing through the two nitrogen and the other plane perpendicular to the first and bisecting the N—N axis. It has a horizontal mirror plane σ_h passing through all the six atoms. There is an inversions centre (i) at the mid point of the N—N bond. In fact, *i* exists in all those molecules of D_{nh} point group, where n is even.

D_{3h} Point Group

A molecule containing a C_3 axis and a σ_h plane and three C_2 axes perpendicular to the C_3 axis belong to the point group D_{3h}. $D_{3h} = C_3 + 3C_2 + \sigma_h$ planar BF_3 molecule, CO_3^{2-} (planar) and other symmetric trigonal planar molecules belong to D_{3h} point groups. In BF_3 molecule one C_3 axis passing through B and perpendicular to the plane of the molecule (σ_h), three C_2 axes passing through each one of three F and the B perpendicular to the C_3 axis, three σ_v planes containing the three C_2 axes and one σ_h containing the planar molecule (Fig. 2.20).

D_{4h} Point Group

This group has principal axis C_4 perpendicular to the molecular plane (σ_h). There are four two fold axes (C_2). These C_2 axes are perpendicular to C_4 axis. Thus $D_{4h} = C_4 + 4 C_2 + \sigma_h$. Examples include XeF_4, $[PtCl_4]^{2-}$, *trans*-$[CoCl_2(Mt_3)_4]^+$, etc. order of the group = 16 (Fig. 2.21).

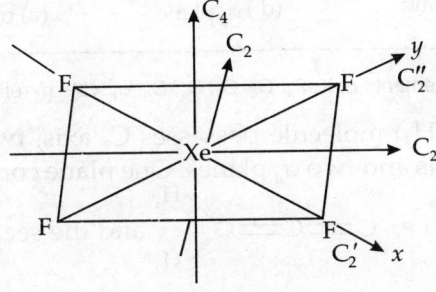

Fig. 2.21

D_{nd} Point Group

Group

When, in addition to C_n and nC_2, there are n vertical planer ($n\sigma_v$) which lie between the C_2's, the symmetry class is D_{nd}. These vertical planes are called dihedral planes and are represented by σ_d.

$$D_{nd} = C_n + nC_2 + n\sigma_d$$

$$D_{2d} = C_2 + 2C_2 + 2\sigma_d$$

$$D_{3d} = C_3 + 3C_2 + 3\sigma_d$$

An example of the group D_{2d} is allene (Fig. 2.22).

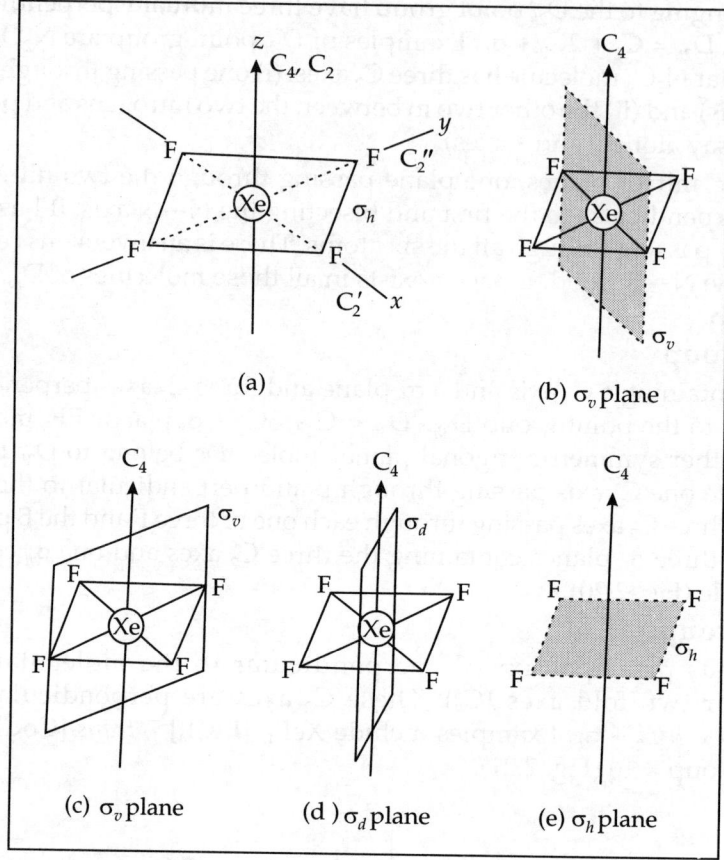

(a)

(b) σ_v plane

(c) σ_v plane

(d) σ_d plane

(e) σ_h plane

Fig. 2.22 Symmetry elements of XeF_4; C_4, C_2, C_2', C_2'', σ_h; (b – d) reflection planes in XeF_4

Allene ($H_2C = C = CH_2$) molecule possesses C_2 axis, two twofold, axes (C_2', C_2'') perpendicular to the C_2 axis and two σ_d planes. One plane contains three carbon atoms ($C = C = C$), H_1 and H_2, i.e. $C = C = C\begin{smallmatrix} \diagup H_1 \\ \diagdown H_2 \end{smallmatrix}$ and the second plane contains three

carbon atoms, H_3 and H_4, i.e. $\begin{smallmatrix} H_3 \diagdown \\ H_4 \diagup \end{smallmatrix} C = C = C$. These two planes are dihedral planes

because these two planes bisect the angle between two successive C_2 axes.

We now summarily construct the axial point group with no further fetish for explanation except for what is given under the 'remarks' column (Table 2.7). The nonaxial point groups, C_1, C_s and C_i ($\equiv S_2$) have already been given in Section 2.4.

Table 2.7 Axial point groups

Point group	Subgroup+ generators	Group elements subgroup elements + coset	Remark
C_n			
C_2	(C_2)	$\{E, C_2\}$	(a) No. of group elements = n
C_3	(C_3)	$\{E, C_3, C_3^2, = \overline{C_3}\}$	
C_4	(C_4)	$\{E, C_4, C_2, C_4^3 = \overline{C_4}\}$	
C_5	(C_5)	$\{E, C_5, C_5^2, C_5^3, C_5^4, \overline{C_5}\}$	
C_6	(C_6)	$\{E, C_6, C_3, C_2, C_3^2, C_6^5 = \overline{C_6}\}$	
C_{nv}			
C_{2v}	$C_2 + (\sigma_v)$	$\{E, C_2, \sigma_v E, \sigma_v C_2\} = \{E, C_2, \sigma_v, \sigma_v\}$	(a) No. of group elements = $2n$
C_{3v}	$C_3 + (\sigma_v)$	$\{E, C_3, C_3^2, \sigma_v E, \sigma_v C_3, \sigma_v C_3^2\}$ $= \{E, C_3, C_3^2, 3\sigma_v\}$	(b) Product of $\sigma_v \, C_n^k$ are a set of σ_v's which form a class for n = odd and two classes for n = even
C_{4v}	$C_4 + (\sigma_v)$	$\{E, C_4, C_2, C_4^3 \, 2\sigma_v, 2\sigma_d\}$	
C_{5v}	$C_5 + (\sigma_v)$	$\{E, C_5, C_5^2 C_5^3, C_5^4, 5\sigma_v\}$	
C_{6v}	$C_6 + (\sigma_v)$	$\{E, C_6, C_3, C_2, C_3^2, C_6^5, 3\sigma_v, 3\sigma_d\}$	
C_{nh}			
C_{2h}	$C_2 + (\sigma_h)$	$\{E, C_2, \sigma_h, i\}$	(a) No. of group elements = $2n$
C_{3h}	$C_3 + (\sigma_h)$	$\{E, C_3, C_3^2, \sigma_h, S_3, S_3^5\}$	(b) For products $\sigma_h C_n^k$
C_{4h}	$C_4 + (\sigma_h)$	$\{E, C_4, C_2, C_4^3, \sigma_h, S_4, i, S_4^3\}$	
C_{5h}	$C_5 + (\sigma_h)$	$\{E, C_5, C_5^2, C_5^3 C_5^4, \sigma_h, S_5, S_5^7 S_5^3 S_5^9\}$	
C_{6h}	$C_6 + (\sigma_h)$	$\{E, C_6, C_3, C_2, C_3^2, C_6^5, \sigma_h, S_6, S_3, i, S_3^5, S_6^5\}$	
D_n			
D_2	$C_2 + (C_2')$	$\{E, C_2, C_2', C_2''\}$	(a) No. of group elements = $2n$
D_3	$C_3 + (C_2')$	$\{E, C_3, C_3^2, 3C_2'\}$	(b) Products of C_2', C_n^k are equal and other C_2's which are in one class when n = odd and in two classes for is even n.
D_4	$C_4 + (C_2')$	$\{E, C_4, C_2, C_4^3, 2C_2', 2C_2''\}$	
D_5	$C_5 + (C_2')$	$\{E, C_5, C_5^2, C_5^3, C_5^4, 5C_2'\}$	
D_6	$C_6 + (C_2')$	$\{E, C_6, C_3, C_2, C_3^2, C_6^5, 3C_2', 3C_2''\}$	
D_{nh}			
D_{2h}	$D_2 + (\sigma_h)$	$\{E, C_2, C_2', C_2'', i, \sigma_v', \sigma_v'', \sigma_v'''\}$	(a) No. of elements = $4n$
D_{3h}	$D_3 + (\sigma_h)$	$\{E, C_3 C_3^2, 3C_2', \sigma_h, S_3, S_3^5, 3\sigma_v\}$	(b) Products $\sigma_h C_2' = \sigma_v$'s which are distributed in one or two classes depending on odd or even values of n.
D_{4h}	$D_4 + (\sigma_h)$	$\{E, C_4, C_2, C_4^3 \, 2C_2', 2C_2'', \sigma_h, S_4, i, S_4^3, 2\sigma_v, 2\sigma_d\}$	
D_{5h}	$D_5 + (\sigma_h)$	$\{E, C_5, C_5^2, C_5^3 C_5^4, 5C_2', \sigma_h S_5 S_5^7, S_5^3, S_5^9, 5\sigma_v\}$	
D_{6h}	$D_6 + (\sigma_h)$	$\{E, C_6, C_3 C_2, C_3^2 C_6^5 \, 3C_2', 3C_2'', \sigma_h, S_6, S_3, i, S_3^5, S_6^5, 3\sigma_v, 3\sigma_d\}$	
D_{nd}			
D_{2d}	$D_2 + (\sigma_d)$	$\{E, C_2, C_2', C_2'', 2\sigma_d, S_4, S_4^3,\} =$ $\{E, C_2, C_2', C_2'', 2\sigma_d, 2S_4\}$	(a) Elements = $4n$
D_{3d}	$D_3 + (\sigma_d)$	$\{E, C_3, C_3^2, 3C_2', 3\sigma_d, S_6, S_6^3, S_6^5\} =$ $\{E, C_3, C_3^2 \, 3C_2', 3\sigma_d, 2S_6, i\}$	(b) $C_n^k \sigma_d = \sigma_d$'s all equivalent
D_{4d}	$D_4 + (\sigma_d)$	$\{E, C_4, C_2, C_4^3 \, 4C_2', 4\sigma_d, S_8, S_8^3, S_8^5, S_8^7\}$	(c) All C_2' are equivalent.

Contd...

Table 2.7 Axial point groups (Contd...)

Point group	Subgroup+ generators	Group elements subgroup elements + coset	Remark
D_{5d}	$D_5 + (\sigma_d)$	$\{E, C_5, C_5^2, C_5^3, C_5^4, 5C_2', 5\sigma_d, S_{10},$ $S_{10}^5 = i, S_{10}^7, S_{10}^9\}$	(d) Products of σ_d and set of C_2'''s lead to series $S_{2n}, S_{2n}^3 \ldots$
D_{6d}	$D_6 + (\sigma_d)$	$\{E, C_6, C_3, C_2, C_3^2, C_6^5, 6C_2', 6\sigma_d,$ $S_{12}, S_{12}^3 = S_4, S_{12}^5, S_{12}^7, S_{12}^9 = S_4^3, S_{12}^{11}\}$	$S_{2n}^5 \ldots$ which can then be put into classes
S_{2n}			
S_2	(S_2)	$\{E, S_2 = i\}$	(a) Elements = $2n$
S_4	(S_4)	$\{E, S_4, C_2, S_4^3\}$	(b) $S_2 \equiv C_i$ already include in the nonaxial group.
S_6	(S_6)	$\{E, S_6, C_3, i, C_3^2, S_6^5\}$	(c) S_n group with n = odd are identical with C_{nh} point groups
S_8	(S_8)	$\{E, S_8, C_4, S_8^3, C_2, S_8^5, C_4^3 S_8^7\}$	

2.6.4 Features of Group Elements—Classes and Products

The group elements recorded in the foregoing section can be presented in somewhat more condensed form by making use of the concept of classes. The class idea has been partially but not uniformly used throughout in the foregoing tables. It is very important to remember the following points while writing the group elements classwise.

1. The inversion i and horizontal reflection plane σ_h always form separate classes by themselves as also E.

2. The vertical reflection planes (σ_v's) either belong to one class or split into two classes. If they are of one class, they are written as $n\sigma_v$'s. If two classes are involved, these elements are grouped under two heads $n\sigma_v$'s and $n\sigma_d$'s.

3. The C_2' rotations follow the same type of grouping as for σ_v's. In D_{nd} point groups, all the C_2's belong to a single class, so also do all the σ_d's.

4. Rotations of C_n point group and also the group elements of higher group C_{nh} and S_{2n} mutually commute. All the elements of these point groups form separate classes.

5. In higher groups confining C_2's and (or) σ_v's as in D_n, D_{nh}, D_{nd} pairs of proper rotations (C_n^k and C_n^{n-k}) form a class as can be verified by the similarity operation $X^{-1} A X$. These C_6 and C_6^5 belong to a single class and are written as $2C_6$. Similarly C_5^2 and C_5^3 of the groups D_5, D_{5h}, D_{5d} are represented in a class as $2C_5^2$. In class notations, the lowest possible numerals are used.

   The convention for grouping into classes, and the improper rotations of the higher point groups (D_{nh}, D_{nd}, O_h, T_d) is the same as for proper rotations of these groups. Thus S_6 and S_6^5 belong to a class ($2S_6$) and so do the group elements S_5^3 and S_5^7 to the class $2S_5^3$.

To illustrate the above classwise grouping, we rewrite the elements of some groups.

$$C_4 \quad \rightarrow \quad \{E, C_4, C_4^2 (= C_2), C_4^3\}$$
$$D_4 \quad \rightarrow \quad \{E, 2 C_4, C_4^2 (= C_2), 2C_2', 2C_2''\}$$

$$C_{5v} \quad \rightarrow \quad \{E, 2C_5, 2C_5^2, 5\sigma_v\}$$
$$C_{6h} \quad \rightarrow \quad \{E, 2C_6, C_6^3 (\equiv C_2), 2C_3, 3C_2', 3C_2'', \sigma_h, 2S_6, i, 2S_3, 3\sigma_d, 3\sigma_v\}$$
$$C_{6d} \quad \rightarrow \quad \{E, 2C_6, 2C_3, C_2, 6C_2', 2S_{12}, 2S_4, 2S_{12}^5, 6\sigma_d\}$$

Attention may be drawn to the product relations under the 'remarks' column of the axial point groups (Table 2.7). The products of $C_2' C_n^k$ (with different k's) are nC_2's grouped into one or two classes and similarly for the generated $n\sigma_v$'s. Although these operations, C_2's and σ_v's are all associated with distinct symmetry axes and symmetry planes, the group elements as indicated above do not reveal their location. This specification of locations, though often unnecessary from the view point of application, can nevertheless be ascertained with a little bit of geometry and coordinate transformation. To acquaint the reader with the principle of specifying the locations of C_2's and σ_v's, etc. We take up the issue in the more difficult cases of the cubic groups discussed in Section 2.4.4.

2.6.5 Cubic Point Groups

These are of two types

  (a) Tetrahedral point group (T_d and)
  (b) Octahedral point group (O_n)

Tetrahedral Point Group (T_d)

Methane is a tetrahedral molecule. It has four three fold rotation axes ($4C_3$) and three two fold rotation axes ($3C_2$). Each C_3 axis passing through one of the hydrogen atoms and carbon atom. Each C_3 generates C_3^1, C_3^2 and C_3^3 ($= E$) operations.

Therefore, number of distinct operations $= 4 \times (C_3^1, C_3^2, C_3^3 (= E)) = 4 + 4 + 1 = 9$. It is to be noted that all the E operations are equivalent to one operations (Fig. 2.23).

Each C_2 axis bisecting the bond angles between the two pairs of bonds, pointing opposite to each other. Each C_2 axis of symmetry can generate only one distinct operation, i.e. C_2^1 because $C_2^2 = E$. Therefore total number of operations on $3C_2$ will be $3 \times 1 = 3$. $3S_4$ axes of symmetry exist collinear with C_2 axes. Each S_4 generates the two new operations S_4^1 and S_4^3 $[S_4^2 (= C_4^2 \sigma^2 = C_4^2 = C_2) S_4^4 (= E)]$. Hence for $3S_4$ axes, total operations on $3S_4$ in $3 \times 2 = 6$.

There exists six σ_d planes. Each of these planes generates only one operation so $6 \times 1 = 6$. The symmetry elements and their locations are shown in Table 2.8.

Thus total 24 symmetry operations are possible for CH_4 molecule and they are:

$1E$, $8C_3$, $3C_2$, $6S_4$, $6\sigma_d$.

It should be noted that molecules may have tetrahedral geometry, but may not possess all the 24 operations and do not belong to T_d point group. Examples: CH_3Cl, $CHCl_3$.

Table 2.8 Symmetry elements and their locations (T_d)

Symmetry elements and their locations	Symmetry operations	No. of distinct operations	Remarks
(a) Four 3-fold axes collinear with the body diagonals of the cube.	$4 \times (C_3, C_3^2, C_3^3 = E)$	9	All the E operations are equivalent to one operation.
(b) Three C_2-axes normal to the faces of the cube and incidentally passing through the mid points of the edges of the tetrahedron.	$3 \times (C_2, \{C_2^2 = E\})$	3	Operations within { } are redundant since E has already been considered under C_4 operations.

Contd...

Table 2.8 Symmetry elements and their locations (T_d) (Contd...)

Symmetry elements and their locations	Symmetry operations	No. of distinct operations	Remarks
(c) Three S_4-axes collinear with C_2 axes	$3 \times (S_4, \{S_4^2 = C_2\}, S_4^3, \{S_4^4 = E\})$	6	Operations within { } are redundant.
(d) Six σ_d's passing through the diametrically opposite edges of the cube. Each such σ_d comprises a pair of vertices and accommodates a C_3 and a C_2 axis.	$6 \times (\sigma_d, \{\sigma_d^2 = E\})$	6	Operations within { } are redundant .

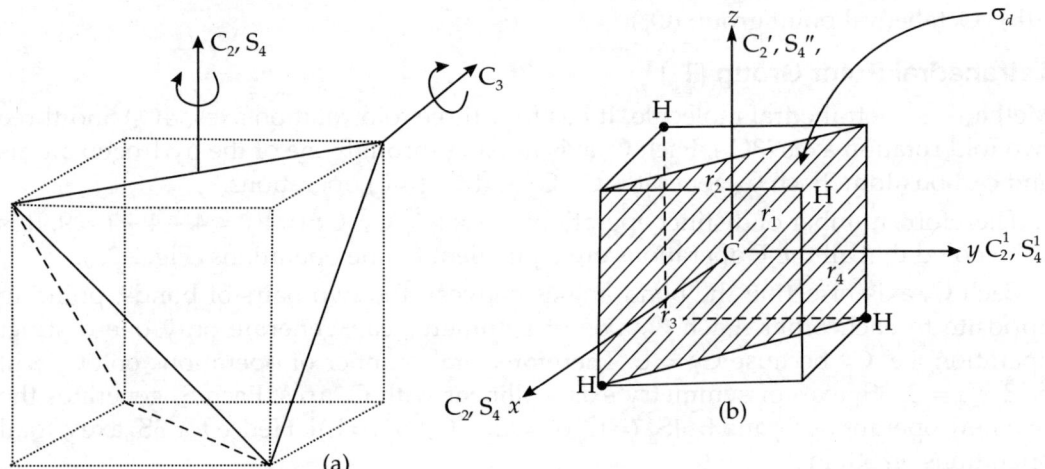

Fig. 2.23 (a) A tetrahedron drawn within a cube (b) coordinate system of CH_4 molecule

Octahedral Point Group (O_h)

An octahedral molecule has three C_4 axes and three C_2 axes (coincident with C_4 axes). Each C_4 axes passes through the opposite point of atoms (Fig. 2.24).

There are $6C_2'$ axes distinct from the $3C_2$ axes. Each C_2' axis passes through the mid-points of pairs of opposite edges.

An octahedral has eight equilateral triangular faces. There are four different C_3 axes, each passing through the mid-point of a pair of triangular faces on the opposite sides of an octahedron.

An octahedral molecule also contains other symmetry elements such as centre of inversion, reflections planes and rotation–reflection axes.

There are three σ_h planes and each σ_h plane contains the central atom and passes through four apical atoms. They bisect the angle between the pairs of C_2' axes.

In an octahedral, each C_4 axis is also an S_4 axis and similarly each C_3 axis is also an S_6 axis. Hence, there are $6S_4$ and $8S_6$ symmetry operations.

In short, we may say

(i) $3C_4$-axes of symmetry pass through opposite vertices of the octahedron, i.e. they pass through the centers of opposite faces of the cube.

(ii) $3C_2$ and $3C_4$ axes of symmetry collinear with the C_4-axis of symmetry.

(iii) $4S_6$ axes coincident with $4C_3$

(iv) $6C_2$ axes bisecting opposite edges

(v)   A centre of symmetry i is present

(vi) $3\sigma_h$ these are due to the presence of three proper axes of symmetry ($3C_4$). These three horizontal planes are perpendicular to C_4 axes each σ_h is $\perp$ to a C_4.

(vii) $3\sigma_h$ planes passing through two apices and bisecting opposite edges.

Thus for a regular octahedral molecule the total symmetry operations will be

$E + 8C_3 + 6C_4 + 36C_2 + 6C_2' + i + 6S_4 + 8S_6 + 3\sigma_h + 6\sigma_d = 48$ (Table 2.9). Molecules with these symmetry operations belong to O_h point symmetry.

Table 2.9 Symmetry elements and their locations (O_h)

	Symmetry elements and their locations	Symmetry operations	No. of distinct operations	Remarks
(a)	There 4-fold axes passing through the midpoints of the opposite faces of the cube	$3 \times (C_4, C_4^2 = C_2, C_4^3, C_4^4 = E)$	10	Four E's are just equivalent to a single E.
(b)	Three S_4 axes collinear with C_4	$3 \times (S_4, \{S_4^2 = C_2\} S_4^3, \{S_4^4 = E\})$	6	Operations within { } are redundant
(c)	Six C_2' passing through the midpoints of the opposite edges of the cube of the cube and also through the center	$6 \times (C_2' \{C_2'^2 = E\})$	6	"
(d)	Six σ_d's each passing through a pair of diametrically opposite edges of the cube	$6 \times (\sigma_d, \{\sigma_d^2 = E\})$	6	"
(e)	Three σ_h perpendicular to the C_4 axes	$3 \times (\sigma_h, \sigma_h^2 = E)$	3	"
(f)	i. inversion center	$(i, \{i_2 = E\})$	1	"
(g)	Four 3-fold symmetry axes collinear with the body diagonals of the cube	$4 \times (C_3, C_3^2, \{C_3^3 = E\})$	8	"
(h)	Four S_6 improper axes collinear with the body diagonals of the cube	$4 \times (S_6, \{S_6^2 = C_3\}), \{S_6^3 = i\}, \{S_6^4 = C_3^2\} S_6^5, \{S_6^6 = E\})$	8	"
	Total number		$\overline{48}$	

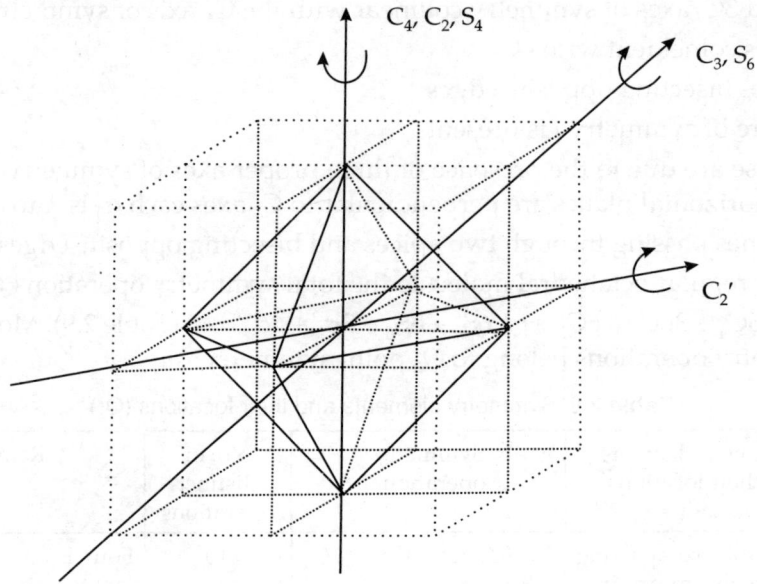

Fig. 2.24 An octahedron drawn within a cube

Molecules may have octahedral geometry but may not belong to O_h point group. Example: $[Co\,(NH_3)_5\,Cl]^{2+}$, *cis* and *trans*-$[Co\,(NH_3)_4\,Cl_3]$

2.6.6 Special Point Group of Linear Molecules

Linear molecules like HCl, CO, CO_2, N_2O, can be classed under two groups $C_{\infty v}$ and $D_{\infty h}$ depending on whether the molecule lacks any inversion centre i (as in CO, N_2O) or possesses the later (e.g. CO_2)

(i) Molecules falling under $C_{\infty v}$ have the internuclear line as the C_∞ axis about which any rotation ranging from an infinitesimally small value to 2π will be a symmetry operation. Such molecules also have an infinite number of σ_v's all containing the C_∞ axis. Examples: HCN, HCl, CO, N_2O, etc. The order of the group $h = \infty$

(ii) If the molecule possesses a centre of symmetry (i) in addition to C_∞ then it will also have a σ_h plane ($\sigma_h \perp C_\infty$) and infinite number of C_2 axis perpendicular to C_∞, there are infinite σ_vs each containing one C_2 axis. In such case, the point symmetry will be $D_{\infty h}$. Examples: CH≡CH, H_2, CO_2, etc.

Icosahedral Point Groups (I_h)

This group contains molecules with either icosahedral or pentagonal dodecahedral shapes and belong to I_h point group. The point group I_h is highly symmetrical group with $h = 120$. These are E, $12C_5$, $12C_5^2$, $20C_3$, $15C_2$, i, $12S_{10}$, $12S_{10}^3$, $20S_6$ and 15σ (Fig. 2.25). Examples include regular icosahedron and dodecahedron and Buck minister fullerene (C_{60}).

Each icosahedron contains six C_5 axes passing through two opposite corners and are coincident with S_{10}. The possible operations are:

$C_5^1 = S_{10}^2$, $C_5^2\,(S_{10}^4)$, $C_5^3\,(S_{10}^6)\,C_5^6\,(S_{10}^8)$ and S_{10}^1, S_{10}^3, S_{10}^7 and S_{10}^9. S_{10}^5 is equal to inversion (i).

There are ten C_3 axes passing through opposite trigonal faces and are collinear with S_6 axes.

The possible operations are:

$C_3^1 (= S_6^2)$, C_3^2, $(= S_6^4)$. The other operations are S_6^1, S_6^5, S_4^2 is equal to inversion i. There are fifteen C_2 axes, bisecting opposite edges and fifteen planes of symmetry. The total number of symmetry operations are 12 and they are

E, 24 C_5, 24 S_{10}, 20 C_3, 20 S_3, 15 C_2, 15σ and i.

Typical boronhydride, dodecaborane ion $B_{12} H_{12}^{2-}$ has I_h symmetry.

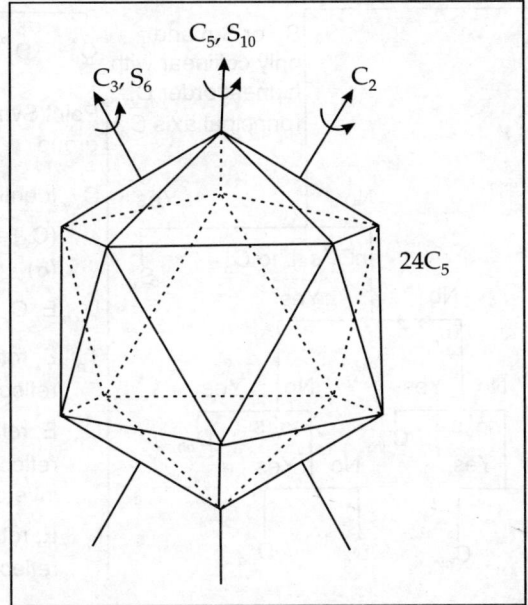

(i) Icosahedral elements (rotational)

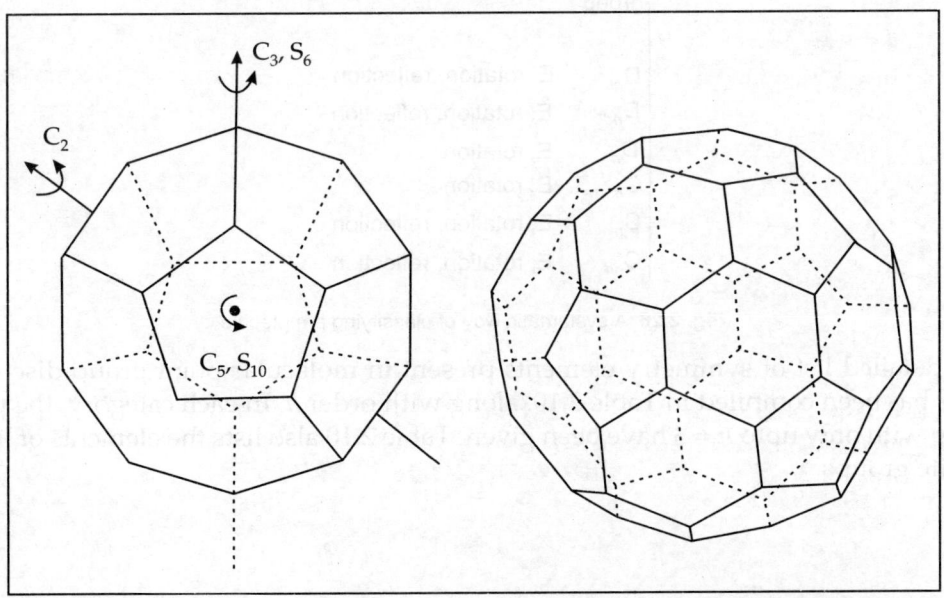

(ii) Dodecahedral elements   (iii) Structure of C_{60}—fullerene

Fig. 2.25 Molecules of I_h point symmetry

A systematic procedure for classifying a molecule in its point group is shown in Fig 2.26. It is self-explanatory.

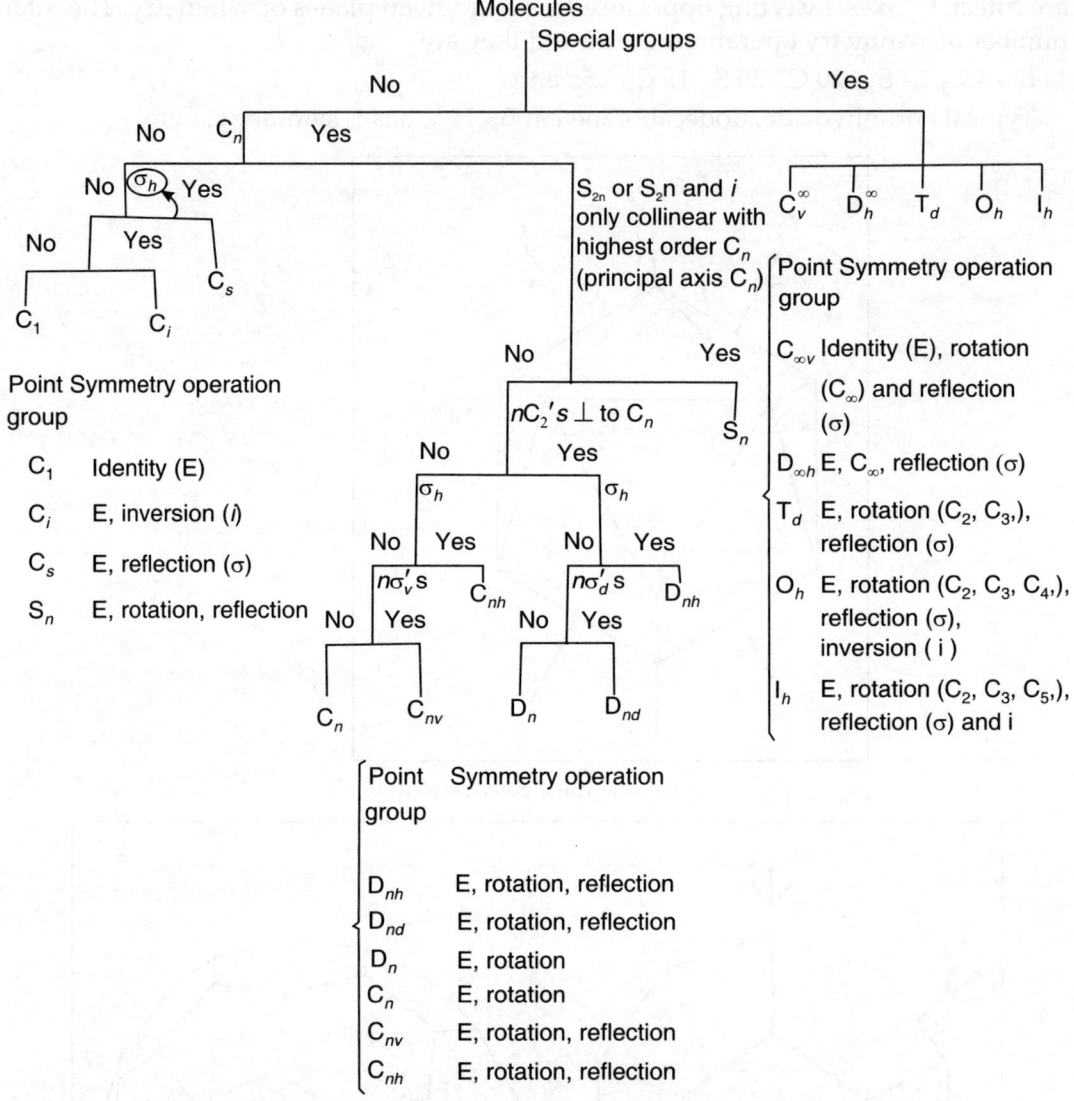

Fig. 2.26 A systematic way of classifying a molecule

A detailed list of symmetry elements present in molecular point group discussed so far has been compiled in Table 2.10. along with order h. In each category, the point group with only upto $n = 3$ have been given. Table 2.10 also lists the elements of linear infinite groups.

Table 2.10 Point groups and their detailed list of symmetry elements

Point group	Order of group, h	Types of symmetry elements	Example
C_1	1	$E(= C_1)$	CHCl Br F, Si Br Cl FI
C_i	2	$E, i(= S_1)$	CHFCl-CHFCl (staggered)
C_s	2	E, σ	HOCl, CH_2Cl Br, POBr Cl_2
C_n groups ($h = n$)			
C_2	2	E, C_2	H_2O_2, Gauchi CH_2Cl-CH_2Cl, N_2H_4,
C_3	3	E, C_3^1, C_3^2	NH_3, P $(C_6H_5)_3$, N $(C_6H_5)_3$
C_4	4	$E, C_4^1, C_4^2, (= C_2), C_4^3$	For C_n point groups, $n > 3$; examples or molecules are rare
C_5	5	$E, C_5^1, C_5^2, C_5^3, C_5^4$	
C_{nv} groups: ($h = 2n$)			
C_{2v}	4	$E, C_2, 2\sigma_v$	H_2O, SO_2, CH_2Cl_2, $BClF_2$, P_y
C_{3v}	6	$E, C_3^1, C_3^2, 3\sigma_v$	NH_3, $POCl_3$, PCl_3, $S_2O_3^{2-}$
C_{4v}	8	$E, C_4^1, C_4^2, (= C_2), C_4^3, 2\sigma_v, 2\sigma_n^i$	$XeOF_4$, square pyramidal ML_5
C_{nh} groups: ($h = 2n$)			
C_{2h}	4	$E, C_2, i (= S_2)\sigma_v$	*trans*-H_2O_2, *trans*-1, 2-dichloro ethylene
C_{3h}	6	$E, C_3^1, C_3^2, S_3^1, S_3^5, S_3^5, \sigma_h$	$H_3 BO_3$
C_{4h}	8	$E, C_4^1, C_4^2, (= C_2), C_4^3, S_4^1, S_4^3, \sigma_h, i (= S_2)$	For C_{nh} ($n > 3$), examples are rare
D_n groups: ($h = 2n$)			
D_2	4	$E, C_2, 2C_2'$,	Ethylene (skew), Pt(en)$_2^{2+}$
D_3	6	$E, C_3^1, C_3^2, 3C_2$	Co(en)$_3^{3+}$, ethane (gauche)
D_4	8	$E, 2C_4, C_2, 4C_2'$,	$n > 3$, examples or molecules are rare
D_{nh} groups: ($h = 4n$)			
D_{2h}	8	$E, C_2, 2C_2', i (= S_2), \sigma_h, 2\sigma_v$	C_2H_4, B_2H_6, N_2O_4
D_{3h}	12	$E, 2C_3, 3C_2, \sigma_h, 3\sigma_v, 2S_3 (S_3^1, S_3^5)$	BF_3, CO_3^{2-}, PCl_5
D_{4h}	16	$E, 2C_4, (C_4^1, C_4^3), C_2(= C_4^2), 2C_2, C_2'' \sigma_h, 2\sigma_v, 2\sigma_d, i, 2S_4 (S_4^1, S_4^3)$	$PtCl_4^{2-}$, XeF_4 + *trans*-$[CoCl_2(NH_3)_4]^+$
D_{nd} groups: ($h = 4n$)			
D_{2d}	8	$E, C_2, 2C_2', 2\sigma_d, 2S_4$	Allene, Spirane
D_{3d}	12	$E, 2C_3 (C_3^1, C_3^2), 3C_2, i, 3\sigma_d, 2S_6 (S_6^1, S_6^5)$	C_2H_6 (staggered) $Cu_2(CO)_6$
D_{4d}	16	$E, 2C_4(C_4^1, C_4^3), C_2(= C_4^2), 4C_2 4\sigma_d 4S_8 (S_8^1, S_8^3, S_8^5 S_8^7)$	All square antibiprism molecules rare example

Contd...

Table 2.10 Point groups and their detailed list of symmetry elements (Contd...)

S_n (n = even) groups (h = n)			
S_4	4	E, S_4^1, S_4^3, C_2	Si, $(OCH_3)_4$
S_6	6	$E, S_6^1, S_4^5, C_3^1, C_3^2, i$	$[(CH_2)_2CH]_3CCl[(CH_2)_2CH]_3$
S_8	8	$E, S_8^1, S_8^3, S_8^5, S_8^7, C_4^1, C_4^3, C_2(=C_4^2)$	rare example

(structures labelled S_4 and S_6)

Infinite-groups: (h = ∞)			
$C_{\infty v}$	∞	$E, \infty C_\infty, C\sigma_v$	HF, CO, HCN, HCl, COS
$D_{\infty h}$	∞	$E, \infty C_\infty, \infty \sigma_v, \sigma_h, i$	$BeCl_2, XeF_2, CO_2, H_2, Cl_2$

Table 2.11 Lists for ready reckoning some important point groups with their respective symmetry elements along with examples.

Table 2.11 Symmetry table of symmetry elements and symmetry operations

Point group	Symmetry operation	Symmetry element	Operation	Example	Structure
C_1	Identity, E	None other than identity E	All atoms unshifted	CH F Cl Br	
C_i	Identity, E Inversion, i	E and inversion centre (point), i	Inversion through the centre	HClBrC—CHClBr (staggered)	
C_1	Identity, E Reflection, σ	E and mirror plane σ	Reflection through a mirror plane	POClBr$_2$ CH$_2$ Br Cl	
C_2	Identity, E Rotation, C_2	E and rotation axis, C_2	Rotation by $360°/2$	p dichloro benzene	
C_{2v}	Identity, E Rotation, C_2 and reflection σ_v	E, one two fold rotation axis (C_2) and two vertical planes, σ_v, σ_v'	Rotation by $360°/2$ and then reflection by σ_v and σ_v' mirror planes	H_2O, SO_2, C_2	

Contd...

Point group	Symmetry operation	Symmetry element	Operation	Example	Structure
C_{3v}	Identity, E rotation, C_3 and reflection, σ_v	E, one three-fold rotation axis (C_3) and three vertical planes, $3\sigma_v$	Rotation by $360°/3$ and then reflection by σ_v mirror planes	NH_3 (trigonal pyramidal), PCl_3, $CHCl_3$ (tetrahedral), $POBr_3$	
C_{2h}	Identity, E rotation, C_2 and reflection, σ_h	E, one two fold rotation axis, (C_2) and a horizontal molecular plane, σ_h, to (= S_2)	Rotation by $360°/2$ and then reflection by σ_h, mirror plane	*trans* $C_2H_2Cl_2$ (planar), N_2F_2	
$C_{\infty v}$	Identity, E rotation, C_∞ and reflection	E, infinite number, of rotation axes ($\infty C\infty$) and infinite number of reflection planes (∞, σ_v)	Rotation and reflection	HCl, CO, HCN	
$D_{\infty h}$	Identity, E rotation, C_∞ and reflection	E, $\infty C\infty$, ∞, σ_v, i, σ_h	Rotation and reflection	CO_2, H_2, N_2,	
D_{2h}	Identity, E rotation, C_2, reflection and inversion	E, C_2, $2C_2'$, $2\sigma_v$, i (= S_2)	Rotation, reflection and inversion	N_2O_4, C_2H_4, B_2H_6	
D_{3h}	Identity, E rotation, C_3 rotation, C_2 and reflection	E, $2C_3$, $3C_2$, σ_h, $3\sigma_v$, S_3 (S_3^1, S_3^5)	Rotation, reflection	BF_3, PF_3	
D_{4h}	Identity, E rotation, C_4 rotation, C_2 and reflection	E, C_4, EC_2, $3\sigma_2 C^{**}\sigma_h$, S_4, i, $2\sigma_v$, $2\sigma_d$	Rotation, reflection, inversion	XeF_4, $[PtCl_4]^{2-}$ *trans*-MA_4B_2	
D_{2d}	Identity, E Rotation, C_2 Dihedral and plane σ_d	E, C_2, C_2', C_2'', σ_d, σ_d'	Rotation, reflection,	allene	
T_d	Identity, E Rotation, C_3 Rotation, C_2 and reflection	E, $3C_2$, $4C_3$, $3S_4$, $6\sigma_d$	Rotation, reflection	CH_4, NH_4^{+4}, $SiCl_4$	

Contd...

Point group	Symmetry operation	Symmetry element	Operation	Example	Structure
C_h	Identity, E Rotation, C_3 Rotation, C_2 Rotation, C_4 reflection and inversion	$E, 6C_2, 4C_3, 3C_4, 4C_6,$ $3S_4, i, 3\sigma_h, 3\sigma_v$	Rotation, refection and inversion	$SF_6, PtCl_6^{2-}, SiF_6^{2-}$	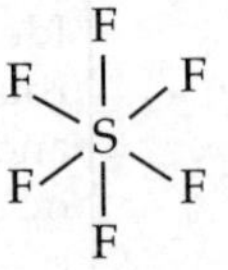
I_h	Identity, E Rotation, C_3 Rotation, C_2 Rotation, C_4 reflection and inversion	$E, 12C_5, 12C_5^2, 20C_3,$ $15C_2, i, 12S_{10}, 12S_{10}^3,$ $20S_6$ and $16s$	Rotation, refection and inversion	icosahedron, dodecahedron, buckminister fullerene (C_{60})	

* σ_d plane: If the symmetry plane bisects the two σ_v planes it is labelled as σ_d.

2.7 DESCENT IN SYMMETRY OF MOLECULES WITH SUBSTITUTION

So far the symmetry of a molecule has been assigned into a point group under 'ideal-ised' circumstances, gradual substitution of the existing groups in the molecules leads to less symmetrical structures of the molecules. The descent in symmetry continues with increased amount of substitution, resulting in lower order point groups from what were initially the higher-order point groups. Thus,

$$C_{nh} \rightarrow C_2$$
$$C_{nv} \rightarrow C_2 \text{ or } C_{n'v} \ (n' < n)$$
$$D_{nh}, D_{nd} \rightarrow C_{nv}, C_{nh}, C_{n'v}, C_{n'h}, Ci \text{ (from } D_{vd})$$

The scheme shown below describes the type of lower in symmetry in AB_n, type ($n = 3 - 6$) molecules with increasing order of substitution.

I. AB_3 molecule

(i) AB_3 (planar)

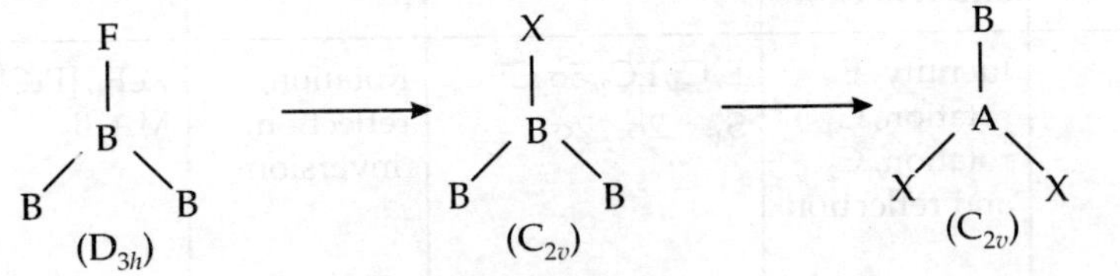

(ii) AB_3 (trigonal pyramidal)

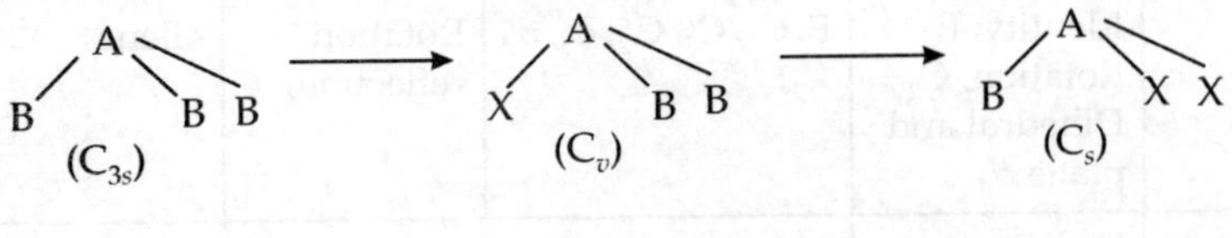

II. AB_4 molecule

(ii) AB_4 (tetrahedral)

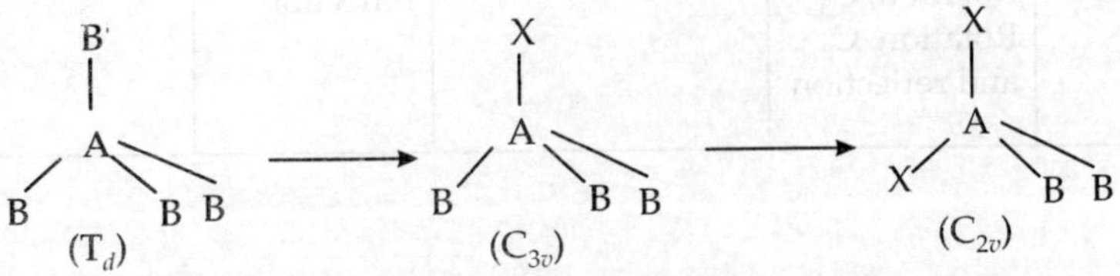

(ii) AB$_4$ (square planar)

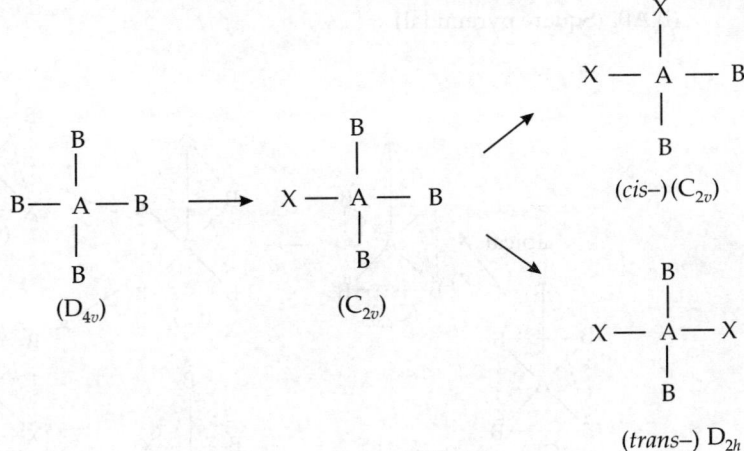

(iii) AB$_4$ (square pyramidal)

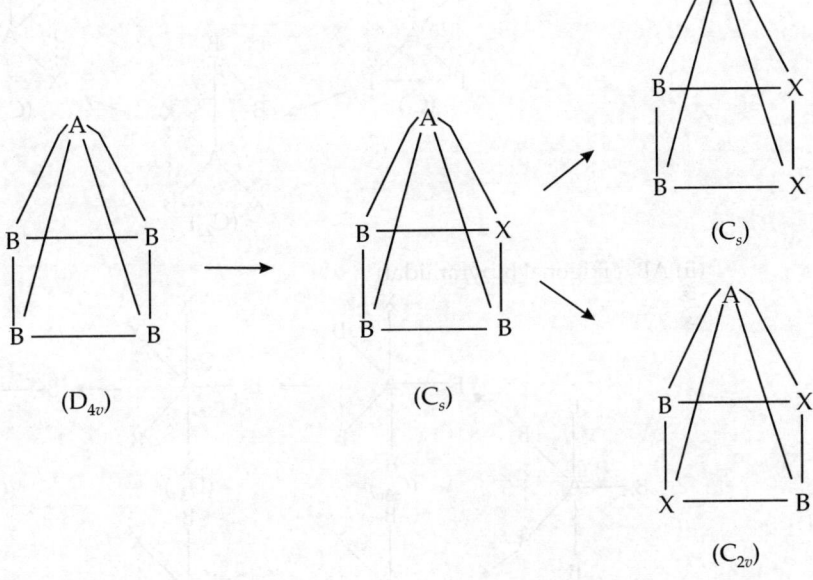

III. AB$_5$ molecule

(i) AB$_5$ (Square pyramidal)

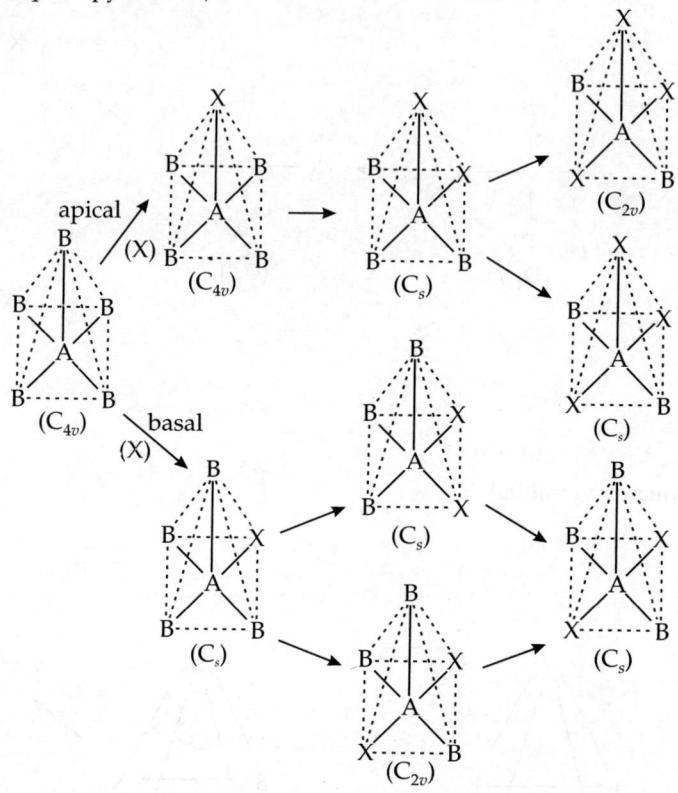

(ii) AB$_5$ (Trigonal bipyramidal)

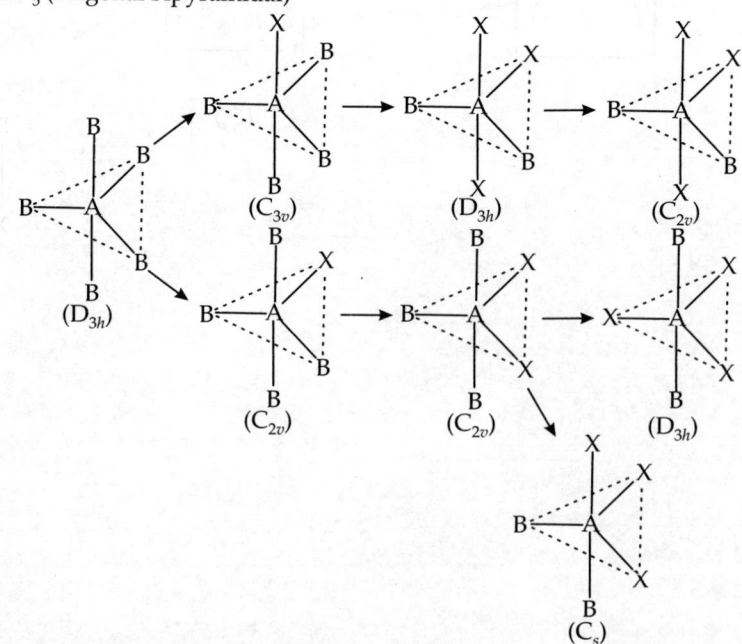

IV. AB_6 (octahedral)

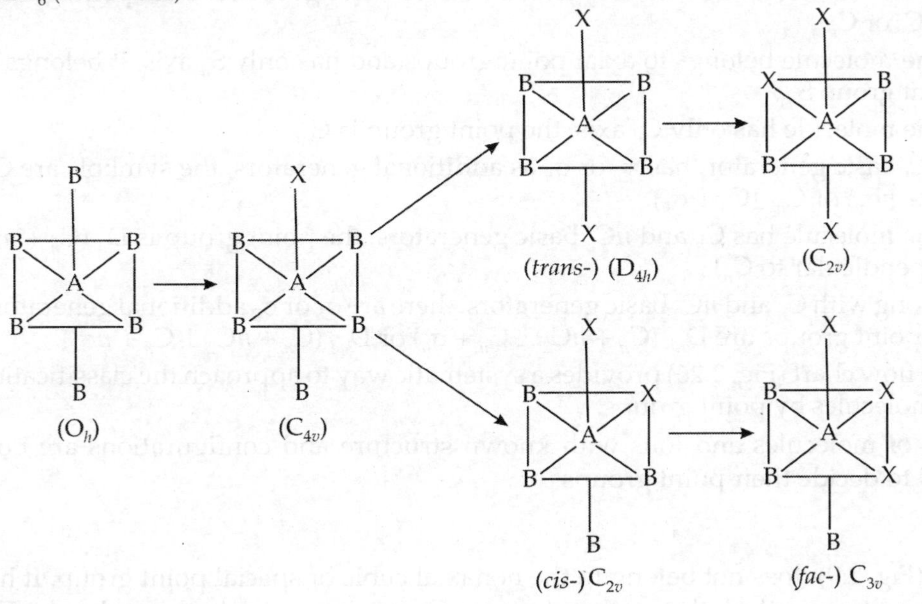

(trans-) (D_{4h})

(C_{2v})

(cis-) C_{2v}

(fac-) C_{3v}

(O_h)

(C_{4v})

The substituted derivatives of benzene with their reduced symmetry point groups are shown below:

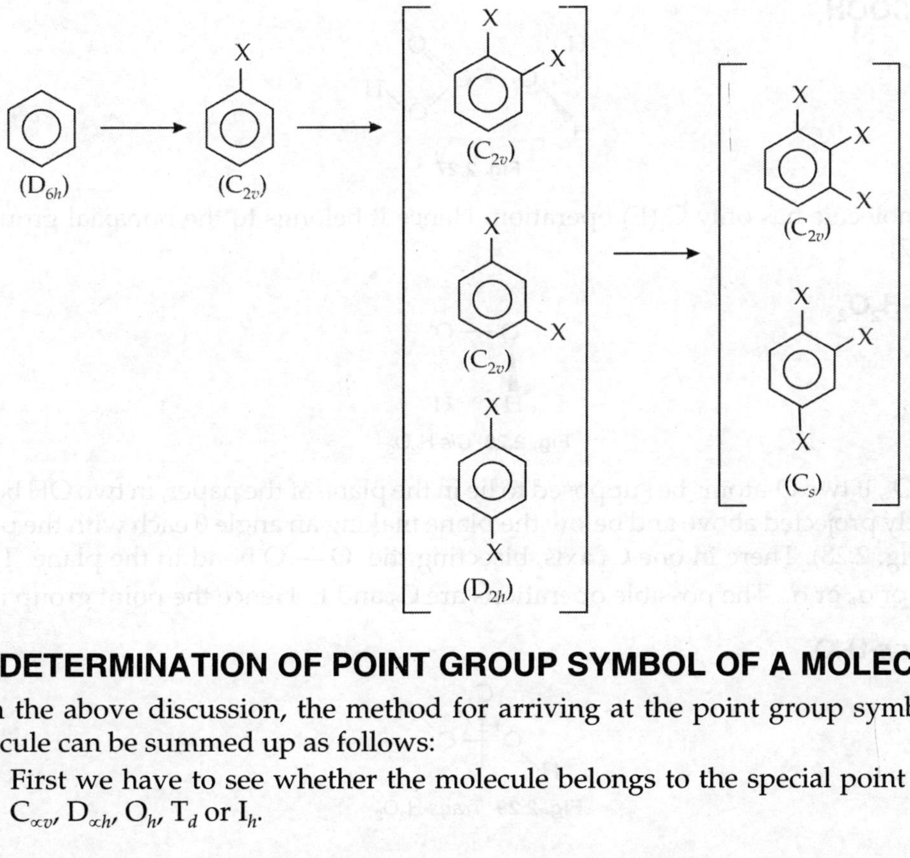

(D_{6h})

(C_{2v})

(C_{2v})

(C_{2v})

(C_{2v})

(C_{2v})

(C_s)

(D_{2h})

2.8 DETERMINATION OF POINT GROUP SYMBOL OF A MOLECULE

From the above discussion, the method for arriving at the point group symbol of a molecule can be summed up as follows:

(i) First we have to see whether the molecule belongs to the special point groups $C_{\infty v}$, $D_{\infty h}$, O_h, T_d or I_h.

(ii) If the molecule has no axis of symmetry then it belongs to non-axial point groups C_1, C_s or C_i.

(iii) If the molecule belongs to axial point groups and has only S_n axis, it belongs to point group S_n.

If the molecule has only C_n axis, the point group is C_n.

If C_n, basic generator, has σ_v or σ_h as additional generators, the symbols are C_{nv} ($C_n + n\sigma_v$) or C_{nh} ($C_n + \sigma_h$).

If the molecule has C_n and nC_2 basic generators, the point group is D_n ($C_n + nC_2$ perpendicular to C_n).

If along with C_n and nC_2 basic generators, there are σ_h or σ_d additional generators, the point groups are D_{nh} ($C_n + nC_2 \perp C_n + \sigma_h$) or D_{nd} ($C_n + nC_2 \perp C_n + n\sigma_d$).

The flow chart (Fig. 2.26) provides a systematic way to approach the classification of molecules by point groups.

A host of molecules and ions with known structure and configurations are now examined to decide their point groups.

(i) H_2O

Water (Fig. 2.9) does not belong to the nonaxial cubic or special point group. It has a C_2 axis and two vertical planes σ_v and σ'_h. No C'_2 axis is present in the molecule. The two σ_v planes are $\perp$ to each other. Looking for σ_h, one does not find it, the point group is therefore C_{2v}.

(ii) CH_3COOH

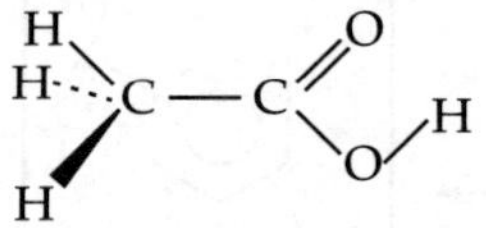

Fig. 2.27

The molecule has only $C_1(E)$ operation. Hence it belongs to the nonaxial group C_1 (Fig. 2.27).

(iii) *Cis*-H_2O_2

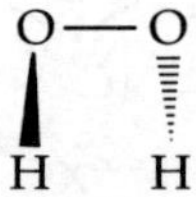

Fig. 2.28 *Cis*-H_2O_2

In H_2O_2, if two O-atoms be supposed to lie in the plane of the paper, in two OH bonds separately projected above and below the plane making an angle θ each with the paper plane (Fig. 2.28). There in one C_2 axis, bisecting the O — O bond in the plane. There is no C'_2 or σ_h or σ_v. The possible operations are C_2 and E. Hence the point group is C_2.

(iv) *Trans*-H_2O_2

Fig. 2.29 *Trans*-H_2O_2

Trans-H_2O_2 has a C_2 axis $\perp$ to the molecular plane and a σ_h along the molecular plane. The point group is C_{2h} (Fig. 2.29).

(v) XeOF₄

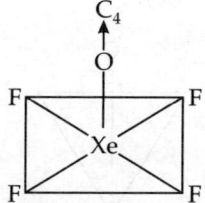

Fig. 2.30 Xenon oxyfluoride

$XeOF_4$ molecule has a square pyramid C_4 axis passes through O—Xe bond and $\perp$ to the square base C_2' axis and σ_h plane is absent. There are $4\sigma_v$ planes, each bisecting the square base. Hence the point group is C_{4v} (Fig. 2.30).

(vi) *Trans* [Pt (NH₃) Cl₂]

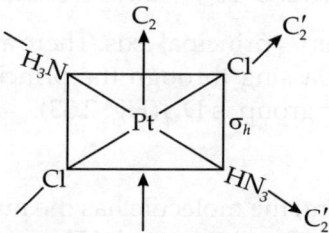

Fig. 2.31 *Trans*-[Pt II (NH₃)₂Cl₂]

Trans-[Pt II $(NH_3)_2Cl_2$] has a C_2 axes perpendicular to the molecular plane (Fig. 2.31). There are two C_2' axis passing through opposite NH_3 and opposite Cl^-. The molecular plane is also an additional generator σ_h. Hence the point group is D_{2h}.

Cyclopentadienide Ion, C₅H₅⁻

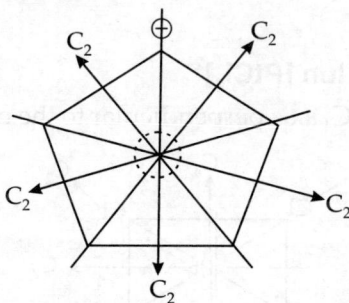

Fig. 2.32 Cyclopentadienide ion $C_5H_5^-$

Cyclopentadienide ion ($C_5H_5^-$) has principal axis C_5 perpendicular to the molecular plane, $5C_2$ axes in the molecular plane and σ_h. Hence the point group is D_{5h} (Fig. 2.32).

Formic acid HCOOH

This molecule, being planar, has just one reflection plane coincident with the molecular plane. The point group is, hence, C_s.

Ethane (Staggered form)

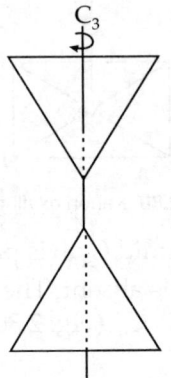

Fig. 2.33 Staggered form of ethane

Staggered form of ethane has C_3 principal axis. There are three subsidiary axes ($3C_2$) also. There are three planes, passing through the principal axis, but in between the subsidiary axes (C_2). The point group is D_{3d} (Fig. 2.33).

Ethane (Eclipsed form)

C_3 and C_2 being being present, the molecule has the minimum symmetry of D_3. The presence of σ_h perpendicular to C_3 axis is noted. Hence it belongs to the group D_{3h}.

Triphenyl Phosphine, $P(C_6H_5)_3$

The compound possesses only a C_3 axis. So it is C_3.

Boron Trifluoride, BF_3

BF_3 has a C_3 axis perpendicular to the σ_h plane (molecular plane) of the molecule. It has also three C_2 axes, each passes through the B atom and an F atom. The overall point group is D_{3h}.

Platinum (II) Tetrachloride Ion $[PtCl_4]^{2-}$

The compound possesses a C_4 axis perpendicular to the molecular plane σ_h (Fig. 2.34).

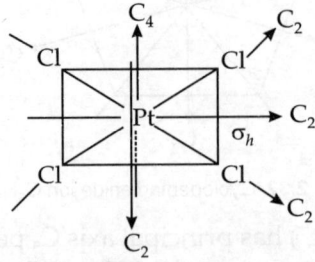

Fig. 2.34 In $[PtCl_4]^{2-}$ ion four-fold axis (C_4) passes through the centre of Pt-atom and $\perp$ to the σ_h plane [C_4 is also $\perp$ to all C_2 axes]

Bis-(Cyclopentadienide) Iron (II) (Ferrocene), Fe (C_5H_5)$_2$

Fe (C_5H_5)$_2$ (Staggered form)

Ferrocene has a C_5 axis passing through the centres of the rings and the iron atom. There are five C_2 axes passing through Fe atom and are perpendicular bisectors of the lines drawn from a carbon atom of one ring to one of the nearest carbon atom of the other ring. Hence the group must be of the D type. There is no σ_h plane but there are five vertical planes of symmetry (σ_d) which pass between adjacent C_2's. Thus the group is D_{5d} (Fig. 2.35(a)).

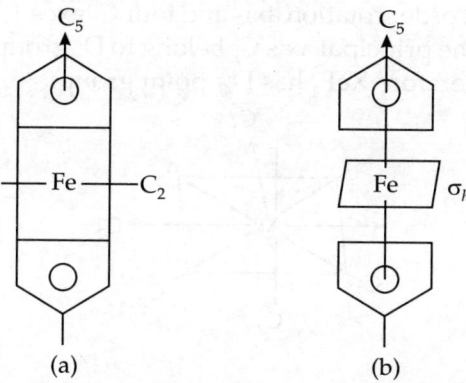

(a)   (b)

Fig. 2.35 Bis-(cyclopentadienide) iron (II) (a) staggered form $5C_2 \perp C_5$ (b) eclipsed form $5C_2 \perp C_5$ [Horizontal plane σ_h contains $5C_2$]

Eclipsed ferrocene has a proper axis C_5. It does not have an improper axis. It has five C_2 axes perpendicular to C_5. It has a horizontal plane σ_h containing five C_2 axes. So the point group of eclipsed ferrocene is D_{5h} (Fig. 2.35(b)).

Os (C_5H_5)$_2$ (Eclipsed form)

Eclipsed Os (C_5H_5)$_2$ has a C_5 axis passing through the centre of the two cyclopentadienyl ring and the Os atom. There are five C_2 axes parallel to the rings and pass through the Os atom. There is one σ_h plane parallel to the rings through the Os atom. Therefore the compound has D_{5h} point group.

Benzene (C_6H_6)

Benzene has a C_6 axis perpendicular to the σ_h plane of the ring, and six C_2 axes in the plane of the ring, three through two carbon atoms and three between the atoms. These are sufficient to make it D_{6h} (Fig. 2.36).

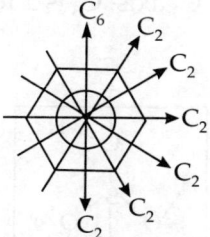

Fig. 2.36

PROBLEMS AND SOLUTIONS

Problem: Determine the point groups of the following molecules and ions.

XeF_4, SF_4, IOF_3, N_2F_2, H_3BO_3, H_2O, PCl_3, BrF_5, HF, Co, HCN, N_2H_4, $P(C_6H_5)_3$, BF_3, $PtCl_4^{2-}$, $Os(C_5H_5)_2$ (eclipsed), C_6H_6, F_2, N_2, C_2H_2, $CH_2 = C = CH_2$ (allene), $Fe(C_5H_5)_2$ (staggered), $[Ru(Cn)_3]^{2-}$.

Solution

XeF_4

XeF_4 has C_4, i.e. highest order rotation axis and four C_2 axes $\perp$ to the C_4 axis. Molecules possessing nC_2 axes $\perp$ to the principal axis C_n belong to D_n groups. XeF_4 has a horizontal plane $\perp$ to the C_4 axis. Therefore XeF_4 has D_{4h} point group.

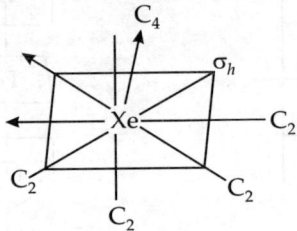

SF_4

The molecule has only one highest order rotation axis C_2. It has no horizontal mirror plane $\perp$ to the C_2 axis but has two vertical mirror planes containing the C_2 axis. So the point group is C_{2v}.

IOF_3

The molecule has no symmetry other then E. Therefore its point is C_1.

N_2F_2

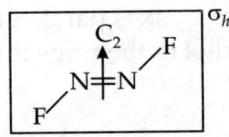

The molecule has σ_h mirror plane and C_2 axis $\perp$ to the σ_h plane. No other symmetry elements are present. Therefore it is C_{2h}.

H_3BO_3

H_3BO_3 has a σ_h mirror plane and C_3 axis. C_3 is $\perp$ to σ_h. There are no other symmetry elements present. So it is C_{3h}.

H₂O

H_2O has a C_2 axis and two vertical planes σ_v and σ_v'. The two vertical planes are $\perp$ to each other. The point group is therefore C_{2v}.

PCl₃

PCl_3 has a C_3 axis through the P atom and equidistant from the three Cl atoms. It has three σ_v planes, each through the P atom and one of the Cl atoms, overall C_{3v}.

BrF₅

BrF_5 has one C_4 axis through the Br atom and the F atom in the plane of the drawing, two σ_v plane each with the Br atom, the F atom in the plane of the drawing, and two of the other F atom, and two σ_d planes between the equatorial F atoms, therefore the overall C_{4v}.

HF, CO, HCN

All are linear molecules with infinite rotation axis through the centre of all the atoms. There are also an infinite number of σ_v planes, all of which contain the C_∞ axis. So the point group of each is $C\infty_v$.

N₂H₄

This molecule contain C_2 axis $\perp$ to the N—N bond. There are no other symmetry elements, so it is C_2.

P(C₆H₅)₃

This compound possesses only a C_3 axis. So it is C_3.

BF₃

BF_3 has a C_3 axis $\perp$ to the σ_h plane of the molecule. It has also three C_2 axes, each passes through the B atom and an F atom. The overall point group is D_{3h}.

PtCl₄²⁻

$PtCl_4^{2-}$ has a C_4 axis $\perp$ to the σ_h plane of the molecule. It also has four C_2 axes in the plane of the molecule, two through opposite Cl atoms and two splitting the Cl—Pt—Cl angles, thus making it D_{4h}.

Four-fold axes (C_4) passes through the centre of Pt-atom and $\perp$ to the plane of the molecule (σ_h). C_4 is also $\perp$ to all C_2 axis.

Os (C$_5$H$_5$)$_2$ (Eclipsed)

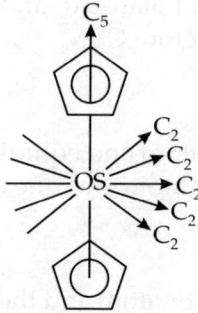

Os (C$_5$H$_5$)$_2$ has a C$_5$ axis through the centre of the two cyclopentadienyle rings and Os, five C$_2$ axes parallel to the rings and pass through the Os atom and a σ_h plane parallel to the rings through the Os atom. Therefore the compound has D$_{5h}$ point group.

C$_6$H$_6$

Benzene has a C$_6$ axis $\perp$ to the σ_h plane of the ring, and six C$_2$ axes in the plane of the ring, three through two C atoms and three between the atom. These are sufficient to make it D$_{6h}$.

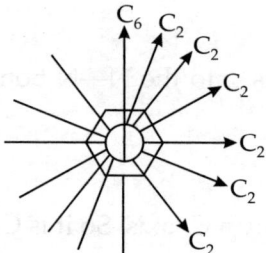

F$_2$, N$_2$ and C$_2$H$_2$

These are all liner molecules with a C$_\alpha$ axis through the atoms. There are also an infinite number of C$_2$ axes $\perp$ to the C$_\alpha$ axes and a σ_h plane $\perp$ to the C$_\alpha$ axes, sufficient to make them D$_{\alpha h}$.

Allene, CH$_2$ $=$ C $=$ CH$_2$

Allene has a C$_2$ axis through the three carbon atoms and two C$_2$ axes $\perp$ to the line of the carbon atoms, both at 45° angles to the planes of the H atoms. Two σ_d mirror planes through each H—C—H combination complete the assignment of D$_{2d}$.

Fe(C$_5$H$_5$)$_2$ (Staggered) and Fe (C$_5$H$_5$)$_2$ (Eclipse)

Ferrocene has a C$_5$ axis through the centres of the rings and the Fe atom. There are five C$_2$ axes $\perp$ to the C$_5$ axis. These C$_2$ axes pass through Fe atoms and are $\perp$ bisectors of lines drawn from a carbon atom of one ring to one of the nearest carbon atoms of the other rings. Hence the group must be of the D type. There is no σ_h. However, there are five vertical planes of symmetry which pass between adjacent C$_2$'s. Thus the group is D$_{5d}$.

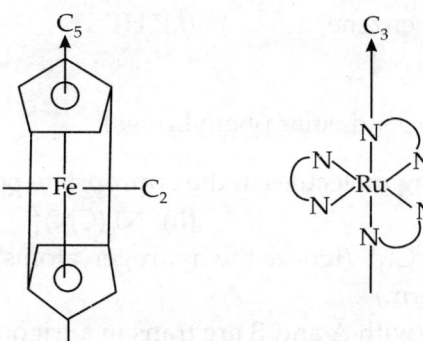

Eclipse ferrocene has a proper axis C_5. It does not have an improper axis. It has five C_2 axes $\perp$ to C_5 axis. It has a horizontal plane containing five C_2 axes. So the point group of eclipse ferrocene is D_{5h}.

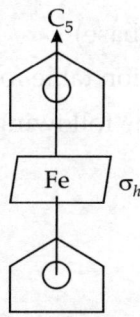

[Ru (en)$_3$]$^{2+}$

[Ru (en)$_3$]$^{2+}$ has a C_3 axis $\perp$ to the drawing through the Ru. There are three C_2 axes in the plane of the paper, each intersecting an on ring at the midpoint and passing through the Ru atom. The group is D_3.

STUDY QUESTIONS

1. Identify the symmetry elements present in the following molecules (ions) and assign them the point group (or determine their symmetry):
   - (a) Pyridine
   - (b) ethane (staggered)
   - (c) CH$_2$Cl Br
   - (d) *cis*-2-butene
   - (e) *trans*-2- butene
   - (f) *trans*-dichlorethane
   - (g) cyclopentadienyl ion
   - (h) *cis*-[Co(en)$_2$ Cl$_2$]$^+$
   - (i) Cyclohexane (boat form)
   - (j) Xe O$_2$ F$_2$
   - (k) TeCl$_4$
   - (l) Cyclopropane
   - (m) Cu(NO$_2$)$_6^{4-}$
   - (n) Cyclopropene
   - (o) Benzene
   - (p) Allene
   - (q) O=C=C=C=O
   - (r) White phosphorus P$_4$
   - (s) C$_{60}$
   - (t) Ferrocene (staggered)
   - (u) Ferrocene (eclipsed)
   - (v) CDCl$_3$

(w) 1, 2, 4-trifluorobenzene   (x) HF

(y) $PtCl_6^{2-}$

(z) ⬡—⬡ (Perpendicular phenyl rings)

2. Classify the following molecules in the appropriate point groups

   (a) $COCl_4^{2-}$ (b) $Ni(CN)_4^{2-}$

   (c) Cis-$[CO(NH_3)_2 Cl_4]^-$ (ignore the hydrogen atoms)

   (d) C_6H_{12} (chain form)

   (e) $Si(CH_3)_3 \cdot A \cdot B$ (with A and B are trans in a trigonal bipyramid)

   (f) PF_3 (g) $(CH_3)_2 B \overset{H}{\underset{H}{<>}} B (CH_3)_2$

   (h) $Cl–I–Cl^-$ (i) planar cis-$PdCl_2 B_2$ (B = base)

   (j) planar trans-$PdCl_2 B_2$ (B = base)

3. Workout the group multiplication table for H_2O.

4. Show that for H_2O molecule, the following relations of operations are correct :

   (i) $\hat{\sigma}_v(xz) \cdot \hat{C}_2 = \hat{\sigma}_v(yz)$

   (ii) $\hat{C}_2 \cdot \hat{\sigma}_v(xz) = \hat{\sigma}_v(yz)$

   (iii) $\hat{\sigma}_v(xz) \cdot \hat{\sigma}_v(yz) = C_2$

5. Workout the products of the elements in C_{3v} point group.

   $\sigma_v C_3^1 \sigma_v, \ \sigma_v'' C_3^1 \sigma_v, \ \sigma_v' C_3^2 \sigma_v, \ \sigma_v C_3^1 \sigma_v''$

6. Show that

   $C_3^2 \sigma_v C_3^1 = \sigma_v''$

   $\sigma_v' \sigma_v \sigma_v' = \sigma_v''$

   $C_3^1 \sigma_v'' C_3^2 = \sigma_v$

3

Representation of Groups

3.1 REPRESENTATION OF GROUPS BY MATRICES

Representation is a set of matrices (of their characters) which represent the operations of a point group. The following four matrices form a representation for the C_{2v} point group. It is denoted by the symbol Γ.

$$E = \begin{pmatrix} 1 & 0 & 0 \\ 0 & 1 & 0 \\ 0 & 0 & 1 \end{pmatrix}, C_2 = \begin{pmatrix} -1 & 0 & 0 \\ 0 & -1 & 0 \\ 0 & 0 & 1 \end{pmatrix}, \sigma_v(xz) = \begin{pmatrix} 1 & 0 & 0 \\ 0 & -1 & 0 \\ 0 & 0 & 1 \end{pmatrix}, \sigma_v(yz) = \begin{pmatrix} -1 & 0 & 0 \\ 0 & 1 & 0 \\ 0 & 0 & 1 \end{pmatrix}$$

$$\Gamma \begin{array}{cccc} E & C_2 & \sigma_v(xz) & \sigma_v(yz) \\ \begin{pmatrix} 1 & 0 & 0 \\ 0 & 1 & 0 \\ 0 & 0 & 1 \end{pmatrix} & \begin{pmatrix} -1 & 0 & 0 \\ 0 & -1 & 0 \\ 0 & 0 & 1 \end{pmatrix} & \begin{pmatrix} 1 & 0 & 0 \\ 0 & -1 & 0 \\ 0 & 0 & 1 \end{pmatrix} & \begin{pmatrix} -1 & 0 & 0 \\ 0 & 1 & 0 \\ 0 & 0 & 1 \end{pmatrix} \end{array}$$

Such set of matrices that multiply together in the same manner as a group multiplication table is said to be the representation of that group. It is

$$C_2 \, \sigma_v(xz) = \begin{pmatrix} -1 & 0 & 0 \\ 0 & -1 & 0 \\ 0 & 0 & 1 \end{pmatrix} \begin{pmatrix} 1 & 0 & 0 \\ 0 & -1 & 0 \\ 0 & 0 & 1 \end{pmatrix} = \begin{pmatrix} -1 & 0 & 0 \\ 0 & 1 & 0 \\ 0 & 0 & 1 \end{pmatrix} = \sigma_v(yz)$$

$$\sigma_v(xz) \, \sigma_v(xz) = \begin{pmatrix} 1 & 0 & 0 \\ 0 & -1 & 0 \\ 0 & 0 & 1 \end{pmatrix} \begin{pmatrix} 1 & 0 & 0 \\ 0 & -1 & 0 \\ 0 & 0 & 1 \end{pmatrix} = \begin{pmatrix} 1 & 0 & 0 \\ 0 & 1 & 0 \\ 0 & 0 & 1 \end{pmatrix} = E$$

Thus each element in the C_{2v} group is its own inverse.

$$C_{2(z)} \, C_{2(z)} = \begin{pmatrix} 1 & 0 & 0 \\ 0 & -1 & 0 \\ 0 & 0 & 1 \end{pmatrix} \begin{pmatrix} 1 & 0 & 0 \\ 0 & -1 & 0 \\ 0 & 0 & 1 \end{pmatrix} = \begin{pmatrix} 1 & 0 & 0 \\ 0 & 1 & 0 \\ 0 & 0 & 1 \end{pmatrix} = E$$

Since the sum of the diagonal term elements of a matrix is called its trace or character, the following characters are observed.

E	C_2	$\sigma_v(xy)$	$\sigma_v(yz)$
3	-1	1	1

This set of characters also forms a representation.

Thus the set of matrices for the symmetry operations of a point group forms a representation. In C_{2h} point group E, C_2, $\sigma_v(xy)$ and i are the four symmetry operations present in this group. The matrix representation for this point group is shown below.

$$E = \begin{pmatrix} 1 & 0 & 0 \\ 0 & 1 & 0 \\ 0 & 0 & 1 \end{pmatrix} \quad C_2 = \begin{pmatrix} -1 & 0 & 0 \\ 0 & -1 & 0 \\ 0 & 0 & 1 \end{pmatrix}$$

$$i = \begin{pmatrix} -1 & 0 & 0 \\ 0 & -1 & 0 \\ 0 & 0 & -1 \end{pmatrix} \quad \sigma_v(xy) = \begin{pmatrix} 1 & 0 & 0 \\ 0 & 1 & 0 \\ 0 & 0 & -1 \end{pmatrix}$$

E	C_2	i	$\sigma_v\,(xy)$
3	−1	−3	1

Representations can be divided into two types:
1. Reducible representations
2. Irreducible representations.

3.2 REDUCIBLE AND IRREDUCIBLE REPRESENTATIONS

A higher dimension representation can be reduced to lower dimension, called reducible representation. Those representations which can not be further reduced to representations of lower dimension is called *irreducible representation*. Each point group can be decomposed into symmetry pattern known as irreducible representations.

Reducible Representations: Let A, B, C, D and P be the matrices in the representation T of a group. Let P be the similarity transformation matrix in this group. By similarity transformation in the matrices A, B, C, D and P are changed into A', B', C', D' and P' as shown below.

$$P^{-1} A P = A'$$
$$P^{-1} B P = B'$$
$$P^{-1} C P = C'$$
$$P^{-1} D P = D'$$
$$P^{-1} P P = P' = P'$$

If the resulting matrices can be blocked into smaller matrices, then the representation T is called a reducible representation. For example, A' can be blocked into $a_1', a_2', a_3', ..., a_4',$ submatrices as

$$A' = \begin{pmatrix} a_1' & 0 & 0 & 0 & 0 & 0 \\ 0 & a_2' & 0 & 0 & 0 & 0 \\ 0 & 0 & a_3' & 0 & 0 & 0 \\ 0 & 0 & 0 & a_4' & 0 & 0 \\ 0 & 0 & 0 & 0 & a_5' & 0 \\ 0 & 0 & 0 & 0 & 0 & a_6' \end{pmatrix}$$

Irreducible Representation: If it is not possible to find a similarity transformation matrix which will reduce the matrices of representation T, then the representation is said to be irreducible. All one-dimensional representations are always irreducible. The example of an irreducible representation is illustrated below.

Let us consider the matrices of transformation for the z-coordinate of a hydrogen atom in H_2 molecule, which is produced by the symmetry operations of $D_{\infty h}$ group.

The operations of $D_{\infty h}$ are E, C_∞, σ_v, C_2, σ_h, S_∞ and i. It is seen from Fig. 3.1 that the z-coordinate is unaffected by E, C_∞ and σ_v operations. The equations and matrices for the transformation of z-coordinate of hydrogen atom by these operations are given below.

$$E \cdot z = 1\, z$$
$$E \text{ matrix} = [1]$$
$$C_\infty \cdot z = 1\, z$$
$$C_\infty \text{ matrix} = [1]$$
$$\sigma_v \cdot z = 1\, z$$
$$\sigma_v \cdot \text{matrix} = [1]$$

All other operations of this group change the coordinate z of hydrogen atom into $-z$ and we get the following equations and matrices for the transformation.

$$C_2 \cdot z = -1\, z \qquad C_2 \text{ matrix} = [-1]$$
$$S_\infty \cdot z = -1\, z \qquad S_\infty \text{ matrix} = [-1]$$
$$\sigma_h \cdot z = -1\, z \qquad \sigma_h \text{ matrix} = [-1]$$
$$i \cdot z = -1\, z \qquad i \text{ matrix} = [-1]$$

The matrix representation thus obtained for the z-coordinate of hydrogen atom in H_2 molecule is given below.

	E	C_∞	σ_v	C_2	S_∞	σ_h	i
T	[1]	[1]	[1]	[−1]	[−1]	[−1]	[−1]

This representation T, is irreducible since it is one-dimensional.

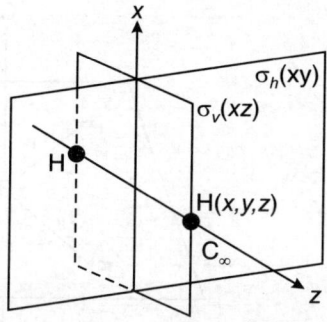

Fig. 3.1 The C_∞ axis, vertical plane $\sigma_v(xz)$ and horizontal plane $\sigma_h(xy)$ in H_2 molecule

A *reducible representation* results when the various symmetry operations are performed on all the sigma-bonds of a molecule. Let us consider the BF_3 molecule which belongs to D_{3h} point group. The twelve symmetry operations of this group are divided into six classes as follows.

$$(E),\ (C_3^1, C_3^2),\ (3\, C_2),\ (3\, \sigma_v),\ (S_3^1, S_3^5) \text{ and } (\sigma_h)$$

The symmetry operations in these classes are carried out on the three sigma bonds in BF_3 molecule. The three sigma bonds in BF_3 are considered as vectors r_1, r_2 and r_3. By a symmetry operation R, the vectors r_1, r_2 and r_3 are changed into r_1', r_2' and r_3' according to matrix Eq. (3.1).

$$\begin{pmatrix} r_1' \\ r_2' \\ r_3' \end{pmatrix} = (R) \begin{pmatrix} r_1 \\ r_2 \\ r_3 \end{pmatrix} \qquad \qquad \dots(3.1)$$

where (R) is the matrix for the operation R. The vectors r_1, r_2 and r_3 remain unaffected by the identity operation (Fig. 3.2). The equations relating r_1', r_2' and r_3' to r_1, r_2 and r_3 are given below.

$$r_1' = 1\,r_1 + 0\,r_2 + 0\,r_3 \qquad \qquad \text{...(3.2)}$$
$$r_2' = 0\,r_1 + 1\,r_2 + 0\,r_3 \qquad \qquad \text{...(3.3)}$$
$$r_3' = 0\,r_1 + 0\,r_2 + 1\,r_3 \qquad \qquad \text{...(3.4)}$$

In matrix form, these equations become

$$\begin{pmatrix} r_1' \\ r_2' \\ r_3' \end{pmatrix} = \begin{pmatrix} 1 & 0 & 0 \\ 0 & 1 & 0 \\ 0 & 0 & 1 \end{pmatrix} \times \begin{pmatrix} r_1 \\ r_2 \\ r_3 \end{pmatrix} \qquad \qquad \text{...(3.5)}$$

From Eqs. (3.1) and (3.5) we get

$$\begin{pmatrix} 1 & 0 & 0 \\ 0 & 1 & 0 \\ 0 & 0 & 1 \end{pmatrix}$$

as the the matrix for identity operation. From Fig. 3.2 it is seen that the C_3^1 operation changes the vectors r_1, r_2 and r_3 in cyclic order. The resulting vectors r_1', r_2' and r_3' are related to r_1, r_2 and r_3 by the equations.

$$r_1' = 0\,r_1 + 1\,r_2 + 0\,r_3 \qquad \qquad \text{...(3.6)}$$
$$r_2' = 0\,r_1 + 0\,r_2 + 1\,r_3 \qquad \qquad \text{...(3.7)}$$
$$r_3' = 1\,r_1 + 0\,r_2 + 0\,r_3 \qquad \qquad \text{...(3.8)}$$

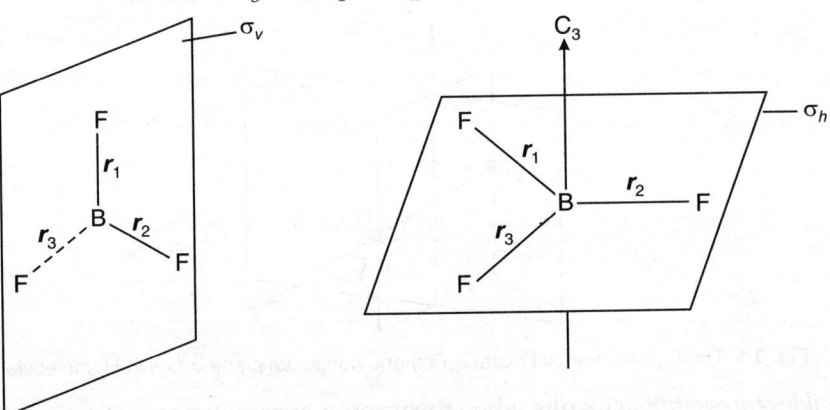

Fig. 3.2 The 3-fold axis, horizontal plane σ_h, and vertical plane σ_v in BF$_3$ molecule, r_1, r_2 and r_3 represent the B–F sigma bonds

There equations can be represented in the matrix form as follows.

$$\begin{pmatrix} r_1' \\ r_2' \\ r_3' \end{pmatrix} = \begin{pmatrix} 0 & 1 & 0 \\ 0 & 0 & 1 \\ 1 & 0 & 0 \end{pmatrix} \times \begin{pmatrix} r_1 \\ r_2 \\ r_3 \end{pmatrix} \qquad \qquad \text{...(3.9)}$$

From Eqs. (3.1) and (3.9), we get the matrix

$$\begin{pmatrix} 0 & 1 & 0 \\ 0 & 0 & 1 \\ 1 & 0 & 0 \end{pmatrix}$$

for C_3' operation. By a similar procedure, the matrices for σ_v, σ_h, S_3' and C_2' operations can be obtained. Thus, we get the following reducible representations T_R for sigma bonds of BF_3 molecule. Reducible representation (T_R):

$$
\overset{E}{\begin{pmatrix} 1 & 0 & 0 \\ 0 & 1 & 0 \\ 0 & 0 & 1 \end{pmatrix}}
\overset{C_3'}{\begin{pmatrix} 0 & 1 & 0 \\ 0 & 0 & 1 \\ 1 & 0 & 0 \end{pmatrix}}
\overset{\sigma_v}{\begin{pmatrix} 1 & 0 & 0 \\ 0 & 0 & 1 \\ 0 & 1 & 0 \end{pmatrix}}
$$

$$
\overset{\sigma_h}{\begin{pmatrix} 1 & 0 & 0 \\ 0 & 1 & 0 \\ 0 & 0 & 1 \end{pmatrix}}
\overset{S_3'}{\begin{pmatrix} 0 & 1 & 0 \\ 0 & 0 & 1 \\ 1 & 0 & 0 \end{pmatrix}}
\overset{C_2'}{\begin{pmatrix} 1 & 0 & 0 \\ 0 & 0 & 1 \\ 0 & 1 & 0 \end{pmatrix}}
$$

We need only the characters of matrices of the reducible representation to solve problems in chemistry. Thus we have

E	C_3'	C_2'	σ_v	σ_h	S_3'
3	0	1	1	3	0

Reducible and irreducible representations play a major role in the application of group theory to solve problems in chemistry. The first step in any application of group theory involves the formation of the reducible representation. This representation is then decomposed into the various irreducible representations of the point group under consideration. This aspect will be discussed in detail in Chapter 6.

We shall now consider, another example, water molecule. It belongs to C_{2v} point group. Each transformation matrix in the C_{2v} set is block diagonalised, i.e. it can be broken down into smaller matrices along the diagonal, with other elements equal to zero.

$$
E: \begin{pmatrix} 1 & 0 & 0 \\ 0 & 1 & 0 \\ 0 & 0 & 1 \end{pmatrix}
\xrightarrow{\text{block diagonalised}}
\begin{pmatrix} [1] & 0 & 0 \\ 0 & [1] & 0 \\ 0 & 0 & [1] \end{pmatrix} \rightarrow
\begin{pmatrix} 1 & & \\ & 1 & \\ & & 1 \end{pmatrix}
$$

$$
C_2: \begin{pmatrix} -1 & 0 & 0 \\ 0 & -1 & 0 \\ 0 & 0 & 1 \end{pmatrix}
\xrightarrow{\text{block diagonalised}}
\begin{pmatrix} [-1] & 0 & 0 \\ 0 & [-1] & 0 \\ 0 & 0 & [1] \end{pmatrix} \rightarrow
\begin{pmatrix} -1 & & \\ & -1 & \\ & & -1 \end{pmatrix}
$$

$$
\sigma_v(xz): \begin{pmatrix} 1 & 0 & 0 \\ 0 & -1 & 0 \\ 0 & 0 & 1 \end{pmatrix}
\xrightarrow{\text{block diagonalised}}
\begin{pmatrix} [1] & 0 & 0 \\ 0 & [-1] & 0 \\ 0 & 0 & [1] \end{pmatrix} \rightarrow
\begin{pmatrix} 1 & & \\ & -1 & \\ & & 1 \end{pmatrix}
$$

$$
\sigma_v(yz): \begin{pmatrix} -1 & 0 & 0 \\ 0 & 1 & 0 \\ 0 & 0 & 1 \end{pmatrix}
\xrightarrow{\text{block diagonalised}}
\begin{pmatrix} [-1] & 0 & 0 \\ 0 & [1] & 0 \\ 0 & 0 & [1] \end{pmatrix} \rightarrow
\begin{pmatrix} -1 & & \\ & 1 & \\ & & 1 \end{pmatrix}
$$

Thus all the nonzero elements become $| x |$ matrices along the principal diagonal. The total representation describing the effect of all the symmetry operations in the

C_{2v} point group on the point with coordinates x, y, and z is

$$
\begin{matrix} & E & & C_2 & & \sigma_v(xz) & & \sigma_v(yz) \end{matrix}
$$

$$
\begin{pmatrix} 1 & & \\ & 1 & \\ & & 1 \end{pmatrix}\begin{pmatrix} 1 & & \\ & -1 & \\ & & 1 \end{pmatrix}\begin{pmatrix} 1 & & \\ & -1 & \\ & & 1 \end{pmatrix}\begin{pmatrix} -1 & & \\ & 1 & \\ & & 1 \end{pmatrix}
$$

Note that each of these matrices is block diagonalised, i.e. the total matrix can be broken up into blocks of smaller matrices with no off-diagonal elements between the blocks.

When the matrices are block diagonalised in this way, the x, y, and z coordinates are also block diagonalised. As a result, the x, y, and z coordinates are independent of each other. The matrix elements in the 1, 1 positions (numbered as row, column) describe the results of the symmetry operations on the x coordinate, those in the 2, 2 positions describe the results of the operations on the y coordinate, and those in the 3, 3 positions describe the results of the operations on the z coordinate.

$$
E: \begin{pmatrix} 1 & & \\ & 1 & \\ & & 1 \end{pmatrix}
$$

The position of this element is 1, 1 (row and column)
The position of this element is 2, 2
The position of this element is 3, 3

3.3 REPRESENTATIONS OF SYMMETRY OPERATIONS OF A POINT GROUP

Earlier, we have seen that the symmetry operation of a molecule can be expressed mathematically in the matrix notations. In this section, a few representations of symmetry operations in a point group are described.

3.3.1 Translational Vectors as a Basis of Representation

We choose water molecule. This molecule has C_{2v} point group. Translational vectors are attached to both oxygen atom and two hydrogen atoms as water molecule (Fig. 3.3) may be taken as a basis for representation of symmetry operations in a point group.

$E(T_y) = (+1)T_y$

σ_{xz}

σ_{yz}

$C_2(T_y) = (-1)T_y$

$\sigma_{xy}(T_y) = (-1)T_y$

$\sigma_{yz}(T_y) = (+1)T_y$

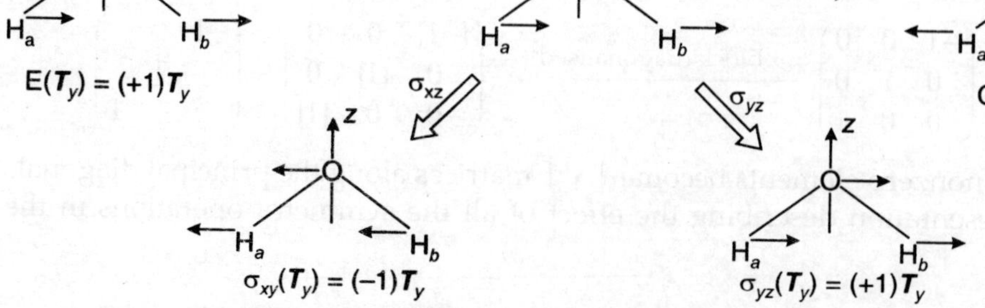

Fig. 3.3 Translational vectors T_y in H_2O molecule which lie in the yz-plane

Suppose a translational vector (T_y) in y direction is attached to each atom of a water molecule. The behaviour of this set of vectors on applying the symmetry operations of the point group C_{2v} is studied. Figure 3.1 shows that the direction of the vectors either remain unchanged (a and d) or show reversal of direction.

$$E\,(T_y) = (+1)\,(T_y),\; +1 \text{ indicates vector direction unchanged}$$
$$C_2\,(T_y) = (-1)\,(T_y),\; -1 \text{ indicates vector direction changed}$$
$$\sigma_v\,(xz)\,(T_y) = (-1)\,(T_y),\; -1 \text{ indicates vector direction changed}$$
$$\sigma_v\,(yz)\,(T_y) = (+1)\,(T_y),\; +1 \text{ indicates vector direction unchanged}$$

The above mentioned facts are shown in the following way.

C_{2v}	E	C_2	$\sigma_v(xz)$	$\sigma_v(yz)$	
	+1	−1	−1	+1	T_y

This is one-dimensional representation of the C_{2v} point group for which translational vectors T_y form the basis.

Similarly, we can get one-dimensional representations based on translational vectors T_x (Fig. 3.4) and T_z.

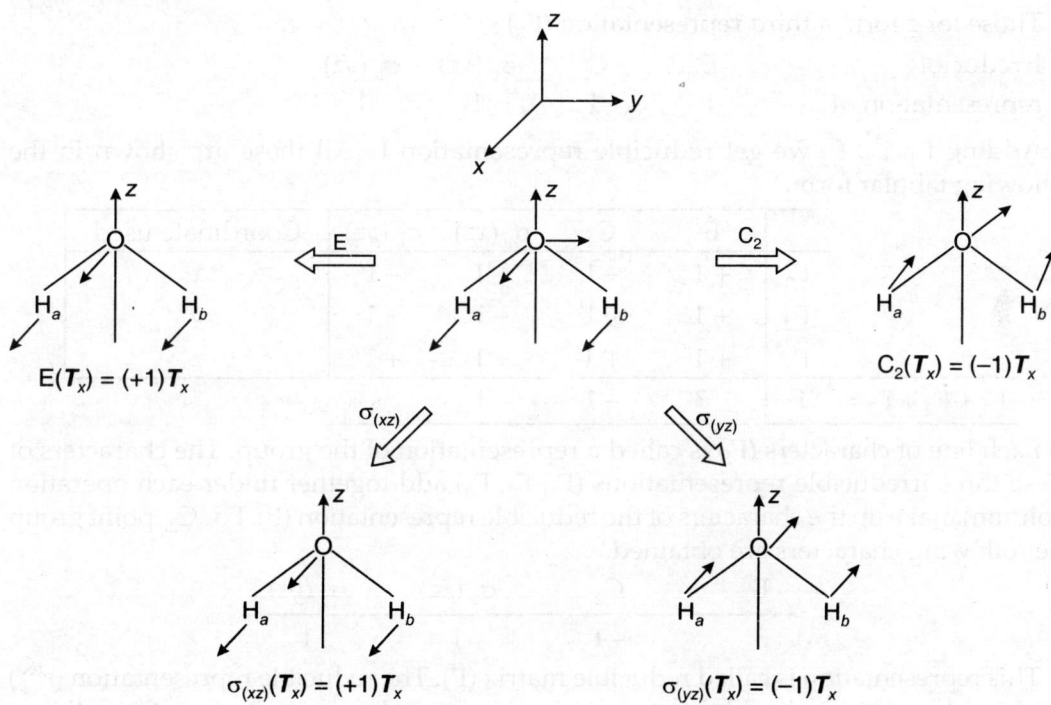

Fig. 3.4 Translational vectors T_x in water molecule

Therefore one-dimensional representations based on T_x are

C_{2v}	E	C_2	$\sigma_v(xz)$	$\sigma_v(yz)$	
	+1	−1	+1	−1	T_x

The one-dimensional representations based on T_x, T_y and T_z vectors are shown in Table 3.1.

Table 3.1 One-dimensional representations based on T_x T_y and T_z

C_{2v}	E	C_2	$\sigma_v(xz)$	$\sigma_v(yz)$	
	+1	−1	+1	−1	T_x
	+1	−1	−1	+1	T_y
	+1	+1	+1	+1	T_z

The four matrix elements for x, i.e. 1 for E, − 1 for C_2, 1 for $\sigma_v(xz)$ and − 1 for $\sigma_v(yz)$ form a representation of the group (Γ_3).

Irreducible	E	C_2	$\sigma_v(xz)$	$\sigma_v(yz)$
representation Γ_3	1	− 1	1	− 1

Those for Y form a second representation (Γ_2)

Irreducible	E	C_2	$\sigma_v(xz)$	$\sigma_v(yz)$
representation Γ_2	1	− 1	− 1	1

Those for z form a third representation (Γ_3)

Irreducible	E	C_2	$\sigma_v(xz)$	$\sigma_v(yz)$
representation Γ_1	1	1	1	1

Adding Γ_1, Γ_2, Γ_3 we get reducible representation Γ. All these are shown in the following tabular form.

	E	C_2	$\sigma_v(xz)$	$\sigma_v(yz)$	Coordinate used
Γ_3	+ 1	− 1	+ 1	− 1	x
Γ_2	+ 1	− 1	− 1	+ 1	y
Γ_1	+ 1	+ 1	+ 1	+ 1	z
$\Gamma_1 + \Gamma_2 + \Gamma_3 =$ Γ	3	− 1	1		

Each line of characters (Γ_i) is called a representation of the group. The characters of these three irreducible representations (Γ_1, Γ_2, Γ_3) add together under each operation (column) make up the characters of the reducible representation (Γ). For C_{2v} point group the following characters are obtained.

E	C_2	$\sigma_v(xz)$	$\sigma_v(yz)$
3	− 1	1	1

This representation is called reducible matrix (Γ). The reducible representation (r^{red}) can be reduced to irreducible representations (r_1, r_2, r_3) by similarity transformation

$$r^{red} = r_1 \oplus r_2 \oplus r_3$$

3.3.2 Rotational Vectors as a Basis of Representation

This is being illustrated by taking the example of water molecule. Let the water molecule (see Fig. 3.3) be rotated about the z-axis. The rotational vectors (R_z) associated with H_a and H_b atoms lie in xy plane. Their behaviour under different symmetry operations of the C_{2v} point group is also shown in Fig. 3.5.

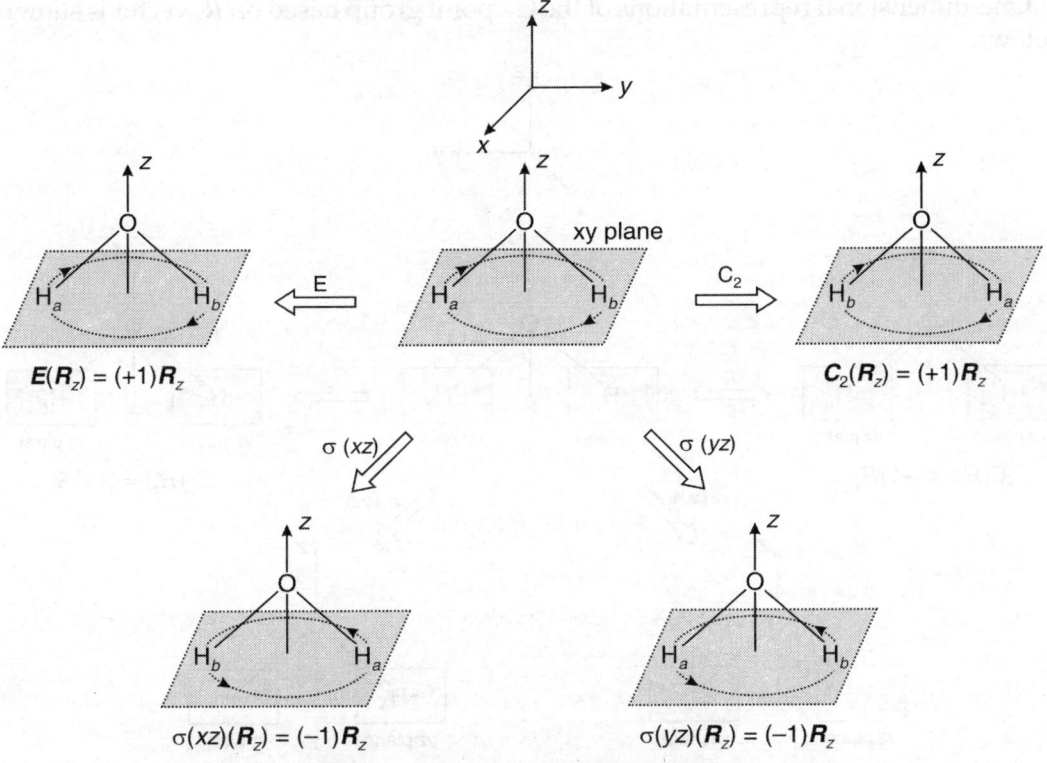

$E(R_z) = (+1)R_z$

$C_2(R_z) = (+1)R_z$

σ (xz)

σ (yz)

$\sigma(xz)(R_z) = (-1)R_z$

$\sigma(yz)(R_z) = (-1)R_z$

Fig. 3.5 Rotational vectors R_z in H_2O molecule

Figure 3.5 reveals that the direction of rotational vectors remain unchanged while the others show reversal of directions. These facts may be expressed as

$$E\,(R_z) = (+1)\,R_z \qquad\qquad \sigma_v\,(xz)\,(R_z) = (-1)\,R_z$$
$$C_2\,(R_z) = (+1)\,R_z \qquad\qquad \sigma_v\,(yz)\,(R_z) = (-1)\,R_z$$

where (+1) represents unchanged vector direction and (–1) represents changed vector direction. The characters are shown in the following way.

C_{2v}	E	C_2	$\sigma_v\,(xz)$	$\sigma_v\,(yz)$	
	1	1	–1	–1	R_z

This is one-dimensional representation of the C_{2v} point group based on the rotational vectors (R_z). We can also find out other one-dimensional representations based on R_x and R_y (Figs 3.4 and 3.5).

Rotational vectors R_x lie in the yz plane. These R_x vectors are attached to H_a and H_b atoms and have identical orientations. They originate tangentially from their respective $-y$ direction. It is to be noted that rotational vectors R_x and the translational vector T_y behave in identical manner under the symmetry operations and thus produce exactly identical one-dimensional representations (Fig. 3.6).

One-dimensional representations of the C_{2v} point group based on R_x vector is shown below:

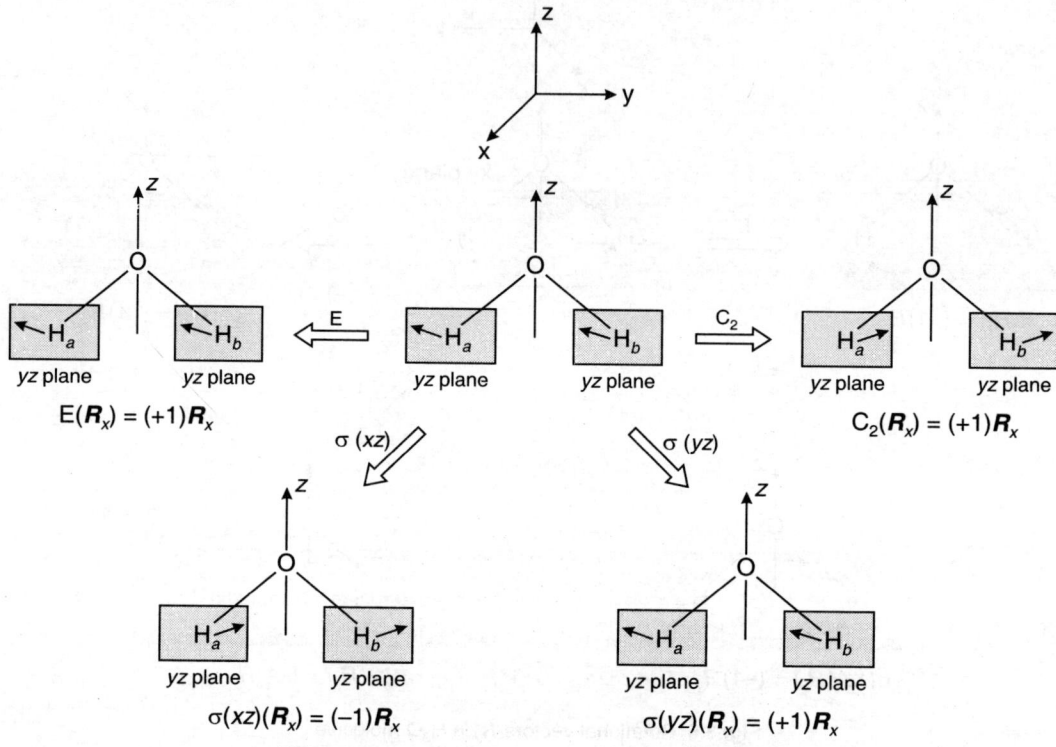

Fig. 3.6 Behaviour of R_x under symmetry operations

Rotational vectors R_y lie in the xz plane. These vectors are attached to H_a and H_b atoms. Both these vectors have identical orientations. They originate tangentially from their respective $+x$ direction. So rotational vectors R_y and translational vectors T_x behave in identical manner under the symmetry operations and thus produce exactly identical one-dimensional representations (Fig. 3.7).

The behaviour of R_y under different symmetry operations are shown below:

C_{2v}	E	C_2	$\sigma_v(xz)$	$\sigma_v(yz)$	
	1	−1	1	−1	R_y (behaves like T_x)

Table 3.2, after compilation of R_x, R_y and R_z vectors under different symmetry operation of C_{2v} point group, displays one-dimensional representations based on the rotational vectors.

Table 3.2 One-dimensional representations based on the rotational vectors

C_{2v}	E	C_2	$\sigma_v(xz)$	$\sigma_v(yz)$	
	1	−1	−1	1	R_x (behaves like T_y)
	1	−1	1	−1	R_y (behaves like T_x)
	1	1	−1	−1	R_z
	1	−1	−1	1	R_x (behaves like T_y)

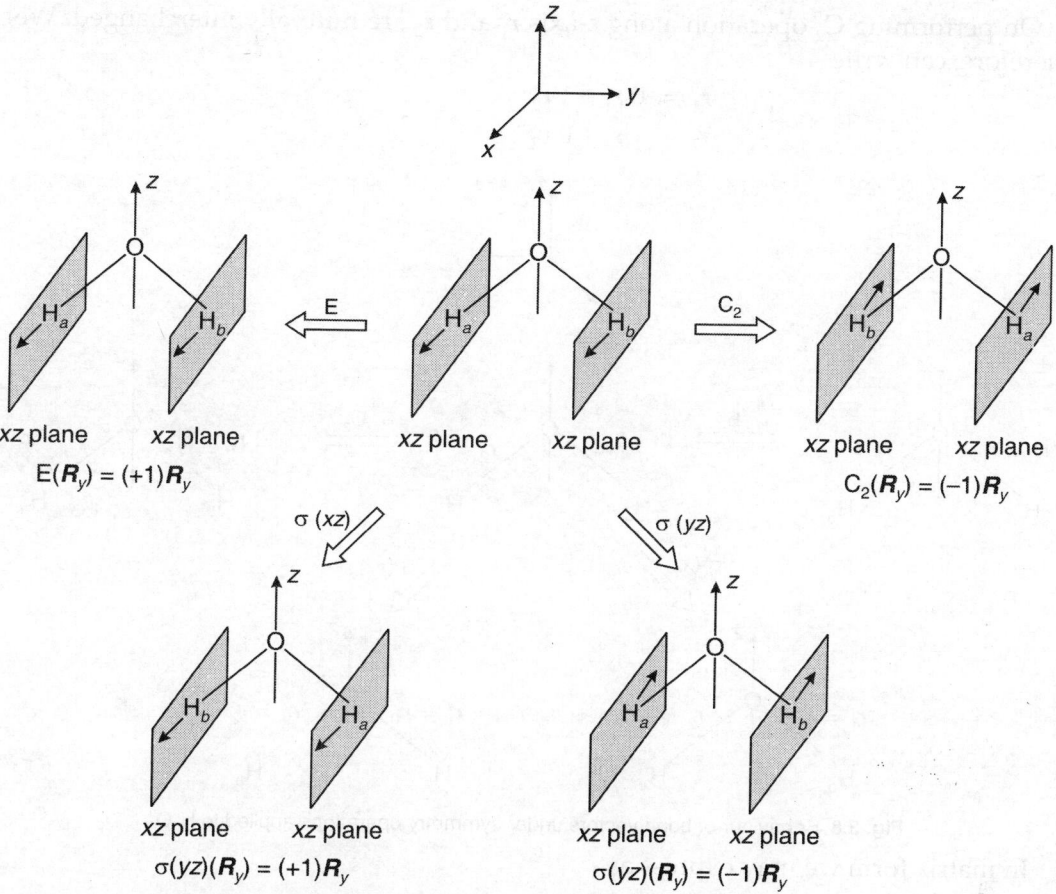

Fig. 3.7 Behaviour of R_y under symmetry operations

3.3.3 Bond Vectors as a Basis of Representations

The set of bond vectors are considered as the basis of representations of symmetry operations of a point group. This is being illustrated by taking the example of water molecule.

The two bond vectors r_1 and r_2 of water molecule are shown in Fig. 3.6. We carry out several symmetry operations of C_{2v} group on these two r_1 and r_2 vectors and get the following result.

The bond vectors r_1 and r_2 remains (Fig. 3.8) unchanged by the identity operation the equations relating to r_1', r_2' to r_1 and r_2 are shown below.

$$r_1' = 1\, r_1 + 0\, r_2$$
$$r_1' = 0\, r_1 + 1\, r_2$$

The above relations may be expressed in the matrix form as

$$\begin{pmatrix} r_1' \\ r_2' \end{pmatrix} = \begin{pmatrix} 1 & 0 \\ 0 & 1 \end{pmatrix} \begin{pmatrix} r_1 \\ r_2 \end{pmatrix}$$

On performing C_2 operation along z-axis r_1 and r_2 are mutually interchanged. We, therefore, can write

$$r_1' = 0\,r_1 + 1\,r_2$$
$$r_2' = 1\,r_1 + 0\,r_2$$

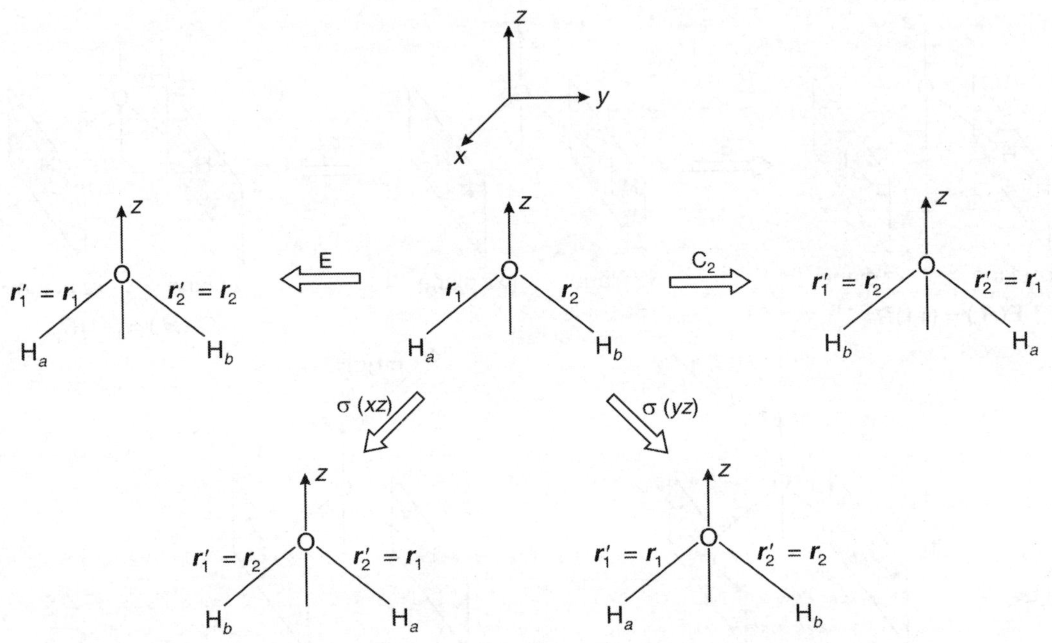

Fig. 3.8 Behaviour of bond vectors under symmetry operations applied to H_2O

In matrix form we may express as

$$\begin{pmatrix} r_1' \\ r_2' \end{pmatrix} = \begin{pmatrix} 0 & 1 \\ 1 & 0 \end{pmatrix}\begin{pmatrix} r_1 \\ r_2 \end{pmatrix}$$

In σ_v (xz) operation r_1 and r_2 mutually interchanged

$$\begin{pmatrix} r_1' \\ r_2' \end{pmatrix} = \begin{pmatrix} 0 & 1 \\ 1 & 0 \end{pmatrix}\begin{pmatrix} r_1 \\ r_2 \end{pmatrix}$$

So

In case of σ_v (yz) operation r_1 and r_2 vectors remain unchanged

$$\begin{pmatrix} r_1' \\ r_2' \end{pmatrix} = \begin{pmatrix} 1 & 0 \\ 0 & 1 \end{pmatrix}\begin{pmatrix} r_1 \\ r_2 \end{pmatrix}$$

Hence, the transformation matrices of the point group C_{2v} are as follows.

C_{2v}	E	C_2	$\sigma_v(xz)$	$\sigma_v(yz)$
	$\begin{pmatrix} 1 & 0 \\ 0 & 1 \end{pmatrix}$	$\begin{pmatrix} 0 & 1 \\ 1 & 0 \end{pmatrix}$	$\begin{pmatrix} 0 & 1 \\ 1 & 0 \end{pmatrix}$	$\begin{pmatrix} 1 & 0 \\ 0 & 1 \end{pmatrix}$

3.3.4 Atomic Wave Function as a Basis of Representations

Atomic orbitals, i.e. atomic wave functions may be taken as a basis of representations of symmetry operations of a point group. Here we shall consider only the angular part of the wave function since radial part remains unchanged by all symmetry operations.

We take water molecule as the example and carry out different symmetry operations on p orbitals, (i.e. p_x, p_y and p_z orbitals) of O-atom in water molecule (Fig. 3.9).

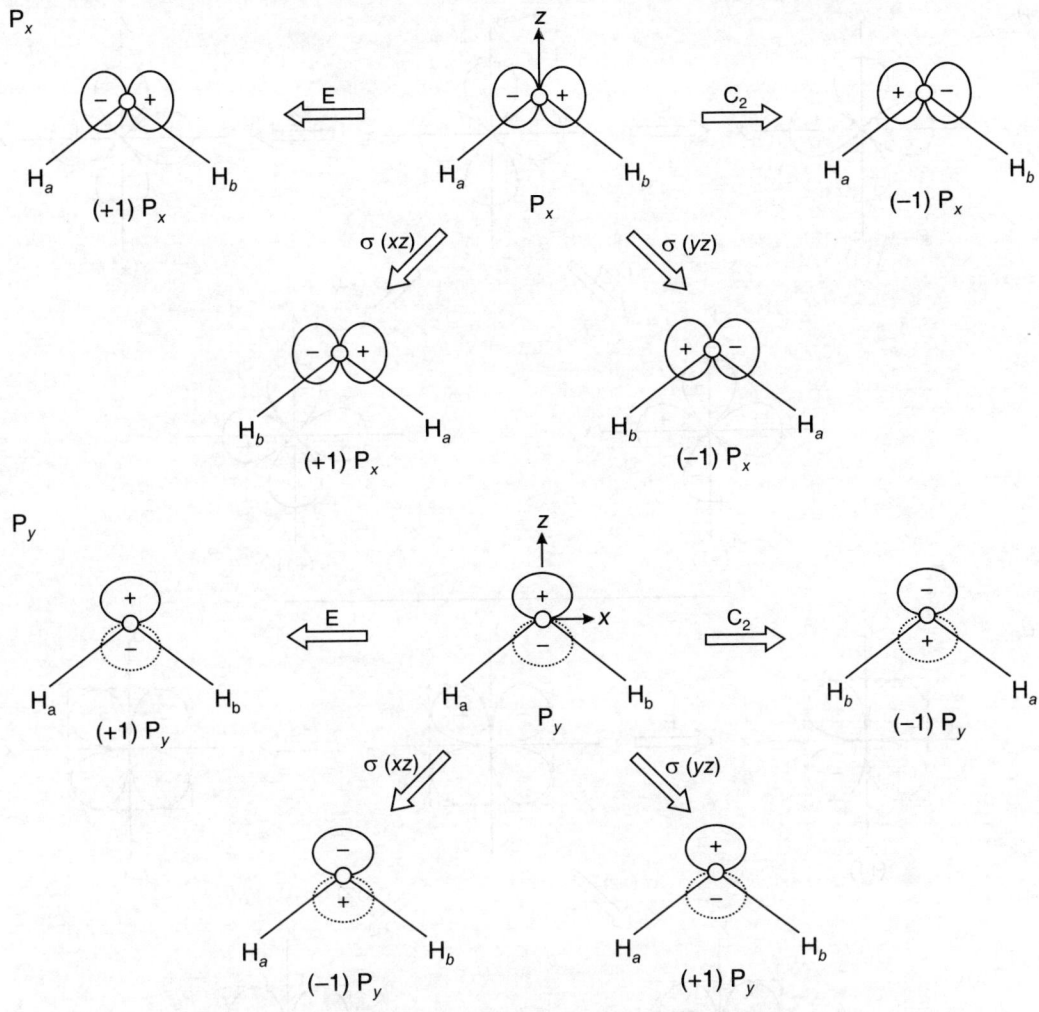

Fig. 3.9 Effect of symmetry operations on p_x and p_y orbitals of oxygen in water

The transformation matrices (1×1 matrix) of the point group C_{2v} based on p orbitals are as follows.

C_{2v}	E	C_2	$\sigma_v (x)$	$\sigma_v (yz)$	
	$+1$	$+1$	$+1$	$+1$	p_z
	$+1$	-1	$+1$	-1	p_x
	$+1$	-1	-1	$+1$	p_y

We shall now carry out different symmetry operation of C_{2v} point group on the d orbital diagrams. Figure 3.10 displays the effects for $d_{x^2-y^2}$, d_{xy} and d_{yz} orbitals of in SO_2.

The transformation matrices (each one is $|\times|$ matrix) of the point group C_{2v} based on d orbitals are as follows.

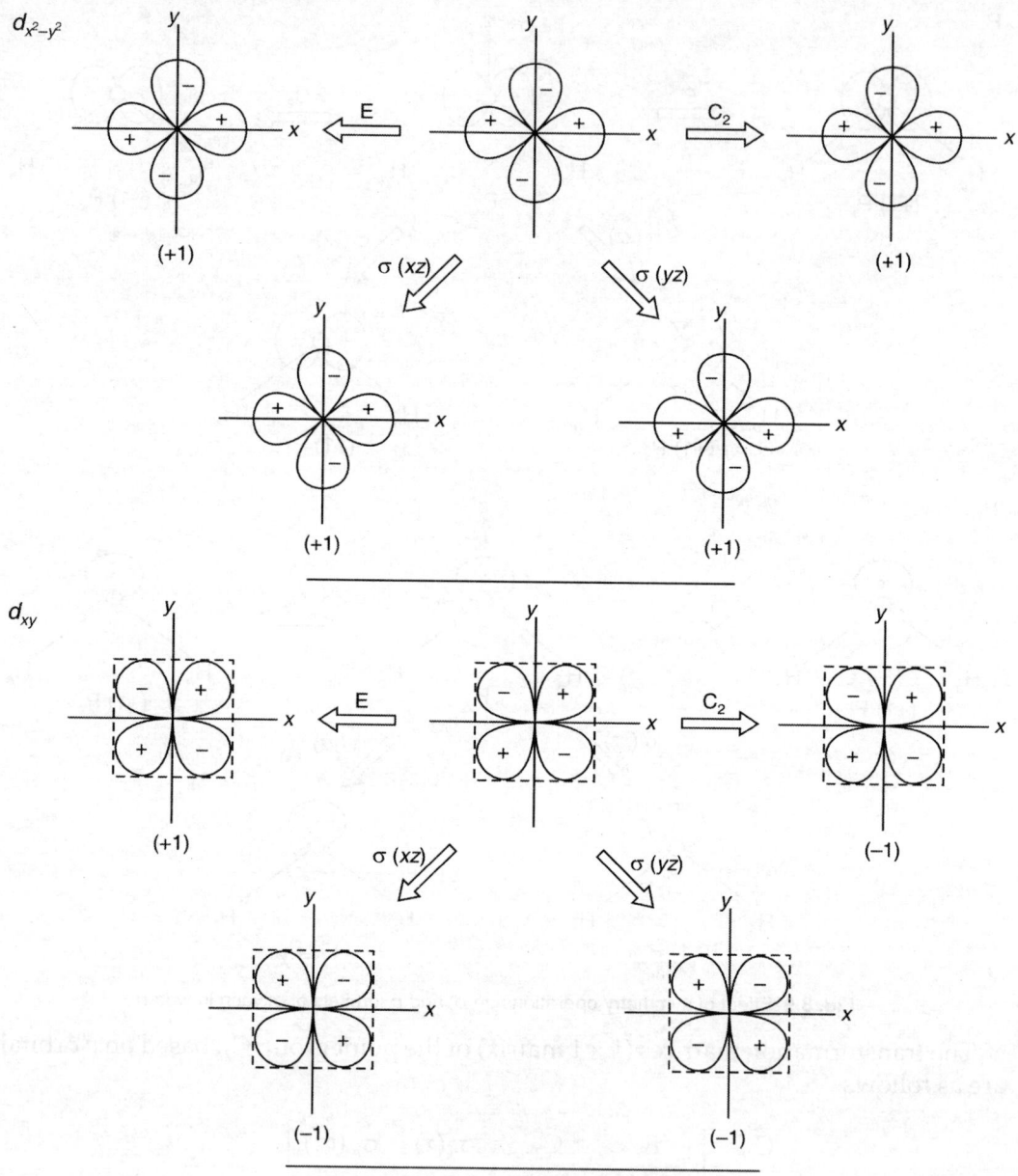

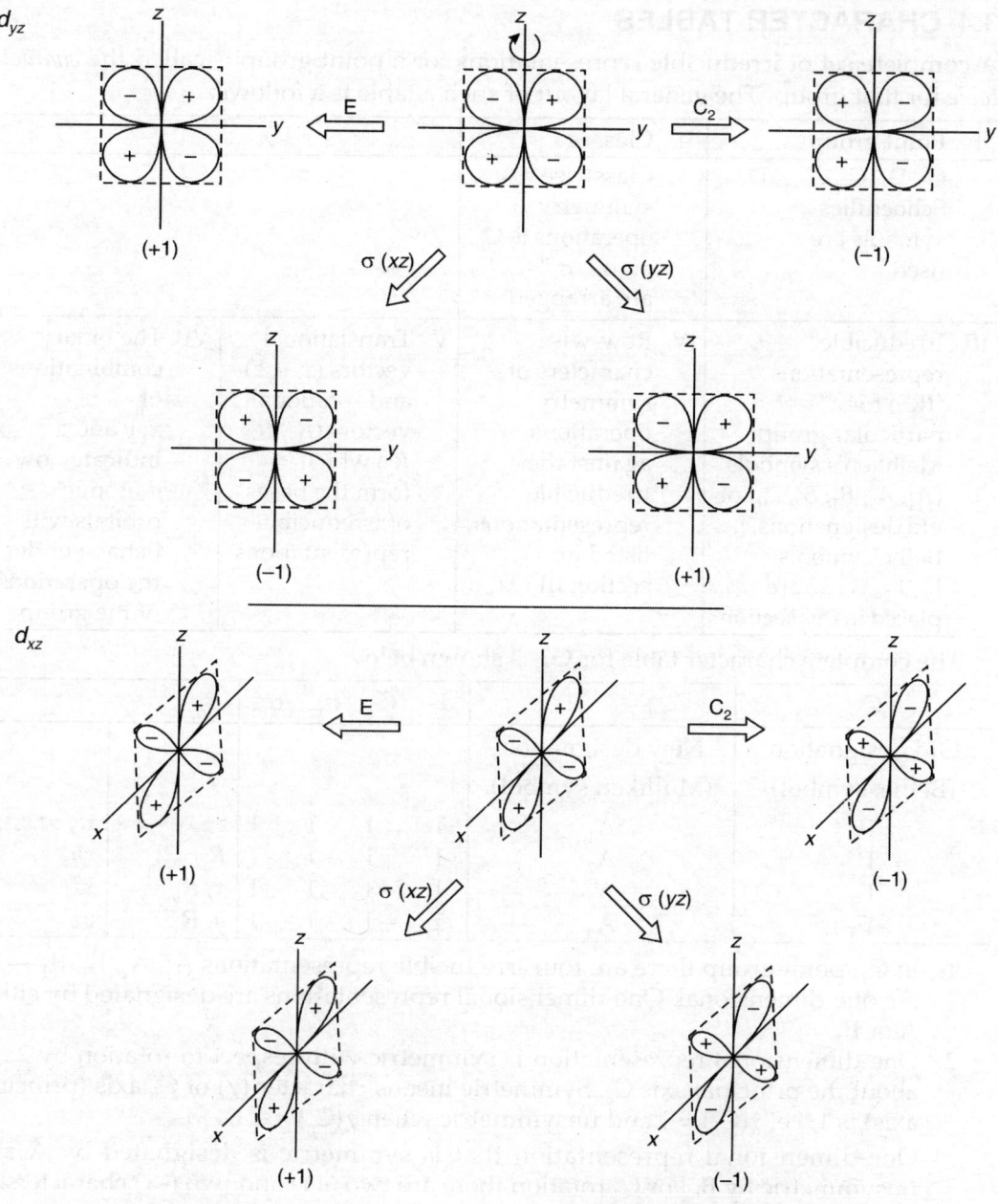

Fig. 3.10 Effect of symmetry operations on d-orbital of sulphur in SO_2

C_{2v}	E	C_2	$\sigma_v(xz)$	$\sigma_v(yz)$	
	+1	+1	+1	+1	$d_{x^2-y^2}, d_{z^2}$
	+1	+1	−1	−1	d_{xy}
	+1	−1	+1	−1	d_{xz}
	+1	−1	−1	+1	d_{yz}

3.4 CHARACTER TABLES

A complete set of irreducible representations for a point group is called the *character table* for that group. The general layout of such a table is a follows:

I **Point group**	II **Classes**		
C_n, D_n, C_{nh}, C_{nv}, D_h... Schoenflies symbols are used	Classwise symmetry operations E, C_n, σ_v, σ_v', σ_v'' are arranged		
III Irreducible representations (IRS) for a particular group Melliken's symbols (A_1, A_2, B_1, B_2, ...,) or old designations, i.e. Bethe symbols Γ_1, Γ_2, Γ_3, ..., are placed in this section.	IV Row-wise characters of symmetry operations against the irreducible representations listed in section III	V Translational vectors (x, y, z) and rotational vectors (R_x, R_y, R_z) which form the bases of irreducible representations	VI The binary combinations of x, y and z indicate how d-atomic orbitals will behave under the operations of the group

The complete character table for C_{2v} is shown below.

C_{2v}		E	C_2	σ_{xz}	σ_{yz}		
Old designation (Bethe, symbol)	New designation (Mulliken symbol)						
Γ_1	A_1	1	1	1	1	z	x^2, y^2, z^2
Γ_2	A_2	1	1	-1	-1	R_z	xy
Γ_3	A_3	1	-1	1	-1	x, R_y	xz
Γ_4	A_4	1	-1	-1	1	y, R_x	yz

1. In C_{2v} point group there are four irreducible representations A_1, A_2, B_1, B_2 — all are one-dimensional. One-dimensional representations are designated by either A or B.

2. One-dimensional representation is symmetric with respect to rotation by $2\pi/n$ about the principal axis C_n. Symmetric means character (χ) of C_n axis (principal axis) is 1, i.e. $\chi(C_n) = 1$ and unsymmetric when $\chi(C_n) = -1$

   One-dimensional representation that is symmetric is designated by A and unsymmetric by B. For C_2 rotation there are two (+1) and two (–1) characters.

   A irreducible representation must have $\chi(C_2) = +1$ and B irreducible must have $\chi(C_2) = -1$. Thus we have

Irreducible representation	Character of C_2, i.e. $\chi(C_2)$
A	
A	+ 1 } symmetric
B	+ 1 } representation
B	– 1 } unsymmetric
	– 1 } or antisymmetric representation

3. Subscripts 1 and 2 are usually attached to A's and B's to designate those which are symmetric and antisymmetric with respect to C_2 perpendicular to the principal axis. If such a C_2 axis is missing, the character of reflection in a σ_v plane should be +1 for symmetric and –1 for antisymmetric.

In C_{2v} point group only one C_2 is present. This C_2 is the principal axis. No other C_2 axis is present. In that case we have to consider the characters of σ_v (xz) plane.

Axis (C_2)	Plane ($\sigma_v(xz)$)	
+1	1	Since $\chi(C_2) = H$, the IR is A since $\chi(\sigma_v(xz)) = +1$, the subscript is 1. So the complete symbol is A_1
+1	–1	As $\chi(C_2) = +1$, the IR is A since $\chi(\sigma_v(xz)) = -1$, the subscript is 2. So the complete symbol is A_2
–1	+1	Since $\chi(C_2) = -1$, the IR is B So the complete symbol is B_1
–1	–1	The symbol is B_2

Thus,	E	C_2	σ_v	σ_v'		
	1	1	1	1	→	A_1
	1	1	–1	–1	→	A_2
	1	–1	1	–1	→	B_1
	1	–1	–1	1	→	B_2

Totally symmetric representation is always designated by A_1. Each line of the character (Γ_i) such as $\Gamma_1 : 1\ 1\ 1\ 1$ is called representation of the group.

3.5 DEFINITION OF SOME USEFUL TERMS

Trace and Character

The sum of the diagonal term elements of a matrix is called its *trace*. The character (χ) is the trace of the matrix. For C_{2v} point group, we have

$$E = \begin{pmatrix} 1 & 0 & 0 \\ 0 & 1 & 0 \\ 0 & 0 & 1 \end{pmatrix}, C_2 = \begin{pmatrix} -1 & 0 & 0 \\ 0 & -1 & 0 \\ 0 & 0 & 1 \end{pmatrix}, \sigma(xz) = \begin{pmatrix} 1 & 0 & 0 \\ 0 & -1 & 0 \\ 0 & 0 & 1 \end{pmatrix}, \sigma(yz) = \begin{pmatrix} -1 & 0 & 0 \\ 0 & 1 & 0 \\ 0 & 0 & 1 \end{pmatrix}$$

and the following character is noted.

E	C_2	$\sigma(xy)$	$\sigma_v(yz)$
3	–1	1	1

Order (h):

The total number of symmetry operations (E, C_n, σ_v, etc. in a group is called the *order (h) of the group*. In C_{2v} point form symmetry operations are E, C_2, σ_v (xz), σ_v (yz). So the order (h) is 4. In C_{3v} six symmetry operations are E, C_3, C_3^2, σ_v, σ_v', σ_v''. So the order is 6.

Class:

A set of symmetry elements which are conjugate to one another is called a *class* of the group. C_{2v} point group contains 4 classes of symmetry operations E, C_2, σ_v (xz), σ_v' (yz)

C_{3v} has 3 classes E, $2C_3$ $(= C_3^1, C_3^2)$ and $3\ \sigma_v$ $(= \sigma_v, \sigma_v', \sigma_v'')$.

Dimension (k):

The character under the identity operation (E) denotes the *dimensionality* of the irreducible representation.

C$_{2v}$ $\longrightarrow$		E	E $\longleftarrow$ D$_4$	
Designation				
Old (Bethe symbol)	New (Mulliken symbol)			
Γ_1	A$_1$	1	1	Γ_1
Γ_2	A$_2$	1	1	Γ_2
Γ_3	A$_3$	1	1	Γ_3
Γ_4	A$_4$	1	1	Γ_4
			2	Γ_5

In C$_{2v}$, the dimension of $\Gamma_1 = 1$

In D$_4$, the dimension of $\Gamma_5 = 2$

3.6 THE GREAT ORTHOGONALITY THEORY

The elements of the matrices constitute the irreducible representations (IR) of a group.

Let Γ_i and Γ_j are the two irreducible representations* (IR) of a point group and K_i and K_j are dimensions** of these representation. The mathematical expression for the great orthogonality theorem is

$$\sum_R [\Gamma_i(R)_{mn}][\Gamma_j(R)_{m'n'}]^* = \frac{h}{\sqrt{K_iK_j}} \delta_{ij} \delta_{mm'} \delta_{nn'} \qquad ...(3.10)$$

---

* Representation is a set of matrices. These matrices represent the operations of a point group. In C$_{2v}$ point group we have

$$E: \begin{pmatrix} 1 & & \to x \\ & 1 & \to y \\ & & 1 \to z \end{pmatrix} ; C_2: \begin{pmatrix} -1 & & \\ & -1 & \\ & & -1 \end{pmatrix} ; \sigma_v(xz) = \begin{pmatrix} 1 & & \\ & -1 & \\ & & 1 \end{pmatrix} ; \sigma'_v = \begin{pmatrix} -1 & & \\ & 1 & \\ & & 1 \end{pmatrix}$$

For x coordinate, $\Gamma_3 \to$ 1 –1 1 –1   form the representation of the group

For y coordinate, $\Gamma_2 \to$ 1 –1 –1 1

For z coordinate, $\Gamma_1 \to$ 1 1 1 1

Reducible representation r^{red}, $\Gamma = \sum$ irreducible representation

$= r_1 (\Gamma_1) \oplus r_2 (\Gamma_2) \oplus r_3 (\Gamma_3)$

$\Gamma = \Gamma_1 + \Gamma_2 + \Gamma_3$

** Dimension (K): The character under the identity operation (E) denotes the dimensionality of the irreducible representation. Consider C$_{2v}$ point group.

C$_{2v}$	E	C$_2$	$\sigma_v(xz)$	$\sigma_v(yz)$
Γ_1	1	1	1	1
Γ_2	1	1	–1	–1
Γ_3	1	–1	1	–1
Γ_4	1	–1	–1	1

In C$_{2v}$, the dimension of $\Gamma_1 = 1$; $\Gamma_2 = 1$; $\Gamma_3 = 1$, $\Gamma_4 = 1$

where

   (i) R is the particular *symmetry operation* of the point group.

   (ii) h is the order of the group

   (iii) $\sum\limits_{R}$ represents the summation of various operations in a group.

   (iv) $[\Gamma_i(R)_{mn}]$ denotes an element in the mth row and nth column of a matrix corresponding to an operation R in the ith irreducible representation.

   (v) $[\Gamma_i(R)m'n']^*$ is the complex conjugate of the element in the m'th row and n'th column of a matrix in the jth irreducible representation. If the element $\Gamma_j(R)^*_{m'n'}$ does not have imaginary or complex numbers, then there is no need of complex conjugate and $\Gamma_j(R)^*_{m'n'}$ is replaced by $\Gamma_j(R)_{m'n'}$ term.

   (vi) K_i and K_j are the dimensions of the ith and jth representations of the group.

   (vii) δ_{ij} is the Krönecker delta

$$\delta_{ij} = 1 \text{ when } i = j$$
$$= 0 \text{ when } i \neq j$$

The complicated mathematical expression for the great orthogonality theorem can be put in the form of three simple equations as

$$\sum_{R}\Gamma_i(R)_{mn}\Gamma_j(R)_{mn} = 0 \text{ if } i \neq j \qquad \qquad ...(3.11)$$

$$\sum_{R}\Gamma_i(R)_{mn}\Gamma_j(R)_{m'n'} = 0 \text{ if } m \neq m' \text{ and/or } n \neq n' \qquad ...(3.12)$$

$$\sum_{R}\Gamma_i(R)_{mn}\Gamma_j(R)_{mn} = \frac{h}{K_i} \qquad \qquad ...(3.13)$$

Equation (3.11) implies that the sum of the products of the corresponding elements of matrices of various symmetry operations belonging to two different irreducible representations is zero ($\Sigma\ \Gamma_i\Gamma_j = 0$). Let us consider irreducible representations of water (C_{2v} point group).

$\Gamma_1(E)\ \Gamma_2(E) + \Gamma_1(C_2)\ \Gamma_2(C_2) + \Gamma_1(\sigma_v(xz)\ \Gamma_2(\sigma_v(xz) + \Gamma_1\ \sigma_v(yz)\ \Gamma_2\sigma_v(yz)$

$= 1 \times 1 + 1 \times 1 + 1 \times (-1) + 1 \times (-1) = 0$

Similarly

$\Gamma_1(E)\ \Gamma_4(E) + \Gamma_1(C_2) \cdot \Gamma_4(C_2) + \Gamma_1\ \sigma_v(xz)\ \Gamma_4\ \sigma_v(xz) + \Gamma_1\ \sigma_v(yz)\ \Gamma_4\ \sigma_v(yz)$

$= 1 \times 1 + 1 \times (-1) + 1 \times (-1) + 1 \times 1 = 0$

3.7 PROPERTIES OF IRREDUCIBLE REPRESENTATIONS (IRs)

No two irreducible representations (IRs) are identical in any point group. The dimensionality (character under E) may be same, but the IRS differ from each other in the nature of their characters. Consider C_{2v} point group.

The irreducible representations Γ_1 : 1 1 1 1 (dimension $K_1 = 1$) and Γ_2 : 1 1 –1 –1, (dimension $K_2 = 1$) have same dimension but have different characters.

We shall now discuss five important rules about irreducible representations and their characters.

Rule 1: The sum of the squares of the dimensions (K) $\sum K_i^2$ (character under E) of each of the irreducible representations equals the *order* (h) of the group.

$$h = \sum K_i^2 = K_1^2 + K_2^2 + \ldots + K_n^2$$

Let us consider C_{2v} and C_{3v} groups.

C_{2v}	E	C_2	$\sigma_v(xz)$	$\sigma_v(yz)$
A_1	1	1	1	1
A_2	1	1	-1	-1
B_1	1	-1	1	-1
B_2	1	-1	-1	1

C_{3v}	E	$2\,C_3$	$3\,\sigma_v$
A_1	1	1	1
A_2	1	1	-1
E	2	-1	0

In C_{2v} group, $h = 1^2 + 1^2 + 1^2 + 1^2 = 4$

In C_{3v} group, $h = 1^2 + 1^2 + 2^2 = 6$

Rule 2: For any irreducible representation, the sum of the squares of the characters $\sum [\chi_i(R)]^2$ multiplied by the number of operations ($n(R)$) in the class equals the order (h) of the group

$$\sum n(R)[\chi_i^2(R)] = h$$

In C_{3v} group

For $A_1, h = 1(1)^2 + 2(1)^2 + 3(1)^2 = 6$

$A_2, h = 1(1)^2 + 2(1)^2 + 3(-1)^2 = 6$

$A_3, h = 1(2)^2 + 2(-1)^2 + 3(0)^2 = 6$

For C_{2v} group

$A_1 = 1 \cdot 1^2 + 1 \cdot 1^2 + 1 \cdot 1^2 + 1 \cdot 1^2 = 4$

$A_2 = 1 \cdot 1^2 + 1 \cdot 1^2 + 1 \cdot (-1)^2 + 1 \cdot (-1) = 4$

$B_1 = 1 \cdot 1^2 + 1 \cdot (-1)^2 + 1 \cdot 1^2 + 1 \cdot (-1)^2 = 4$

$B_2 = 1 \cdot 1^2 + 1(-1)^2 + 1(-1)^2 + 1 \cdot 1^2 = 4$

Rule 3: All irreducible representations are orthogonal to one another, i.e. for a pair of representations, the sum of the products of the characters in each column multiplied by the number of operations (n) in the class is zero.

$$\sum n(R)\,\chi_i(R)\,\chi_j(R) = 0 \quad \text{(when } i \neq j)$$

where n is the number of operations $\chi_i(R)$ is the character of the symmetry operation in the ith irreducible representation.

In C_{2v} point group, B_1 and B_2 are orthogonal

$(1 \times 1 \times 1) + (1 \times -1 \times -1) + (1 \times 1 \times -1) + (1 \times -1 \times 1) = 0$

$\quad$|E $\qquad\qquad$ |C_2 $\qquad\qquad$ |$\sigma_v(xz)$ $\qquad$ |$\sigma_v(yz)$

In C_{3v} point group, A_2 and E are orthogonal

$(1 \times 1 \times 2) + (2 \times 1 \times -1) + (3 \times -1 \times 0) = 0$

$\quad$|E $\qquad\qquad$ $2C_3$ $\qquad\qquad$ $3\sigma_v$

Rule 4: In a given representation (reducible/irreducible), the characters of all matrices belonging to operations in the same class are identical.

In C_{3v}, all σ_v belong to the same class and have identical characters. So three σ_v have $A_1 = 1$, $A_2 = -1$, E = 0.

Rule 5: The number of irreducible representation of a group is equal to the number of classes in the group.

C_{2v} point group has 4 class (E, C_2, σ_v (xz), σ_v (yz)) and 4 IRs (Γ_1, Γ_2, Γ_3, Γ_4).

C_{3v} point group has 3 classes (E, C_3 and σ_v) and 3 IRs (Γ_1, Γ_2, Γ_3).

3.8 CONSTRUCTION OF CHARACTER TABLES OF A POINT GROUPS

For the construction of a character table of a point group, we make use of the five rules derived from the *great orthogonality theorem*.

3.8.1 Character Table for C_{2v} Point Group

In C_{2v} point group there are *four* symmetry classes or *four* symmetry operations. There (symmetry classes/symmetry operations) are E, C_2, σ_v (xz), σ_v (yz).

Since the order of the group (h) = total number of symmetry operations or classes, we may write $h = 4$.

i. **Number of irreducible representations (IRs) in a group**

Recall **rule 5** which states that

No. of IRs in a group = No. of classes in a group = No. of symmetry operations

So C_{2V} group contains 4 IRs Γ_1, Γ_2, Γ_3, Γ_4 (capital gamma).

ii. **Dimensionality of IRs**

We know that the character under the identity operation (E) denotes the dimensionality of the irreducible representation.

Rule 1 states that sum of the squares of the dimensions $\sum_i k_i^2$ (character under E) of each of the IRs equals the order (h) of the group. Let the dimensions of four IRs be k_1, k_2, k_3, k_4, i.e., dimension of $\Gamma_1 = k_1$, that of $\Gamma_2 = k_2$, etc.

$$\sum k_i^2 \, (E) = k_1^2 + k_2^2 + k_3^2 + k_4^2 = h = 4$$

By error and trial method, we can determine the dimensions of the IRs as

$$k_1 = k_2 = k_3 = k_4 = 1$$

The minimum value of $k = +1$ and never 0 or –ve value. Thus all the IRs Γ_1, Γ_2, Γ_3 and Γ_4 of C_{2v} group are one-dimensional. Following the rule we may write.

$$\sum \chi_i^2 \, (E) = 1^2 + 1^2 + 1^2 + 1^2 = 4$$

So the character of Γ_1, Γ_2, Γ_3, Γ_4 under E will be +1 each. Therefore, we may write

C_{2v}	E
Γ_1	1
Γ_2	1
Γ_3	1
Γ_4	1

iii. **Character set of each IRs**

The characters under other classes for a given IR are such that

$$\sum n(R)\chi_i^2 \, (R) = h \text{ (rule 2)}$$

For example, the characters of Γ_i are given by

$$n(E) \, \chi_1^2 \, (E) + n(C_2) \, \chi_1^2(C_2) + n(\sigma_v(xz)) \cdot x_1^2 \, (\sigma_v(xz)) + n(\sigma_v)(yz) \, \chi_1^2 \, (\sigma_v(yz)) = 4$$

Here $n(E) = n(C_2) = n(\sigma_v(xz)) = n(\sigma_v(yz)) = 1$

Thus $\chi_1^2(E) + \chi_1^2(C_2) + \chi_1^2(\sigma_v(xz)) + \chi_1^2(\sigma_v(yz)) = 4$

The above relation is justified only when we put +1 values for characters, i.e.

$$\chi(E) = \chi(C_2) = \chi(\sigma_v(xz)) = \chi(\sigma_v(yz)) = 1$$

$\therefore \quad \Gamma_1 : 1 \cdot 1^2 + 1 \cdot 1^2 + 1 \cdot 1^2 + 1 \cdot 1^2 = 4$

So one of the representations for $\chi(R) = +1$ for each operation (R), i.e.

	E	C_2	$\sigma_v(xz)$	$\sigma_v(yz)$
Γ_1	1	1	1	1

At this stage, partially constructed C_{2v} table has the following form:

C_{2v}	E	C_2	$\sigma_v(xz)$	$\sigma_v(yz)$
Γ_1	1	1	1	1
Γ_2	1			
Γ_3	1			
Γ_4	1			

Orthogonality of IRs

From **rule 3** we have $\sum n(R)\chi_i(R)\chi_j(R) = 0 (i \neq j)$

This implies that each of the remaining three IRs (Γ_2, Γ_3, Γ_4) has two +1's and two −1's as the character. Since the character of identity operation (E) is always +1, we have to distribute one +1 and two − 1's in the remaining three operations (C, $\sigma_v(xz)$, $\sigma_v(yz)$). Therefore, we will have

C_{2v}	E	C_2	$\sigma_v(xz)$	$\sigma_v(yz)$
Γ_1	1	1	1	1
Γ_2	1	−1	1	−1
Γ_3	1	−1	1	−1
Γ_4	1	−1	−1	1

All these representations are orthogonal to one another. For example Γ_1/Γ_2, Γ_2/Γ_3 and Γ_3/Γ_4 pairs are orthogonal to one another:

$(1)(1)(1) + (1)(1)(1) + (1)(1)(-1) + (1)(1)(-1) = 0$

$(1)(1)(1) + (1)(1)(-1) + (1)(-1)(1) + (1)(-1)(-1) = 0$

$(1)(1)(1) + (1)(-1)(-1) + (1)(1)(-1) + (1)(-1)(1) = 0$

Therefore, these are the 4 irreducible representations of C_{2v} group. The complete table is given below:

C_{2v}	E	C_2	$\sigma_v(xz)$	$\sigma_v(yz)$
Γ_1	1	1	1	1
Γ_2	1	1	−1	−1
Γ_3	1	−1	1	−1
Γ_4	1	−1	−1	1

3.8.2 Character Table for C_{3v} Point Group

C_{3v} group contains E, C_3^1, C_3^2, σ_v, σ_v', σ_v'' symmetry operations. So the order (h) of the group is 6.

In C_{3v} group (all three σ_v classes are identical. So the classes are E, $2C_3$ and $3\sigma_v$.

i. **Number of irreducible representations in a group**

Since there are 3 classes, the number of IRs is also 3. They are $\Gamma_1, \Gamma_2, \Gamma_3$.

ii. **Dimensionality of IRs**

Let the dimensions of three IRs be k_1, k_2, k_3. Following **rule 1** we may write

$$k_1^2 + k_2^2 + k_3^2 = 6 \qquad \qquad ...(3.14)$$

If we put $k_1 = k_2 = 1$ and $k_3 = 2$, Eq. (3.14) is satisfied, i.e. $1^2 + 1^2 + 2^2 = 6$

So there are two one-dimensional (Γ_1, Γ_2) and one two-dimensional (Γ_3) irreducible representations.

So the characters of Γ_1 and Γ_2 under E class will be +1 each and Γ_3 will be +2. At this stage, the partially constructed C_{3v} table has the following form.

C_{3v}	E
Γ_1	1
Γ_2	1
Γ_3	2

iii. **Character set of each IR**

Rule 2 states that

$$\sum n(R)\chi_i^2(R) = h$$

The set of characters of the IR (Γ_1) is a row of +1 so that

$$n(E)\,\chi_1^2(E) + n(C_3)\,\chi_1^2(C_3) + n(\sigma_v)\,\chi_1^2(\sigma_v) = 6$$
$$1 \cdot 1^2 + 2 \cdot 1^2 + 3 \cdot 1^2 = 6$$

At this stage, partially constructed C_{3v} table has the following form.

C_{3v}	E	$2C_3$	$3\sigma_v$
Γ_1	1	1	1
Γ_2	1		
Γ_3	2		

iv. **Orthogonality of IRs**

Rule 3 states that

$$\sum n(R)\,\chi_i(R)\,\chi_j(R) = 0; \quad \text{when } i \neq j$$

Γ_2 can be worked out by using +1 and –1 characters for $\chi(C_3)$ and $\chi(\sigma_v)$ so that

$$\Gamma_1 \Gamma_2 = n(E)\,\chi_1(E)\,\chi_2(E) + n(C_2)\,\chi_1(C_3)\,\chi_2(C_3) + n(\sigma_v)\,\chi_1(\sigma_v)\,\chi_2(\sigma_v) = 0$$
$$1 \cdot 1 \cdot 1 + 2 \cdot 1 \cdot 1 + 3 \cdot 1 \cdot (-1) = 0$$

So we may write

C_{3v}	E	$2C_3$	$3\sigma_v$
Γ_1	1	1	–1

Updated C_{3v} table now appears as

C_{3v}	E	$2C_3$	$3\sigma_v$
Γ_1	1	1	1
Γ_2	1	1	–1
Γ_3	2	x	y

The values of x and y can be determined by taking the cross-products of $\Gamma_1 \cdot \Gamma_3$ and $\Gamma_2 \cdot \Gamma_3$. Thus

$$\Gamma_1 \cdot \Gamma_3 : 1 \cdot 1 \cdot 2 + 2 \cdot 1 \cdot x + 3 \cdot 1 \cdot y = 0 \quad \text{or} \quad 2x + 3y = -2 \qquad \qquad ...(3.15)$$

$$\Gamma_2 \cdot \Gamma_3 : 1 \cdot 1 \cdot 2 + 2 \cdot 1 \cdot x + 3(-1)\, y = 0 \quad \text{or} \quad 2x - 3y = -2 \qquad \qquad ...(3.16)$$

Solving Eqs (3.15) and (3.16), we get $x = -1, y = 0$

The complete table can be written by substituting the values of x and y.

C_{3v}	E	$2C_3$	$3\sigma_v$
Γ_1	1	1	1
Γ_2	1	1	−1
Γ_3	2	−1	0

In order to check the correctness of Γ_3 we may follow **rule (2),**

i.e. $$\sum n(R)\, \chi_i^2(R) = h$$

which is the order (h) of this group.

3.8.3 Character Table for C_{4v} Point Group

In the point C_{4v} there are five symmetry classes E, $2C_4$, C_3, $2\sigma_v$ and $2\sigma_v'$. So the number of irreducible representation (Γ_i) = number of classes = 5 (**rule 5**). The dimensionalities of the irreducible representations are determined as follows (**rule 1**)

$$k_1^2 + k_2^2 + k_3^2 + k_4^2 + k_5^2 = 8 \qquad \qquad ...(3.17)$$

So the order of the group is also 8. If we put $k_1 = k_2 = k_3 = k_4 = 1$ and $k_5 = 2$, Eq. (3.17) is satisfied. This means that there are four single dimensional representations and only one two-dimensional representation, i.e.

$$\chi_1(E) = \chi_2(E) = \chi_3(E) = \chi_4(E) = 1 \quad \text{and} \quad \chi_5(E) = 2$$

At this point, we have the following part of the character table.

C_{4v}	E
Γ_1	1
Γ_2	1
Γ_3	1
Γ_4	1
Γ_5	2

The first IR(Γ_1) will have only +1 in each class. Therefore

$$1 \cdot \chi_1^2(E) + 2 \cdot \chi_1^2(C_4) + 1 \cdot \chi_1^2(C_2) + 2 \cdot \chi_1^2(\sigma_v) + 2 \cdot \chi_1^2(\sigma_v') = h = 8$$

$$1 \cdot 1^2 + 2 \cdot 1^2 + 1 \cdot 1^2 + 2 \cdot 1^2 + 2 \cdot 1^2 = 8$$

So we may write

E	$2C_4$	C_2	$2\sigma_v$	$2\sigma_v'$
1	1	1	1	1
1				
1				
1				
1				

The characters of the remaining three one-dimensional representations may be obtained by the orthogonality requirements (**rule 3**). To make a representation orthogonal to the first one, we assign +1 to the operator C_2 and to those in one of the other classes, with –1 being assigned to the remaining two classes. Thus, we have

C_{4v}	E	$2C_4$	C_2	$2\sigma_v$	$2\sigma_v'$
Γ_1	1	1	1	1	1
Γ_2	1	1	1	–1	–1
Γ_3	1	–1	1	1	–1
Γ_4	1	–1	1	–1	1
Γ_5	2	a	b	c	d

In order to determine the values of a to d in the two-dimensional representation Γ_5. We follow **rule 3** (i.e. orthogonality relationship $\Sigma\chi_i(R)\,\chi_j(R) = 0$)

$$2 + 2a + b + 2c + 2d = 0$$
$$2 + 2a + b - 2c - 2d = 0$$
$$2 - 2a + b + 2c - 2d = 0$$
$$2 - 2a + b - 2c + 2d = 0$$

On solving, we get $a = 0$, $b = -2$, $c = 0$ and $d = 0$. Putting the values of a, b, c, d in the above table we get the complete character table for the C_{4v} point group as shown below in table.

C_{4v}	E	$2C_4$	C_2	$2\sigma_v$	$2\sigma_v'$
Γ_1	1	1	1	1	1
Γ_2	1	1	1	–1	–1
Γ_3	1	–1	1	1	–1
Γ_4	1	–1	1	–1	1
Γ_5	1	0	–2	0	0

Table 3.3 records the properties of characters of irreducible representations (IRs) in C_{2v} and C_{3v} point groups.

Table 3.3 Properties of characters of irreducible representations in C_{2v} and C_{3v} point groups

Property	Example: C_{2v}	Example: C_{3v}
1. Order: Total number of symmetry operations in the group is called the **order** (h).	In C_{2v} point group there are four symmetry operations: E, C_2, $\sigma_v(xz)$, $\sigma_v'(yz)$. So the order = 4.	In C_{3v} point group there are six symmetry operations: E, C_3^1, C_3^2, $\sigma_v(xz)$, $\sigma_v'(yz)$ and $\sigma_v''(zx)$. So the order = 6.
2. Classes: Symmetry operations are arranged in classes. All operations in a class have identical characters.	C_{2v} group has 4 classes E, C_2, σ_v, σ_v'. Each of symmetry operations is in separate class. There are 4 columns in the character table.	C_{3v} has 3 classes: E, $2C_3 (= C_3, C_3^2)$, $3\sigma_v(= \sigma_v, \sigma_v', \sigma_v'')$.

C_{2v}		E	C_2	σ_v	σ_v'
A_1	Γ_1	1	1	1	1
A_2	Γ_2	1	1	–1	1
B_1	Γ_3	1	–1	1	–1
B_2	Γ_4	1	–1	–1	1

C_{3v}		E	$2C_2$	σ_v
A_1	Γ_1	1	1	1
A_2	Γ_2	1	1	–1
E	Γ_3	2	–1	0

Contd...

	Property	Example: C_{2v}	Example: C_{3v}
3.	The number of irreducible representations (IRs) is equal to the number of classes. This means that the character tables have the same number of rows and columns.	Since C_{2v} group has four classes so there must be 4 IRs present ($\Gamma_1, \Gamma_2, \Gamma_3, \Gamma_4$).	C_{3v} has three classes. So there must be there IRs ($\Gamma_1, \Gamma_2, \Gamma_3$)
4.	The character under identity operation (E) denotes the dimension (k) of the IR. The sum of the squares of dimension of each IRs is equal to the *order* of the group.	Order of C_{2v}, $= (1)^2 + (1)^2 + (1)^2 + (1)^2 = 4 = h$	Order of $C_{3v} = (1)^2 + 2(1)^2 + (2)^2 = 6 = h$
5.	For any IRs, the sum of the squares of the characters multiplied by the number of operations in the class is equal to the *order* of the group $$h = \sum n[\chi_i(R)]^2$$	For $A_1 : 1(1)^2 + 1(1)^2 + 1(1)^2 + 1(1)^2 = 4$ For $A_2 : 1(1)^2 + 1(1)^2 + 1(-1)^2 + 1(1)^2 = 4$	For $A_1 : 1(1)^2 + 2(1)^2 + 3(1)^2 = 6$ For $A_2 : 1(1)^2 + 2(1)^2 + 3(-1)^2 = 6$ For $E : 1(2)^2 + 2(-1)^2 + 3(0)^2 = 6$
6.	Irreducible representations (IRs) are *orthogonal* to each other. The sum of the characters (multiplied together for each class) of any pair of irreducible representation is zero $\sum n\chi_i(R)\chi_j(R) = 0$ when $i \neq j$	B_1 and B_2 are orthogonal: $\underset{E}{(1)(1)} + \underset{C_2}{(-1)(-1)} + \underset{\sigma_v}{(1)(-1)}$ $+ \underset{\sigma_v'}{(-1)(1)} = 0$	A_1 and E are orthogonal: $\underset{E}{1(1)(2)} + \underset{C_3}{2(1)(-1)} + \underset{\sigma_v}{3(-1)(0)} = 0$
7.	A totally symmetric representation is included in all groups with characters of 1 for all operations.	C_{2v} has A_1, which has all characters = 1.	C_{3v} has A_1, with all characters = 1.

3.9 RESOLUTION OF REDUCIBLE REPRESENTATION IN TERMS OF IRREDUCIBLE REPRESENTATIONS

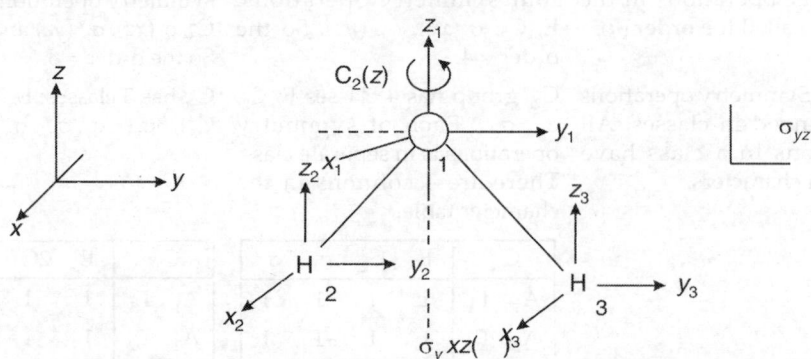

Fig. 3.11 Each atom in H_2O displayed with Cartesian coordinates; the symmetry elements are also shown

Water molecule (C_{2v} point group) contains four symmetry elements E, C_2, σ_v (xz) and σ_v (yz). So the order of the group $h = 4$. Number of classes is also 4 (Fig. 3.11).

The atoms of water molecule are numbered. The coordinates of O_1 is x_1, y_1, z_1, those of H_2 is x_2, y_2, z_2 and H_3 is x_3, y_3, z_3. Each atom is taken as the origin. We have fixed three vectors along the Cartesian coordinates.

Water is a nonlinear triatomic molecule. Total number of degrees of freedom is $3N = 3 \times 3 = 9$ of which 3 are translational, 3 are rotational and the remaining 3 are vibrational.

We shall use transformation matrices to determine the symmetry of all new motions.

E Operations: Under E operation all the 9 vectors are unchanged so the character is 9.

$$
\begin{matrix} O_1 \left\{ \begin{matrix} x_1' \\ y_1' \\ z_1' \end{matrix} \right. \\ H_a \left\{ \begin{matrix} x_2' \\ y_2' \\ z_2' \end{matrix} \right. \\ H_b \left\{ \begin{matrix} x_3' \\ y_3' \\ z_3' \end{matrix} \right. \end{matrix}
=
\begin{bmatrix} 1 & & & & & & & & \\ & 1 & & & & & & & \\ & & 1 & & & & & & \\ & & & 1 & & & & & \\ & & & & 1 & & & & \\ & & & & & 1 & & & \\ & & & & & & 1 & & \\ & & & & & & & 1 & \\ & & & & & & & & 1 \end{bmatrix}
\begin{matrix} \left. \begin{matrix} x_1 \\ y_1 \\ z_1 \end{matrix} \right\} O_1 \\ \left. \begin{matrix} x_2 \\ y_2 \\ z_2 \end{matrix} \right\} H_a \\ \left. \begin{matrix} x_3 \\ y_3 \\ z_3 \end{matrix} \right\} H_b \end{matrix}
$$

| (New coordinate) | (Transformation matrix) | (Old coordinate) |

So the character of E, $\chi(E) = 9$

During symmetry operation, the following points should be noted:

1. If the atom changes position during symmetry operation, one zero (O) is entered.

2. If the atom remains in its original location and vector direction is unchanged, one 1 is entered.

3. If the atom remains in its original location but vector direction is reversed, one –1 is entered.

C_2 Operation:

During C_2 operation O_1 atom remains in its original position but vectors direction along x_1 and y_1 axes are reversed. So –1 is entered for each vector. The vector direction along z axis unchanged. So 1 is entered for each vector. Again due to C_2 operation, H_2 and H_3 atoms are interchanged their positions. So zero(0) is entered in each vector. So the matrix may be expressed as follows.

σ_v **(xz) Operation:** Here O-atom does not change its original position. Only the vector along y axis is reversed. So 1, −1, 1 are entered.

$$O_1 \left\{ \begin{matrix} x_1' \\ y_1' \\ z_1' \end{matrix} \right. \\ H_2 \left\{ \begin{matrix} x_2' \\ y_2' \\ z_2' \end{matrix} \right. \\ H_3 \left\{ \begin{matrix} x_3' \\ y_3' \\ z_3' \end{matrix} \right. = \begin{bmatrix} -1 & & & & & & & & \\ & -1 & & & & & & & \\ & & 1 & & & & & & \\ & & & 0 & & & & & \\ & & & & 0 & & & & \\ & & & & & 0 & & & \\ & & & & & & 0 & & \\ & & & & & & & 0 & \\ & & & & & & & & 0 \end{bmatrix} \begin{bmatrix} x_1 \\ y_1 \\ z_1 \\ x_2 \\ y_2 \\ z_2 \\ x_3 \\ y_3 \\ z_3 \end{bmatrix}$$

$$\chi\,(C_2) = -1$$

Due to $\sigma_v(xz)$ operation H_a and H_b atoms changed their original positions. So 0, 0, 0 are entered.

$$\begin{bmatrix} x_1' \\ y_1' \\ z_1' \\ x_2' \\ y_2' \\ z_2' \\ x_3' \\ y_3' \\ z_3' \end{bmatrix} = \begin{bmatrix} 1 & & & & & & & & \\ & -1 & & & & & & & \\ & & 1 & & & & & & \\ & & & 0 & & & & & \\ & & & & 0 & & & & \\ & & & & & 0 & & & \\ & & & & & & 0 & & \\ & & & & & & & 0 & \\ & & & & & & & & 0 \end{bmatrix} \begin{bmatrix} x_1 \\ y_1 \\ z_1 \\ x_2 \\ y_2 \\ z_2 \\ x_3 \\ y_3 \\ z_3 \end{bmatrix}$$

$$\chi\,(\sigma_v(xz)) = 1$$

σ_v **(yz) Operation:** Here the direction of all the x-axis vectors of O, H_A and H_B atoms are only changed but the other vectors along and z-axes remain unchanged.

$$\begin{bmatrix} x_1' \\ y_1' \\ z_1' \\ x_2' \\ y_2' \\ z_2' \\ x_3' \\ y_3' \\ z_3' \end{bmatrix} = \begin{bmatrix} -1 & & & & & & & & \\ & 1 & & & & & & & \\ & & 1 & & & & & & \\ & & & -1 & & & & & \\ & & & & 1 & & & & \\ & & & & & 1 & & & \\ & & & & & & -1 & & \\ & & & & & & & 1 & \\ & & & & & & & & 1 \end{bmatrix} \begin{bmatrix} x_1 \\ y_1 \\ z_1 \\ x_2 \\ y_2 \\ z_2 \\ x_3 \\ y_3 \\ z_3 \end{bmatrix}$$

$$\chi\,(\sigma_v(yz)) = 3$$

The table characters of the reducible representation Γ_{total} (or Γ_{3N}) are

C_{2v}	E	C_2	$\sigma_v(xz)$	$\sigma_v(yz)$
Γ_{total} or Γ_{3N}	9	−1	1	3

The above discussions may be presented in a tabular form

C_{2v}	E	C_2	$\sigma_v(xz)$	$\sigma_v(yz)$
Unshifted atom	3 (all 3 atom)	1 (O-atom)	1 (O-atom)	3 (all 3 atom)
Contribution/atom	3	−1	+1	+1
Total character Γ_{total} or Γ_{3N}	$3 \times 3 = 9$	$1 \times -1 = -1$	$1 \times 1 = 1$	$3 \times 1 = 3$

3.10 RESOLUTION OF REDUCIBLE REPRESENTATION IN TERMS OF IRREDUCIBLE REPRESENTATION–REDUCTION FORMULA

The number of times (n), an irreducible representation Γ_i occurs in a reducible representation is determined by using the *standard reduction formula*.

$$n\,(\Gamma_i) \;=\; \frac{1}{h}\left[\sum_R n(R)\,\chi_{RR}(R)\,\chi_{IR}(R)\right]$$

where

$n\,(\Gamma_i)$ = Number of times (n) the ith irreducible representation (Γ_i) occur in the reducible representation

h = Order of the group

$n\,(R)$ = No. of operations (R) in the class to which symmetry operation (R) belongs

$\chi_{RR}\,(R)$ = The character of reducible representation (RR) under operation R

$\chi_{IR}\,(R)$ = The character of irreducible representation (IR) under operation R

Thus

$$\begin{pmatrix} \text{No. of} \\ \text{irreducible} \\ \text{representation } (n) \\ \text{of a given type } \Gamma_i \end{pmatrix} = \frac{1}{\text{order }(h)} \sum \left[\begin{pmatrix} \text{No. of} \\ \text{operations} \\ \text{in the class} \end{pmatrix} \begin{pmatrix} \text{Character} \\ \text{of RR, i.e.} \\ \chi_{RR}\,(R) \end{pmatrix} \begin{pmatrix} \text{Character} \\ \text{of IR, i.e.} \\ \chi_{IR}(R) \end{pmatrix} \right]$$

Therefore, if a reducible representation of a group is given, then by the use of the character table and applying the reduction formula, one can determine the irreducible representations of which the reducible representation is made up.

Example: The reducible representation (total representation) in C_{2v} point group is given below. Establish the relationship between the reducible and irreducible representations

C_{2v}	E	C_2	$\sigma_v(xz)$	$\sigma_v(yz)$
Γ_r	9	−1	1	3

Solution. Let us consider C_{2v} point group character table. The four irreducible representation with their characters under the operations are shown in the character table

C_{2v}	1E	$1C_2$	$1\sigma_v(xz)$	$1\sigma_v(yz)$
A_1	1	1	1	1
A_2	1	1	−1	−1
B_1	1	−1	1	−1
B_2	1	−1	−1	1

The order (total number of symmetry operations) of the group $h = 4$. Order of each class $(E, C_2, \sigma_v (xz), \sigma_v (yz)) = 1$.

We now apply the reduction formula for each irreducible representation of the group. The number of times A_1 will appear in the reducible representation is given by

IR character   IR character   IR character
↑   ↑   ↑

$$n(A_1) = \frac{1}{4}\left[(1)\cdot(9)\cdot(1) + (1)\cdot(-1)\cdot(1) + (1)(1)(1) + (1)(3)(1)\right] = 3$$

↑   ↑   ↑   ↑

Character	for C_2,	for $\sigma_v(xz)$,	for $\sigma_v(yz)$,
χ_{RR} (E)	total	total	total
	character	character	character
	= –1	= 1	= 3

This means that A_1 will appear 3 times. Similarly A_2, B_1 and B_2 will appear as

$$n(A_2) = \frac{1}{4}\left[(1)(9)(1) + (1)(-1)(1) + (1)(1)(-1) + (1)(3)(-1)\right] = 1$$

$$n(B_1) = \frac{1}{4}\left[(1)(9)(1) + (1)(-1)(-1) + (1)(1)(1) + (1)(3)(-1)\right] = 2$$

$$n(B_2) = \frac{1}{4}\left[(1)(9)(1) + (9)(-1)(-1) + (1)(1)(-1) + (1)(3)(1)\right] = 3$$

$$\therefore \qquad \Gamma_{red} = 3\,A_1 + 1\,A_2 + 2\,B_1 + 3\,B_2$$

This is the relationship for the given reducible representation and irreducible representation of C_{2v} group.

Note that total no. of IRs = 3 + 1 + 2 + 3 = 9 and total RRs = 9.

Example: For C_{3v} point group, a reducible representation is given below:

C_{3v}	E	$2\,C_3$	$3\,\sigma_v$
Γ	6	0	0

Deduce the relationship between this reducible representation and the irreducible representations of the group.

Solution. The main part of the character table of the C_{3v} point group is show below:

C_{3v}	E	$2\,C_3$	$3\,\sigma_v$
A_1	1	1	1
A_2	1	1	–1
A_3	2	–1	0

We use the standard reduction formula

$$n(\Gamma_i) = \frac{1}{h}\left[\sum_R n(R)\,\chi_{RR}(R)\,\chi_{IR}(R)\right]$$

Here $\qquad h = 6$ (total symmetry operations of the group)

$$n(A_1) = \frac{1}{6}\left[(1)(6)(1) + (2)(0)(1) + (3)(0)(1)\right] = 1$$

$$n(A_2) = \frac{1}{6}\left[(1)(6)(1) + (2)(0)(1) + (3)(0)(-1)\right] = 1$$

$$n \, (A_1) \; = \; \frac{1}{6} \; [(1)(6)(2) + (2)(0)(-1) + (3)(0)(0)] = 2$$

Thus reducible representation $\Gamma_{red} = 1 \, A_1 + 1 \, A_2 + 2E$

3.11 DIRECT PRODUCT REPRESENTATION

Let there be two irreducible representations Γ_i and Γ_j of a group. The direct product representation is obtained by multiplying the characters of the two representations under each symmetry operation.

The irreducible representation of C_{2v} point group is shown below.

C_{2v}	E	C_2	$\sigma_v(xz)$	$\sigma_v(yz)$
A_1	1	1	1	1
A_2	1	1	−1	−1
B_1	1	−1	1	−1
B_2	1	−1	−1	1

The direct product representation $A_1 \times A_2$ can be obtained by multiplying the characters of these representations under each operations.

$$
\begin{array}{cccc}
 & E & C_2 & \sigma_v(xz) & \sigma_v(yz) \\
\text{i.e.} \quad A_1 \times A_2 \rightarrow & 1 \times 1 & 1 \times 1 & 1 \times -1 & 1 \times -1 \\
 & = 1 & = 1 & = -1 & = 1
\end{array}
$$

The result $1, 1, -1, -1$ is identical to the irreducible representation A_2.

Thus $B_1 \times B_2 = A_2$

In direct product only characters are to be multiplied, the order of the class $n(R)$ is not to be multiplied. For example, in C_{3v} point group we have many direct product representations such as $A_1 \times A_2$, $A_1 \times E$, $A_2 \times E$ and $E \times E$.

C_{3v}	E	$2 \, C_3$	$3 \, \sigma_v$	
A_1	1	1	1	
A_2	1	1	−1	
E	2	−1	0	
$A_2 \times A_2$	1	1	−1	A_2
$A_1 \times E$	2	−1	0	E
$A_2 \times E$	2	−1	0	E
$E \times E$	4	1	0	

New representation of C_{3v} group having characters 4, 1, 0. This is not a irreducible representation but must be reducible into a sum of irreducible representation. It is reducible to $A_1 + A_2 + E$.

A_1	1	1	1
A_2	1	1	−1
E	2	−1	0
$A_1 + A_2 + E =$	4	1	0

Thus $E \times E = A_1 + A_2 + E$

In general, the direct product of two or more irreducible representations will be a reducible representation. For example, the direct products of some irreducible representations of he C_{4v} group are shown in a tabular from. In practice, it is not necessary to have to multiply lines of characters together. There are some rules obtained below which unable us to obtain the products automatically. Rules for the determination of direct products.

(i) Dimensions of IRs

$A \times A = A$; $B \times B = A$; $A \times B = B \times A = B$

$A \times E = E \times A = E$; $B \times E = E \times B = E$

$A \times T = T \times A = T$; $B \times T = T \times B = T$; $E \times T = T \times E = T_1 + T_2$

(ii) Subscripts and superscripts of IRs.

$g \times g = g$	$1 \times 1 = 1$	$'\times' = '$
$g \times u = u$	$1 \times 2 = 2$	$'\times'' = ''$
$u \times u = g$	$2 \times 2 = 1$	$''\times'' = '$
$u \times g = u$	$2 \times 1 = 2$	$''\times' = ''$

C_{3v}	E	$2\,C_3$	$2\,\sigma_v$	
A_1	1	1	1	
A_2	1	1	−1	
E	2	−1	0	
$A_1 \times A_2$	1	1	−1	A_2
$A_1 \times E$	2	−1	0	E
$A_2 \times A_2$	1	1	1	A_1
$A_2 \times E$	2	−1	0	E
$E \times E$	4	1	0	$A_1 + A_2 + E$

C_{4v}	E	C_2	$2C_2$	$2\sigma_v$	$2\sigma_d$	
A_1	1	1	1	1	1	
A_2	1	1	1	−1	−1	
B_1	1	1	−1	1	−1	
B_2	1	1	−1	−1	1	
E	2	−2	0	0	0	
$A_1 \times A_1$	1	1	1	1	1	A_1
$A_1 \times A_2$	1	1	1	−1	−1	A_2
$A_1 \times B_2$	1	1	−1	−1	1	B_2
$B_1 \times E$	2	−2	0	0	0	E
$B_2 \times E$	2	−2	0	0	0	E
$A_1 \times E \times B_2$	2	−2	0	0	0	E
$E \times E$	4	4	0	0	0	$A_1 + A_2 + B_1 + B_2$

The direct products written above are infact reducible, some simple and the other complex. The complex ones can be reduced by using the standard reducible formula.

PROBLEMS AND SOLUTIONS

Problem 1. Find the matrix representation for the symmetry operations present is orthoboric acid.

Solution

H_3BO_3 belongs to C_{3h} point group. The matrices for the operations E, C_3, C_3^2, S_3, S_3^5 and σ_h are as follows.

$$E = \begin{pmatrix} 1 & 0 & 0 \\ 0 & 1 & 0 \\ 0 & 0 & 1 \end{pmatrix}, \quad \sigma_h = \begin{pmatrix} 1 & 0 & 0 \\ 0 & 1 & 0 \\ 0 & 0 & -1 \end{pmatrix}$$

$$C_n \text{ matrix} = \begin{pmatrix} \cos\theta & \sin\theta & 0 \\ -\sin\theta & \cos\theta & 0 \\ 0 & 0 & 1 \end{pmatrix}$$

For C_3 operation $\theta = 120°$

$$\therefore \quad C_3 = \begin{pmatrix} \cos 120° & \sin 120° & 0 \\ -\sin 120° & \cos 120° & 0 \\ 0 & 0 & 1 \end{pmatrix} = \begin{pmatrix} -\dfrac{1}{2} & \dfrac{\sqrt{3}}{2} & 0 \\ -\dfrac{\sqrt{3}}{2} & -\dfrac{1}{2} & 0 \\ 0 & 0 & 1 \end{pmatrix}$$

For C_3^2 operation, $\theta = 240°$

$$\therefore \quad C_3^2 = \begin{pmatrix} \cos 240° & \sin 240° & 0 \\ -\sin 240° & \cos 240° & 0 \\ 0 & 0 & 1 \end{pmatrix} = \begin{pmatrix} -\cos 60° & -\sin 60° & 0 \\ \sin 60° & -\cos 60° & 0 \\ 0 & 0 & 1 \end{pmatrix}$$

$$= \begin{pmatrix} -\dfrac{1}{2} & -\dfrac{\sqrt{3}}{2} & 0 \\ \dfrac{\sqrt{3}}{2} & -\dfrac{1}{2} & 0 \\ 0 & 0 & 1 \end{pmatrix}$$

$$S_n \text{ matrix} = \begin{pmatrix} \cos\theta & \sin\theta & 0 \\ -\sin\theta & \cos\theta & 0 \\ 0 & 0 & -1 \end{pmatrix}$$

θ assumes the following values for S_3 and S_3^5 operations

$$\theta (S_3) = 120°, \quad \theta (S_5) = 240°.$$

Problem 2. The square planar $[PtCl_4]^{2-}$ belongs to the D_{4h} point group. Consider the Pt–Cl sigma bonds in this complex ion and obtain the matrices of the reducible representation.

Solution

$PtCl_4^{2-}$ belongs to D_{4h} symmetry. The symmetry operations are E, C_4, C_2, σ_h. The reducible representations are as follows.

$$E = \begin{pmatrix} 1 & 0 & 0 & 0 \\ 0 & 1 & 0 & 0 \\ 0 & 0 & 1 & 0 \\ 0 & 0 & 0 & 1 \end{pmatrix}, \quad C_4 = \begin{pmatrix} 0 & 1 & 0 & 0 \\ 0 & 0 & 1 & 0 \\ 0 & 0 & 0 & 1 \\ 1 & 0 & 0 & 0 \end{pmatrix}, \quad C_2 = \begin{pmatrix} 1 & 0 & 0 & 0 \\ 0 & 0 & 0 & 1 \\ 0 & 0 & 1 & 0 \\ 0 & 1 & 0 & 0 \end{pmatrix}$$

Similarly other matrices can be obtained.

STUDY QUESTIONS

1. Explain the terms reducible and irreducible representations.
2. Taking an example of H_2O molecule, set up one-dimensional representations of the point group C_{2v} based on the translational vectors T_x, T_y, T_z.
3. Taking an example of H_2O molecule, set up the one-dimensional representations of the point group C_{2v} based on rotational R_x, R_y and R_z.
4. Based on p_x, p_y and p_z orbitals of oxygen atom in H_2O molecule, set up the one-dimensional representations of the point group C_{2v}.
5. Based on five d orbitals of sulphur atom in H_2S molecule, set up the one-dimensional representations of the point group C_{2v}.
6. Set up a two-dimensional representation of the point group C_{2v} based on the bond vectors associated with the two O—H bonds of water molecule.
7. Construct the group multiplication table for the following.
   (i) point group C_{2v} (ii) point group C_{3v}
8. State the 'great orthogonality theorem'.
9. Determine irreducible representations contained in the direct product of the following irreducible representations.
   (i) $A_1 \times A_1$; $A_1 \times A_2$ for C_{3v}
   (ii) $E' \times E'$, $A_1'' \times A_2''$ for D_{3h}
10. Based on the characteristics of irreducible representation, set up the irreducible representations of the point groups C_{2v} and C_{3v}.
11. Reduce the following representations to their irreducible representations in the point group indicated

C_{2h}	E	C_2	i	σ_h
Γ	4	0	2	2

12. Analysis of x, y and z coordinates of each atom in NH_3 gives the following representation.

C_{3v}	E	$2C_3$	$3\sigma_v$
Γ	12	0	2

   (a) Reduce Γ to its irreducible representations
   (b) Classify the irreducible representations into translational, rotational and vibrational modes.
   (c) Show that the total number of degrees of freedom = 3N.
   (d) Which vibrational modes are infrared active?
13. Explain the following general layout of a character table.

I	II		
III	IV	V	VI

14. Determine from character tables, the representations to which s, p and d orbitals of an atom in a molecule belonging to
   (i) O_h (ii) T_d (iii) O_{4h} (iv) D_3 groups (v) C_{2v} and (vi) C_{3v} group.

4

Symmetry-Adapted Linear Combinations

4.1 LINEAR COMBINATIONS—AN INTRODUCTION

Molecular wave function is built by the appropriate combination of atomic wave functions and the resulting hybrid orbital wave functions are called *symmetry adapted linear combinations* (SALCs), the SALCs of a molecule is built by using the projection operator defined as

$$\hat{P}_i \left(\text{or } P^{\Gamma_i} \right) = \frac{l_i}{h} \sum_R \chi_i(R) \, \hat{R} \qquad \qquad ...(4.1)$$

where $\hat{P}^i$ is the projection operator of ith irreducible representation (Γ_i) of a point group, h is the order of the group*, l_i is the dimension of the irreducible representations**, $\chi_i(R)$ is the character of each class of representations in the ith irreducible representations, $\hat{R}$ is the symmetry operator for the symmetry operations R, the summation is taken over all the symmetry operations (h in total) of a group.

To construct the molecular orbitals for an AX_n molecule, we need to combine atomic orbitals on X and then match the resultant combinations with the atomic orbitals on the central atom A, and with group theory, we can derive the linear combinations systematically.

4.2 WORKED OUT EXAMPLES OF SALCs AND MOs

4.2.1 A Bent AX₂ Molecule (Example: H₂O)

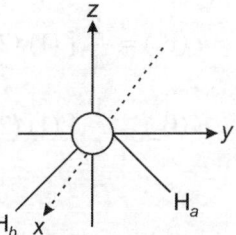

The coordinate system we adopt for H_2O molecule is shown in Fig. 4.1. It is a bent molecule with C_{2v} point group.

We assume that the $2s$ and $2p$ orbitals of oxygen atom and the s orbitals of the two hydrogen atoms take part in the bonding. From the C_{2v} character table, it can be seen that the $2p_x$, $2p_y$ and $2p_z$ orbitals on oxygen atom have B_1, B_2 and A_1 symmetric, respectively, while the oxygen $2s$ orbital, leaving totally symmetric, has A_1 symmetry.

Fig. 4.1 Coordinate system for H₂O molecule [Note that the x-axis points towards the reader]

---

* Sum of the squares of dimensions equals to the order of the group, i.e. $h = \sum (\chi_i(E))^2$

**Sum of the squares of the characters $\sum (\chi_i(R))^2$ multiplied by the number of operations in the class ($n(R)$) equals to the order of the group, i.e. $l_i = h \sum n(R) [(\chi_i(R))^2$. Thus $l_i = h$.

133

The reducible representation obtained on the basis of $1s$ (H_a) and $1s$ (H_b) is

C_{2v}	E	C_2	$\sigma_v(xz)$	$\sigma_v(yz)$
Γ_H	2	0	0	2
	(both orbitals remain unshifted)	(orbitals exchange positions)	(orbitals exchange positions)	(both orbitals remain unshifted)

The number of lines an irreducible representation (IR) occurs in a reversible representation (RR) is determined by using the standard reduction formula.

$$n(\Gamma_i) = \frac{1}{h}\left[\sum n(R)\,\chi_{RR}(R)\,\chi_{IR}(R)\right] \qquad \text{...(4.2)}$$

where $n(\Gamma_i)$ = number of times (n) the ith IR occurs in RR

          $n(R)$ = number of operations in the class

          $\chi_{RR}(R)$ = character of RR under operation R

          $\chi_{IR}(R)$ = character of IR under operation R

The character table of C_{2v} is shown below.

C_{2v}	E	C_2	$\sigma_v(xz)$	$\sigma_v(yz)$
A_1	1	1	1	1
A_2	1	1	−1	−1
B_1	1	−1	1	−1
B_2	1	−1	−1	1

$$n(A_1) = \frac{1}{4}\,[\,(1)(2)(1) + (1)(0)(1) + (1)(0)(1) + (1)(2)(1)\,] = 1$$

$$n(A_2) = \frac{1}{4}\,[\,(1)(2)(1) + (1)(0)(1) + (1)(0)(-1) + (1)(2)(-1)\,] = 0$$

$$n(B_1) = \frac{1}{4}\,[\,(1)(2)(1) + (1)(0)(-1) + (1)(0)(1) + (1)(2)(-1)\,] = 0$$

$$n(B_2) = \frac{1}{4}\,[\,(1)(2)(1) + (1)(0)(-1) + (1)(0)(-1) + (1)(2)(1)\,] = 1$$

$$\Gamma = A_1 + B_2 \qquad \text{...(4.3)}$$

SALC construction from 1s orbitals of hydrogen atoms

SALC of $1s$ (H_a) and $1s$ (H_b) orbitals can be constructed by using the projection operator (Eq. 4.2).

$$P_i\left(\text{or } P^{\Gamma_i}\right) = \frac{l_i}{h}\sum \chi_i(R)\,\hat{R}$$

The two 1s orbitals ($1s_a$ and $1s_b$) will form two linear combinations, one with A_1 symmetry and the other with B_2 symmetry. The linear combinations with A_1 symmetry.

C_{2v}	E	C_2	$\sigma_v(xz)$	$\sigma_v(yz)$	
A_1	1	1	1	1	SALCs (summation of functions*)
$1s_a$	$1s_a$	$1s_b$	$1s_b$	$1s_a$	$1s_a + 1s_b + 1s_b + 1s_a = 2(1s_a + 1s_b)$

*SALCs are obtained by summation of the functions obtained by the operations and multiplied by the character of the irreducible representation (A_1), i.e. $\sum \chi_i(R)\ \hat{R}$

$P^{A_1}(1s_a) = [1.E + 1.C_2 + 1.\sigma_v(xz) + 1.\ \sigma_v(yz)]\,1s_a$

$\qquad = 1s_a.E + 1s_a.C_2 + 1s_a\,\sigma_v(xz) + 1s_a.\sigma_v(yz)$

$\qquad = 1s_a + 1s_b + 1s_b + 1s_a$

$\qquad = 2(1s_a) + 2(1s_b)$

$\Rightarrow 2^{-\frac{1}{2}}(1s_a + 1s_b)$ (after normalisation)

The linear combination with B_2 symmetry

C_{2v}	E	C_2	$\sigma_v(xy)$	$\sigma_v(yz)$	
B_2	1	−1	−1	1	SALCs
$1s_a$	$1s_a$	$-1s_b$	$-1s_b$	$1s_a$	$1s_a - s_b - 1s_b + 1s_a = 2(1s_a - 1s_b)$

$P^{B_2}(1s_a) = [1.E + (-1)\,C_2 + (-1)\,\sigma_v(xz) + 1.\ \sigma_v(yz)]1s_a$

$\qquad = 1s_a - 1s_b - 1s_b + 1s_a = 2(1s_a) - 2(1s_b)$

$\Rightarrow 2^{-\frac{1}{2}}(1s_a + 1s_b)$ (after normalisation)

The MOs of water can be constructed by mixing atomic orbitals of oxygen with the correct SALC of $1s$ orbitals of hydrogen atoms. In Table 4.1, we summaries the way in which the molecular orbitals are formed in H_2O. A schematic energy diagram for this molecule is shown in Fig. 4.2. From this diagram, we can see that there are two bonding orbitals ($1a_1$ and $1b_2$) two other orbitals essentially nonbonding ($2a_1$ and $1b_1$), two antibonding orbitals ($2b_2$ and $3a_1$). All the bonding and nonbonding orbitals are filled, giving rise to a ground electronic configuration of $(1a_1)^2 (1b_2)^2 (2a_1)^2 (1b_1)^2$ and an electronic sates of $^1A_1$. This bonding picture indicates that H_2O has two σ bonds and two filled nonbonding orbitals. Such a result is in qualitative agreement with the familiar valence bond description for this molecule.

Table 4.1 Formation of the molecular orbitals in H_2O

Symmetry	Orbital on O	Orbital on H	Molecular in H_2O
A_1	$2s$	$(2)^{\frac{1}{2}}(1s_a + 1s_b)$	$1a_1, 2a_1, 3a_1$
	$2p_z$		$1b_1$
B_1	$2p_x$	-----	$1b_1$
B_2	$2p_y$	$(2)^{\frac{1}{2}}(1s_a + 1s_b)$	$1b_2, 2b_2$

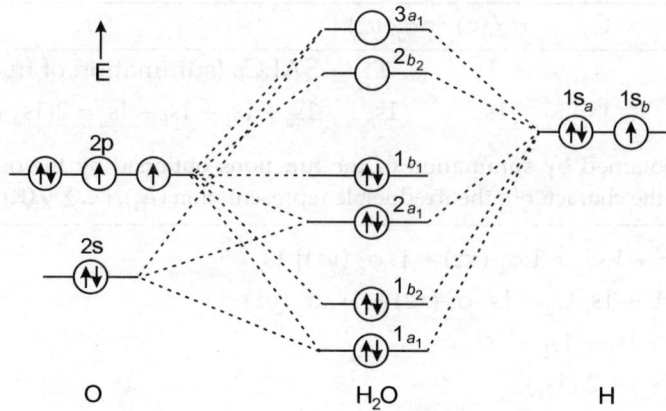

Fig. 4.2 A schematic of energy level diagram for H_2O

4.2.2 A Planar AX₃ Molecule (Example: BH₃)

AX₃ molecule belongs to D_{3h} $(D_3 + \sigma_n)$ point group

Point Group	–	Generator	–	Group Elements
D_{3h}		$D_3 + \sigma_n$ $\underbrace{E, C_3, C_3^2, 3C_2^1}$		$\sigma_h E, \sigma_h C_3, \sigma_h C_3^2, 3\sigma_h C_2^1$

$$= E, C_3, C_3^2, 3C_2^1, \sigma_n, S_3^1, S_3^5, 3\sigma_v$$

Note that $S_3^5 = C_3^5 \sigma_n^5 = C_3^3 C_3^2 \sigma_n^4 \sigma_n = C_3^2 \sigma_n$

$\sigma_n C_2^1 = \sigma_v$ and S_3 ($S_3^1, S_3^2, S_3^3 \equiv E$) is collinear with C_3. So there are two S_3 (S_3^1, S_3^2).

Thus D_{3h} contains C_3, C_3^2 and $C_3^3 (= E), 3C_2, S_3^1, S_3^5, \sigma_n$ (molecular plane) and $3\sigma_v$ (Fig. 4.3).

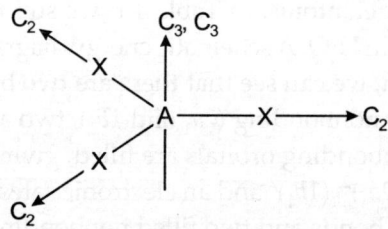

Fig. 4.3 Three C_2 axes perpendicular to C_3 axis. C_3 is collinear with S_3

Here we shall discuss a slightly more complicated system, that of BH₃ will D_{3h} symmetry. BH₃ is not a stable species; it dimerises spontaneously to form diborane, B_2H_6. A convenient coordinate system for BH₃ is shown is Fig. 4.4.

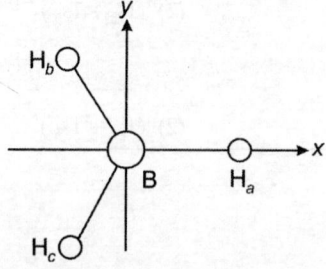

Fig 4.4 Coordinate system for BH₃

The reducible representation obtained on the basis of $1s$ (H_a), $1s$ (H_b) and $1s$ (H_c)

D_{3h}	E	$2C_3$	$3C_2$	σ_n	$2S_3$	$3\sigma_v$
Γ_H	3 (3 orbitals remain unshfited)	0 (All three orbitals suited)	1 (One of the orbitals remain unshfited)	3 (Three orbitals unshfited)	0 (All three orbitals shfited)	1 (One of the orbitals remain unshfited)

This reducible representation Γ_H may be reduced to irreducible representations following the standard reduction equation.

$$n\left(\Gamma_i\right)\frac{1}{h}\left[\sum n\left(R\right)\chi_{RR}\left(R\right)\chi_{IR}\left(R\right)\right]$$

where

$$n\left(R\right) = \text{Number of operations in the class}$$
$$h = \text{Order of the group}$$
$$\chi_{RR}(R) = \text{Character of reducible representation}$$
$$\chi_{IR}(R) = \text{Character of irreducible representation}$$

The character table of D_{3h} point group is shown below.

D_{3h}	E	$2C_3$	$3C_2$	σ_n	$2S_3$	$3\sigma_v$
A'_1	1	1	1	1	1	1
A''_2	1	1	−1	1	1	−1
E'	2	−1	0	2	−1	0
A''_1	1	1	1	−1	−1	−1
A''_2	1	1	−1	−1	−1	1
E''	2	−1	0	−2	1	0

$$n\left(A'_1\right) = \frac{1}{12}\left[(1)\,(3)\,(1) + (2)\,(0)\,(1) + (3)\,(1)\,(1) + (1)\,(3)\,(1) + (2)\,(0)\,(1) + (3)\,(1)\,(1)\right] = 1$$

$$n\left(A'_2\right) = \frac{1}{12}\left[(1)\,(3)\,(1) + (2)\,(0)\,(1) + (3)\,(1)\,(-1) + (1)\,(3)\,(1) + (2)\,(0)\,(1) + (3)\,(1)\,(-1)\right] = 0$$

$$n\left(E'\right) = \frac{1}{12}\left[(1)\,(3)\,(2) + (2)\,(0)\,(-1) + (3)\,(1)\,(0) + (1)\,(3)\,(2) + (2)\,(0)\,(-1) + (3)\,(1)\,(0)\right] = 1$$

$$n\left(A''_1\right) = \frac{1}{12}\left[(1)\,(3)\,(1) + (2)\,(0)\,(1) + (3)\,(1)\,(1) + (1)\,(3)\,(-1) + (2)\,(0)\,(-1) + (3)\,(1)\,(-1)\right] = 0$$

$$n\left(A''_2\right) = \frac{1}{12}\left[(1)\,(3)\,(1) + (2)\,(0)\,(1) + (3)\,(1)\,(-1) + (1)\,(3)\,(-1) + (2)\,(0)\,(-1) + (3)\,(1)\,(1)\right] = 0$$

$$n\left(E''\right) = \frac{1}{12}\left[(1)\,(3)\,(2) + (2)\,(0)\,(-1) + (3)\,(1)\,(0) + (1)\,(3)\,(-2) + (2)\,(0)\,(1) + (3)\,(1)\,(0)\right] = 0$$

Thus we have $\Gamma_H \equiv A'_1 + E'$. To obtain the explicit functions, we require the symmetry operation results.

D_{3h}	E	$2C_3$	$3C_2$	σ_n	$2S_3$	$3\sigma_v$
$1s_a$	$1s_a$	$1s_b, 1s_c$	$1s_a, 1s_b, 1s_c$	$1s_a$	$1s_b, 1s_c$	$1s_a, 1s_b, 1s_c$

D_{3h} contains C_3, C_3^2 and C_3^3 ($= E$), i.e. $2C_3$ operations, $3C_2$ operation, $2S_3$ (S_3^1 and S_3^5) three, operations, one σ_n and three σ_v planes. Therefore, $2C_3$ and $2S_3$ operations on $1s_a$ give the same result, i.e. $1s_b, 1s_c$. To derive the linear combinations of $1s$ orbitals of hydrogen with A'_1 symmetry, we apply $P^{A'_1}$ on the $1s$ orbitals on H_a, denoted as $1s_a$.

$$P^{A'_1}(1s_a) = \sum \chi_i \left(\chi_i(R) \right) \hat{R}$$

$\hat{R}$ is the symmetry operation, $\chi_i(R)$ is the character of that operation, for A_1 irreducible representation, we have (from D_{3h} character table)

E = 1, C_3 = 1, C_2 = 1, σ_n = 1, S_3 = 1, σ_v = 1

$$
\begin{aligned}
P^{A'_1}(1s_a) &= (1)(1s_a) + (1)(1s_b + 1s_c) + (1)(1s_a + 1s_b + 1s_c) + (1)(1s_a) + (1)(1s_b + 1s_c) \\
&\quad + (1)(1s_a + 1s_b + 1s_c) \\
&= 4(1s_a + 1s_b + 1s_c) \hspace{3cm} \text{...(4.4)}
\end{aligned}
$$

$\Rightarrow (3)^{-\frac{1}{2}} (1s_a + 1s_b + 1s_c)$ (after normalisation)

The expression of normalised SALC is

$$\phi_1 = \frac{1}{\sqrt{3}} (1s_a + 1s_b + 1s_c)$$

The operation of $P^{E'}$ on $1s_a$ (H_a) gives

$$P^{E'}(1s_a) = \sum \chi_i(R) \, \hat{R}$$

The characters of E' (from D_{3h} characters table)

E = 2, C_3 = – 1, C_2 = 0, σ_n = 2, S_3 = – 1, σ_v = 0.

$$
\begin{aligned}
P^{E'}(1s_a) &= (2)(1s_a) + (-1)(1s_b + 1s_c) + (0)(1s_a + 1s_b + 1s_c) \\
&\quad + (2)(1s_a) + (-1)(1s_b + 1s_c) + (0)(1s_a + 1s_b + 1s_c) \\
&= 4(1s_a) - 2(1s_b) - 2(1s_c) \\
&= 2[2(1s_a) - (1s_b + 1s_c)] \\
&= (6)^{-\frac{1}{2}} [2(1s_a) - 1s_b - 1s_c] \text{ (after normalisation)} \hspace{1.5cm} \text{...(4.5)}
\end{aligned}
$$

$$\phi_2 = -\frac{1}{\sqrt{6}} [2(1s_a) - 1s_b - 1s_c]$$

[Note that the SALC ϕ_2 is orthogonal to ϕ_1]

One more combination is required to complete the E' set. When we operate $P^{E'}$ on $1s_b$, we get

$$P^{E'}(1s_b) = \phi_3 = -\frac{1}{\sqrt{6}} [2(1s_b) - 1s_a - 1s_c] \text{ and when we operate } P^{E'} \text{ on } 1s_c \text{ we get}$$

$$\text{...(4.6)}$$

$$P^{E'}(1s_b) = \phi_4 = -\frac{1}{\sqrt{6}} [2(1s_c) - 1s_a - 1s_b] \hspace{3cm} \text{...(4.7)}$$

Neither ϕ_3 nor ϕ_4 is orthogonal to ϕ_2. Summation of Eqs (4.6) and (4.7) yields Eq. (4.5)

$$\phi_3 + \phi_4 = \frac{1}{\sqrt{6}} \left[(1s_b + 1s_c) - 2\,(1s_a) \right].$$

$$= -\frac{1}{\sqrt{6}} \left[2\,(1s_a) - (1s_b - 1s_c) \right] = -\phi_2$$

This is excluded. The difference of Eqs (4.5) and (4.6) yields

$$\phi_3 - \phi_4 \;= \frac{1}{\sqrt{6}} \left[2\,(1s_b) - 1s_a - 1s_c - 2\,(1s_c) + 1s_a + 1s_b \right]$$

$$= \frac{1}{\sqrt{6}} \left[3\,(1s_a - 1s_c) \right]$$

$$= \frac{3}{\sqrt{6}} \left[(1s_b - 1s_c) \right]$$

$$\Rightarrow \frac{1}{\sqrt{2}} \left[(1s_b - 1s_c) \right] \text{ (after normalisation)} \qquad \ldots (4.8)$$

Equations (4.5) and (4.8) are chosen because these functions overlap with boron $2p_x$ and $2p_y$ respectively as shown in Fig. 4.5.

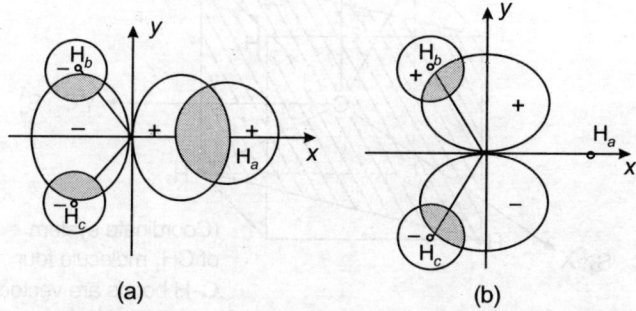

Fig. 4.5 (a) Overlap between the boron $2p_x$ orbital with combination $(6)^{-1/2}[2(1s_a) - 1s_b - 1s_c]$
(b) Overlap between the boron $2p_y$ orbital with the combination $(2)^{-1/2}[1s_b - 1s_c]$
[By symmetry, the total overlaps in (a) and (b) are the same]

Table 4.2 summaries how molecular orbitals in BH_3 are formed. A schematic energy level diagram for BH_3 is shown in Fig. 4.6.

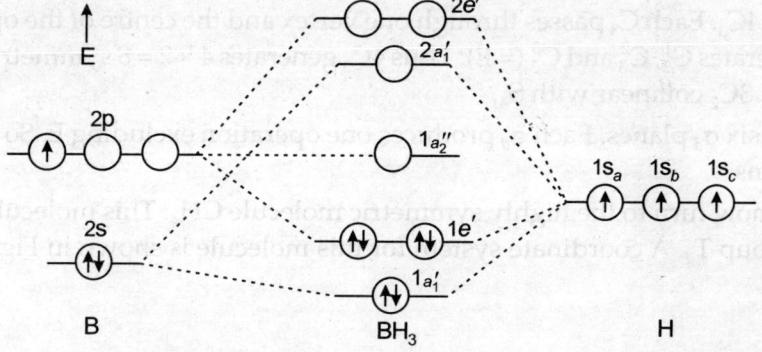

Fig. 4.6 A schematic energy level diagram for BH_3

Table 4.2 Summary of the formation of the molecular orbitals in BH_3

Symmetry	Orbitals on B	Orbitals of H	Molecular orbitals
A_1'	$2s$	$(3)^{-\frac{1}{2}}(1s_a + 1s_b + 1s_c)$	$1a_1', 2a_1'$
E'	$\begin{cases} 2p_x \\ 2p_y \end{cases}$	$\begin{cases} (6)^{-\frac{1}{2}}[(1s_a) - 1s_b - 1s_c] \\ (2)^{-\frac{1}{2}}[1s_b - 1s_c] \end{cases}$	$1e', 2e'$
A_2''	$2p_z$	—	$1a''_2$

4.2.3 A Tetrahedral AX₄ Molecule (Example: CH₄)

The table symmetry operations of a tetrahedral group are 24 of which 9 from $3S_4$, 8 from $4C_3$, 6 from σ_d and 1 from E (Fig. 4.7).

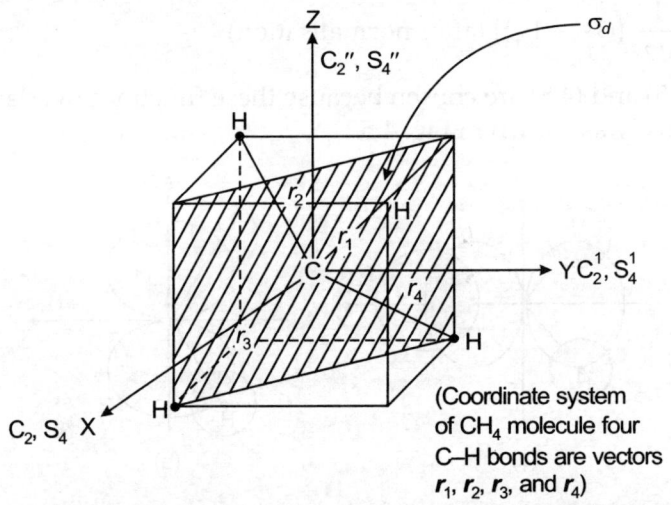

Fig 4.7 A tetrahedral AX₄ molecule

$3S_4$ collinear with C_2 axes along x, y, z axes. Each S_4 generates S_4^1, S_4^2 ($= C_4^2 \sigma^2 = C_4^2 = C_2$), $S_4^3 S_4^4 (= E)$.

Thus, we get S_4^1, S_4^3, C_2 and E from each S_4. Thus $37S_4$ generates $6S_4$ and $3C_2$ operations.

There are $4C_3$. Each C_3 passes through one vertex and the centre of the opposite face. Each C_3 generates C_3^1, C_3^2 and $C_3^3 (= E)$. Thus $4C_3$ generates $4 \times 2 = 8$ symmetry operations excluding E. $3C_2$ collinear with S_4.

There are six σ_d planes. Each σ_d produces one operation excluding E. So $6\sigma_d$ produce $6\sigma_d$ operations.

We shall now turn to the highly symmetric molecule CH_4. This molecule belongs to the point group T_d. A coordinate system for this molecule is shown in Fig. 4.8.

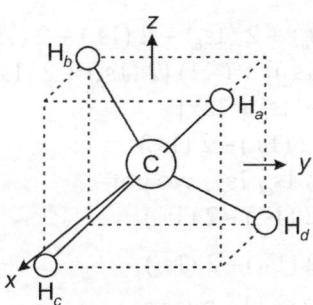

Fig 4.8 Coordinate system for CH$_4$

For CH$_4$ molecule there are eight valance atomic orbitals: $2s$ and $2p$ orbitals of carbon and the $1s$ orbitals are of the hydrogens. $2s$ orbital of carbon has A$_1$ symmetry while $2p_x$, $2p_y$ and $2p_z$ orbitals from a T$_2$ set.

The reducible representation χ_{RR} obtained on the basis of four hydrogen atoms is shown in the following tabular form.

T$_d$	E	8C$_3$	3C$_2$	6S$_4$	6σ$_d$
Γ$_H$	(All the orbitals remain unshifted)	(One orbital remain unshifted)	(All orbitals change positions)	(All orbitals change positions)	(Two orbitals remain unshifted)
	4	1	0	0	2
A$_1$	1	1	1	1	1
T$_2$	3	0	– 1	– 1	1

So the four hydrogen $1s$ orbitals form one linear combination with. A$_1$ symmetry and three other combinations that make up a T$_2$ set. To obtain these combinations, we make use of the symmetry operation results.

T$_d$	E	8C$_3$	3C$_2$	6S$_4$	6σ$_d$
$1s_a$	$1s_a$	2 ($1s_a$), 2 ($1s_b$) 2 ($1s_c$), 2 ($1s_d$)	$1s_b$, $1s_c$, $1s_d$	2 ($1s_b$), 2 ($1s_c$) 2 ($1s_d$)	3 ($1s_a$) $1s_b$, $1s_c$, $1s_d$

Now it is easy to obtain the linear combinations.

$$P^{A_1}(1s_a) = \sum_R \chi_i(R)\,\hat{R}$$

$$= \Sigma\,(\text{character}) \times \text{symmetry operation } \hat{R}$$

Here we drop the numerical factor (l_i/h) in Eq. (4.1) since we are interested only in the functional form of the SALC]

The characters of A$_1$ for E = 1, C$_3$ = 1, C$_2$ = 1, S$_4$ = 1, σ$_d$ = 1 and the characters of T$_2$ are E = 3, C$_3$ = 0, C$_2$ = – 1, S$_4$ = – 1, σ$_d$ = 1.

$P^{A_1}(1s_a) = (1)\,(1s_a) + (1)\,[2\,(1s_a) + 2\,(1s_b) + 2\,(1s_c) + 2\,(1s_d)]$

$\qquad + (1)\,(1s_b + 1s_c + 1s_d) + (1)\,[2\,(1s_b) + 2\,(1s_c) + 2\,(1s_d)]$

$\qquad + (1)\,[3\,(1s_a) + 1s_b + 1s_c + 1s_d\,]$

$\Rightarrow \dfrac{1}{2}\,(1s_a + 1s_b + 1s_c + 1s_d)$ (after normalisation) ...(4.9)

$P^{T_2} (1s_a) = 3 (1s_a) + (0) [2 (1s_a) + 2 (1s_b) + 2 (1s_c) + 2 (1s_d)]$

$\qquad + (-1) [1s_b + 1s_c + 1s_d)] + (-1) [2 (1s_b) + 2 (1s_c) + 2 (1s_d)]$

$\qquad + (1) [3 (1s_a) + 1s_b + 1s_c + 1s_d]$

$\qquad = 6 (1s_a) - 2 (1s_b) - 2 (1s_c) - 2 (1s_d)$ \qquad\qquad ...(4.10)

When we operate P^{T_2} on $1s_b$, $1s_c$, $1s_d$, we get

$P^{T_2} (1s_b) = 6 (1s_b) - 2 (1s_a) - 2 (1s_c) - 2 (1s_d)$ \qquad\qquad ...(4.11)

$P^{T_2} (1s_c) = 6 (1s_c) - 2 (1s_a) - 2 (1s_b) - 2 (1s_d)$ \qquad\qquad ...(4.12)

$P^{T_2} (1s_d) = 6 (1s_d) - 2 (1s_a) - 2 (1s_b) - 2 (1s_c)$ \qquad\qquad ...(4.13)

To obtain the three linear combinations of the T_2 set, we combine Eqs (4.10) to (4.13) in the following manner.

Sum of Eqs (4.10) and (4.11) :

$4 (1s_b) + 4 (1s_b) - 4 (1s_c) - 4 (1s_d)$

$= 4 (1s_a + 1s_b - 1s_c - 1s_d)$

$= \dfrac{1}{2} (1s_a - 1s_b - 1s_c - 1s_d)$ (after normalisation) \qquad\qquad ...(4.14)

Sum of Eqs (4.10) and (4.12) :

$4 (1s_a) - 4 (1s_b) + 4 (1s_c) - 4 (1s_d)$

$= 4 (1s_a - 1s_b + 1s_c - 1s_d)$

$= \dfrac{1}{2} (1s_a - 1s_b - 1s_c + 1s_d)$ (after normalisation) \qquad\qquad ...(4.15)

Sum of Eqs (4.9) and (4.11) :

$\dfrac{1}{2} (1s_a - 1s_b - 1s_c + 1s_d)$ (after normalisation) \qquad\qquad ...(4.16)

There are many ways of combining Eqs (4.10) to (4.13) to arrive at the three combinations that form the T_2 set. We choose those ones given by Eqs (4.14) to (4.16) as these functions overlap effectively with the $2p_z$, $2p_x$, and $2p_y$ orbitals respectively as shown in Fig. 4.9.

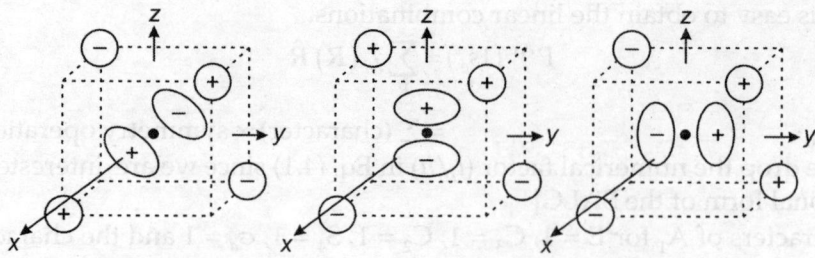

Fig. 4.9 The overlap between $2p_x$, $2p_z$, and $2p_y$ orbitals of carbon with the $1s$ orbitals of hydrogen in CH_4

Table 4.3 summarises the formation of the molecular orbitals in CH_4. A schematic energy level diagram for CH_4 is shown in Fig. 4.10. According to this diagram, the ground configuration is simply $(1a_1)^2 (1t_2)^6$ and the ground state is $^1A_1$.

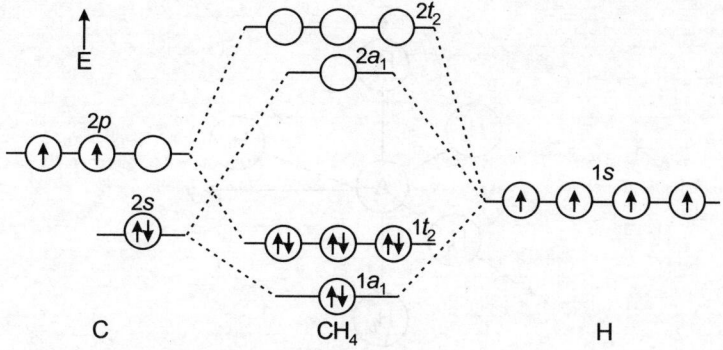

Fig. 4.10 A schematic energy level diagram for CH_4

Table 4.3 Formation of the molecular orbitals in CH_4

Symmetry	Orbital on C	Orbitals on H	Molecular orbitals
A_1	$2s$	$\frac{1}{2}(1s_a + 1s_b + 1s_c + 1s_d)$	$1a_1, 2a_1$
T_2	$2p_x$ $2p_y$ $2p_z$	$\frac{1}{2}(1s_a + 1s_b + 1s_c + 1s_d)$ $\frac{1}{2}(1s_a + 1s_b + 1s_c + 1s_d)$ $\frac{1}{2}(1s_a + 1s_b + 1s_c + 1s_d)$	$1t_2, 2t_2$

$$\begin{cases} (12)^{-\frac{1}{2}} [2(1s_e) + 2(1s_f) \\ \quad - 1s_a - 1s_b - 1s_c - 1s_d] \\ \frac{1}{2}[1s_a + 1s_b - 1s_c - 1s_d] \end{cases}$$

4.2.4 A H_5 Trigonal Bipyramidal Molecule (Hypothetical Molecule with D_{3h} Symmetry)

So far we have illustrated the method for constructing the linear combinations of atomic orbitals, using H_2O, BH_3, and CH_4 as examples. In these simple system, all the ligand orbitals are equivalent to each other.

For molecules, with nonequivalent ligand sites, we first linearly combine the orbitals on the equivalent atoms. Then, if the need arises, we can further combine the combinations that have the same symmetry. For example, let us consider a hypothetical structure (D_{3h} symmetry), Fig. 4.11 shows a convenient coordinate system for this molecule.

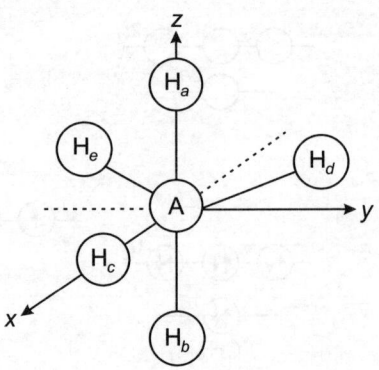

Fig 4.11 Coordinate system for AR_5

From Fig. 4.11 it is clear that there are two sets of H-atoms:

(i) equatorial hydrogens H_c, H_d, H_e and

(ii) axial hydrogens H_a and H_b

when we linearly combine the orbitals on H_a and H_b, two (unnormalised) combinations are obtained.

$1s_a + 1s_b$ (A$'_1$ symmetry)

$1s_a - 1s_b$ (A$''_2$ symmetry)

The combinations for the orbitals on the equatorial hydrogens are

$1s_c + 1s_d + 1s_e$ (A$'_1$ symmetry)

$2(1s_c) - 1s_d - 1s_e$ and $1s_d - 1s_c$ (E$'$ symmetry)

It we assume central atom A contributes n_s and n_p orbitals to bonding, we can easily arrive at the results summarized in Table 4.4.

Table 4.4 Formation of the molecular orbitals in AX_5

Symmetry	orbital on A	orbital on H	Molecular orbitals
A$'_1$	ns	$(2)^{-\frac{1}{2}}[1s_a + 1s_b]$ $(3)^{-\frac{1}{2}}[1s_c + 1s_d + 1s_e]$	$1a'_1$, $2a'_1$, $3a'_1$
E$'$	$\begin{cases} np_x \\ np_y \end{cases}$	$(6)^{-\frac{1}{2}}[2(1s_c) - 1s_d - 1s_e]$	$1e'$, $2e'$
A$''_2$	np_z	$(2)^{-\frac{1}{2}}[1s_d + 1s_e]$ $(2)^{-\frac{1}{2}}[1s_a - 1s_b]$	$1a''_2$, $2a''_2$

Note that we can further combine the two ligand linear combinations with A$'_1$ symmetry by taking their sum and difference.

4.2.5 Octahedral Molecule (AH₆)

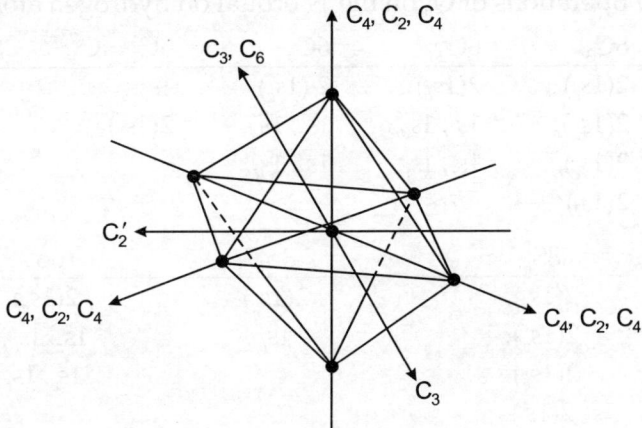

Fig 4.12 Some symmetry elements in AH₆

In the O_h point group, the following symmetry elements are present.

(i) Three S_4 axes of rotation, each passing through opposite apices. Each generates $S_4^1, S_4^2 (= C_2), S_4^3, S_4^4 (\equiv E)$, i.e. two S_4 (S_4^1, S_4^3) rotational axis. Thus $3S_4$ generate six S_4 axis of rotation (Fig. 4.12).

(ii) $3C_2$ and $3C_4$ collinear with S_4.

(iii) $6C'_2$ passing through the centers of opposite edges.

(iv) $4S_6$, collinear with C_3, passing though the center of opposite triangular faces.

(v) $4C_3$, collinear with S_6.

(vi) i

(vii) $3\sigma_h$, each containing 4 vertices in a basal plane.

(viii) $6\sigma_d$ planes, each passing through two apices's and biseeling two opposite edges.

Thus, we have the following set of sym. operation in the point group O_h.

$E, 8C_3, 6C_4, 6C_2, 3C_2 (= C_4^2), i, 6S_4, 8S_6, 3\sigma_n, 6\sigma_d$.

We now consider the highly symmetrical octahedral system for this molecule is shown Fig. 4.13.

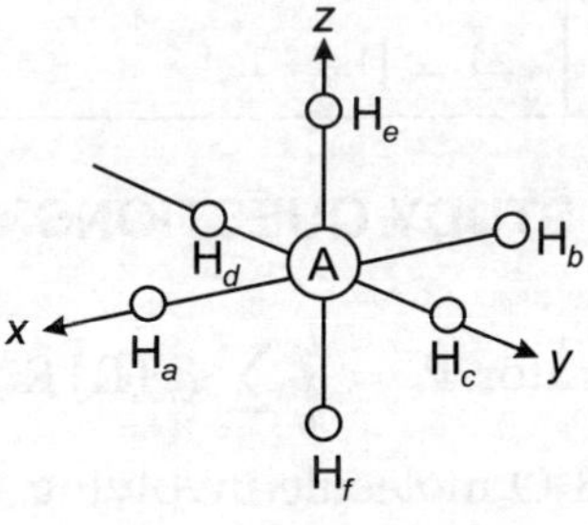

Fig 4.13 Coordinate system for AH₆

We assume the central atom A contributes n_s, n_p and n_d orbitals to bonding. To obtain the symmetry of the six $1s$ orbital linear combinations.

O_h	E	$8C_3$	$6C_2$	$6C_4$	$3C_2$	$\equiv$	C_4^2	i	$6S_4$	$8S_6$	$3\sigma_n$	σ_d		
Γ_H	6	0	0	2	2		0	0	0	4	2	$\equiv$	$A_{1g} + E_g^{-1} T_{1u}$	

To derive the combinations, we need the following tabulations listing the effect of various symmetry operations of O_h on the $1s$ orbital on hydrogen atom H_a.

O_h	E	$8C_3$	$6C_2$	$6C_4$	$3C_2 = C_4^2$	i	$6S_4$
$1s_a$	$1s_a$	$2(1s_c)$,	$2(1s_b)$,	$2(1s_a)$,	$1s_{a'}$	$1s_b$	$2(1s_b)$,
		$2(1s_d)$,	$1s_c, 1s_d$,	$1s_c, 1s_d$,	$2(1s_b)$		$1s_c, 1s_d$,
		$2(1s_e)$,	$1s_e, 1s_f$	$1s_e, 1s_f$			$1s_e, 1s_f$
		$2(1s_f)$					

	$8S_6$	$3\sigma_h$	$6\sigma_d$
	$2(1s_c)$,	$2(1s_a)$	$2(1s_a)$
	$2(1s_d)$,	$1s_b$	$1s_c, 1s_d$,
	$2(1s_e)$,		$1s_e, 1s_f$
	$2(1s_f)$		

Now it is straight forword to derive the results summarised in Table 4.5.

Table 4.5 Formation of the molecular orbitals in AH_6

Symmetry	orbital on A	orbital on H	Molecular orbitals
A_1'	n_s	$(6)^{-\frac{1}{2}}[1s_a + 1s_b + 1sc + 1sd + 1se + 1sf]$	$1a_1, 2a_1,$
E_g	$\begin{cases} nd_{z^2} \\ nd_{x^2-y^2} \end{cases}$		$1e_g, 2e_g$
T_{2g} T_{1u}	$\begin{cases} nd_{xy} \\ nd_{yz} \\ nd_{xz} \end{cases}$ $\begin{cases} np_x \\ np_y \\ np_z \end{cases}$	_____ _____ $(2)^{-\frac{1}{2}}[1s_a + 1s_b]$ $(2)^{-\frac{1}{2}}[1s_c + 1s_d]$ $(2)^{-\frac{1}{2}}[1s_e + 1s_f]$	$1t_{2g}$ $t_{1u}, 2t_{1u}$

STUDY QUESTIONS

1. Set up the projection operator $\hat{P}_i = \dfrac{l_i}{h} \sum_R \chi_i(R)\hat{R}$

2. Workout the SALCs of H_2O molecule involving $1s$ orbitals of hydrogen atoms and also form molecular orbitals of H_2O molecule.

3. Workout the SALSs of BeH_2 molecule involving $1s$ orbitals of hydrogen atoms and also form molecular orbitals of BeH_2 molecule.

4. Workout the SALSs of BH_3 molecule involving $1s$ orbitals of hydrogen atoms and also form molecular orbitals of BH_3 molecule.

5. Workout the SALSs of NH_3 molecule involving $1s$ orbitals of hydrogen atoms and also form molecular orbitals of NH_3 molecule.

5

Application of Group Theory to Atomic Orbitals

Schrödinger equation [Eq. (5.1)] describes accurately the atomic systems.

$$\frac{\partial^2 \psi}{\partial x^2} + \frac{\partial^2 \psi}{\partial y^2} + \frac{\partial^2 \psi}{\partial z^2} + \frac{8\pi^2 m}{h^2}\left(E + \frac{e^2}{r}\right)\psi = 0 \qquad \qquad ...(5.1)$$

In order to solve Eq. (5.1), we have to express in terms of spherical polar coordinates r, θ, ϕ. The relationship between the Cartesian and spherical coordinates is given by a study of Fig. 5.1.

$$x = r \sin \theta \cos \phi, \; y = r \sin \theta \sin \phi, \; z = r \cos \theta, \; r = \sqrt{x^2 + y^2 + z^2}$$

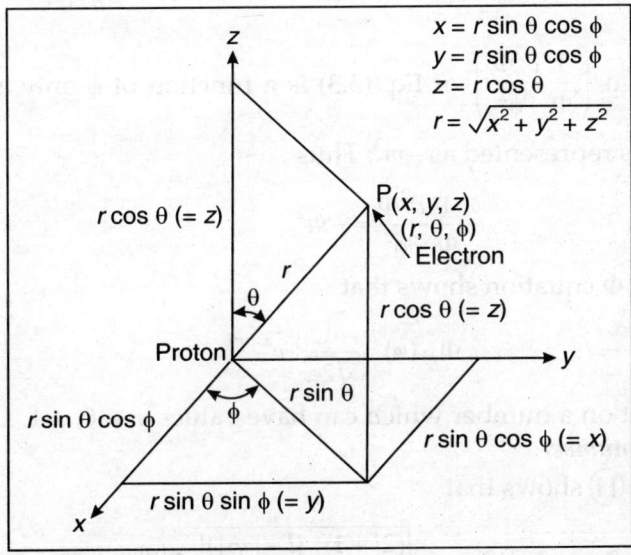

Fig. 5.1 The spherical coordinate system and its relation to the Cartesian coordinate system

Utilisation of the values of x, y, z and substitution of the operator

$$\left(\frac{\partial^2}{\partial x^2} + \frac{\partial^2}{\partial y^2} + \frac{\partial^2}{\partial z^2}\right)$$

into the corresponding polar form enable us to write Eq. (5.1) in the follow form:

$$\frac{1}{r^2}\frac{\partial}{\partial r}\left(r^2\frac{\partial\psi}{\partial r}\right)+\frac{1}{r^2\sin^2\theta}\frac{\partial^2\psi}{\partial\phi^2}+$$

$$\frac{1}{r^2\sin^2\theta}\cdot\frac{\partial}{\partial\theta}\left(\sin\theta\frac{\partial\psi}{\partial\theta}\right)+\frac{8\pi^2 m}{h^2}\left(E+\frac{e^2}{r}\right)\psi=0 \quad ...(5.2)$$

Equation (5.2) contains three variables r, θ and ϕ and can be written as the product of three simple functions:

R(r) dependent only on r (r = electron nucleus separation)

Ⓗ(θ) dependent only on θ and

$\Phi(\phi)$ dependent only on ϕ. Then

$$\psi(r,\theta,\phi)=\mathbb{R}(r)\,Ⓗ\,\Phi(\phi)$$

Equation (5.2) transforms in the following form

$$\frac{r^2\sin^2\theta}{R}\frac{\partial}{\partial r}\left(r^2\frac{\partial R}{\partial r}\right)+\frac{1}{\Phi}\cdot\frac{\partial^2\Phi}{\partial\phi^2}+\frac{\sin\theta}{Ⓗ}\cdot\frac{\partial}{\partial\theta}\left(\sin\theta\frac{\partial Ⓗ}{\partial\theta}\right)$$

$$+\,r^2\sin^2\theta\cdot\frac{8\pi^2 m}{h^2}\left(E+\frac{Ze^2}{r}\right)=0 \quad ...(5.3)$$

The second term $\left(\dfrac{1}{\Phi}\dfrac{\partial^2\Phi}{\partial\phi^2}\right)$ of Eq. (5.3) is a function of ϕ only and this term is a

constant term. It is represented as $-m^2$. Thus

$$\frac{1}{\Phi}\frac{\partial^2\Phi}{\partial\phi^2}=-m^2$$

The solution of Φ equation shows that

$$\Phi_m(\phi)=\frac{1}{\sqrt{2\pi}}\cdot e^{\pm im\phi}, \qquad\qquad ...(5.4)$$

i.e. it is dependent on a number which can have values $m=0,\pm1,\pm2$. m is known as *magnetic quantum number*.

The solution of Ⓗ shows that

$$\theta_{l,m}=Y_l^m(\theta)=\epsilon\sqrt{\frac{(2l+1)}{2}\cdot\frac{(l-|m|)!}{(l+|m|)!}}\,P_l^{|m|}(\cos\theta) \qquad ...(5.5)$$

where $P_l^{|m|}$ is the associated Legendre function (Table 5.1).

The θ part of the wave function is dependent on two numbers l and m. The possible values of m are $0,\pm1,\pm2,...,\pm l$.

The radial part of the wave does not affect the shape of the orbital. The shape is dependent on the angular parts ϕ and θ. The normalised wave functions $Y_l^m(\theta,\phi)$ are called *spherical harmonics*.

Table 5.1 Some associated Legendre functions, $P_l^{|m|}$ (cos θ) (functional form)

$P_0^0 = 1$	$P_2^0 = \dfrac{1}{2}\ (3 \cos^2 θ - 1)$
$P_1^1 = \sin θ$	$P_3^3 = 15 \sin θ\ (1 - \cos^2 θ) = 15 \sin^3 θ$
$P_1^0 = \cos θ$	$P_3^2 = 15 \sin^2 θ \cos θ$
$P_2^2 = 3 \sin^2 θ$	$P_3^1 = \dfrac{3}{2}\ \sin θ\ (5 \cos^2 θ - 1)$
$P_2^1 = 3 \sin θ \cos θ$	$P_3^0 = \dfrac{1}{2}\ (5 \cos^3 θ - 3 \cos θ)$

$$Y_l^m(θ, φ) = Y_l^m(θ)\ Φ(φ)$$

$$= ∈ \sqrt{\frac{(2l+1)}{2} \cdot \frac{(l-|m|)!}{(l+|m|)!}}\ P_l^{|m|}(\cos θ) \cdot \frac{1}{\sqrt{2π}}\ e^{imφ}$$

$$= ∈ \sqrt{\frac{(2l+1)}{4π}\ \frac{(l-|m|)!}{(l+|m|)!}}\ e^{imφ}\ P_l^{|m|}(\cos θ) \qquad \text{...(5.6)}$$

where $∈ = (-1)^m$ for $m ≥ 0$ and $∈ = 1$ for $m ≤ 0$. Few spherical harmonics are tested in Table 5.2.

Table 5.2 The first few spherical harmonics, $Y_l^m(θ, φ)$

$Y_0^0 = \left(\dfrac{1}{4π}\right)^{1/2}$	$Y_2^{\pm 2} = \left(\dfrac{15}{32π}\right)^{1/2} \sin^2 θ\ e^{\pm 2iφ}$
$Y_1^0 = \left(\dfrac{3}{4π}\right)^{1/2} \cos θ$	$Y_3^0 = \left(\dfrac{7}{16π}\right)^{1/2} (5 \cos^3 θ - 3 \cos θ)$
$Y_1^{\pm 1} = \mp \left(\dfrac{3}{8π}\right)^{1/2} \sin θ\ e^{\pm iφ}$	$Y_3^{\pm 1} = \mp \left(\dfrac{21}{64π}\right)^{1/2} \sin θ\ (5 \cos^2 θ - 1)\ e^{\pm iφ}$
$Y_2^0 = \left(\dfrac{5}{16π}\right)^{1/2} (3 \cos^2 θ - 1)$	$Y_3^{\pm 2} = \left(\dfrac{105}{32π}\right)^{1/2} \sin^2 θ \cos θ\ e^{\pm 2iφ}$
$Y_2^{\pm 1} = \mp \left(\dfrac{15}{8π}\right)^{1/2} \sin θ \cos θ\ e^{\pm iφ}$	$Y_3^{\pm 3} = \mp \left(\dfrac{35}{64π}\right)^{1/2} \sin^3 θ\ e^{\pm 3iφ}$

Angular Wave Functions

The angular function of an orbital does not depend on the value of n, but is governed by l and m.

5.1 SHAPE OF s-ORBITAL

For s-orbital, $l = 0$ and $m = 0$, we use Eq. (5.5), i.e.

$$Y_l^m(θ) = ∈ \sqrt{\frac{(2l+1)}{2}\ \frac{(l-|m|)!}{(l+|m|)!}}\ P_l^{|m|}(\cos θ)$$

$$Y_0^0(\theta) = 1 \cdot \sqrt{\frac{(2.0+1)}{2} \cdot \frac{(0-0)!}{(0+0)!}} \; P_0^0 (\cos \theta) \; [\text{for } m = 0, \in = 1]$$

$$Y_0^0(\theta) = \sqrt{\frac{1}{2}} \left[\because 0! = 1; P_0^0 (\cos \theta) = 1 \right]$$

$$\Phi_m(\phi) = \frac{1}{\sqrt{2\pi}} e^{im\phi}$$

$$\Phi_0(\phi) = \frac{1}{\sqrt{2\pi}} \cdot e^{i0\phi} = \frac{1}{\sqrt{2\pi}}$$

$$\therefore Y_l^m(\theta, \phi) = Y_0^0(\theta) \cdot \Phi_0(\phi) = \frac{1}{\sqrt{2}} \cdot \frac{1}{\sqrt{2\pi}} = \frac{1}{2\sqrt{\pi}}$$

$$\therefore Y_0^0 = \frac{1}{2\sqrt{\pi}} = \left(\frac{1}{4\pi} \right)^{1/2} \quad \text{(refer to Table 5.1)}$$

Thus for s-orbital ($l = 0$, $m = 0$), the angular wave function $Y_l^m(\theta, \phi)$ is independent of the angle θ and ϕ. At equal distance from the nucleus, the probability of finding the electron is same. The region showing the maximum probability (orbital) is also equidistant from the nucleus. Thus the orbital is spherically symmetrical (Fig. 5.2).

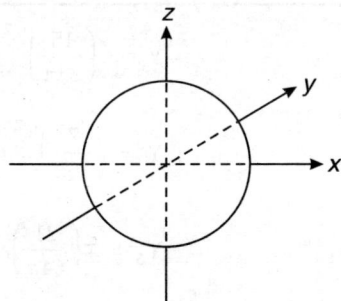

Fig. 5.2 Angular probability distribution of an s-orbital

5.2 SHAPE OF p-ORBITAL

For p-orbitals, $l = 1$ and m (or m_l) = + 1, 0, – 1. So three solutions are possible following Eq. (5.5).

1. When $l = 1$ and m (or m_l) = 0

$$Y_1^0(\theta) = 1 \cdot \sqrt{\frac{(2.1+1)}{2} \cdot \frac{(1-0)!}{(1+0)!}} \cdot P_1^0 (\cos \theta)$$

$$= \frac{\sqrt{3}}{2} \cdot \cos \theta \left[\because P_1^0 (\cos \theta) = \cos \theta, \text{ for } m = 0, \in = 1 \right]$$

$$\Phi_0(\phi) = \frac{1}{\sqrt{2\pi}} e^{i0\phi} = \frac{1}{\sqrt{2\pi}}$$

Thus the orbitals for which $l = 1$ and $m = 0$ is independent of ϕ but dependent on θ. $2p_z$ orbital attains maximum value when $\cos \theta = 1$, i.e. when $\theta = 0$. $2p_z$ orbital assumes positive and negative values for $\theta = 0$ to $\dfrac{\pi}{2}$, and $\dfrac{\pi}{2}$ to π respectively. The positive and negative signs represent the phases of the wave.

2. When $l = 1$ and m (or m_l) $= +1$, $\in = -1$ ($\because m > 0$)

$$Y_1^0(\theta) = (-1)\sqrt{\frac{(2.1+1)}{2} \cdot \frac{(1-1)!}{(1+1)!}} \; P_1^1 (\cos \theta)$$

$$= (-1)\sqrt{\frac{3}{4}} \, \sin \theta \; \left[P_1^1 (\cos \theta) = \sin \theta) \right]$$

$$\Phi_1(\phi) = \frac{1}{\sqrt{2\pi}} e^{i1\phi} = \frac{1}{\sqrt{2\pi}} e^{i\phi}$$

$$\therefore \; Y_1^1 = 2p_x = (-1)\sqrt{\frac{3}{4}} \, \sin \theta \cdot \frac{1}{\sqrt{2\pi}} e^{i\phi} = -\sqrt{\frac{3}{8\pi}} \, \sin \theta \, e^{i\phi}$$

$$= -\frac{\sqrt{3}}{2} \cdot \frac{1}{\sqrt{\pi}} \cdot \sin \theta \cos \phi \left[\frac{e^{i\phi}}{2} = \cos \phi \right]$$

3. When $l = 1$, m (or m_l) $= -1$, $\in = 1$, since $m < 0$

$$Y_1^{-1}(\theta) = 1 \cdot \sqrt{\frac{(2.1+1)}{2} \cdot \frac{(1-|-1|)!}{(1+|-1|)!}} \; P_1^{1-|1|}(\cos \theta)$$

$$= \sqrt{\frac{3}{2} \cdot \frac{1}{2}} \; P_1^1 (\cos \theta) = \sqrt{\frac{3}{4}} \, \sin \theta \; \left[\because P_1^1 (\cos \theta) = \sin \theta \right]$$

$$= \frac{\sqrt{3}}{2} \sin \theta$$

$$\Phi_{-1}(\phi) = \frac{1}{\sqrt{2\pi}} e^{i(-1)\phi} = \frac{1}{\sqrt{2\pi}} e^{-i\phi}$$

$$\therefore \; Y_1^{-1} = 2py = \frac{\sqrt{3}}{2} \sin \theta \cdot \frac{1}{\sqrt{2\pi}} e^{-i\phi} = \sqrt{\frac{3}{8\pi}} \, \sin \theta \, e^{-i\phi}$$

$$= \frac{\sqrt{3}}{2} \cdot \frac{1}{\sqrt{\pi}} \sin \theta \sin \phi \left[\frac{e^{-i\phi}}{\sqrt{2}} = \sin \phi \right]$$

Both $Y_1^1 (\theta)$ and $Y_1^{-1} (\theta)$ functions involve the imaginary quantity (i). This i can be removed by a linear combination of the two wave functions which is also a solution of the wave function.

The two resulting wave functions obtained by linear combination are

$$\frac{1}{\sqrt{2}}(Y_1^1 + Y_1^{-1}) = \frac{1}{\sqrt{2}}\left[\left(\frac{3}{4}\right)^{1/2} \sin\theta\,(2\pi)^{-\frac{1}{2}}\,e^{i\phi} + \left(\frac{3}{4}\right)^{1/2} \sin\theta\,(2\pi)^{-\frac{1}{2}}\,e^{i\phi}\right]$$

$$= \left(\frac{3}{4}\right)^{1/2} \sin\theta\cos\phi$$

$$= \text{constant term} \cdot \sin\theta\cos\phi$$

As $\sin\theta \cdot \cos\theta$ describes the angular dependence of the x component of r, the combination $\frac{1}{\sqrt{2}}(Y_1^1 + Y_1^{-1})$ is hence called the p_x orbital. Therefore

$$\left(\frac{3}{4}\right)^{1/2} \sin\theta\cos\phi = \text{constant} \cdot \sin\theta\cos\phi \sim p_x$$

Similarly,

$$\frac{1}{\sqrt{2}}(Y_1^1 - Y_1^{-1}) = \frac{1}{\sqrt{2}}\left(\frac{3}{4}\right)^{1/2} \cdot \frac{1}{(2\pi)^{1/2}} \sin\theta\,[e^{i\phi} - e^{-i\phi}]$$

$$= \frac{1}{2} \cdot \left(\frac{3}{4\pi}\right)^{1/2} \sin\theta \cdot 2i\sin\phi$$

$$= \left\{\left(\frac{3}{4\pi}\right)^{1/2} i\right\} \sin\theta\sin\phi$$

$$= \text{constant term} \cdot \sin\theta\sin\phi$$

Since $\sin\theta \cdot \sin\phi$ describes the angular dependence of the y component of r, the combination $\frac{1}{\sqrt{2}}(Y_1^1 - Y_1^{-1})$ is p_y orbital.

5.3 SHAPE OF d-ORBITAL

For d-orbitals, $l = 2$ and m (or m_l) = +2, +1, 0, -1, -2. By using Eq. (5.6) we can directly determine $Y_l^m(\theta, \phi)$ values.

For $l = 2$, m or $m_l = 0$, $\epsilon = 1$

$$Y_2^0 = 1 \cdot \sqrt{\frac{(2.2+1)}{4\pi} \cdot \frac{(2-|0|)!}{(2+|0|)!}}\, e^{i0\phi} \cdot P_2^0(\cos\theta)$$

$$= \sqrt{\frac{5}{16\pi}}(3\cos^2\theta - 1)\left[\because P_2^0(\cos\theta) = \frac{1}{2}(3\cos^2\theta - 1); \text{see Table 5.1}\right]$$

For $l = 2$, m (or m_l) = 1, $\epsilon = -1$

$$Y_2^1 = (-1)\sqrt{\frac{(2.2+1)}{4\pi} \cdot \frac{(2-1)!}{(2+1)!}}\, e^{i1\phi} \cdot P_2^1(\cos\theta)$$

$$= -\sqrt{\frac{15}{8\pi}} \sin\theta\cos\theta\, e^{i\phi}\left[P_2^1\cos\theta = 3\sin\theta \cdot \cos\theta\right]$$

For $l = 2$, m (or m_l) $= -1$, $\in = 1$

$$Y_2^{-1} = 1 \sqrt{\frac{(2.2+1)}{4\pi} \cdot \frac{(2-|-1|)!}{(2+|-1|)!}} \; e^{i(-1)\phi} \; P_2^{|-1|}(\cos\theta)$$

$$= \sqrt{\frac{15}{8\pi}} \sin\theta \cos\theta \, e^{-i\phi}$$

For $l = 2$, m (or m_l) $= 2$, $\in = (-1)^2 = 1$

$$Y_2^2 = (-1)^2 \sqrt{\frac{(2.2+1)}{4\pi} \cdot \frac{(2-|2|)!}{(2+|2|)!}} \; e^{i2\phi} \; P_2^2(\cos\theta)$$

$$= \sqrt{\frac{15}{32\pi}} \sin^2\theta \, e^{2i\phi} \left[P_2^2(\cos\theta) = 3\sin^2\theta \right]$$

For $l = 2$, m (or m_l) $= 2$, $\in = 1$

$$Y_2^{-2} = \sqrt{\frac{(2.2+1)}{4\pi} \cdot \frac{(2-|-2|)!}{(2+|-2|)!}} \; e^{i(-2)\phi} \; P_2^{|-2|}(\cos\theta)$$

$$= \sqrt{\frac{15}{32\pi}} \sin^2\theta \, e^{-2i\phi}$$

Thus, we have

$$3d^2 = Y_2^0 = \sqrt{\frac{5}{16\pi}} (3\cos^2\theta - 1) = 4^{-1} \cdot 5^{1/2} \cdot \pi^{-1/2} (3\cos^2\theta - 1)$$

$$3d_{xz} = Y_2^1 = \sqrt{\frac{15}{8\pi}} \sin\theta \cos\theta \, e^{i\phi}$$

$$= 2^{-3/2} \cdot 30^{1/2} \cdot \pi^{-1/2} \sin\theta \cos\theta \cos\phi$$

$$3d_{yz} = Y_2^{-1} = \sqrt{\frac{15}{8\pi}} \sin\theta \cos\theta \, e^{i\phi} = 2^{-3/2} \cdot 30^{1/2} \cdot \pi^{-1/2} \sin\theta \cos\theta \cos\phi$$

$$3d_{x^2-y^2} = Y_2^2 = \sqrt{\frac{15}{32\pi}} \sin^2\theta \, e^{2i\phi} = 4^{-1} \cdot 15^{1/2} \cdot \pi^{-1/2} \sin^2\theta \cos 2\phi$$

$$3d_{xy} = Y_2^{-2} = \sqrt{\frac{15}{32\pi}} \sin^2\theta \, e^{-2i\phi} = 4^{-1} \cdot 15^{1/2} \cdot \pi^{-1/2} \sin^2\theta \cos 2\phi$$

The angular wave functions $(H)_{l,\,m}$, $\Phi_m(\phi)$ and $Y_l^m(\theta, \phi)$ of hydrogen like atom are given separately in Table 5.3. Table 5.4 records the total wave functions for 1s, 2s, 2p, 3s, 3p and 3d-orbitals.

Table 5.3 The angular wave functions Y_l^m for (l = 0, 1, 2, 3, 4) hydrogen like atoms

$$Y_l^m(\theta, \phi) = Y_l^m(\theta)\,\Phi_m(\phi) = \Theta(\theta)\,\Phi(\phi)$$

$$Y_l^m(\theta) = \Theta^{(0)} = \Theta_{l,m}^{(0)} = \sqrt{\frac{(2l+1)}{2} \cdot \frac{(l-|m|)!}{(l+|m|)!}}\; P_l^{|m|}(\cos\theta) \qquad \Phi_m(\phi) = \frac{e^{im\phi}}{\sqrt{2\pi}}$$

Quantum numbers			Orbital symbol	Spherical harmonics Y_l^m	$Y_l^m(\theta)=\Theta^{(0)}=\Theta_{l,m}^{(0)}$	$\Phi_m(\phi)=\dfrac{e^{im\phi}}{\sqrt{2\pi}}$	$Y_l^m(\theta,\phi)=Y_l^m(\theta)\Phi_m(\phi)=\Theta(\theta)\Phi(\phi)$
n	l	m (or m_l)					
1	0	0	$1s$	Y_0^0	$\sqrt{\dfrac{1}{2}}$	$\sqrt{\dfrac{1}{2\pi}}$	$\sqrt{\dfrac{1}{4\pi}}$ or $(2\sqrt{\pi})^{-1}$ or $\dfrac{1}{2\sqrt{\pi}}$
2	0	0	$2s$	Y_0^0	$\sqrt{\dfrac{1}{2}}$	$\sqrt{\dfrac{1}{2\pi}}$	$\sqrt{\dfrac{1}{4\pi}}$ or $(2\sqrt{\pi})^{-1}$ or $\dfrac{1}{2\sqrt{\pi}}$
2	1	0	$2p_z$	Y_1^0	$\sqrt{\dfrac{3}{2}}\cos\theta$	$\sqrt{\dfrac{1}{2\pi}}$	$\sqrt{\dfrac{3}{4\pi}}\cos\theta$ or $2^{-1}\sqrt{3}\,\pi^{-1/2}\cos\theta$ or $\dfrac{\sqrt{3}}{2}\cdot\dfrac{\cos\theta}{\sqrt{\pi}}$
2	1	+1	$2p_x$	Y_1^1	$\dfrac{\sqrt{3}}{2}\sin\theta$	$\sqrt{\dfrac{1}{2\pi}}\,e^{i\phi}$	$-\sqrt{\dfrac{3}{8\pi}}\sin\theta\, e^{i\phi}$ or $-2^{-1}\sqrt{3}(\pi)^{-1/2}\sin\theta\cos\phi$ or $-\dfrac{\sqrt{3}}{2}\cdot\dfrac{\sin\theta\cos\phi}{\sqrt{\pi}}$
2	1	−1	$2p_y$	Y_1^{-1}	$\dfrac{\sqrt{3}}{2}\sin\theta$	$\sqrt{\dfrac{1}{2\pi}}\,e^{-i\phi}$	$\sqrt{\dfrac{3}{8\pi}}\sin\theta\, e^{-i\phi}$ or $2^{-1}\sqrt{3}\cdot(\pi)^{-1/2}\sin\theta\sin\phi$ or $\dfrac{\sqrt{3}}{2}\cdot\dfrac{\sin\theta\sin\phi}{\sqrt{\pi}}$
3	0	0	$3s$	Y_0^0	$\sqrt{\dfrac{1}{2}}$	$\sqrt{\dfrac{1}{2\pi}}$	$\sqrt{\dfrac{1}{4\pi}}$ or $(2\sqrt{\pi})^{-1}$ or $\dfrac{1}{2\sqrt{\pi}}$

(The orbitals $2p_x$ and $2p_y$ are grouped together as $2p$.)

Contd...

$$Y_l^m(\theta,\phi) = \Theta_{l,m}^{(\theta)} = \sqrt{\frac{(2l+1)}{2}\cdot\frac{(l-|m|)!}{(l+|m|)!}}\,P_l^{|m|}(\cos\theta) \qquad \Phi_m(\phi) = \frac{e^{im\phi}}{\sqrt{2\pi}} \qquad Y_l^m(\theta,\phi) = Y_l^m(\theta)\,\Phi_m(\phi) = \Theta(\theta)\,\Phi(\phi)$$

Quantum numbers			Orbital symbol	Spherical harmonics Y_l^m	$Y_l^m(\theta) = \Theta_{l,m}^{(\theta)}$	$\Phi_m(\phi) = \dfrac{e^{im\phi}}{\sqrt{2\pi}}$	$Y_l^m(\theta,\phi) = Y_l^m(\theta)\,\Phi_m(\phi) = \Theta(\theta)\,\Phi(\phi)$
n	l	$m\ (\text{or } m_l)$					
3	1	0	$3p_z$	Y_1^0	$\sqrt{\dfrac{3}{2}}\cos\theta$	$\sqrt{\dfrac{1}{2\pi}}$	$\sqrt{\dfrac{3}{4\pi}}\cos\theta$ or $2^{-1}\cdot 3^{1/2}\cdot\pi^{-1/2}\cos\theta$ or $\dfrac{\sqrt{3}}{2\sqrt{\pi}}\cos\theta$
3	1	1	$3p_x$	Y_1^1	$\dfrac{\sqrt{3}}{2}\sin\theta$	$\sqrt{\dfrac{1}{2\pi}}\,e^{i\phi}$	$-\sqrt{\dfrac{3}{8\pi}}\sin\theta\,e^{i\phi}$ or $-2^{-1}\cdot\sqrt{3}\pi^{-1/2}\sin\theta\cos\phi$ or $-\dfrac{\sqrt{3}}{2\sqrt{\pi}}\sin\theta\cos\phi$
3	1	-1	$3p_y$	Y_1^{-1}	$\dfrac{\sqrt{3}}{2}\sin\theta$	$\sqrt{\dfrac{1}{2\pi}}\,e^{-i\phi}$	$\sqrt{\dfrac{3}{8\pi}}\sin\theta\,e^{-i\phi}$ or $2^{-1}\cdot\sqrt{3}(\pi)^{-1/2}\sin\theta\cos\phi$ or $\dfrac{\sqrt{3}}{2\sqrt{\pi}}\sin\theta\sin\phi$
3	2	0	$3d_{z^2}$	Y_2^0	$\dfrac{\sqrt{5}}{2}\cdot\dfrac{1}{2}(3\cos^2\theta-1)$	$\sqrt{\dfrac{1}{2\pi}}$	$\sqrt{\dfrac{5}{16\pi}}(3\cos^2\theta-1)$
3	2	2	$3d_{x^2-y^2}$	Y_2^2	$\dfrac{1}{4}\sqrt{15}\sin^2\theta$	$\sqrt{\dfrac{1}{2\pi}}\,e^{2i\phi}$	$\sqrt{\dfrac{15}{32\pi}}\sin^2\theta\,e^{2i\phi}$ or $4^{-1}\cdot15^{1/2}\cdot\pi^{-1/2}\sin^2\theta\cos2\phi$
3	2	-2	$3d_{xy}$	Y_2^{-2}	$\dfrac{1}{4}\sqrt{15}\sin^2\theta$	$\sqrt{\dfrac{1}{2\pi}}\,e^{-2i\phi}$	$\sqrt{\dfrac{15}{32\pi}}\sin^2\theta\,e^{-2i\phi}$ or $4^{-1}\cdot15^{1/2}\cdot\pi^{-1/2}\sin^2\theta\sin2\phi$
3	2	1	$3d_{xz}$	Y_2^1	$\dfrac{1}{2}\sqrt{15}\sin\theta\cos\theta$	$\sqrt{\dfrac{1}{2\pi}}\,e^{i\phi}$	$-\sqrt{\dfrac{15}{8\pi}}\sin\theta\cos\theta\,e^{i\phi}$ or $-2^{-3/2}\cdot30^{1/2}\cdot\pi^{-1/2}\sin\theta\cos\theta\cos\phi$
3	2	-1	$3d_{yz}$	Y_2^{-1}	$\dfrac{1}{2}\sqrt{15}\sin\theta\cos\theta$	$\sqrt{\dfrac{1}{2\pi}}\,e^{-i\phi}$	$\sqrt{\dfrac{15}{8\pi}}\sin\theta\cos\theta\,e^{-i\phi}$ or $2^{-3/2}\cdot30^{1/2}\cdot\pi^{-1/2}\sin\theta\cos\theta\sin\phi$

(Orbital symbols $3p$ and $3d$ group the respective rows.)

Table 5.4 Orbitals of a hydrogen-like atom

Quantum number			Orbital symbols	Wave functions		
n	l	m (or m_l)		Radial part of the wave function	Angular part of the wave function	Total wave function = Radial part of the wave function × angular part of the wave function
1	0	0	$1s$	$2\left(\dfrac{Z}{a_0}\right)^{3/2} e^{-Zr/a_0}$	$\dfrac{1}{2\sqrt{\pi}}$	$\psi_{1s} = \dfrac{1}{\sqrt{\pi}}\left(\dfrac{4\pi^2 mZe^2}{h^2}\right)^{3/2} e^{-Zr/a_0}$
2	0	0	$2s$	$\dfrac{1}{2\sqrt{2}}\left(\dfrac{Z}{a_0}\right)^{3/2}\left(2-\dfrac{Zr}{a_0}\right)e^{-Zr/2a_0}$	$\dfrac{1}{2\sqrt{\pi}}$	$\psi_{2s} = \dfrac{1}{4\sqrt{2\pi}}\left(\dfrac{4\pi^2 mZe^2}{h^2}\right)^{3/2}\left(2-\dfrac{Zr}{a_0}\right)e^{-Zr/a_0}$
2	1	0	$2p_z$	$\dfrac{1}{2\sqrt{6}}\left(\dfrac{Z}{a_0}\right)^{3/2}\left(\dfrac{Zr}{a_0}\right)e^{-Zr/2a_0}$	$\dfrac{\sqrt{3}}{2}\cdot\dfrac{\cos\theta}{\sqrt{\pi}}$	$\psi_{2p_x} = \dfrac{1}{4\sqrt{2\pi}}\left(\dfrac{4\pi^2 mZe^2}{h^2}\right)^{3/2}\left(\dfrac{Zr}{a_0}\right)e^{-Zr/2a_0}\cos\theta$
2	1	1	$2p_x$	$\dfrac{1}{2\sqrt{6}}\left(\dfrac{Z}{a_0}\right)^{3/2}\left(\dfrac{Zr}{a_0}\right)e^{-Zr/2a_0}$	$\dfrac{\sqrt{3}}{2}\cdot\dfrac{\sin\theta\cos\phi}{\sqrt{\pi}}$	$\psi_{2p_x} = \dfrac{1}{4\sqrt{2\pi}}\left(\dfrac{4\pi^2 mZe^2}{h^2}\right)^{3/2}\left(\dfrac{Zr}{a_0}\right)e^{-Zr/2a_0}\sin\theta\cos\phi$
2	1	–1	$2p_y$	$\dfrac{1}{2\sqrt{6}}\left(\dfrac{Z}{a_0}\right)^{3/2}\left(\dfrac{Zr}{a_0}\right)e^{-Zr/2a_0}$	$\dfrac{\sqrt{3}}{2}\cdot\dfrac{\sin\theta\sin\phi}{\sqrt{\pi}}$	$\psi_{2p_y} = \dfrac{1}{4\sqrt{2\pi}}\left(\dfrac{4\pi^2 Ze^2}{h^2}\right)^{3/2}\left(\dfrac{Zr}{a_0}\right)e^{-Zr/2a_0}\sin\theta\sin\phi$
3	0	0	$3s$	$\dfrac{2}{81}\dfrac{1}{\sqrt{3}}\left(\dfrac{Z}{a_0}\right)^{3/2}$ $\times\left(27-\dfrac{18Zr}{a_0}+\dfrac{2Z^2 r^2}{a_0^2}\right)e^{-Zr/3a_0}$	$\dfrac{1}{2\sqrt{\pi}}$	$\psi_{3s} = \dfrac{2}{81\sqrt{3\pi}}\left(\dfrac{4\pi^2 mZe^2}{h^2}\right)^{3/2}\left(27-18\dfrac{Zr}{a_0}+2\dfrac{Z^2 r^2}{a_0^2}\right)e^{-Zr/3a_0}$

Contd...

Table 5.4 Orbitals of a hydrogen-like atom (Contd...)

Quantum number			Orbital symbols	Wave functions		
n	l	m (or m_l)		Radial part of the wave function	Angular part of the wave function	Total wave function = Radial part of the wave function × angular part of the wave function
3	1	0	$3p_z$	$\frac{4}{81\sqrt{6}}\left(\frac{Z}{a_0}\right)^{3/2}\left(6\frac{Zr}{a_0}-\frac{Z^2r^2}{a_0^2}\right)e^{-Zr/3a_0}$	$\frac{\sqrt{3}}{2\sqrt{\pi}}\cdot\cos\theta$	$\psi_{3p_z}=\frac{2}{81\sqrt{\pi}}\left(\frac{4\pi^2mZe^2}{h^2}\right)^{3/2}\left(6\frac{Zr}{a_0}-\frac{Z^2r^2}{a_0^2}\right)e^{-Zr/3a_0}\cos\theta$
3	1	1	$3p_x$	$\frac{4}{81\sqrt{6}}\left(\frac{Z}{a_0}\right)^{3/2}\left(6\frac{Zr}{a_0}-\frac{Z^2r^2}{a_0^2}\right)e^{-Zr/3a_0}$	$\frac{\sqrt{3}}{2\sqrt{\pi}}\sin\theta\cos\phi$	$\psi_{3p_x}=\frac{2}{81\sqrt{\pi}}\left(\frac{4\pi^2mZe^2}{h^2}\right)^{3/2}\left(6\frac{Zr}{a_0}-\frac{Z^2r^2}{a_0^2}\right)e^{-Zr/3a_0}\sin\theta\cos\phi$
3	1	-1	$3p_y$	$\frac{4}{81\sqrt{6}}\left(\frac{Z}{a_0}\right)^{3/2}\left(6\frac{Zr}{a_0}-\frac{Z^2r^2}{a_0^2}\right)e^{-Zr/3a_0}$	$\frac{\sqrt{3}}{2\sqrt{\pi}}\sin\theta\sin\phi$	$\psi_{3p_y}=\frac{2}{81\sqrt{\pi}}\left(\frac{4\pi^2mZe^2}{h^2}\right)^{3/2}\left(6\frac{Zr}{a_0}-\frac{Z^2r^2}{a_0^2}\right)e^{-Zr/3a_0}\sin\theta\cos\phi$
3	2	0	$3d_{z^2}$	$\frac{4}{81\sqrt{30}}\left(\frac{Z}{a_0}\right)^{3/2}\left(\frac{Zr}{a_0}\right)^2 e^{-Zr/3a_0}$	$\frac{\sqrt{5}}{4\sqrt{\pi}}(3\cos^2\theta-1)$	$\psi_{3d_{z^2}}=\frac{1}{81\sqrt{6\pi}}\left(\frac{4\pi^2mZe^2}{h^2}\right)^{3/2}\frac{Z^2r^2}{a_0^2}e^{-Zr/3a_0}(3\cos^2\theta-1)$
3	2	1	$3d_{xz}$	$\frac{4}{81\sqrt{30}}\left(\frac{Z}{a_0}\right)^{3/2}\left(\frac{Zr}{a_0}\right)^3 e^{-Zr/3a_0}$	$\frac{\sqrt{30}}{2\sqrt{2}}\cdot\frac{1}{\sqrt{\pi}}\sin\theta\cos\theta\cos\phi$	$\psi_{3d_{xz}}=\frac{\sqrt{2}}{81\sqrt{\pi}}\left(\frac{4\pi^2mZe^2}{h^2}\right)^{3/2}\frac{Z^2r^2}{a_0^2}e^{-Zr/3a_0}\sin\theta\cos\theta\cos\phi$
3	2	-1	$3d_{yz}$	$\frac{4}{81\sqrt{30}}\left(\frac{Z}{a_0}\right)^{3/2}\left(\frac{Zr}{a_0}\right)^2 e^{-Zr/3a_0}$	$\frac{\sqrt{30}}{2\sqrt{2}}\cdot\frac{1}{\sqrt{\pi}}\sin\theta\cos\theta\sin\phi$	$\psi_{3d_{yz}}=\frac{\sqrt{2}}{81\sqrt{\pi}}\left(\frac{4\pi^2mZe^2}{h^2}\right)^{3/2}\frac{Z^2r^2}{a_0^2}e^{-Zr/3a_0}\sin\theta\cos\theta\sin\phi$

Contd...

Table 5.4 Orbitals of a hydrogen-like atom (Contd...)

Quantum number			Orbital symbols	Wave functions		
n	l	m (or m_l)		Radial part of the wave function	Angular part of the wave function	Total wave function = Radial part of the wave function × angular part of the wave function
3	2	2	$3d_{x^2-y^2}$	$\dfrac{4}{81\sqrt{30}}\left(\dfrac{Z}{a_0}\right)^{3/2}\left(\dfrac{Zr}{a_0}\right)^2 e^{-Zr/3a_0}$	$\dfrac{\sqrt{15}}{4}\cdot\dfrac{1}{\sqrt{\pi}}\cdot\sin^2\theta\cos 2\phi$	$\psi_{3d_{x^2-y^2}}=\dfrac{1}{81\sqrt{2\pi}}\left(\dfrac{4\pi^2 mZe^2}{h^2}\right)^{3/2}\dfrac{Z^2 r^2}{a_0^2}e^{-Zr/3a_0}\sin^2\theta\cos 2\phi$
3	2	−2	$3d_{xy}$	$\dfrac{4}{81\sqrt{30}}\left(\dfrac{Z}{a_0}\right)^{3/2}\left(\dfrac{Zr}{a_0}\right)^2 e^{-Zr/3a_0}$	$\dfrac{\sqrt{15}}{4}\cdot\dfrac{1}{\sqrt{\pi}}\cdot\sin^2\theta\sin 2\phi$	$\psi_{3d_{xy}}=\dfrac{1}{81\sqrt{2\pi}}\left(\dfrac{4\pi^2 mZe^2}{h^2}\right)^{3/2}\dfrac{Z^2 r^2}{a_0^2}e^{-Zr/3a_0}\sin^2\theta\sin 2\phi$

$Z = 1$ for hydrogen; $a_0 = \dfrac{h^2}{4\pi^2 mZe^2}$ = Polar radians = 0.529 Å

Since Y_2^2 and Y_2^{-2} are complex, so correct linear combinations of the imaginary functions are of real form and are commonly used by the chemists in their description of d-orbitals.

Linear combination of Y_2^2 and Y_2^{-2} gives

$$\frac{1}{\sqrt{2}}(Y_2^2 + Y_2^{-2}) = \frac{1}{\sqrt{2}} \cdot \sqrt{\frac{15}{16}} \cdot \sqrt{\frac{1}{2\pi}} \sin^2\theta \, (e^{2i\phi} + e^{-2i\phi})$$

$$= \sqrt{\frac{15}{16\pi}} \sin^2\theta \cos 2\phi$$

$$= \sqrt{\frac{15}{16\pi}} \sin^2\theta \, (\cos^2\phi - \sin^2\phi)$$

$$= \sqrt{\frac{15}{16\pi}} \, r^2 \sin^2\theta \, \frac{(\cos^2\phi - \sin^2\phi)}{r^2} \cdot$$

$$= \sqrt{\frac{15}{16\pi}} \cdot \frac{(x^2 - y^2)}{r^2} \qquad \left[\begin{array}{l} x = r\sin\theta\cos\phi \\ y = r\sin\theta\sin\phi \\ x^2 - y^2 = r^2\sin^2\phi\,(\cos^2\phi - \sin^2\phi) \end{array} \right]$$

$$\sim d_{x^2-y^2}$$

The d_{z^2} orbital can be resolved into a linear combination of two orbitals, $d_{z^2-x^2}$ and $d_{z^2-y^2}$, each of which is equivalent to the $d_{x^2-y^2}$ orbital.

$$d_{z^2} = d_{z^2-x^2} + d_{z^2-y^2} = d_{2z^2-x^2-y^2}$$

The trigonometric function of $d_{2z^2-x^2-y^2}$ is $3\cos^2\theta - 1$

$$3\cos^2\theta - 1 = 3\cos^2\theta - (\sin^2\theta + \cos^2\theta) = 2\cos^2\theta - \sin^2\theta$$

$$= 2\cos^2\theta - \sin^2\theta \, (\cos^2\phi + \sin^2\phi)$$

$$= 2\cos^2\theta - \sin^2\theta\cos^2\phi - \sin^2\theta\sin^2\phi$$

$$= 2\left(\frac{z}{r}\right)^2 - \left(\frac{x}{r}\right)^2 - \left(\frac{y}{r}\right)^2 \qquad \left[\begin{array}{l} \cos\theta = \dfrac{z}{r} \\[6pt] \sin\theta\cos\phi = \dfrac{x}{r} \\[6pt] \sin\theta\sin\phi = \dfrac{y}{r} \end{array} \right]$$

$$= \frac{1}{r^2}(2z^2 - x^2 - y^2)$$

Thus $d_{z^2} = d_{2z^2-x^2-y^2} = \dfrac{\sqrt{5}}{4\sqrt{\pi}}(3\cos^2\theta - 1)$

$$= \sqrt{\frac{5}{8}}(3\cos^2\theta - 1)\cdot(2\pi)^{-\frac{1}{2}}$$

It is to be noted that in the z direction (where $x = y = 0$), the expression $2z^2 - x^2 - y^2$ has the value $2z^2$, and so it is positive in both the $\pm z$ direction. For this direction $(z = 0, y = 0)$, the expression reduces to $-x^2$ and so is negative for both $\pm x$ directions and, more generally, for the xy plane. Similarly,

$$\frac{1}{i\sqrt{2}}(Y_2^2 - Y_2^{-2}) = \frac{1}{i\sqrt{2}}\sqrt{\frac{15}{16}} \cdot \sqrt{\frac{1}{2\pi}} \sin^2\theta\,(e^{2i\phi} - e^{-2i\phi})$$

$$= \frac{1}{i\sqrt{2}} \sqrt{\frac{15}{32\pi}} \sin^2 \theta \, 2i \sin 2\phi$$

$$= \frac{\sqrt{15}}{8\sqrt{\pi}} \cdot 2 \sin^2 \theta \sin 2\phi$$

$$= \frac{\sqrt{15}}{\sqrt{16\pi}} \cdot \frac{(r \sin \theta \cos \phi)(r \sin \theta \sin \phi)}{r^2}$$

$$= \frac{\sqrt{15}}{\sqrt{16\pi} \, r^2} (r \sin \theta \cos \phi)(r \sin \theta \sin \phi)$$

$$\sim d_{xy}$$

Linear combinations of Y^1_2 and Y^{-1}_2 give the following results

$$\frac{1}{\sqrt{2}}(Y^1_2 + Y^{-1}_2) = \frac{1}{\sqrt{2}} \sqrt{\frac{15}{4}} \cdot \sqrt{\frac{1}{2\pi}} \sin \theta \cos \theta (e^{i\phi} + e^{-i\phi})$$

$$= \frac{1}{\sqrt{2}} \sqrt{\frac{15}{8\pi}} \sin \theta \cos \theta \, 2 \cos \phi$$

$$= \frac{\sqrt{15}}{2\sqrt{\pi}} \cos \theta \sin \theta \cos \phi$$

$$= \frac{\sqrt{15}}{2\sqrt{\pi}} \frac{(r \cos \theta)(r \sin \theta \cos \phi)}{r^2}$$

$$= \left(\frac{\sqrt{15}}{2\sqrt{\pi} r^2} \right) d_{xz}$$

$$\sim d_{xz}$$

$$\frac{1}{i\sqrt{2}}(Y^1_2 - Y^{-1}_2) = \frac{1}{i\sqrt{2}} \cdot \sqrt{\frac{15}{4}} \cdot \sqrt{\frac{1}{2\pi}} \sin \theta \cos \theta (e^{i\phi} - e^{-i\phi})$$

$$= \frac{\sqrt{15}}{\sqrt{4\pi}} \cdot \frac{(r \cos \theta)(r \sin \theta \sin \phi)}{r^2} = \left(\sqrt{\frac{15}{4\pi}} \cdot \frac{1}{r^2} \right) d_{yz} \sim d_{yz}$$

Finally, it is necessary to make sure that the new orbital is normalised. To normalise a wave function ψ, a new wave function of the following is formed.

$$\psi \Big/ \left(\int \psi^* \, \psi \, d\tau \right)^{1/2}$$

For example, let $\psi = [(2) + (-2)] \quad \therefore \psi^* = [(2) + (-2)]^*$

$$\therefore \quad \int \psi^* \, \psi \, d\tau = \int [(2) + (-2)]^* (2) + (-2)] \, d\tau$$

$$= \int [(2)^* (2) + (2)^* (-2) + (-2)^* (2) + (-2)^* (-2)] \, d\tau$$

$$= 1 + 0 + 0 + 1 = 2$$

Hence the real function, which is called $d_{x^2-y^2}$, is

$$d_{x^2-y^2} = [(2)+(-2)] \Big/ \left\{ \int [(2)+(-2)]^* [(2)+(-2)]\, d\tau \right\}^{1/2}$$

$$= 2^{-\frac{1}{2}} [(2)+(-2)]$$

By making other suitable combinations the five d wave functions in their real forms are obtained as

$$d_{z^2} = 0$$

$$d_{yz} = 2^{-\frac{1}{2}} [(1)-(-1)]$$

$$d_{xz} = 2^{-\frac{1}{2}} [(1)+(-1)]$$

$$d_{xy} = 2^{-\frac{1}{2}} [(2)-(-2)]$$

$$d_{x^2-y^2} = 2^{-\frac{1}{2}} [(2)+(-2)]$$

A wave function ψ is said to be normalised if $\int \psi^* \psi\, d\tau = 1$.

For example,

$$\int d_{yz}^* d_{yz}\, d\tau = \int 2^{-\frac{1}{2}}[(1)-(-1)]^* \, 2^{-\frac{1}{2}} [(1)-(-1)]\, d\tau$$

$$= 2^{-1} \int [(1)^* (1) - (1)^* (-1) - (-1)^* (1) + (-1)^* (-1)]\, d\tau$$

$$= 2^{-1} [1-0-0+1] = 2^{-1} \cdot 2 = 1$$

Similarly,

$$\int d_{xy}^* d_{xy}\, d\tau = \int 2^{-\frac{1}{2}} [(2)-(-2)]^* \, 2^{-\frac{1}{2}}[(2)-(-2)]\, d\tau$$

$$= 2^{-1} \int [(2)^*(2) - (2)^*(-2) - (-2)^*(2) + (-2)^*(-2)]\, d\tau$$

$$= 2^{-1} [1-0-0+1] = 1$$

Again two different wave functions ψ_1 and ψ_2, are said to be orthogonal if

$$\int \psi_1^* \psi_2\, d\tau = 0$$

For example,

$$\int d_{xy}^* d_{x^2-y^2}\, d\tau = \int 2^{-\frac{1}{2}} [(2)-(-2)]^* \, 2^{-\frac{1}{2}} [(2)+(-2)]\, d\tau$$

$$= 2^{-1} \int [(2)^*(2) + (2)^*(-2) - (-2)^*(2) - (-2)^*(-2)]\, d\tau$$

$$= 2^{-1} [1+0-0-1] = 0$$

5.4 APPLICATION OF VARIOUS SYMMETRY OPERATIONS ON DIFFERENT ATOMIC ORBITALS

We shall carry out symmetry operations of the group on each orbital and thus determine the characters and the representation.

5.4.1 *s*-orbital

s-orbital is spherical. All operations such as E, C_{2z}, C_{2x}, C_{2y}, i, σ_{xz}, σ_{yz}, σ_{zx} done on this orbital leave it unchanged, i.e. no change of character (+1) (Fig. 5.3).

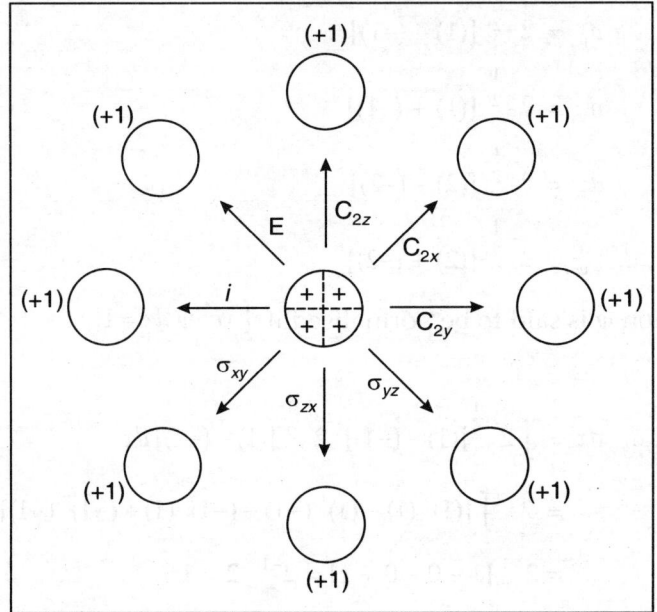

Fig. 5.3 The operations of the group D_{2h} upon the *s*-orbital

$$
\begin{array}{ccccccccc}
E & C_{2z} & \Gamma & C_{2y} & C_{2x} & i & \sigma(xy) & \sigma(yz) & \sigma(xz) \\
| & | & | & | & | & | & | & | & | \quad - A_{1g}
\end{array}
$$

This corresponds to transformation properties of A_{1g} irreducible representation. The transformation properties of the orbtials can be better appreciated when considered with respect to the operations of a particular group. We shall perform the symmetry operations E, C_2, $\sigma_v(xz)$, $\sigma_v(yz)$ on the atomic orbitals and see the effect if any. If the character remains unchanged it is assigned +1 and if it is changed then the character is assigned to –1.

As seen in Fig. 5.3, *s*-orbital remains unchanged on performing all the symmetry operations. We take C_{2v} operations

$$
\begin{array}{cccc}
E & C_2 & \sigma_v(xz) & \sigma_v(yz) \\
| & | & | & |
\end{array}
$$

The transformation properties correspond to A_1 or A_{1g} irreducible representation.

5.4.2 *p*-orbital

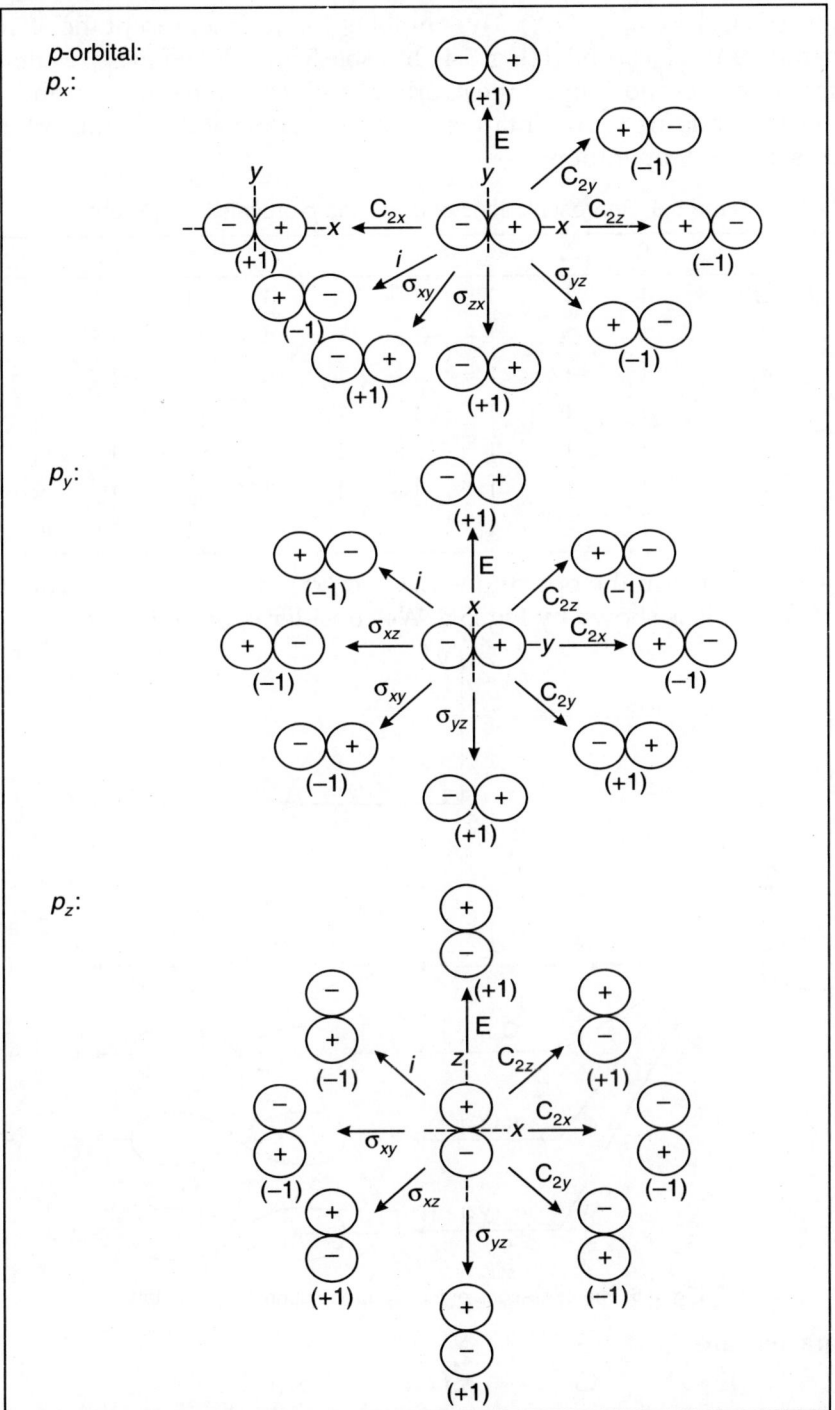

Fig. 5.4 The operation of the group D_{2h} upon the p_x, p_y and p_z orbitals

Let us consider p_x orbital. The identity operation (E) leaves the p_x orbital unchanged. Thus $E(p_x) = p_x = 1(p_x)$. On the other hand, the C_{2z} operation changes the sign of the p_x orbital, hence $C_{2z}(p_x) = -p_x = -1(p_x)$. The numbers 1 and -1 represent the effect of the E and C_{2z} operations upon p_x orbital (Fig. 5.4). In Table 5.5, are listed these results and those for the remaining operations upon the p_x orbital and upon the other s, p and d-orbitals. In fact these numerals are representative of the behaviour of the atomic orbitals under the various symmetry operations.

Table 5.5 Representations of d-, s- and p-orbitals in D_{2h} group

	E	C_{2z}	C_{2y}	C_{2x}	i	σ_{xy}	σ_{xz}	σ_{yz}	Γ
$d_{x^2-y^2}, d_{z^2}, S$	1	1	1	1	1	1	1	1	a_g
d_{xy}	1	-1	1	-1	1	1	-1	-1	b_{1g}
d_{xz}	1	-1	1	-1	1	-1	1	-1	b_{2g}
d_{yz}	1	-1	-1	1	1	-1	-1	1	b_{3g}
p_x	1	-1	-1	1	-1	1	1	-1	b_{3u}
p_y	1	-1	1	-1	-1	1	-1	1	b_{2u}
p_z	1	1	-1	-1	-1	-1	1	1	b_{1u}

We shall now perform the operations E, C_2, $\sigma_v(xz)$, $\sigma_v(yz)$ on the p-orbitals and see the effect if any. This is shown by Fig. 5.5. We consider p_x orbital.

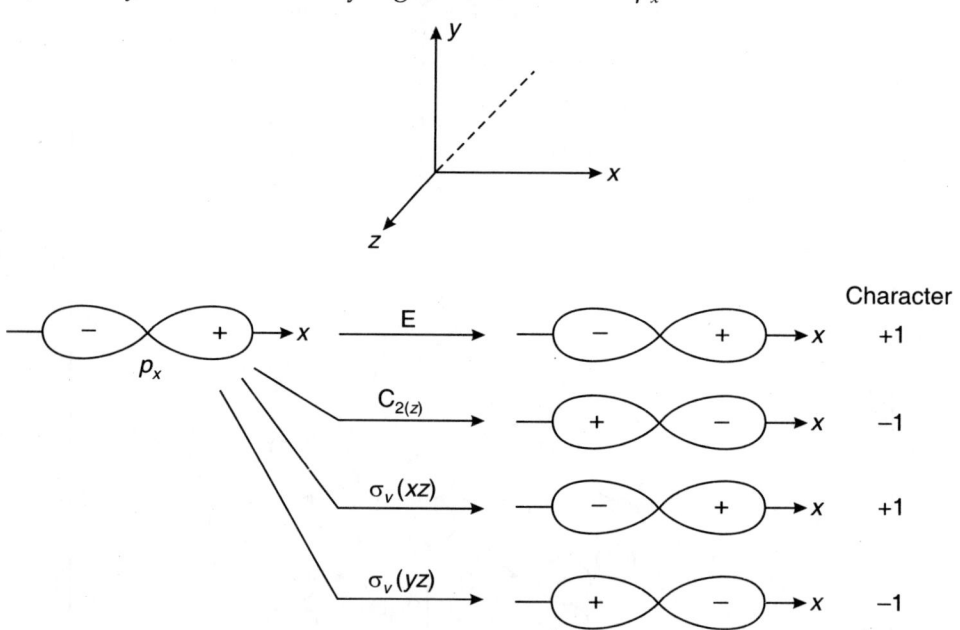

Fig. 5.5 The operations of the C_{2v} group upon the p_x orbital

The characters are

E	C_2	$\sigma_v(xz)$	$\sigma_v(yz)$
1	-1	1	-1

This corresponds to transformation properties of B_1 irreducible representation.

Similarly it can be shown that the p_y and p_z orbitals have B_2 and A_1 irreducible representations.

	E	C_2	$\sigma_v(xz)$	$\sigma_v(yz)$	
p_y	1	-1	-1	1	B_2
p_z	1	1	1	1	A_1

Thus the p-orbitals (p_x, p_y, p_z) which belong to a three-dimensional irreducible representation in free atom to be splitted up into three separate one-dimensional irreducible representations (B_1, B_2, A_1) in the C_{2v} point group. Total character can be determined by adding all the characters of E, C_2, $\sigma_v(xz)$, $\sigma_v(yz)$ columnwise in C_{2v} point group.

	E	C_2	$\sigma_v(xz)$	$\sigma_v(yz)$
p_x	1	-1	1	-1
p_y	1	-1	-1	1
p_z	1	1	1	1
Total character, $\Gamma =$	3	-1	1	1

Worked out in the following way:

$$\begin{Bmatrix} p_x \\ p_y \\ p_z \end{Bmatrix} \xrightarrow{\text{E}} \begin{Bmatrix} p_x \\ p_y \\ p_z \end{Bmatrix}, \qquad E = \begin{pmatrix} 1 & 0 & 0 \\ 0 & 1 & 0 \\ 0 & 0 & 1 \end{pmatrix} \qquad \text{Character} = 3$$

$$\xrightarrow{C_2(z)} \begin{Bmatrix} -p_x \\ -p_y \\ p_z \end{Bmatrix}, \qquad C_2(z) = \begin{pmatrix} -1 & 0 & 0 \\ 0 & -1 & 0 \\ 0 & 0 & 1 \end{pmatrix} \qquad \text{Character} = -1$$

$$\xrightarrow{\sigma_v(xz)} \begin{Bmatrix} p_x \\ -p_y \\ p_z \end{Bmatrix}, \qquad \sigma_v(xz) = \begin{pmatrix} 1 & 0 & 0 \\ 0 & -1 & 0 \\ 0 & 0 & 1 \end{pmatrix} \qquad \text{Character} = 1$$

$$\xrightarrow{\sigma_v(yz)} \begin{Bmatrix} -p_x \\ p_y \\ p_z \end{Bmatrix}, \qquad \sigma_v(yz) = \begin{pmatrix} -1 & 0 & 0 \\ 0 & 1 & 0 \\ 0 & 0 & 1 \end{pmatrix} \qquad \text{Character} = 1$$

	E	$C_2(z)$	$\sigma_v(xz)$	$\sigma_v(yz)$
$\therefore$ Total character, $\Gamma =$	3	-1	1	1

By using reduction formula it can be shown

$$\Gamma = A_1 + B_1 + B_2$$

Thus p-orbitals belong to three different irreducible representations in C_{2v} point group.

5.4.3. d-orbitals

The transformation of the $d_{x^2 - y^2}$ orbital by the operations of the point group D_{2h} is shown below (Fig. 5.6).

Identical characters are observed for $3d_{z^2}$ orbital (Table 5.5). For d_{xy} orbital we have the following observations (Fig. 5.7).

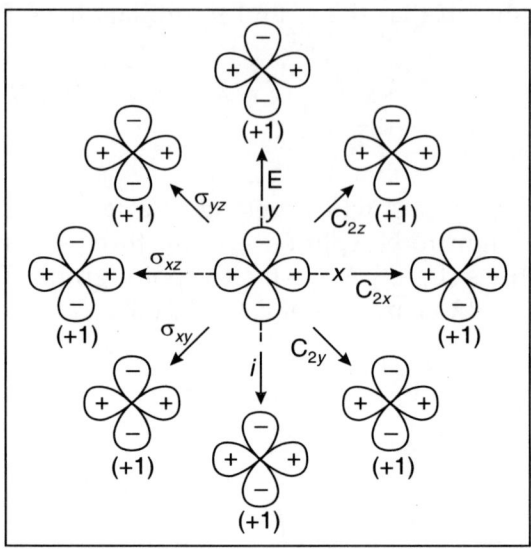

Fig. 5.6 Transformation of the $d_{x^2-y^2}$ orbital by the operations of the point group D_{2h}

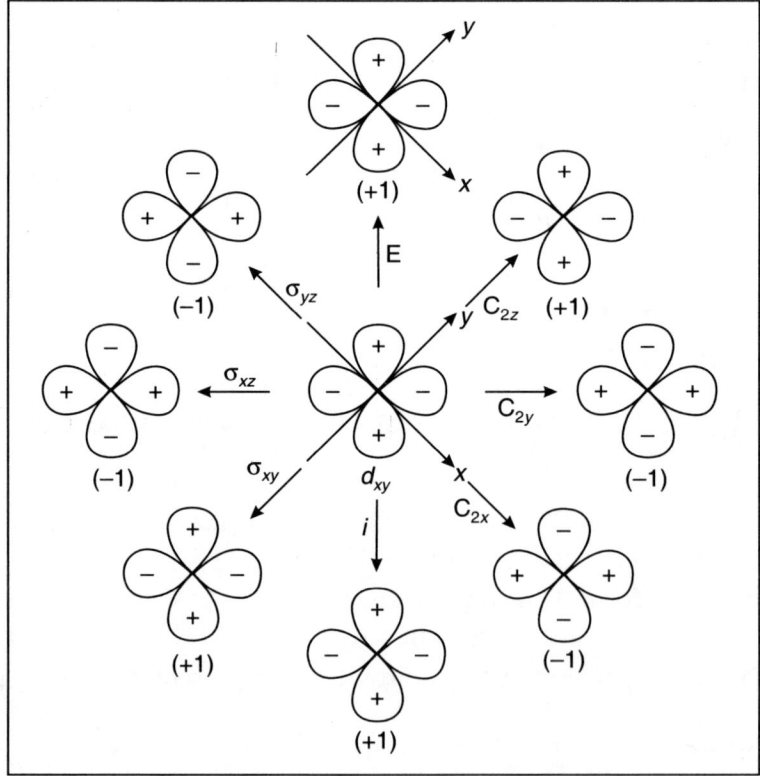

Fig. 5.7 The operations of the group D_{2h} upon the d_{xy} orbital

By comparison of these lines of numbers, or characters, with those in the character table for the D_{2h} group we can immediately give a symmetry label to each of the atomic orbitals, thus:

$$d_{x^2-y^2}, d_{z^2} \text{ and } s \qquad \text{belongs to } a_g$$
$$d_{xy} \qquad\qquad\qquad \text{belongs to } b_{1g}$$
$$d_{xz} \qquad\qquad\qquad \text{belongs to } b_{2g}$$
$$d_{yz} \qquad\qquad\qquad \text{belongs to } b_{3g}$$
$$p_z \qquad\qquad\qquad \text{belongs to } b_{1u}$$
$$p_y \qquad\qquad\qquad \text{belongs to } b_{2u}$$
$$\text{and} \quad p_x \qquad\qquad\qquad \text{belongs to } b_{3u}$$

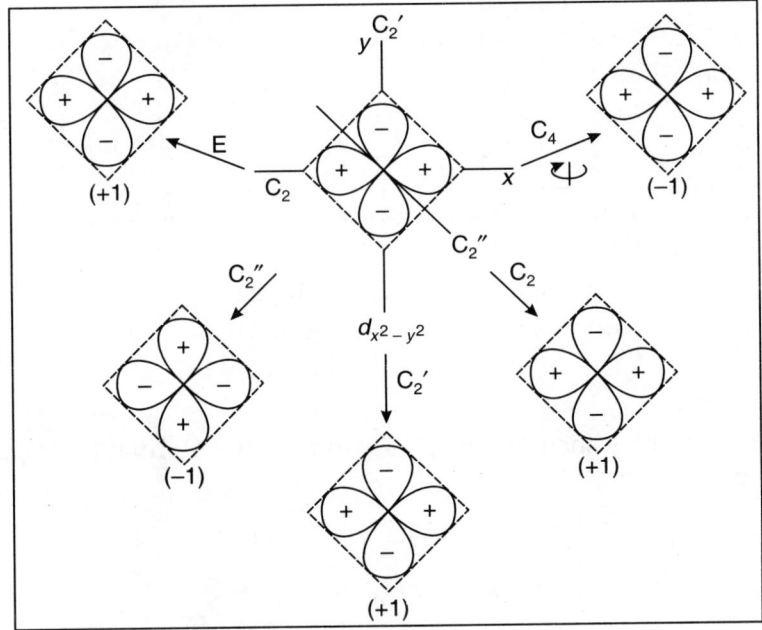

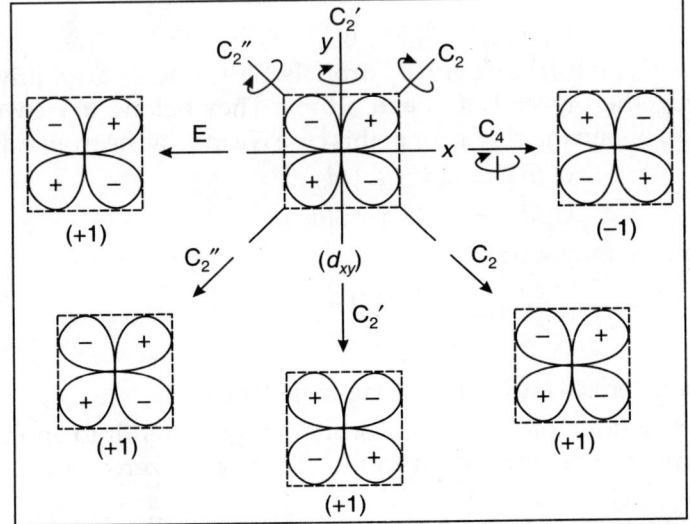

Fig. 5.8 The operations of the group D_{4h} upon the $d_{x^2-y^2}$ and d_{xy} orbitals

When single electron wave functions are considered, small letters are used to describe the symmetry representations, whilst capital letters are used for the total electronic wave function of the atom or molecule.

We shall now consider the square planar geometry. The molecule belongs to the group D_{4h}. For the moment D_4 symmetry is used to see how the orbitals behave under the following symmetry operation.

Let us consider $d_{x^2-y^2}$ and d_{xy} orbitals separately (Fig. 5.8).

Thus we see

	E	C_4	C_2	C_2'	C_2''	
s	1	1	1	1	1	a_1
p_z	1	1	1	−1	−1	a_2
d_{z^2}	1	1	1	1	1	a_1
d_{xy}	1	−1	1	−1	1	b_2
$d_{x^2-y^2}$	1	−1	1	1	−1	b_1

Since the s and d-orbitals are even and p-orbitals are odd wrt inversion, we conclude that in D_{4h} symmetry

$$
\begin{array}{ll}
s \text{ and } d_{z^2} & \text{belong to } a_{1g} \\
d_{x^2-y^2} & \text{belong to } b_{1g} \\
d_{xy} & \text{belong to } b_{2g} \\
\text{and} \quad p_z & \text{belong to } a_{2u}
\end{array}
$$

So far we have not yet included p_x, p_y, d_{xz} and d_{yz}. If a C_4 rotation is carried out on d_{xy}, we get as before

$$C_4(d_{xy}) = -d_{xy} \qquad \text{...(5.7)}$$

but with d_{xz}

$$C_4(d_{xz}) = d_{yz} = 1(d_{yz}) \qquad \text{...(5.8)}$$

and with d_{yz}

$$C_4(d_{yz}) = -d_{xz} = -(d_{xz}) \qquad \text{...(5.9)}$$

The operations C_4 mixes the d_{xz} and d_{yz} orbitals. So d_{xz} and d_{yz} are equivalent and must be considered together instead of one at a time. They belong to e (two-dimensional) representation. To obtain the character in this case we may write from Eqs (5.8) and (5.9):

$$
\begin{aligned}
C_4(d_{xz}) &= 0(d_{xz}) + 1(d_{yz}) \\
C_4(d_{yz}) &= -1(d_{xz}) + 0(d_{yz})
\end{aligned} \qquad \text{...(5.10)}
$$

In matrix form we may write

$$C_4 \begin{pmatrix} d_{xz} \\ d_{yz} \end{pmatrix} = \begin{pmatrix} 0 & 1 \\ -1 & 0 \end{pmatrix} \begin{pmatrix} d_{xz} \\ d_{yz} \end{pmatrix} \qquad \text{...(5.11)}$$

Therefore character is 0 (sum of the diagonal elements).

It is important to note that when one orbital is converted into another orbital by a symmetry operation, its contribution to the character is zero. In the transformation matrix $\begin{pmatrix} 0 & 1 \\ -1 & 0 \end{pmatrix}$ the sum of the diagonal is 0.

Therefore, we may write

$$\chi \, C_4(d_{xz}, d_{yz}) = 0 + 0 = 0 \qquad \qquad ...(5.12)$$

Continuing in this way $\chi \, E(d_{xz}, d_{yz}) = 2$.

Since

$$C_2(d_{xz}) = -d_{xz} = -1(d_{xz}) + 0(d_{yz}) \qquad ...(5.13)$$

and

$$C_2(d_{yz}) = -d_{yz} = 0(d_{xz}) + -1(d_{yz}) \qquad ...(5.14)$$

In matrix form we may write

$$C_2 \begin{pmatrix} d_{xz} \\ d_{yz} \end{pmatrix} = \begin{pmatrix} -1 & 0 \\ 0 & -1 \end{pmatrix} \begin{pmatrix} d_{xz} \\ d_{yz} \end{pmatrix} \qquad ...(5.15)$$

$\therefore$ Character $\chi \, (C_2) = (-1) + (-1) = -2$ (sum of the diagonal elements of the matrix).

Also

$$C_2'(d_{xz}) = -1(d_{xz}) + 0(d_{yz})$$
$$C_2'(d_{yz}) = 0(d_{xz}) + 1(d_{yz})$$

i.e.

$$C_2' \begin{pmatrix} d_{xz} \\ d_{yz} \end{pmatrix} = \begin{pmatrix} -1 & 0 \\ 0 & -1 \end{pmatrix} \begin{pmatrix} d_{xz} \\ d_{yz} \end{pmatrix}$$

Therefore the character $\chi \, (C_2) = -1 + 1 = 0$.

The transformation matrix for C_2'' is $\begin{pmatrix} 0 & -1 \\ -1 & 0 \end{pmatrix}$ and the character $\chi \, (C_2'') = 0 + 0 = 0$.

Hence summarising

	E	C_4	C_2	C_2'	C_2''
$\Gamma_{d_{xz}, d_{yz}}$	2	0	-2	0	0

Confirming that the orbitals (d_{xz}, d_{yz}) belong to e, in D_4 and since they are even orbitals to e_g in D_{4h}.

Following the same producer it reveals that p_x and p_y belong to e_u. In case of C_2' and C_2'' above matrix notation is unnecessary since these operations do not mix the orbitals. This means

$$C_2'(d_{xz}) = -1(d_{xz})$$
$$C_2'(d_{yz}) = -1(d_{yz})$$

and the character is $-1 + (-1) = -2$.

We shall now consider a trigonal bipyramidal molecule $[CuCl_5]^{3-}$ (Fig. 5.9). The molecule belongs to D_{3h} group. The principal axis C_3 is perpendicular to the trigonal plane which we call the z axis and three C_2 axes passing through each of the three Cl and Cu perpendicular to C_3 axis. By carrying out the group operations on the s, d_{z^2} and p_z orbitals, their symmetry species can be easily determined.

	E	C_3	C_2	σ_h	S_3	σ_v	
s	1	1	1	1	1	1	a_1'
d_{z^2}	1	1	1	1	1	1	a_1'
p_z	1	1	-1	-1	-1	1	a_2''

The remaining three pairs of orbitals (d_{xz}, d_{yz}), (p_x, p_y) and $(d_{x^2-y^2}, d_{xy})$ are mixed by the C_3 operation and they belong to an e species.

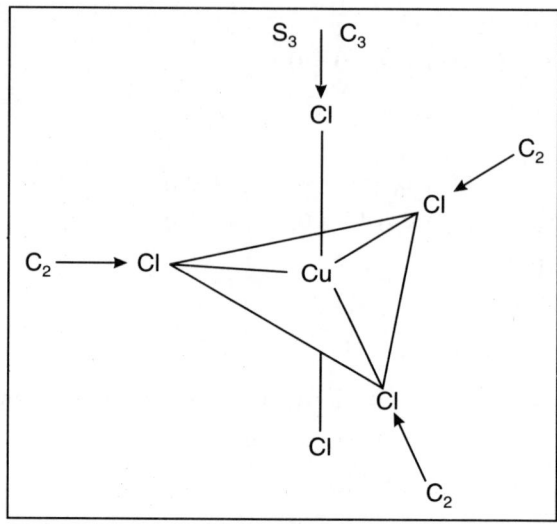

Fig. 5.9 Elements of symmetry of the trigonal bipyramidal $[CuCl_5]^{3-}$ ion (D_{3h}). The σ_v planes contain the copper atom, the apical chlorine atoms and one equatorial chlorine atom

It is obvious that a C_3 rotation on the d_{xy} orbital will convert it into a mixture of the d_{xy} and $d_{x^2-y^2}$ orbitals. It is to be noted that in case of O_h symmetry $d_{x^2-y^2}$ and d_{xy} are not degenerate while they degenerate and form e pair in D_{3h} symmetry (Fig. 5.10).

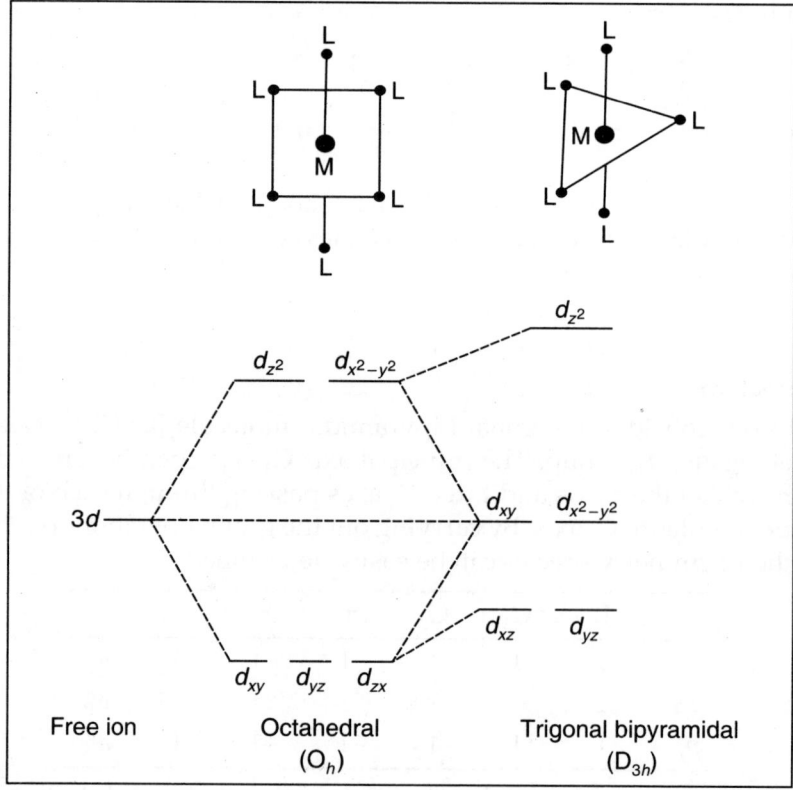

Fig. 5.10 Qualitative d-orbital ordering in trigonal bipyramidal (D_{3h}) crystal field relative to octahedral field (O_h)

The remaining three pairs of orbitals are (d_{xz}, d_{yz}), (p_x, p_y) and $(d_{x^2-y^2}, d_{xy})$. Each of the pairs are mixed by the operation C_3. So each of these pairs must belong to an e species. By considering whether the e orbitals are symmetric or antisymmetric with respect to the trigonal plane, the assignment of the symmetry species of the atomic orbitals in D_{3h} can be completed, viz.

	s and d_{z^2}	belong to a_1'
	p_z	belong to a_2''
	p_x, p_y	belong to e'
	d_{xy}, d_{xy}	belong to e''
and	$d_{x^2-y^2}, p_{xy}$	belong to e'

It is to be noted that in this case the $d_{x^2-y^2}, d_{xy}$ orbitals form a degenerate e pair, whilst in octahedral symmetry they do not (Fig. 5.10). It is obvious that a C_3 rotation on the d_{xy} orbital will convert it into a mixture of d_{xy} and $d_{x^2-y^2}$ orbitals, but the exact form of this mixture is by no means clear.

It has already been shown that the wave function consists of a radial part and an angular part, the former being invariant part under all the symmetry operations. The angular part will be affected by the various operations. The angular wave functions for the d_{xy} and $d_{x^2-y^2}$ orbitals are shown below.

$$d_{xy} = (15/16\pi)^{1/2} \sin^2\theta \sin 2\phi = \frac{\sqrt{15}}{4} \cdot \frac{1}{\sqrt{\pi}} \cdot \sin^2\theta \sin 2\phi$$

$$d_{x^2-y^2} = (15/16\pi)^{1/2} \sin^2\theta \cos 2\phi = \frac{\sqrt{15}}{4} \cdot \frac{1}{\sqrt{\pi}} \cdot \sin^2\theta \cos 2\phi$$

If the rotation is carried out about the z axis, i.e. the wave function is quantised about the z axis, the angle θ will be unaffected by rotation. A rotation of $\in$ degrees about the z axis will convert ϕ into $\phi + \in$.

Now
$$d_{xy} = \frac{\sqrt{15}}{4} \cdot \frac{1}{\sqrt{\pi}} \sin^2\theta \sin 2\phi$$

$$= K \sin 2\phi \qquad \qquad ...(5.16)$$

where K is invariant under all rotation about the z axis.

It is to be noted that in the octahedral group this problem is more difficult since the C_3 axis does not correspond with the axis (z) of quantisation. Thus rotation will alter both ϕ and θ, K is not invarient. However, in D_{3h} the procedure is relatively simple.

Similarly
$$d_{x^2-y^2} = K \cos 2\phi \qquad \qquad ...(5.17)$$

By carrying out the group operations on $d_{xy}, d_{x^2-y^2}$ orbitals we have

$$E(d_{xy}) = d_{xy}$$
$$E(d_{x^2-y^2}) = d_{x^2-y^2}$$

$\therefore$ Character $\chi(E) = 1 + 1 = 2$

$$C_3(d_{xy}) = C_3(K \sin 2\phi) = K \sin 2(\phi + 120°) = K[\sin(2\phi + 240°)]$$
$$= K[\sin 2\phi \cdot \cos 240° + \cos 2\phi \sin 240°]$$

$$= K\left[\sin 2\phi\left(-\frac{1}{2}\right) + \cos 2\phi\left(-\frac{\sqrt{3}}{2}\right)\right] = -\frac{1}{2}K\sin 2\phi - \frac{\sqrt{3}}{2}K\cos 2\phi$$

i.e. $C_3(d_{xy}) = -\dfrac{1}{2}d_{xy} - \dfrac{\sqrt{3}}{2}d_{x^2-y^2} \quad \left[\begin{array}{c} \because\ K\sin 2\phi = d_{xy} \\ K\cos 2\phi = d_{x^2-y^2} \end{array}\right]$ ...(5.18)

$$C_3(d_{x^2-y^2}) = C_3(K\cos 2\phi) = K\cos 2\,(\phi + 120°) = K\,[\cos(2\phi + 240°)]$$

$$= K\,[\cos 2\phi \cos 240° - \sin 2\phi \sin 240°] = -\frac{1}{2}K\cos 2\phi + \frac{\sqrt{3}}{2}K\sin 2\phi$$

i.e. $C_3(d_{x^2-y^2}) = -\dfrac{1}{2}d_{x^2-y^2} + \dfrac{\sqrt{3}}{2}d_{xy} = \dfrac{\sqrt{3}}{2}d_{xy} - \dfrac{1}{2}d_{x^2-y^2}$ ...(5.19)

Expressing Eqs (5.18) and (5.19) into matrix form, we have

$$C_3\begin{pmatrix} d_{xy} \\ d_{x^2-y^2} \end{pmatrix} = \begin{pmatrix} -\dfrac{1}{2} & -\dfrac{\sqrt{3}}{2} \\ +\dfrac{\sqrt{3}}{2} & -\dfrac{1}{2} \end{pmatrix}\begin{pmatrix} d_{xy} \\ d_{x^2-y^2} \end{pmatrix}$$

$\therefore$ Character $\chi(C_3) = -\dfrac{1}{2} + \left(-\dfrac{1}{2}\right) = -1$

Since $C_2(d_{xy}) = -1(d_{xy})$ and $C_2(d_{x^2-y^2}) = +1(d_{x^2-y^2})$

$$\therefore\ \chi C(d_{xy}, d_{x^2-y^2}) = -1 + (+1) = 0$$

Therefore in D_3 the characters are:

	E	$2C_3$	$3C_2$
$d_{xy},\ d_{x^2-y^2}$	2	−1	0

confirming that these orbitals belong to the e representation in D_3, and by inspection the e' representation in D_{3h}.

6

Applications of Group Theory to Valence Bond Theory

6.1 HYBRIDISATIONS INVOLVING s-, p- AND d-ORBITALS

Hybrid orbitals of an atom means mixing of different atomic orbitals of the same atom having their energies very close. If ϕ_1 and ϕ_2 are two combining atomic orbitals (wave functions), the resulting two (the number of hybrid orbitals will be equal to the number of combining pure atomic orbitals) hybrid orbitals will be

$$\psi_{h_1} = C_1\,\phi_1 + C_2\,\phi_2$$
$$\psi_{h_2} = C_3\,\phi_1 + C_4\,\phi_2$$

The coefficients C_1, C_2, C_3, C_4 indicate the contribution of each orbital to the formation of hybrid orbitals. The values of the coefficients should be such that each of the hybridised orbital wave function is normalized and two hybrid orbitals should also be orthogonal to each other.

$$\int \psi_{h_1}^2\, d\tau = \int \psi_{h_2}^2\, d\tau = 1 \quad \text{and}$$

$$\int \psi_{h_1}\,\psi_{h_2}\, d\tau = 0$$

If we apply the normalisation condition for ψ_{h_1}, we may write

$$\int \psi_{h_1}^2\, d\tau = 1$$

$$\int (C_1\,\phi_1 + C_2\,\phi_2)^2\, d\tau = 1$$

or $\quad C_1^2 \int \phi^2\, d\tau + 2C_1 C_2 \int \phi_1\phi_2\, d\tau + C_2^2 \int \phi_2^2\, d\tau = 1$

$$C_1^2 \cdot 1 + 2\,C_1 C_2 \cdot 0 + C_2^2 \cdot 1 = 1$$
$$C_1^2 + C_2^2 = 1$$

Similarly for ψ_{h_2} if normalization is applied, we have

$$C_3^2 + C_4^2 = 1$$

Since ψ_{h_1} and ψ_{h_2} hybrid orbitals are orthogonal, we may write

$$\int \psi_{h_1}\,\psi_{h_2}\, d\tau = 0$$

$$\int (C_1\,\phi_1 + C_2\,\phi_2)(C_3\,\phi_4 + C_4\,\phi_2)d\tau = 0$$

$$C_1 C_3 \int \phi_1^2\, d\tau + C_1 C_4 \int \phi_1\phi_2\, d\tau + C_2 C_3 \int \phi_2\phi_1\, d\tau + C_2 C_4 \int \phi_2^2\, d\tau = 0$$

$$C_1 C_3 \cdot 1 + C_1 C_4 \cdot 0 + C_2 C_3 \cdot 0 + C_2 C_4 \cdot 1 = 0$$

$$C_1 C_3 + C_2 C_4 = 0$$

As the hybrid orbitals are orthogonal wave functions they form the basis for reducible representation of the group to which the molecule belongs.

Hybrid orbitals provide better overlap with the incoming atomic orbitals and result in the formation of more stable bonds and a state of lower energy. The energy liberated is called *hybridisation energy*. This energy is partly used for the excitation of elections from the lower energy orbital to higher energy orbitals in the atom.

We shall now consider the compositions of different hybrid orbitals.

6.1.1 Linear BeCl$_2$ Molecule

In BeCl$_2$ the formation of two σ bonds take place by the overlapping of two hybrid orbitals (*sp* hybrid orbitals) of Be with the p_z orbitals of chlorine atom. The hybrid orbitals are directed at an angle of 180° along the internuclear axis (say z-axis) as shown in the Fig. 6.1.

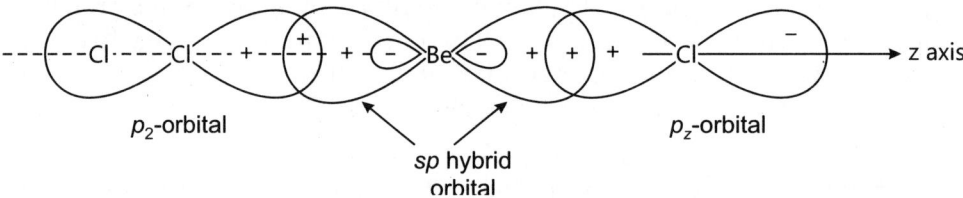

p$_z$-orbital   *sp* hybrid orbital   *p$_z$*-orbital

Fig. 6.1 Formation of two σ-bonds by the overlapping of two hybrid orbitals (*sp* hybrid orbitals) of Be with the *p$_z$*-orbitals of chlorine

The 2s orbital of Be atom is spherically symmetrical and hence contributes to the formation of both the hybrid orbitals. 2p_z orbital is also directed in the direction in which the hybrid orbitals are formed and hence can contribute to their formation.

But 2p_x and 2p_y orbitals of Be atom are directed along x and y axis. This means that p_x and p_y orbitals have nodes along the z-axis and hence can not contribute to the formation of the hybrid orbitals along that direction. Thus the two hybrid orbitals have contributions only from s and p_z orbitals. Hence they are *sp* hybrid orbitals. The compositions of the two hybrid orbitals can be expressed as follows.

$$\psi_{h_1} = C_1 \psi_s + C_2 \psi_{pz}$$

$$\psi_{h_2} = C_3 \psi_s + C_4 \psi_{pz}$$

The coefficients C_1 and C_3 can be easily evaluated by knowing that s-orbital is spherically symmetrical, it contributes equally to the formation of two hybrid orbitals. So $C_1 = C_3$. In terms of probability contribution it is $1/2$; $C_1^2 = C_2^2 = \dfrac{1}{2}$

Hence the coefficient of the s orbital wave function in making the hybrid orbitals in each case shall be $\dfrac{1}{\sqrt{2}}$. So $C_1 = C_3 = \dfrac{1}{\sqrt{2}}$

Hence,

$$\psi_{h_1} = \frac{1}{\sqrt{2}} \psi_s + C_2 \psi_{p_z}$$

$$\psi_{h_2} = \frac{1}{\sqrt{2}} \psi_s + C_4 \psi_{p_z}$$

Since each hybrid orbital is normalised, we have

$$C_1^2 + C_2^2 = 1$$

$$\frac{1}{2} + C_2^2 = 1 \qquad\qquad [\because C_1 = \frac{1}{\sqrt{2}}]$$

$$\therefore \qquad\qquad C_2 = \frac{1}{\sqrt{2}}$$

Hence, $\qquad\qquad \psi_{h_1} = \frac{1}{\sqrt{2}}\,\psi_s + \frac{1}{\sqrt{2}}\,\psi_{p_z} = \frac{1}{\sqrt{2}}\,(\psi_s + \psi_{p_z})$

This is the composition of the first hybrid orbital as the two hybrid orbitals are orthogonal to each other, so we may write

$$C_1 C_3 + C_2 C_9 = 0$$

$$\frac{1}{\sqrt{2}} \cdot \frac{1}{\sqrt{2}} + \frac{1}{\sqrt{2}} \cdot C_4 = 0 \qquad\qquad [\because C_1 = C_2 = C_3 = \frac{1}{\sqrt{2}}]$$

$$\therefore \qquad\qquad C_4 = -\frac{1}{\sqrt{2}}$$

Hence, $\qquad\qquad \psi_{h_2} = \frac{1}{\sqrt{2}}\,\psi_s - \frac{1}{\sqrt{2}}\,\psi_{p_z}$

$$= \frac{1}{\sqrt{2}}\,(\psi_s - \psi_{p_z})$$

The two sp hybrid orbitals are thus equivalent, one having contribution from positive phase of p_z orbital and another from the negative phase of p_z orbital. Their formation is illustrated in Fig. 6.2 using the boundary lines of the constituent orbitals.

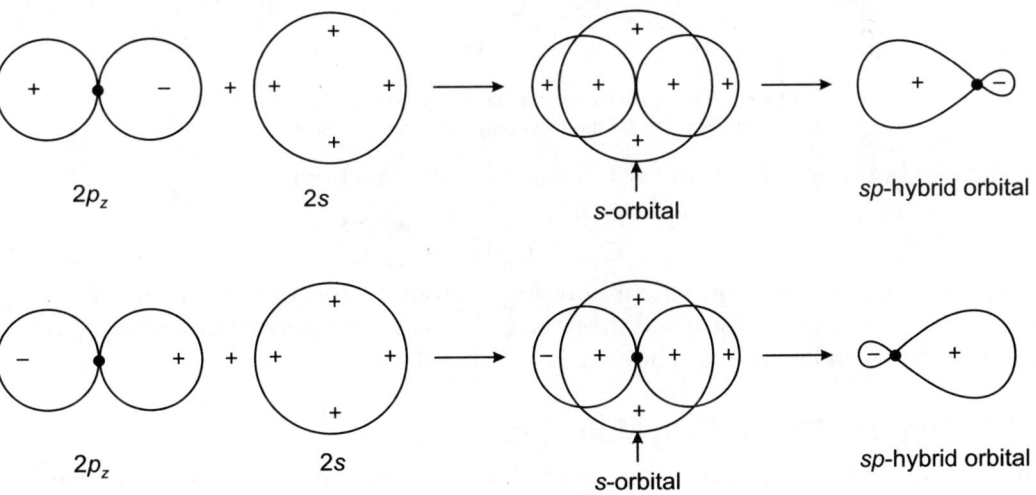

Fig. 6.2 Combination of $2s$ and $2p_z$ atomic orbitals forming two sp hybrid orbitals

The wave functions are $\quad \psi_{sp}(1) = \frac{1}{\sqrt{2}}\,(\psi_{2s} + \psi_{2p_z})$

$$\psi_{sp}(2) = \frac{1}{\sqrt{2}}\,(\psi_{2s} + \psi_{2p_z})$$

Hybridisation scheme for linear triatomic molecule AX_3, we linearly combine the atomic orbitals on atom A in such a way that the resultant combinations (called hybrid orbitals) point toward the X atoms. In BeH_2 molecule, two equivalent colinear hybrid orbitals are constructed from the $2s$ and $2p_z$ orbitals on Be, which can overlap with the two $1s$ hydrogen orbitals to form two Be–H single bonds (the $2p_x$ and $2p_y$ orbitals do not take part in the hybridisation scheme, otherwise the resultant hybrid orbitals would not point directly at the hydrogens). If we combine the $2s$ and $2p_z$ orbitals in the following manner:

$$h_1 = (2)^{-1/2} (2s + 2p_z)$$
$$h_2 = (2)^{-1/2} (2s - 2p_z)$$

The hybrid orbitals h_1 and h_2 would overlap nicely with the $1s$ orbitals on H_a and H_b respectively, as shown in Fig. 6.3.

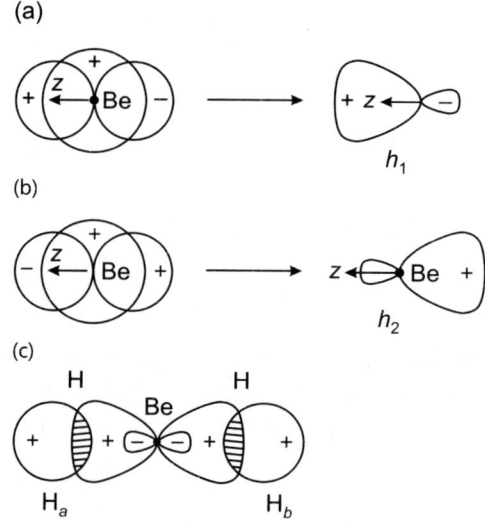

Fig. 6.3 The formation of the two sp hybrid orbitals in BeH_2 (a) and (b), (c) the two equivalent bonds in BeH_2

The two bonding orbitals in BeH_2 have the wave functions

$$\psi_1 = C_5 h_1 + C_6 1s_a, \quad C_6 > C_5$$
$$\psi_2 = C_5 h_2 + C_6 1s_b, \quad C_6 > C_5$$

So now we have two equivalent bonding orbitals ψ_1 and ψ_2 with the same energy. Moreover, ψ_1 and ψ_2 are localised orbitals. ψ_1 is localised between Be and H_a and ψ_2 is localised between Be and H_b. They are $2c - 2e$ bonds.

6.1.2 Trigonal Planar BCl_3 Molecule

The hybrid orbitals $\psi_{h_1}, \psi_{h_2}, \psi_{h_3}$, are directed to the corners of a trigonal plane (Fig. 6.4).

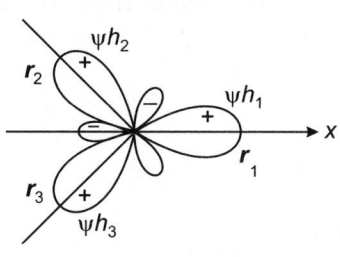

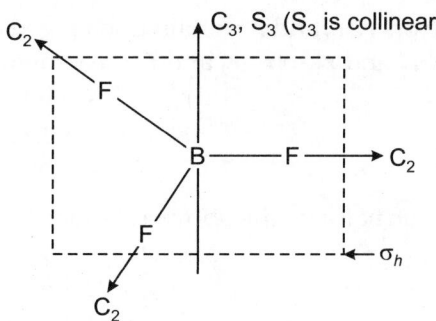

ψ_{h_1} is directed along x-axes

ψ_{h_2} is directed along x and y-axes

ψ_{h_3} is directed along x and y-axes

Fig. 6.4 The hybrid orbitals ψ_{h_1}, ψ_{h_2}, ψ_{h_3} are the bond vectors and BCl_3 belongs to D_{3h} point group

Note: D_{3h} means $D_3 + \sigma_h$; $D_3 \rightarrow E, C_3, C_3^2, 3C_2'$

$\quad\quad D_{3h} \rightarrow E, C_3, C_3^2, 3C_2'$, $\sigma_h E, \sigma_h C_3, \sigma_h C_3^2, 3\,\sigma_h C_2'$

$\quad\quad \rightarrow E, \underbrace{C_3, C_3^2}_{2C_3}\ \underbrace{3C_3'}_{C_2}\ , \sigma_h\ \underbrace{S_3', S_3^s}_{2S_3}, 3\sigma_v\ [\sigma_h C_2' = \sigma_v]$

On performing the different operations (E, C_3, C_2, σ_h, S_3, σ_v), the following total characters (Γ_{total}) are obtained.

D_{3h}	E	$2C_3$*	$3C_2$**	σ_h	$2S_3$#	$3\sigma_v$
	3	0	1	3	0	1
Γ_{total}	All the 3 bond vectors remain unshifted	All the 3 bond vectors are shifted	One bond vector along the rotation axis remains unshifted	All 3 bond vectors remain unshifted	All the 3 bond vectors are shifted	One bond vector lying in the symmetry plane, remains unshifted

The same result can be obtained by working out the matrix for each operation.

Let us consider the hybrid orbitals ψ_{h_1}, ψ_{h_2}, ψ_{h_3} as vectors r_1, r_2 and r_3. By a symmetry operation R, the vectors r_1, r_2 and r_3 changed to r_1', r_2', r_3' according to matrix equation.

$$\begin{pmatrix} r_1' \\ r_2' \\ r_3' \end{pmatrix} = \hat{R} \begin{pmatrix} r_1 \\ r_1 \\ r_1 \end{pmatrix}$$

where $\hat{R}$ is the matrix operator.

---

* C_3^1 and C_3^2 produce indistinguishable configurations. So $2C_3$ comes.

**Three C_2 axes are said to constitute a set of equivalent symmetry elements.

S_3 (S_3^1, S_3^2, S_3^3 ($\equiv$ E)) is collinear with C_3. So two S_3 (S_3^1, S_3^2) are present.

The vectors r_1, r_2 and r_3 remain unchanged by the identity operation (E). The equation relating r_1' r_2' and r_3' to r_1, r_2 and r_3 are shown below.

$$r_1' = 1\,r_1 + 0\,r_2 + 0\,r_3$$
$$r_2' = 0\,r_1 + 1\,r_2 + 0\,r_3$$
$$r_3' = 0\,r_1 + 0\,r_2 + 1\,r_3$$

In the matrix form, the above relations become

$$\begin{pmatrix} r_1' \\ r_2' \\ r_3' \end{pmatrix} = \begin{pmatrix} 1 & 0 & 0 \\ 0 & 1 & 0 \\ 0 & 0 & 1 \end{pmatrix} \begin{pmatrix} r_1 \\ r_1 \\ r_1 \end{pmatrix}$$

$\uparrow$
(Transformation matrix)

Here the transformation matrix is the identity operation (E). Thus

$$E = \begin{pmatrix} 1 & 0 & 0 \\ 0 & 1 & 0 \\ 0 & 0 & 1 \end{pmatrix}$$

Therefore the character of E = 1 + 1 + 1 = 3

On performing C_3 operation, r_1 changes to r_2, r_2 changes to r_3 and r_3 changes to r_1. The resulting vectors r_1', r_1' and r_3' are related to r_1, r_2 and r_3 by the following equations.

$$r_1' = 0\,r_1 + 1\,r_2 + 0\,r_3$$
$$r_2' = 0\,r_1 + 0\,r_2 + 1\,r_3$$
$$r_3' = 1\,r_1 + 0\,r_2 + 0\,r_3$$

In matrix form, we may write

$$\begin{pmatrix} r_1' \\ r_2' \\ r_3' \end{pmatrix} = \begin{pmatrix} 0 & 1 & 0 \\ 0 & 0 & 1 \\ 1 & 0 & 0 \end{pmatrix} \begin{pmatrix} r_1 \\ r_1 \\ r_1 \end{pmatrix}$$

Here $C_3 = \begin{pmatrix} 0 & 1 & 0 \\ 0 & 0 & 1 \\ 1 & 0 & 0 \end{pmatrix}$, the character of C_3, $\chi(C_3) = 0 + 0 + 0 = 0$

On performing C_2 along r_1, r_1 remains unchanged whereas r_2 and r_3 mutually interchange:

$$r_1' = 1\,r_1 + 0\,r_2 + 0\,r_3$$
$$r_2' = 0\,r_1 + 0\,r_2 + 1\,r_3$$
$$r_3' = 0\,r_1 + 1\,r_2 + 0\,r_3$$

$$\begin{pmatrix} r_1' \\ r_2' \\ r_3' \end{pmatrix} = \begin{pmatrix} 1 & 0 & 0 \\ 0 & 0 & 1 \\ 0 & 1 & 0 \end{pmatrix} \begin{pmatrix} r_1 \\ r_1 \\ r_1 \end{pmatrix}$$

Thus $C_2 = \begin{pmatrix} 1 & 0 & 0 \\ 0 & 0 & 1 \\ 0 & 1 & 0 \end{pmatrix}$ $\chi(C_2) = 1 + 0 + 0 = 1$

On performing σ_v along r_1, r_1 remains unchanged but r_2 and r_3 mutually interchanged.

$$\begin{pmatrix} r_1' \\ r_2' \\ r_3' \end{pmatrix} = \begin{pmatrix} 1 & 0 & 0 \\ 0 & 0 & 1 \\ 0 & 1 & 0 \end{pmatrix} \begin{pmatrix} r_1 \\ r_1 \\ r_1 \end{pmatrix} \qquad \therefore \ \chi(\sigma_v) = 1 + 0 + 0 = 1$$

Thus $$\sigma_v = \begin{pmatrix} 1 & 0 & 0 \\ 0 & 0 & 1 \\ 0 & 1 & 0 \end{pmatrix}$$

On performing σ_h operation none of the vectors change their positions.

$$\begin{pmatrix} r_1' \\ r_2' \\ r_3' \end{pmatrix} = \begin{pmatrix} 1 & 0 & 0 \\ 0 & 1 & 0 \\ 0 & 0 & 1 \end{pmatrix} \begin{pmatrix} r_1 \\ r_2 \\ r_3 \end{pmatrix} \qquad \therefore \ \sigma_h = \begin{pmatrix} 1 & 0 & 0 \\ 0 & 1 & 0 \\ 0 & 0 & 1 \end{pmatrix}$$

$$\therefore \ \chi(\sigma_n) = 3$$

On performing S_3 (= $C_3 \cdot \sigma_h$) all the vectors change positions and hence the matrix is the same as that of C_3.

$$S_3 = \begin{pmatrix} 0 & 1 & 0 \\ 0 & 0 & 1 \\ 1 & 0 & 0 \end{pmatrix} \qquad \therefore \ \chi(S_3) = 0$$

D_{3h}	E	$2C_3$	$3C_2$	σ_h	$2S_3$	$3\sigma_v$	
Γ_σ	3	0	1	3	0	1	$\equiv A_1' + E'$

Look at the character table of D_{3h} group.

D_{3h}	E	$2C_3$	$3C_2$	σ_h	$2S_3$	$3\sigma_v$
A_1'	1	1	1	1	1	1
A_2'	1	1	−1	1	1	−1
E'	2	−1	0	2	−1	0
A_1''	1	1	1	−1	−1	−1
A_2''	1	1	−1	−1	−1	1
E''	2	−1	0	−2	1	0

Thus for D_{3h}		E	$2C_3$	$3C_2$	σ_h	$2S_3$	$3\sigma_v$
Reducible representation RR, Γ_σ		3	0	1	3	0	1
Irreducible representation IR$_1$	A_1'	1	1	1	1	1	1
	E'	2	−1	0	2	−1	0

For AB_3 molecule, the order (h) of D_{3h} point group is 12. The number of irreducible representations (IRs) are obtained as given in the following relation.

$$n(\Gamma_i) = \frac{1}{h} \left[\sum n(R) \, \chi_{RR}(R) \, \chi_{IR}(R) \right]$$

where $n(R)$ is the no. of operations in the class R

χ_{RR} (R) is the character of representation $\chi_{(RR)}$ (R)

χ_{IR} (R) is the character of irreducible representation χ_{IR} (R)

$\therefore \quad n(A_1') = \frac{1}{2}[(1)(3)(1) + (2)(0)(1) + (3)(1)(1) + (1)(3)(1) + (2)(0)(1) + (3)(1)(1) = 1$

$n(E') = \frac{1}{2}[(1)(3)(2) + (2)(0)(-1) + (3)(1)(0) + (1)(3)(2) + (2)(0)(-1) + (3)(1)(0) = 1$

Hence $\quad \Gamma_\sigma = 1A_1' + 1\,E' = A_1' + E'$

The atomic orbitals which form the bases of A', and E' are

$\quad A_1' = s$ orbital

$\quad E' = (p_x, p_y)$ or $(d_{x^2-y^2}, d_{xy})$

Hence the three hybrid orbitals can be formed by sp^2 (s, p_x, p_y) or sd^2 $(s, d_{x^2-y^2}, d_{xy})$ combination. $2s$ and $2p$ orbitals are closer in energy whereas in $2s$ and $3d$ orbitals, energy separation is very large. Hence the three hybrid orbitals in BCl_3 are formed by sp_2 hybridisation and not by sd^2 hybridisation.

The composition of the three hybrid orbitals can be shown as follows.

$$\psi_{h_1} = \psi_{sp_1^2} = a_1\,\psi_{2s} + b_1\,\psi_{2p_x} + C_1\,\psi_{2p_y}$$
$$\psi_{h_2} = \psi_{sp_2^2} = a_2\,\psi_{2s} + b_2\,\psi_{2p_x} + C_2\,\psi_{2p_y}$$
$$\psi_{h_3} = \psi_{sp_3^1} = a_3\,\psi_{2s} + b_3\,\psi_{2p_x} + C_3\,\psi_{2p_y}$$

s orbital is spherically symmetrical. It contributes equally to the making of the three hybrid orbitals. Thus

$$a_1 = a_2 = a_3 = \frac{1}{\sqrt{3}} \text{ (because } a_1^2 + a_2^2 + a_3^2 = 1)$$

ψ_{h_1} is formed along x-axis and hence can not have any contribution from p_y. Therefore C_1 will be zero.

Thus

$$\psi_{h_1} = \frac{1}{\sqrt{3}}\,\psi_{2s} + b_1\,\psi_{2p_x} + 0 \cdot \psi_{2p_y}$$

$$= \frac{1}{\sqrt{3}}\,\psi_{2s} + b_1\,\psi_{2p_x}$$

ψ_{h_1} is normalised wave function and hence

$$a_1^2 + b_1^2 = 1$$
$$\left(\frac{1}{\sqrt{3}}\right) + b_1^2 = 1$$
$$b_1^2 = 1 - \frac{1}{3} = \frac{2}{3} \quad \therefore \ b_1 = \sqrt{\frac{2}{3}}$$

Therefore $\qquad \psi_{h_1} = \frac{1}{\sqrt{3}}\,\psi_{2s} + \sqrt{\frac{2}{3}}\,\psi_{2p_x} \hspace{2cm} ...(6.1)$

Now considering that ψ_{h_1} and ψ_{h_2} are orthogonal to each other, so we may write

$$a_1a_2 + b_1b_2 + c_1c_2 = 0$$
$$a_1a_2 + b_1b_2 + 0 = 0 \hspace{2cm} [c_1c_2 = 0; c_2 = 0, \text{ since } c_1 = 0]$$

$$a_1 a_2 + b_1 b_2 = 0$$

$$\frac{1}{\sqrt{3}} \cdot \frac{1}{\sqrt{3}} + \sqrt{\frac{2}{3}} \cdot b_2 = 0 \qquad\qquad \left[a_1 = a_2 = \frac{1}{\sqrt{3}} \text{ and } b_1 = \sqrt{\frac{2}{3}} \right]$$

$$\therefore \qquad\qquad b_2 = \frac{1}{\sqrt{6}}$$

ψ_{h_2} is normalised. Normalisation condition requires

$$a_2^2 + b_2^2 + c_2^2 = 1$$

$$\frac{1}{3} + \frac{1}{6} + C_2^2 = 1 \qquad\qquad \left[a_1 = a_2 = a_3 = \frac{1}{\sqrt{3}} ; b_2 = \frac{-1}{\sqrt{6}} \right]$$

$$C_2^2 = \frac{1}{2} \qquad \therefore \quad C_2 = \frac{1}{\sqrt{2}}$$

Hence the composition of ψ_{h_2} will be

$$\psi_{h_2} = \frac{1}{\sqrt{3}} \psi_{2s} + \left(\frac{-1}{\sqrt{6}} \right) \psi_{2p_x} + \frac{1}{\sqrt{2}} \psi_{2p_y} \qquad\qquad ...(6.2)$$

Now we shall consider the orthogonality of ψ_{h_1} and ψ_{h_3}

$$a_1 a_3 + b_1 b_3 + c_1 c_3 = 0$$

$$a_1 a_3 + b_1 b_3 = 0 \qquad\qquad\qquad [\because c_1 = 0]$$

$$\frac{1}{\sqrt{3}} \cdot \frac{1}{\sqrt{3}} + \sqrt{\frac{2}{3}} b_3 = 0$$

$$\therefore \qquad\qquad b_3 = - \frac{1}{\sqrt{6}}$$

The orthogonality condition of ψ_{h_2} and ψ_{h_3} will result

$$a_2 \cdot a_3 + b_2 \cdot b_3 + c_2 \cdot c_3 = 0$$

$$\frac{1}{\sqrt{3}} \cdot \frac{1}{\sqrt{3}} + \left(\frac{-1}{\sqrt{6}} \right)\left(\frac{-1}{\sqrt{6}} \right) + \frac{1}{\sqrt{2}} \cdot C_3 = 0$$

$$\frac{1}{3} + \frac{1}{6} + \frac{1}{\sqrt{2}} \cdot C_3 = 0$$

$$\frac{C_3}{\sqrt{2}} = 1 - \left(\frac{1}{3} + \frac{1}{6} \right) = \frac{1}{2}$$

$$\therefore \qquad\qquad C_3 = \frac{-1}{\sqrt{2}}$$

Substituting the values of a_3, b_3 and c_3, we get the composition of ψ_{h_3} as

$$\psi_{h_3} = \frac{1}{\sqrt{3}} \psi_{2s} + \left(\frac{-1}{\sqrt{6}} \right) \psi_{2p_x} + \left(-\frac{1}{\sqrt{2}} \right) \psi_{2p_y}$$

$$= \frac{1}{\sqrt{3}} \psi_{2s} - \frac{1}{\sqrt{6}} \psi_{2p_x} - \frac{1}{\sqrt{2}} \psi_{2p_y} \qquad\qquad ...(6.3)$$

In Eqs (6.1), (6.2) and (6.3), the meaning of signs of the coefficients can be understood by physical considerations.

$\psi_{h_1}\left(=\dfrac{1}{\sqrt{3}}\,\psi_{2s}+\sqrt{\dfrac{2}{3}}\,\psi_{2p_z}\right)$ is formed by the combination from $2s$ orbital and

positive phase of $2p_x$ orbital. $\psi_2\left(=\dfrac{1}{3}\,\psi_{2s}-\dfrac{1}{\sqrt{6}}\,\psi_{2p_x}+\dfrac{1}{\sqrt{2}}\,\psi_{2p_y}\right)$ is formed from the

combination of $2s$ orbital and positive phase of $2p_y$ and negative phase of $2p_x$ orbital.

$\psi_{h_3}\left(=\dfrac{1}{\sqrt{3}}\,\psi_{2s}-\dfrac{1}{\sqrt{6}}\,\psi_{2p_x}-\dfrac{1}{\sqrt{2}}\,\psi_{2p_y}\right)$ has contributions from $2s$ orbital, negative phase

of $2p_x$ orbital and negative phase of $2p_y$ orbital. All these hybrid orbitals are actually orthonormal functions.

6.1.3 Tetrahedral AB$_4$ Molecule

Tetrahedral AB$_4$ type molecules are CH_4, NH_4^+, etc. We apply the same procedure as we adopt in AB$_3$ type of molecules. Here we are required to construct a set of four σ-hybrid orbitals on the central atom A, which are pointing along A—B bond directions. Each hybrid orbitals may be represented by the bond vectors. The four bond vectors r_1, r_2, r_3 and r_4 are shown in Fig. 6.5.

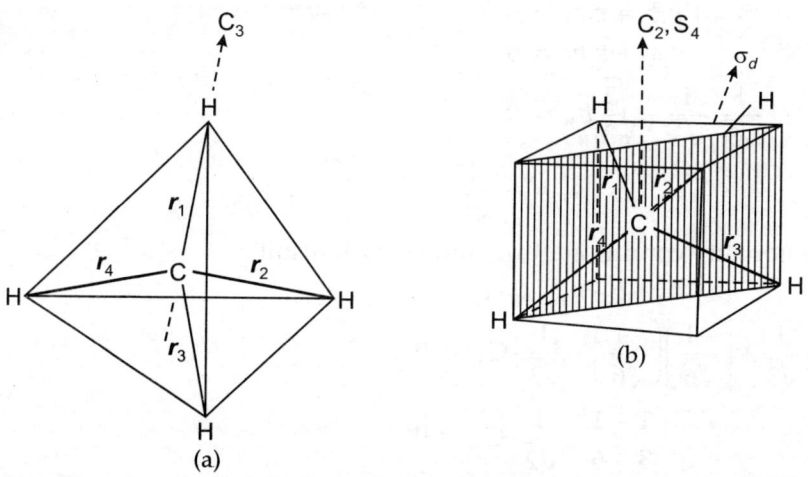

Fig. 6.5 (a) The molecular geometry of methane molecule showing the C–H sigma bonds
(b) The four vectors representing C–H sigma bonds and some symmetry elements in methane

Methane belongs to the T$_d$ point group and possesses the following symmetry operations.

$$E,\ 8\,C_3,\ 3\,C_2,\ 6\,S_4,\ 6\,\sigma_d$$

All the four vectors (C–H) in CH_4 molecule remain unshifted by identity operation (E).

$$r_1 \longrightarrow r_1 + 0\,r_2 + 0\,r_3 + 0\,r_4$$
$$r_2 \longrightarrow 0\,r_1 + r_2 + 0\,r_3 + 0\,r_4$$
$$r_3 \longrightarrow 0\,r_1 + 0\,r_2 + r_3 + 0\,r_4$$
$$r_4 \longrightarrow 0\,r_1 + 0\,r_2 + 0\,r_3 + 0\,r_4$$

In terms of matrix, we may write

$$\begin{pmatrix} r_1 \\ r_2 \\ r_3 \\ r_4 \end{pmatrix} = \begin{pmatrix} 1 & 0 & 0 & 0 \\ 0 & 1 & 0 & 0 \\ 0 & 0 & 1 & 0 \\ 0 & 0 & 0 & 1 \end{pmatrix} \begin{pmatrix} r_1 \\ r_2 \\ r_3 \\ r_4 \end{pmatrix}$$

(New   (Transformation   (Old
coordinates)   matrix)   coordinates)

$\therefore$ Characters of E, $\chi(E) = 1 + 1 + 1 + 1 = 4$

A rotation operation about C_3 axis passing through vector r_1 leaves it unchanged but other vectors will change.

$$r_1 \longrightarrow r_1 \text{ (unchanged)}$$
$$r_2 \longrightarrow r_3$$
$$r_3 \longrightarrow r_4$$
$$r_4 \longrightarrow r_2$$
$$r_1 \longrightarrow r_1 + 0\,r_2 + 0\,r_3 + 0\,r_4$$
$$r_2 \longrightarrow 0\,r_1 + 0\,r_2 + r_3 + 0\,r_4$$
$$r_3 \longrightarrow 0\,r_1 + 0\,r_2 + 0\,r_3 + r_4$$
$$r_4 \longrightarrow 0\,r_1 + r_2 + 0\,r_3 + 0\,r_4$$

In terms of matrix

$$\begin{pmatrix} r_1 \\ r_2 \\ r_3 \\ r_4 \end{pmatrix} = \begin{pmatrix} 1 & 0 & 0 & 0 \\ 0 & 0 & 1 & 0 \\ 0 & 0 & 0 & 1 \\ 0 & 1 & 0 & 0 \end{pmatrix} \begin{pmatrix} r_1 \\ r_2 \\ r_3 \\ r_4 \end{pmatrix} \qquad \therefore \ \chi(C_3) = 1$$

In C_2 operation all the bond vectors (r_1, r_2, r_3, r_4) are interchanged. So the character $\chi(C_2) = 0$

S_4 generates $S_4^1, S_4^2, S_4^3, S_4^4 \ (\equiv E)$.

$$S_4^1 = C_4^1\,\sigma$$
$$S_4^2 = C_4^2\,\sigma^2 = C_2$$
$$S_4^3 = C_4^3\,\sigma^3 = C_4^3\,\sigma$$

In S_4 operation all the bond vectors are shifted

$\therefore$ $\chi(S_4) = 0$

In σ_d operation the vectors r_1 and r_2 remain unshifted whereas r_3 and r_4 change. Thus $\chi(\sigma_d) = 2$

T_d	E	$8\,C_3$	$3\,C_2$	$6\,S_4$	$6\,\sigma_d$	
$\Gamma_\sigma^{(R)}$ or Γ_{tetra}	4	1	0	0	2	$\equiv A_1 + T_2$

The character table of T_d group is given below.

T_d	E	C_3	C_2	S_4	σ_d
A_1	1	1	1	1	1
A_2	1	1	-1	-1	-1
E	2	-1	2	2	0
T_1	3	0	-1	-1	-1
T_2	3	0	-1	-1	1

Thus for T_d	E	$8C_3$	$3C_2$	$6S_4$	$6\sigma_d$
Reducible representation (RR) Γ_σ	4	1	0	0	2
Irreducible representation IR, $\qquad$ A_1	1	1	1	1	1
T_2	3	0	−1	−1	1

In tetrahedral AB_4 molecule; the order (h) of the group = 24. The number of operations $E = 1$, $C_3 = 8$, $C_2 = 3$ S, $S_4 = 6$ and $\chi\sigma_d = 6$.

The character of RR, $\chi(E) = 4$, $\chi(C_3) = 1$, $\chi(C_2) = 0$, $\chi(S_4) = 0$, $\chi(\sigma_d) = 2$.

$$n(A_1) = \frac{1}{24}\left[(1)(4)(1) + (8)(1)(1) + (3)(0)(1) + (6)(0)(1) + (6)(2)(1)\right] = 1$$

$$n(T_2) = \frac{1}{24}\left[(1)(4)(3) + (8)(1)(0) + (3)(0)(-1) + (6)(0)(-1) + (6)(2)(1)\right] = 1$$

Thus RR is reduced to $A_1 + T_2$ $\quad \therefore \Gamma_6 = \Gamma_{\text{tetrahedral bond vectors}} = A_1 + T_2$

The atomic orbitals for A_1 and T_2 orbitals are as follows:

A_1 orbitals	T_2 orbitals
s	$(p_x, p_y, p_z)(d_{xy}, d_{xz}, d_{yz})$

Thus the hybrid orbitals can be constructed either from

(i) one s and three p orbitals to give sp^3 hybrid orbitals

$$s + p_x + p_y + p_z = sp^3 \quad \text{or}$$

(ii) from an s and three d orbitals to yield sd^3 hybrid orbitals

$$s + d_{xy} + d_{xz} + d_{yz} = sd^3$$

From symmetry point of view both are equally possible. To decide which mode of hybridisation is most likely in *a* given molecule or ion, orbital energies must be taken into account.

For tetrahedral CH_4 molecule, $2s$ and $2p$ orbitals of carbon atom are close in energy and hence sp^3 hybridisation takes place. Carbon atom has no $2d$ orbitals but has $3d$ orbitals which are much higher in energy and can not combine with $2s$ orbital.

In cases where s and d orbitals are close in energy, sd^3 hybridisation is possible. For tetrahedral species involving transition metals, such as MnO_4^-, MnO_4^{2-}, CrO_4^{2-}, sd^3 hybridisation may be more important since the $3d$ orbitals of Mn and Cr are somewhat lower in energy then those of $4p$ orbitals. Therefore, Mn and other transition metals would use three $3d$ orbitals instead of the three higher energy $4p$ orbitals for the formation of the hybrids.

6.1.4 Square Planar AB_4 Molecule (XeF$_4$ ion) (PtCl$_4^{-2}$, AvCl$_4^-$, [Ni(CN)$_4$]$^{2-}$)

Square planar AB_4 molecule belongs to D_{4h} point group.

D_{4h} point group has principal axis $C_4 \perp$ to the molecular plane (σ_h). There are four two fold symmetry axes to C_2' and two $C_2'' \perp$ to the C_4 axis (Fig. 6.6).

The molecule has S_4 axis. S_4 generated S_4^1, S_4^2, S_4^3, S_4^4 ($\equiv E$)

$$S_4^1 = C_4^1 \cdot \sigma \; ; S_4^2 = C_4^2 \cdot \sigma^2 = C_2 \; ; S_4^3 = C_4^3 \sigma_3 = C_4^3 \cdot \sigma$$

So there are two S_4 (S_4^1 and S_4^3).

Thus   along the z-axis – C_4, C_2, S_4 are present

        along the along the x-axis – C_2'

        along the y-axis – C_2'

In between x and y axes – 2 C_2'' present

There are 2 σ_v plane and 1 σ_h (molecular plane)

i is also present in the molecular plane

Thus square planar molecule AB_4 has the following operations:

E, $2C_4$, C_2, $2C_2'$, $2C_2''$, i, $2S_4$ (S_4^1, S_4^3) σ_h, $2\sigma_v$, ..., $2\sigma_d$.

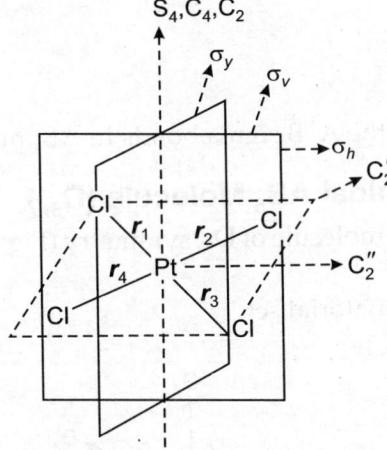

Fig. 6.6 The four vectors representing the Pt–Cl sigma bond and some symmetry element in $[PtCl_4]^{2-}$

The reducible representation Γ_σ is obtained by finding the number of unshifted vectors by each symmetry operation of the point group. The representation, based on bond vectors, is given below.

Operation		E	$2C_4$	C_2	$2C_2'$	$2C_2''$	i	$2S_4$	σ_h	$2\sigma_v$	$2\sigma_d$
	Γ_σ	4	0	0	2	0	0	0	4	2	0
IRs	A_{1g}	1	1	1	1	1	1	1	1	1	1
	B_{1g}	1	–1	1	1	–1	1	–1	1	1	–1
	E_u	2	0	–2	0	0	–2	0	2	0	0

The no. of times (n) an irreducible representation (IR) Γ_i occurs in a reducible representation (RR) is determined by using the reducible formula:

$$n(\Gamma_i) = \frac{1}{h}\left[\sum C(R)\, \chi_{RR}(R)\, \chi_{IR}(R)\right]$$

where n is the no. of IR a given type h is the order

        $C(R)$ is the no. of operations in the class $C(R)$

        $\chi_{RR}(R)$ is the character of reducible representation

        $\chi_{IR}(R)$ is the character of irreducible representation

$$n(A_{1g}) = \frac{1}{16}\Big[(1)(4)(1) + (2)(0)(1) + (1)(0)(1) + (2)(2)(1) + (2)(0)(1) + (1)(0)(1) + (2)(0)(1)$$
$$+ (1)(4)(1) + (2)(2)(1) + (2)(0)(1)\Big]$$
$$= 1$$

$$n(B_{1g}) = \frac{1}{16}\Big[(1)(4)(1) + (2)(2)(1) + (1)(4)(1) + (2)(2)(1)\Big] = 1$$

$$n(E_u) = \frac{1}{16}\Big[(1)(4)(2) + (1)(4)(2)\Big] = 1$$

Hence $\Gamma_{bond} = \Gamma_{A_{1g}} + \Gamma_{B_{1g}} + \Gamma_{E_u}$

From the character table of D_{4h}, we find that

$\Gamma_{A_{1g}} : s, d_{z^2}$

$\Gamma_{B_{1g}} : d_{x^2 - y^2}$

$\Gamma_{E_u} : p_x, p_y$

The possible hybrids for the A–B sigma bonds in AB_4 molecule are dsp^2 and d^2p^2.

6.1.5 Trigonal Bipyramidal AB₅ Molecule (D₃ₕ)

In trigonal bipyramidal AB_5 molecule of D_{3h} symmetry (Fig. 6.7), it is possible to classify the σ-vectors into two sets.

(i) the axial set and (ii) equatorial set

We treat two sets separately.

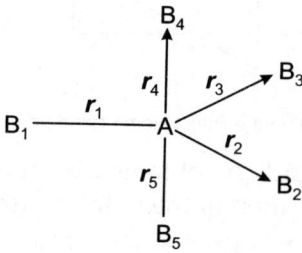

Fig. 6.7 Trigonal bipyramidal AB₅ molecule with a set of r_1, r_2, r_3, r_4, r_5 vectors for σ-hybrid orbitals.

The reducible representations (Γ_σ) corresponding to Γ_{axial} and $\Gamma_{equatorial}$ can be tabulated following Fig. 6.7.

D_{3h}		E	$2\,C_3$	$3\,C_2$	σ_h	$2\,S_3$	$3\,\sigma_v$
Γ_{ax}		2	2	0	0	0	2
Γ_{eq}		3	0	1	3	0	1
Γ_σ		5	2	1	3	0	3
IRS							
Γ_{ax}	A_1'	1	1	1	1	1	1
	A_2''	1	1	−1	−1	−1	1
Γ_{eq}	A_1'	1	1	1	1	1	1
	E'	2	−1	0	2	−1	0

Now $\qquad\qquad \Gamma_\sigma = \Gamma_{ax} + \Gamma_{eq} = A_1' + A_2'' + A_1 \, E' = 2A_1' + A_2'' + E'$

The atomic orbitals (AOs) corresponding to these IRs are:

IRs	AOs
A_1'	s, d_{z^2}
A_2''	p_z
E'	$p_x, p_y, d_{xy}, d_{x^2-y^2}$

For hybridisation, we require two A_1' type orbitals (s and d_{z^2}), one A_2'' (p_z orbital) and a pair of E type orbitals (either p_x, p_y or $d_{xy}, d_{x^2-y^2}$). Therefore the possible hybrids are

(i) dsp^3 ($d_{z^2}, s, p_x, p_y, p_z$)

(ii) d^3sp ($d_{z^2}, d_{xy}, d_{x^2-y^2}, s, p_z$)

Since p orbitals are lower in energy than d orbitals, the dsp^3 hybrid combinations appear to be more realistic. In case of gaseous molecule $MoCl_5$, the appropriate hybridisation must be a mixture of both the schemes: dsp $(p-d)^2$ since the energies of $4d$ orbitals of Mo are comparable to its $5p$ orbitals.

6.1.6 Square Pyramidal AB_5 Molecule [(C_{4v}) ($Ni(CN)_5^{3-}$, $InCl_5^{2-}$)]

For the square pyramidal AB_5 molecule, the transformation of the vectors for the axial (r_5) and equatorial (r_1, r_2, r_3, r_4) can be worked out identically using the Fig. 6.8.

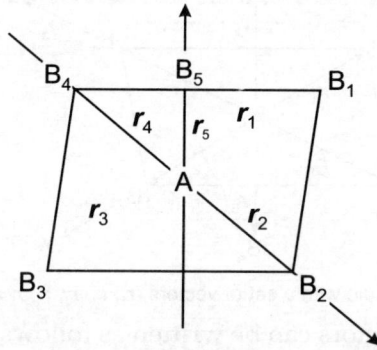

Fig. 6.8 Square pyramidal AB_5 molecule with a set of vectors $r_1 \rightarrow r_5$ for s hybrid orbitals

C_{4v}	E	$2\,C_4$	C_2	$2\,\sigma_v$	$2\,\sigma_d$
Γ_σ^{ax}	1	1	1	1	1
Γ_σ^{base}	4	0	0	2	0
Γ_{total}^σ	5	1	1	3	2
IRS					
Γ_{ax} ; A_1	1	1	1	1	1
$\Gamma_{eq}\begin{cases} A_1 \\ B_1 \\ E \end{cases}$	1	1	1	1	1
	1	−1	1	1	−1
	2	0	−2	0	0

Therefore $\qquad\qquad \Gamma_{total}^\sigma \rightarrow 2\,A_1 + B_1 + E$

The AOs belonging to these IRs are

IRs	AOs
A_1	s, p_z, p_{z^2}
B_1	$d_{x^2-y^2}$
E	p_x, p_y, d_{xz}, d_{yz}

For the possible hybridisation, combination will be from two A_1, one B_1 and a pair of E type orbitals.

dsp^3	$(d_{x^2-y^2}, s, p_z, p_x, p_y)$
d^2sp^2	$(d_{z^2}, d_{x^2-y^2}, s, p_x, p_y)$
d^3sp	$(d_{x^2-y^2}, d_{xz}, d_{yz}, s, p_z)$
d^2p^3	$(d_{z^2}, d_{x^2-y^2}, p_x, p_y, p_z)$
sd^4	$(s, d_{z^2}, d_{x^2-y^2}, d_{xz}, d_{yz})$

of these dsp^3 ($d = d_{x^2-y^2}$) may be more realistic than the other combinations. This is due to the fact that the s and p orbitals are lower in energy than the d orbitals.

6.1.7 Octahedral AB_6 Molecule (O_h)

In octahedral compounds, all the six σ bonds are identical (Fig. 6.9).

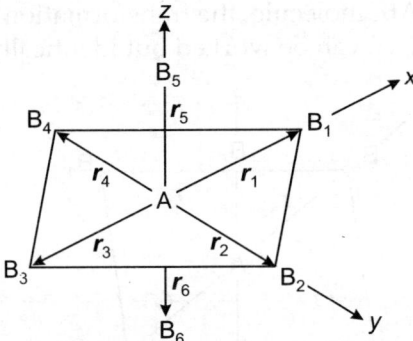

Fig. 6.9 Octahedral AB_6 molecule with a set of vectors, r_1, r_2, r_3, r_4, r_5 and r_6 for σ-hybrid orbitals

The Γ_{toal}^{σ} for all the six σ vectors can be written as follows.

O_h		E	$8\,C_3$	$6\,C_2$	$6\,C_4$	$3\,C_2$	i	$6\,S_4$	$8\,S_6$	$3\,\sigma_h$	$6\,\sigma_d$
	Γ_{total}^{σ}	6	0	0	2	2	0	0	0	4	2
IRS											
	A_{1g}	1	1	1	1	1	1	1	1	1	1
	E_g	2	−1	0	0	2	2	0	−1	2	0
	E_{1u}	3	0	−1	1	−1	−3	−1	0	1	1

Therefore $\quad \Gamma_{total}^{\sigma} \rightarrow A_{1g} + E_g + T_{1u}$

IRs	AOs
A_{1g}	s
E_g	$d_{z^2}, d_{x^2-y^2}$
T_{1u}	p_x, p_y, d_z

This result unequivocally leads to d^2sp^3 hybridisation (d^2 means dz^2 and $d_{x^2-y^2}$ orbitals).

6.2 HYBRIDISATION INVOLVING π-ORBITALS

6.2.1 Trigonal Planar AB₃ Molecule

For each B atom in the planar AB_3 molecule, there are two $2p$ orbitals, viz. $2p_x$ and $2p_y$ of π-nature of these two p orbitals **one is in the molecular plane perpendicular to A–B bond** and the second one is perpendicular to the molecular plane (Fig. 6.10).

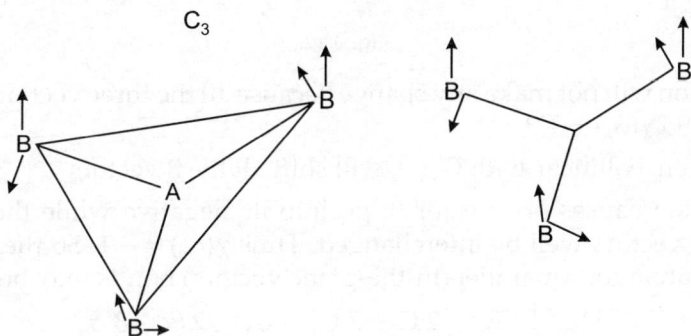

Fig. 6.10 Vectors representing π-orbtials of B atoms in the planar AB₃ molecule

These may be represented by vectors pointing in the appropriate directions and are referred to as π (∥) and π (⊥) orbitals respectively.

The two sets of π-orbitals may be treated separately as no vectors of one set is interchanged with the vectors of the second set under the symmetry operations of the group D_{3h}.

The character of a symmetry operation may be determined based on the following facts.

  (i)   A displaced vector contributes zero

 (ii)   An undisplaced vector without changing direction contributes + 1

(iii)   An undisplaced vector with changing direction contributes – 1.

By performing the operation of each class of the point group, we can get the total character for π-bond representation, Γ_Π.

For Γ_Π (∥) :

• The E operation will not make any change of the three vectors in the molecular plane. So $\chi(E) = + 3$ (Fig. 6.11).

• $C_3(z)$ operation shifts one vector into another

So $\chi(C_3) = 0$

• C_2 operation shifts each in-the-plane vector into its own negative sign. So $\chi(C_2) = -1$, the other two C_2 operations will contribute zero to the character because by these operations other two in-the-plane vectors will be interchanged (Fig. 6.12).

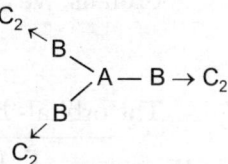

Fig. 6.11

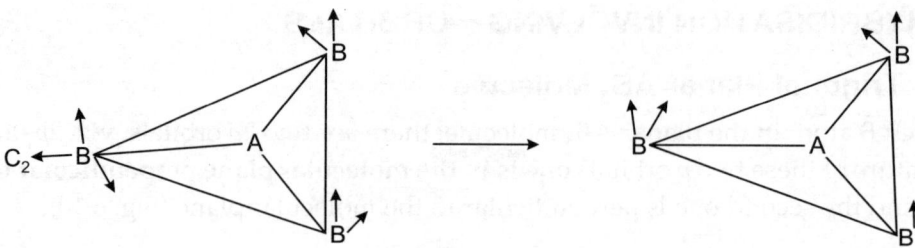

Fig. 6.12

- σ_h operation will not make any change because all the three vectors lie in the same plane σ_h. So $\chi(\sigma_h) = +3$
- S_3 operation, collinear with $C_3 (z)$, will shift all the 3 vectors $\therefore$ $\chi(S_3) = 0$
- σ_v operation causes one vector to go into its negative while the other two in-the-plane vectors well be interchanged. Thus $\chi(\sigma_v) = -1$. So the total reducible representation for π-parallel (in-the-plane vectors) bonds may be expressed as

D_{3h}	E	$2\,C_3$	$3\,C_2$	σ_h	$2\,S_3$	$3\,\sigma_v$
$\Gamma_{\Pi(\|)}$	3	0	−1	3	0	−1

Following the same manner, we can determine the total reducible representation for π-perpendicular (out of plane) bonds ($\pi\,(\perp)$). Here, the σ_h operation will invert the three out of plane vectors at their own position. Thus $\chi(\sigma_h) = -3$

σ_v operation causes one vector unshifted

$\therefore$ $\chi(\sigma_v) = +1$. The results are

D_{3h}	E	$2\,C_3$	$3\,C_2$	σ_h	$2\,S_3$	$3\,\sigma_v$
$\Gamma_{\pi(\perp)}$	3	0	−1	−3	0	1

Adding both representations, we get total π-orbital representation $\Gamma_\pi(R) = \Gamma_{\pi(\|)} + \Gamma_\pi(I)$

D_{3h}	E	$2\,C_3$	$3\,C_2$	σ_h	$2\,S_3$	$3\,\sigma_v$
$\Gamma_{\pi(R)}$	6	0	−2	0	0	0
$\Gamma_{\pi(\|)}$	3	0	−1	3	0	−1
$\Gamma_{\pi(\perp)}$	3	0	−1	−3	0	1

Since $\Gamma_\Pi\,(\perp)$ and $\Gamma_\Pi\,(\|)$ are independent reducible representation for π-hybrid orbitals, we can apply reduction formula to them separately to get the results as

$$\Gamma\pi\,(\perp) = A_2'' + E''$$
$$\Gamma\pi\,(\|) = A_2' + E'$$

The orbitals belonging to the above irreducible representation are as follows:

$\Gamma_\Pi(\perp)$ representation (out-of-plane vectors)		$\Gamma_\|$ (|) representation (in-the-plane vectors)	
AOs of A_2'' symmetry	AOs of E'' symmetry	AOs A_2' symmetry	AOs of E' symmetry
p_z	(d_{xz}, d_{yz})	None	(i) (p_x, p_y) (ii) $d_{x^2-y^2}, d_{xy}$

From the above discussion, the following conclusions can be drawn.

1. The orbitals $\pi(\perp)$ can form bonds with A atom involving pd^2 hybridisation $p_z + (d_{xz}, d_{yz}) = pd^2$.

2. The orbitals π (||) can form two bonds with A atom involving p_x and p_y (or $d_{x^2-y^2}$, d_{xy}) orbitals. These two bonds are shared equally among the three B atoms.

6.2.2 π-Bonding in Tetrahedral AB_4 Molecule

We shall now consider a tetrahedral AB_4 molecule, total $2 \times 4 = 8\pi$ type hybrid orbitals on atom A, two for each B-atom. So a maximum of two π atomic orbitals on each B atom is permitted for π-bonding in AB_4. The atomic orbitals are represented by vectors at right angles to each other and be in plane perpendicular to the A—B bond axis.

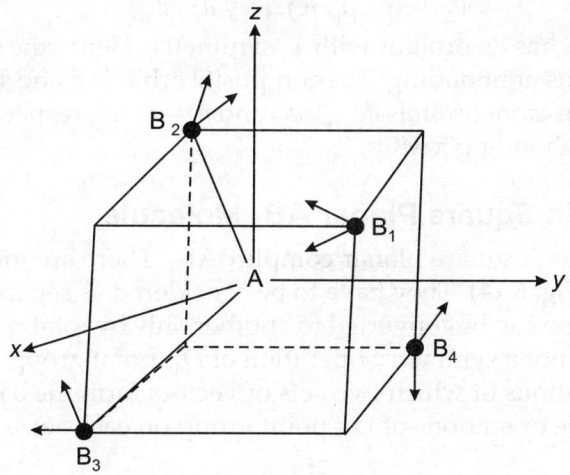

Fig. 6.13

So there are eight possible A—B π bonds. We may attach two vectors on each B atoms along the direction perpendicular to the bond axis and they are also $\perp$ to each other. There vectors form the bases for the representation of the group (Fig. 6.13).

Thus we have a set of 8 vectors, i.e. 2 on each bonded atom suitable for π bonds in T_d group.

The total character the π-hybrid representation can be determined by performing the symmetry operation (of each class of the group) on the vectors and counting the unshifted vectors. We thus get the results as follows.

T_d	E	$8\,C_3$	$3\,C_2$	$6\,S_4$	$6\,\sigma_d$
$\Gamma_\pi(R)$	8	−1	0	0	0

Under C_3 rotation each arrow (vector) on the ligand (B) $\perp$ to A—B bond is rotated by 120°, while the arrows (vectors) on the other ligands get shifted. The transformation matrix expressions is as follows.

$$\begin{pmatrix} x' \\ y' \end{pmatrix} = \begin{pmatrix} \cos\dfrac{2\pi}{3} & -\sin\dfrac{2\pi}{3} \\ \sin\dfrac{2\pi}{3} & \cos\dfrac{2\pi}{3} \end{pmatrix} \begin{pmatrix} x \\ y \end{pmatrix} = \begin{pmatrix} -\dfrac{1}{2} & -\dfrac{\sqrt{3}}{2} \\ \dfrac{\sqrt{3}}{2} & -\dfrac{1}{2} \end{pmatrix} \begin{pmatrix} x \\ y \end{pmatrix}$$

Therefore the character of C_3, $\chi(C_3) = -\dfrac{1}{2} + \left(-\dfrac{1}{2}\right) = -1$

By applying the reduction formula, the above reducible representation $\Gamma_\Pi (R)$ can be reduced to the sum of irreducible representations as

$$\Gamma_\pi (R) = E + T_1 + T_2$$

Thus there are eight π group orbitals formed.

Inspection of the T_d character table shows that A atom has the following AOs are available.

$$E \ : \ (d_{z^2}, d_{x^2 - y^2})$$
$$T_1 \ : \ \text{None}$$
$$T_2 \ : \ (p_x, p_y, p_z), (d_{xy}, d_{yz}, d_{zx})$$

The central atom A has no orbitals with T_1 symmetry. Hence the group orbitals with T_1 symmetry remain as nonbonding. The composite orbitals E and T_2 combine with the corresponding central atom orbitals $d_{x^2 - y^2}$, d_{z^2} and d_{xy}, d_{yz}, d_{zx} respectively, forming five bonding and five antibonding π MOs.

6.2.3 π-Bonding in Square Planar AB$_4$ Molecule

We shall now consider a square planar complex AB$_4$. There are four vectors in $\perp$ and four in $\parallel$ directions (Fig. 6.14). They have to be considered as separate sets because the vectors in one direction can be converted to another only by rotation through 90° along x- and y-axes. This is not a symmetry operation of D_{4h} point group. The total character of the two representations of which two sets of vectors form the bases, can be worked out by performing the operations of D_{4h} point group on each set.

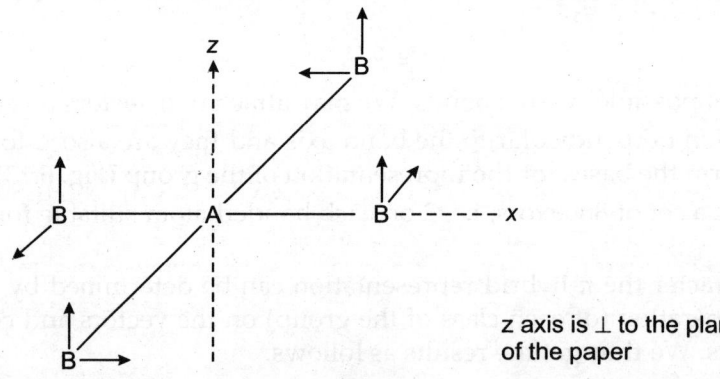

z axis is $\perp$ to the plane of the paper

Fig. 6.14

Now, the reducible representation for these two sets are to be worked out by performing symmetry operations of each class and determining the character. The results are given below:

D_{4h}	E	$2\,C_4$	C_2	$2\,C_2'$	$2\,C_2''$	i	$2\,S_4$	σ_h	$2\,\sigma_v$	$2\,\sigma_d$
$\Gamma_\pi (\perp)$	4	0	0	-2	0	0	0	-4	2	0
$\Gamma_\parallel (\parallel)$	4	0	0	-2	0	0	0	4	-2	0

The total π-representation $(\Gamma_\Pi (R)) = \Gamma_\pi (\perp) + \Gamma_\pi (\parallel)$

	E	$2C_4$	C_2	$2C_2'$	$2C_2''$	i	$2S_4$	σ_h	$2\sigma_v$	$2\sigma_d$
$\therefore \quad \Gamma_\pi(R) =$	8	0	0	-4	0	0	0	0	0	0

By applying the standard reduction formula on both π-representations, we get the following irreducible representations:

$$\Gamma_\pi(\perp) = A_{2u} + B_{2u} + E_g$$
$$\Gamma_\pi(\parallel) = A_{2g} + B_{2g} + E_u$$

The atomic orbitals corresponding to the irreducible representations are as follows:

IR	AO		IR	AO
A_{2u}	p_z		A_{2g}	None
B_{2u}	None		B_{2g}	d_{xy}
E_g	d_{xz}, d_{yz}		E_u	p_x, p_y

Thus to form the π bonds in perpendicular direction p_z, d_{xz} and d_{yz} orbitals can be used and thus three π bonds can be formed between four positions in AB_4.

For the formation of in-plane π bond, the orbitals d_{xy}, p_x and p_y can be used. However, p_x and p_y orbitals are already used up in the making of σ bonds. Hence only d_{xy} orbital of the central atom A can part in the formation of one π bond in the plane of σ bond.

6.2.4 π Bonding in Octahedral AB_6 Molecule

AB_6 type octahedral molecule belongs to O_h point group. In that case, the vectors are equivalent because they are interchangeable by C_4 operation. Operations of the O_h point group are applied on the 12 vectors and the total character is obtained (Fig. 6.15).

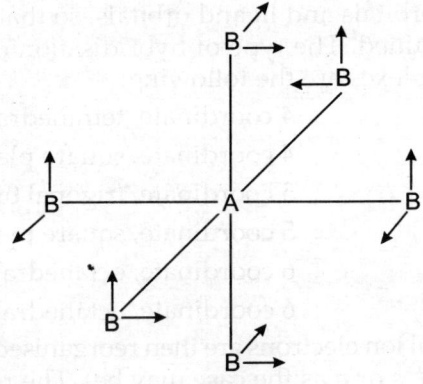

Fig. 6.15

O_h	E	$8\,C_3$	$6\,C_2$	$6\,C_4$	$3C_2(=C_4^2)$	i	$6\,S_4$	$8\,S_6$	$3\,\sigma_h$	$6\,\sigma_d$
Γ_Π	12	0	0	0	-4	0	0	0	0	0

The total character (Γ_π) can be broken down to the following irreducible representations.

$$\Gamma_\pi = T_{1g} + T_{2g} + T_{1u} + T_{2u}$$

IRs	AOs
T_{1g}	None
T_{2g}	d_{xy}, d_{xz}, d_{yz}
T_{1u}	p_x, p_y, p_z
T_{2u}	None

Since there are no central atom orbitals corresponding to T_{1g} and T_{2u} symmetry, all the 12π bonds cannot be formed. The central atom orbitals p_x, p_y, p_z and d_{xy}, d_{xz}, d_{yz} can take part in the formation of π orbitals. As p_x, p_y and p_z orbitals have already been utilised for the formation of σ orbitals in octahedral complexes, only d_{xy}, d_{xz} and d_{yz} orbitals are available for π bonding and only three π bonds can be formed between A and six B atoms.

6.3 VALENCE BOND THEORY IN TRANSITION METAL COMPLEXES

The valence bond theory of L Pauling and JL Slater is the oldest and is based on the concept of hybridisation of suitable orbitals of the metal ion. This VB theory has served inorganic chemist very well during 1930–1950 and has provided a useful understanding of the valence and structure of coordination compounds.

Metal complex formation between a metal ion (M^{n+}) and certain number of ligands (nL) is assumed to occur in the following manner.

(a) First the metal atom (**M**) loses the requisite number of electrons (ne) to form the cation (M^{n+}). The number of electrons the metal atom loses (ne) is the oxidation state of the resulting cation formed.

$$\mathbf{M} - ne \rightleftharpoons \mathbf{M}^{n+}$$

(b) A metal complex a has a definite geometry. Pure metal ion orbitals (s, p or d) cannot generate on their own such stereochemistries as tetrahedral, octahedral, square pyramidal square planar, etc. Pauling showed that a suitable combination of the pure atomic orbitals can lead to highly directed hybrid orbitals. These hybrid orbitals house the ligand lone pair electrons. As such, there is a significant overlap of the metal hybrid orbitals and ligand orbitals, so that desired stereochemistry of the complex is attained. The type of hybridisation occurring in the first row transition metal complexes are the following:

$4s4p^3$ (sp^3)	: 4 coordinate, tetrahedral
$3d_{x^2-y^2}4s4p_x4p_y$ (dsp^2)	: 4 coordinate, square planar
$3d_{z^2}4s4p^3$ (dsp^3)	: 5 coordinate, trigonal bipyramidal
$3d_{z^2}3d_{x^2-y^2}4s4p^2$ (d^2sp^2)	: 5 coordinate, square pyramidal
$4s4p^34d_{z^2}4d_{x^2-y^2}$ (sp^3d^2)	: 6 coordinate, octahedral (outer orbital)
$3d_{z^2}3d_{x^2-y^2}4s4p^3$ (d^2sp^3)	: 6 coordinate, octahedral (inner orbital)

(c) The nonbonding metal ion electrons are then reorganised to occupy the remaining metal orbitals (pure d, s or p as the case may be). The regrouping of electrons is achieved obeying Hund's rule, i.e. with maximum possible unpaired spins.

(d) A coordination complex is termed as an *inner orbital complex* (covalent) or *outer orbital complex* (ionic) depending upon whether the hybridisation involves the inner d orbitals or outer d orbitals.

The outer orbital octahedral complexes are given by weak ligand fields while the inner orbital octahedral complexes are formed by strong ligand fields.

(e) In addition to the σ bond, a π bond may be formed by overlap of a filled, suitable metal d orbital with a vacant ligand orbital: $M \rightleftharpoons L$. Such double bonding usually occurs in complexes of metal ions in low oxidation state.

A magnetic study reveals the number and rearrangement of the nonbonding electrons. These two factors indicate the type of hybridisation. Table 6.1 gives different hybridisation patterns. The hybridisation scheme with respect to Cr^{2+} and Mn^{3+} (d^4 configuration) is shown in Table 6.2. It should be noted that a complex with d^4 configuration gives the same number of unpaired electrons, i.e. four, for each of the sp^3, dsp^2, sp^3d^2 hybridisations. Also, the number of unpaired electrons is two for the d^2sp^3 hybridisation.

Table 6.1 Hybrid orbitals and directional properties

Number of covalent bonds	Hybrid orbitals	Stereochemistry	Example(s)
2	sp	Linear (180°)	$[Ag(NH_3)_2]^+$, $[CuCl_2]^-$ $[Hg(CN)_2]$, $HgCl_2$
3	s, p_x, p_y	Trigonal planar (120°)	BCl_3
4	sp^3	Tetrahedral (109)°	$Cs_2[CoCl_4]$, $[NiCl_2(PPh_3)_2]$ $[NiCl_2(OPPh_3)_2]$
	$d_{x^2-y^2}, s, p_x, p_y$	Square planar (90°)	$[Ni(dmg)_2]$, $K_2[PtCl_4]$, $[Cu(en)_2](NO_3)_2$, $[PtCl_4]^{2-}$, $[Pt(NH_3)_2Cl_2]$, $[Pt(NH_3)_4]^{2+}$
5	d_{z^2}, s, p^3	Trigonal bipyramidal (90°, 120° and 180°)	$[CuI(dipyridyl)_2]$ I, $[NiBr_3(P(C_2H_5)_3)_2]$
	$d_{z^2}, d_{x^2-y^2}, s, p_x, p_y$	Square pyramidal (90°)	$[VO(acac)_2]$, $[Fe(CO)_5]$
6	$s, p^3, d_{z^2}, d_{x^2-y^2}$	Octahedral (outer orbital) (90° and 180°)	$K_3[FeF_6]$, $Na_3[CoF_6]$
	$d_{z^2}, d_{x^2-y^2}, s, p^3$	Octahedral (inner orbital) (90° and 180°)	$[Co(NH_3)_6]$ Cl_3, $[Co(en)_3]$ Cl_3

acacH = acetylacetone; dmgH = dimethylglyoxime

Table 6.2 Hybridisation scheme as applied to d^4 and d^5 systems (↑, denotes metal ion electron; ×, bonding ligand electron; n, number of unpaired electrons)

	3d				4s	4p			4d			n
Cr²⁺, Mn³⁺ (d^4 system)	↑	↑	↑	↑	—	— — — —			— — — —			$\underline{4}$
Tetrahedral sp^3	↑	↑	↑	↑	—	×× ×× ×× ××			— — — —			$\underline{4}$
Sq. planar dsp^2	↑	↑	↑	↑	××	×× ×× ××			— — — —			$\underline{4}$
Outer orbital	↑	↑	↑	↑	—	×× ×× ×× ××			×× ×× —			$\underline{4}$
Octahedral sp^3d^2 Inner orbital	↑↓	↑	↑	×	×	×× ×× ×× ××			— — —			$\underline{2}$
Octahedral d^2sp^3 Mn²⁺, Fe³⁺ (d^5 system)	↑	↑	↑	↑	↑	— — — —			— — — —			$\underline{5}$
Tetrahedral sp^3	↑	↑	↑	↑	↑	×× ×× ×× ××			— — — —			$\underline{5}$
Octahedral sp^3d^2	↑	↑	↑	↑	↑	×× ×× ×× ××			×× ×× —			$\underline{5}$

| Octahedral d^2sp^3 | ↕ ↕ ↑ ×× | ×× | ×× ×× ×× ×× | — — — — — | **1** |
| Sq. planar dsp^2 | ↕ ↑ ↑ ↑ | ×× | ×× | ×× ×× — | — — — — — | **3** |

Further, the Mn^{2+} and Fe^{3+} ions (d^5 configuration) will give five unpaired electrons for each of the sp^3 and sp^3d^2 hybridisation, three unpaired electrons for the dsp^2 hybridisation, and only one unpaired electron for the d^2sp^3 hybridisation (Table 6.2). Table 6.3 lists the magnetic moments of some compounds of the first transition series metal ions.

Electrons in an incompletely filled shell give rise to a resultant angular momentum. The orbital moment is largely quenched by the surrounding ligands. Thus ignoring orbital contributions, the magnetic moment of a complex will depend upon the actual number of unpaired electrons (Table 6.3).

We shall now work out the applications of valence bond theory to coordination compounds to find out the number of unpaired electrons arising out of a certain hybridisation model.

Hexaminecobalt(III) ion [Co(NH₃)₆]³⁺

The $3d$ electron distribution of cobalt(III) ion is

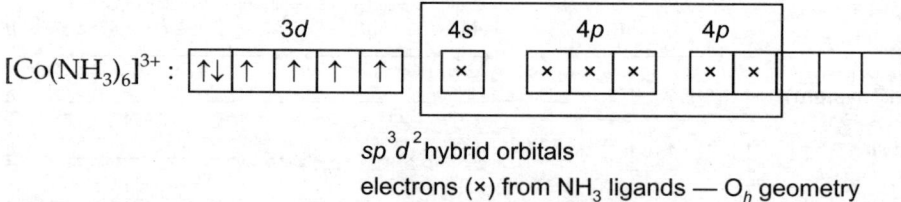

Cobalt(III) ion forms six equivalent linkages to the six nitrogen donor atoms of six ammonia molecules. Valence bond gives six equivalent bonds via sp^3d^2 or d^2sp^3 hybrid orbitals using the outer $4d$ orbitals or inner $3d$ orbitals. These hybrid orbitals will be occupied by ligand lone pairs of electrons. The metal ion will then be forced to rearrange the pairing of its own electrons in the available d orbitals either of the following ways.

With sp^3d^2 hybridisation of cobalt(III) orbital with ligand electrons (marked ×):

If sp^3d^2 hybridisation is accepted for the formation of [Co(NH₃)₆]³⁺ ion then the orbital diagram depicting the VB description of the cobalt-ammonia bonds is shown below:

[Co(NH₃)₆]³⁺ :

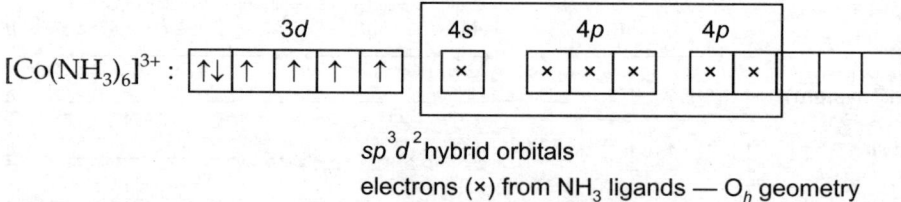

sp^3d^2 hybrid orbitals
electrons (×) from NH₃ ligands — O_h geometry

But of d^2sp^3 hybridisation is taken into consideration then cobalt-ammonia bond will be of the following type (× denotes ligand electrons):

[Co(NH₃)₆]³⁺ :

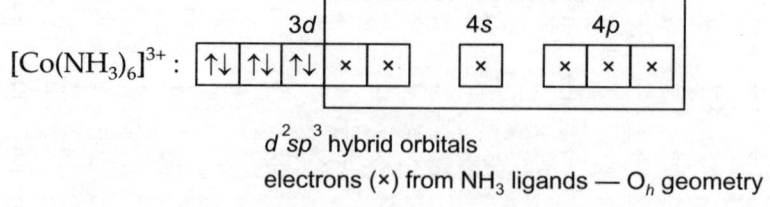

d^2sp^3 hybrid orbitals
electrons (×) from NH₃ ligands — O_h geometry

Experimental determination shows $[Co(NH_3)_6]$ Cl_3 is *diamagnetic,* i.e. with all the spins paired. Thus of the two spin arrangements d^2sp^3 (inner orbital hybridisation) is the correct one.

Sodium hexafluorocobaltate(III), Na$_3$ [CoF$_6$]

The magnetic moment for Na_3 $[CoF_6]$ is 5.39 BM. This indicates that there are four unpaired electrons present in the compound and the compound has the outer orbital hybridisation, (i.e. sp^3d^2) leaving all the five $3d$ orbitals for occupation by the six $3d$ electrons.

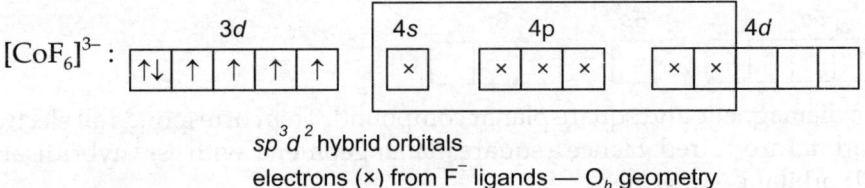

sp^3d^2 hybrid orbitals
electrons (×) from F⁻ ligands — O$_h$ geometry

Potassium tetracyanonickelate(II), K$_2$ [Ni(CN)$_4$]

The electron distribution of nickel(II) ion is given below.

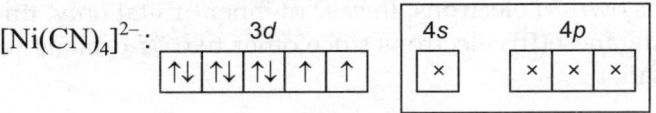

Following VB model, the four equivalent bonds may be either sp^3 or dsp^2 hybridisation and such hybrid orbitals of nickel(II) ion will be occupied by CN⁻ electrons. The d orbitals left out of the hybridisation scheme will be used by nickel(II) to reorganise the spins of its d electrons in either of the following ways.

For sp^3 nickel(II) with CN⁻ electrons pair (×)

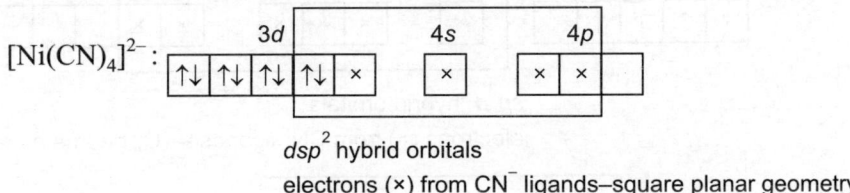

sp^3 hybrid orbitals
electrons (×) from CN⁻ ligands – T$_d$ geometry

For dsp^2 nickel(II) with CN⁻ electron pair (×):

$[Ni(CN)_4]^{2-}$:

dsp^2 hybrid orbitals
electrons (×) from CN⁻ ligands–square planar geometry

$K_2[Ni(CN)_4]$ is a diamagnetic compound, i.e. no unpaired electron is present. This supports square planar structure and rejects the tetrahedral structure containing two unpaired electrons.

Tetraethylammonium tetrachloronickelate(II) [(C$_2$H$_5$)$_4$ N]$_2$ [NiCl$_4$]

The compound is paramagnetic (μ_{eff} = 3.89 BM) corresponding to two unpaired electrons. Hence a tetrahedral geometry with sp^3 hybridisation of nickel(II) orbital is indicated.

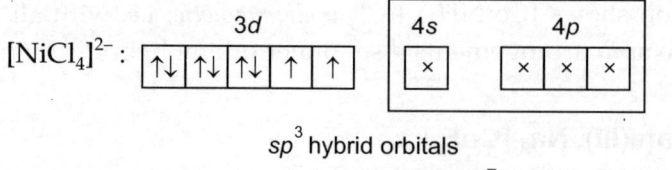

$$sp^3 \text{ hybrid orbitals}$$

electrons (×) from Cl⁻ ligands –T_d geometry

Tetrachloroplatinate(II) ion, [PtCl₄]²⁻

The electron distribution in bipositive platinum(II) is given below.

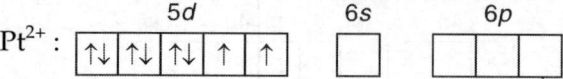

[PtCl₄]²⁻ is a diamagnetic and square planar compound. This corresponds all electrons present in $5d$ orbital are paired. Hence a square planar geometry with dsp^2 hybridisation of platinum(II) orbital is suggested.

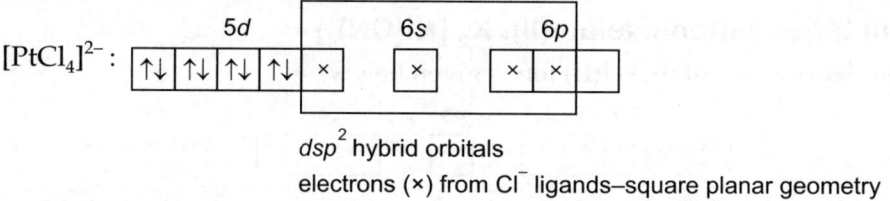

$$dsp^2 \text{ hybrid orbitals}$$

electrons (×) from Cl⁻ ligands–square planar geometry

Potassium hexacyanomanganate(III), K₃ [Mn(CN)₆]

Manganese(III) ion is simply a $3d^4$ ion. The outer orbitals arrangement is sp^3d^2 while inner orbitals arrangement is d^2sp^3. These orbitals will be occupied by the cyanide electrons. In case of the outer orbital hybridisation all five $3d$ orbitals available to manganese(III) for its own $3d$ electrons. In case of inner orbital only, three $3d$ orbitals are available for manganese(III) electrons since other two $3d$ orbitals will have to be vacated for hybridisation.

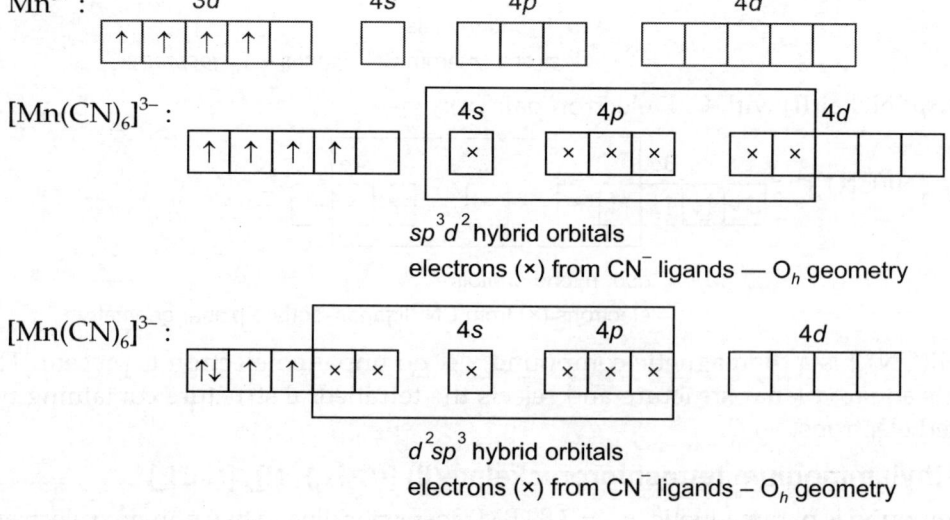

$$sp^3d^2 \text{ hybrid orbitals}$$

electrons (×) from CN⁻ ligands — O_h geometry

$$d^2sp^3 \text{ hybrid orbitals}$$

electrons (×) from CN⁻ ligands – O_h geometry

Experimental value of $K_3[Mn(CN)_6]$ indicates two unpaired electrons and thus rejects the outer orbital hybridisation but supports inner orbital octahedral case.

Pentacarbonyliron(O), Fe(CO)₅

Ion(O) is a $3d^8$ system. In $Fe(CO)_5$ the coordination number of iron(O) is five. Valence bond theory predicts either a dsp^3 hybridisation for trigonal bipyramidal structure or a d^2sp^2 hybridisation for square pyramidal geometry. These hybrid orbitals will be occupied by carbonyl electrons. For dsp^3 and d^2sp^2 hybridisation, the following arrangement will emerge.

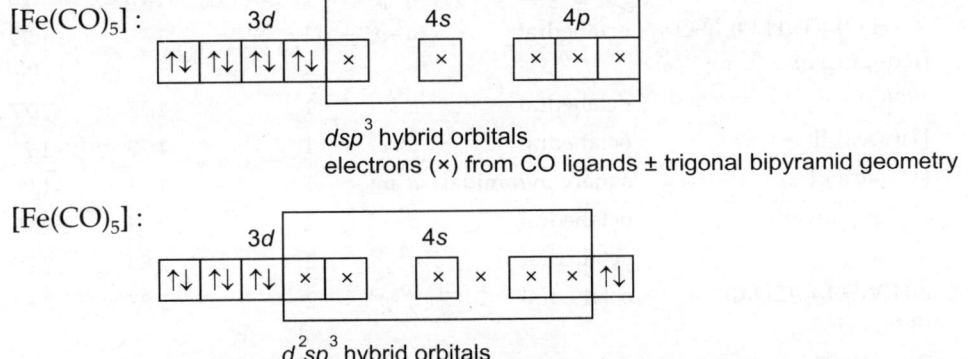

In case of dsp^3 hybridisation, all the eight electrons will remain paired in the form of $3d$ orbitals of iron supporting diamagnetic character of compound. However, in d^2sp^2 hybridisation enough $3d$ orbitals are not available for the accommodation of eight electrons of iron. Thus, two electrons have to be promoted to higher energy outer orbitals. But this is an unfavourable process as promotion to higher energy orbitals will make the electrons vulnerable to oxidation. Thus, a trigonal bipyramidal stereochemistry is accepted.

6.3.1 Shortcomings of Valence Bond Theory

While VB approach, as outlined above, has proved to be extremely useful, it has several inherent limitations also because of which its use is now restricted to a fairly elementary level.

1. Most of the first transition series complexes respond to spin only values for their magnetic moments. But cobalt(II) and nickel(II) in tetrahedral and octahedral geometries register magnetic moments substantially higher than the spin only values. For example, tetrahedral $[CoCl_4]^{2-}$ has $\mu_{SO} = 3.88$ BM while the experimental value is 4.60 BM. Similarly, octahedral $[Ni(NH_3)_6]Cl_2$ complex has $\mu_{SO} = 2.83$ BM while its experimental value is 3.32 BM. Such enhanced magnetic moment values are not explained by VB theory.

2. Magnetic properties of the d^1, d^2 and d^3 configurations are nonresponsive to hybridisation schemes, i.e. to stereochemistries (Tables 6.3 and 6.4).

   For outer orbital, octahedral sp^3d^2 and tetrahedral sp^3 the same number of unpaired electrons are indicated, so that magnetic moment would not distinguish between the two geometries.

3. Both the colour and the magnetic moments of transition metal complexes are due to their possessing d orbital electrons. A correlation between them must exist. Unfortunately VB theory could not establish any connection between them.

Table 6.3 Magnetic moments of some complexes of first row transition metals

Configu-ration	Example	Stereochemistry	Hybrid orbitals	No. of unpaired electrons(n)	μ (BM) Spin only	μ (BM) Et_3p
d^1	$(NH_4)_3(TiF_6)$(purple)	octahedral	d^2sp^3	1	1.73	1.78
	K_3TiF_6	octahedral	d^2sp^3	1	1.73	1.70
	$[C_s(H_2O)_6][Ti(H_2O)_6](SO_4)_2$ (red–purple)	octahedral	d^2sp^3	1	1.73	1.79
	$VO(Q)_2$	octahedral	d^2sp^3	1	1.73	1.77
	$[Ti(urea)_6]I_3$	octahedral	d^2sp^3	1	1.73	1,77
	$[VO(acac)_2]$	square pyramidal	d^2sp^2	1	1.73	1.70
d^2	$K_3[VF_6]$ (green)	octahedral	d^2sp^3	2	2.83	2.79
	$V(acac)_3$	octahedral	d^2sp^3	2	2.83	2.80
	$NH_4V(SO_4)_2.12H_2O$ (Blue – violet)	octahedral	d^2sp^3	2	2.83	2.80
	$[VCl_3(THF)_3]$ (orange)	octahedral	d^2sp^3	2	2.83	2.80
d^3	$[Cr(NH_3)_6]Br_3$(yellow)	octahedral	d^2sp^3	3	3.88	3.77
	$[Cr(bigH)_3]Cl_3$(yellow)	octahedral	d^2sp^3	3	3.87	3.86
	$[Cr(o\text{-}phen) (bigH)_2]Cl_3$	octahedral	d^2sp^3	3	3.87	3.79
	$[Cr(en)_3]I_3H_2O$(yellow)	octahedral	d^2sp^3	3	3.87	3.84
	$K_3[Cr(CN)_6]$(yellow)	octahedral	d^2sp^3	3	3.87	3.87
	$K_3[Cr(C_2O_4)_3].3H_2O$ (reddish – violet)	octahedral	d^2sp^3	3	3.87	3.84
		octahedral	d^2sp^3	3	3.87	3.84
	$K[Cr(H_2O)_6](SO_4)_2.6H_2O$ (violet)	octahedral	d^2sp^3	3	3.87	3.84
d^4	$[Cr(H_2O)_6]SO_4$	octahedral	sp^3d^2	4	4.90	4.80
d^5	$Na_3[FeF_6]$	octahedral	sp^3d^2	5	5.92	5.85
	$K_3[Fe(CN)_6]$	octahedral	d^2sp^3	1	1.73	2.25
	$[Mn(py)_6]Br_2$	octahedral	sp^3d^2	5	5.92	6.00
	$K_3[Mn(CN)_6]3H_2O$	octahedral	d^2sp^3	1	1.73	2.18
	$(Et_4N)_2 [MnCl_4]$	tetrahedral	sp^3	5	5.92	5.94
d^6	$[Fe(NH_3)_6]Cl_2$	octahedral	sp^3d^2	4	4.90	5.45
	$K_4[Fe(CN)_6]$	octahedral	d^2sp^3	0	0	0
	$[Fe(Ph_3P)_2] I_2$	tetrahedral	sp^3	4	4.90	5.10
	Fe(II) phthalocyanine	planar	dsp^2	2	2.83	3.96
	$[Co(NH_3)_6]Cl_3$	octahedral	d^2sp^3	0	0	0
	$[Et_4N]_2 [FeCl_4)$	tetrahedral	sp^3	4	4.90	5.40

Contd...

Configuration	Example	Stereochemistry	Hybrid orbitals	No. of unpaired electrons(n)	μ (BM)	
					Spin only	Et_3p
	$Na_3[CoF_6]$	octahedral	d^2sp^3	4	4.90	5.39
d^7	$[Co(en)_3]SO_4$	octahedral	d^2sp^3	3	3.88	4.56
	$[Co(py)_2Cl_2]$(violet)	octahedral	sp^3d^2	3	3.88	5.15
	$[Co(py)_2Cl_2]$(blue)	tetrahedral	sp^3	3	3.88	4.42
	Co(II)phthalocyanine	planar	dsp^2	1	1.73	2.73
	$Cs_2[CoCl_4]$	tetrahedral	sp^3	3	3.88	4.60
	$K_2Ba[Co(NO_2)_6]$	octahedral	d^2sp^3	1	1.73	1.88
d^8	$[Ni(NH_3)_6]Cl_2$	octahedral	sp^3d^2	2	2.83	3.32
	$[Ni(diars)_3]$ (ClO_4)	octahedral	d^2sp^3	0	0	0
	$(Et_4N)_2[NiCl_4]$	tetrahedral	sp^3	2	2.83	3.89
	$[Ni(Et_3P)_2Cl_2]$	planar	dsp^2	0	0	0
	$Ni(big(H)_2)Cl_2$	planar	dsp^2	0	0	0
	$K_2[Ni(CN)_4]$	planar	dsp^2	0	0	0
d^9	$[Cu(o-phen)_3]$ $(ClO_4)_2$	octahedral	sp^3d^2	1	1.73	1.96
	$[Cu(big\ H)_2]Cl_2$ (rose red)	planar	dsp^2	1	1.73	1.79
	$Cu(dmg)_2Cl_2$	planar	dsp^2	1	1.73	1.85
	$Cs_2[CuCl_4]$	tetrahedral	sp^3	1	1.73	2.0

acac H = acetylacetone; big H = protonated biguanide; QH = quinaldinic acid; $H_2C_2O_4$ = oxalic acid; THF = tetrahydrofuran
diars = o-phenylenelr's dimethylarsine; dmg H = dimethyl-glyoxime; o–phen = o-phenanthroline; Et_3p = triethylphosphine

Table 6.4 Values of n (unpaired electrons) for different stereochemistry and configurations

Number of d electrons	1	2	3	4	5	6	7	8	9
Octahedral (outer orbital sp^3d^2)	1	2	3	4	5	4	3	2	1
Octahedral (inner orbital d^2sp^3)	1	2	3	2	1	0	1	0	1
Tetrahedral, sp^3	1	2	3	4	5	4	3	2	1
Square planar, dsp^2	1	2	3	4	3	2	1	0	1

PROBLEMS

1. Workout the hybrid orbitals of atom A in the molecule AB_3.
2. Workout the hybrid orbitals of C atom in the molecule CH_4.
3. Workout the hybrid orbitals of atom A in a planar molecule AB_4.
4. Discuss the hybrid orbitals in π bonding in the planar AB_3 and AB_4 molecules.

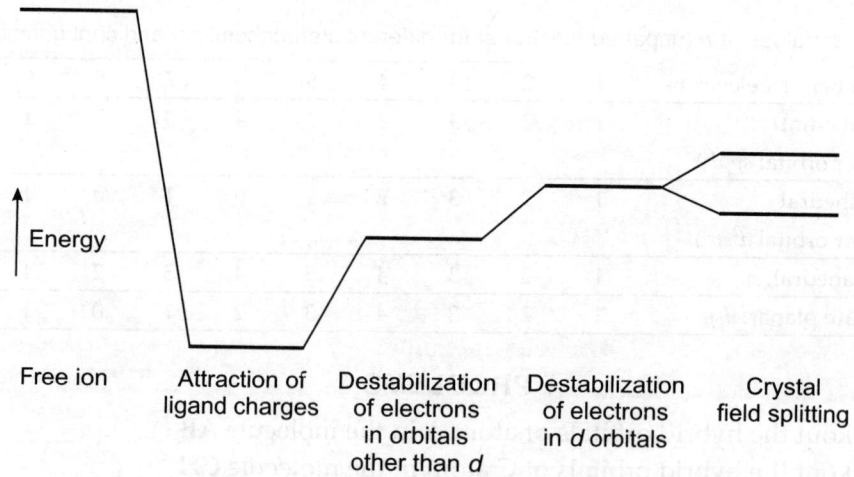

Application of Group Theory to Crystal Field Theory

In crystal field theory, the metal ion and ligands are considered as point charges and the ligands get arranged around the metal ion due to electrostatic attraction, preferring a structure with least energy. Due to this electrostatic attraction, the system attains stability and the liberated energy contributes significantly in the formation of the complex.

This theory goes further in considering that the ligand ions create an electrical field around the metal ion and thus perturb the energies of the metal orbitals. If the electrical field is spherical, it raises the energies of s, p and d orbitals of the metal ion uniformly. In other words, due to the vicinity of the negatively charged ligand field the electrons in the metal orbitals feel repulsion and hence the energy is raised. But spherical shape still retains the triple degeneracy of p orbitals or pentadegeneracy of d orbitals.

However, in case of an octahedral complex, the ligands are at the corners of an octahedron. They still affect the p orbitals equally since p_x, p_y and p_z orbitals are along the axis. Thus, the triple degeneracy of p orbitals is retained.

However, the orientation of five d orbitals are different. The $d_{x^2-y^2}$ are d_{z^2} orbitals are along the axis and d_{xy}, d_{yz} and d_{zx} orbitals are in between the axes. So the d orbital of the metal ion which are oriented directly along the axes (i.e. $d_{x^2-y^2}$ and d_{z^2}) will be raised in energy due to repulsion between ligand electrons and metal d electrons (Fig. 7.1).

Fig. 7.1 The energy of a coordination cluster on the crystal field model

Hence, we conclude that in case of nonspherical field the effect of the ligand field is different on different d orbitals and their degeneracy is removed, i.e. splitting of d orbitals.

The splitting of the d orbitals can be understood in terms of group theory. The complexes of the type ML_6 belong to O_h point group. All the d orbitals are symmetrical with respect to inversion and hence the inversion operation (i) in O_h point group can be excluded (i.e. i operation is not considered here). This leaves us with the point group O with identity operation E and rotation operations C_4, C_3 and C_2 (O is the subgroup of O_h and the point group O_h is the direct product group, $O \times C_i$).

The operations of the O group are performed on the d orbitals and the total character of the reducible representation of which the d orbitals form the basis, is obtained.

E operation: Identity operation leaves all the five d orbitals unchanged. Hence the total character is 5.

C_4 operation: On performing C_4 operation some d orbitals change their positions while the position of one d orbital remains same, i.e. unchanged.

Orbital	Position	Character χ
d_{z^2}	Orbital signs unchanged	+1
$d_{x^2-y^2}$	Orbital signs changed	−1
d_{xy}	Orbital sign changed	−1
d_{xz}	Changes position	0
d_{yz}	Changes position	0

∴ Total character for C_4 rotation $= 1 + (-1) + (-1) + 0 + 0 = -1$

However, it is very difficult to imagine the changes in the d orbitals on performing all the rotation operations, as done in the case of C_4. Hence, the matrix of the rotation operation is worked out.

7.1 GROUP THEORETICAL TREATMENT OF CRYSTAL FIELD SPLITTING

Since the operations are carried out on the d orbital wave functions, the orbital wave function can be expressed as hydrogen like wave function, i.e.

$$\psi_{n, l, m_l} (r, \theta, \phi) = R_{n, l}(r) \cdot (H)_{l, m_l}(\theta) \cdot \Phi_{m_l}(\phi)$$

where n, l, m_l are the quantum numbers are r, θ, ϕ are polar coordinates. $R_{n, l}(r)$ is the radial part of the wave function which depends on r but is independent of the angles θ and ϕ. $(H)_{l, m_l}(\theta) \, \Phi_{m_l}(\phi)$ is the angular part of the wave function. $(H)_{l, m_l}$ depends on the angle θ and $\Phi_{m_l}(\phi)$ depends on the angle ϕ.

During the rotation, r and θ parts of the d orbital wave functions remain unchanged, i.e. $R_{n, l}(r)$ and $(H)_{l, m_l}(\theta)$ remain unchanged. We, therefore, consider only $\Phi_{m_l}(\phi)$ part of the d orbital wave functions in working out the matrix for rotation operation since the angle ϕ is varied during the rotation about the z-axis (Fig. 7.2).

The unnormalised angular wave function is expressed as

$$\Phi_{m_l}(\phi) = e^{im_l\phi}$$

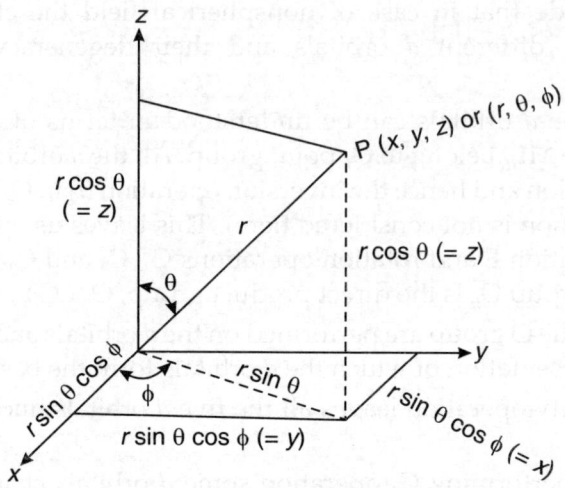

Fig. 7.2 Relation between rectangular and spherical polar coordinates

For d orbitals $m_l = +2, +1, 0, -1, -2$. If we take the function $e^{im_l\phi}$ and rotate anticlockwise by an angle α, the function becomes $e^{im_l(\phi + \alpha)}$. The set of $\Phi_{m_l}(\phi)$ wave functions, I, becomes II on rotation by α.

$$
\begin{array}{l}
l = 2, m_l = 2 \\
l = 2, m_l = 1 \\
l = 2, m_l = 0 \\
l = 2, m_l = -1 \\
l = 2, m_l = -2
\end{array}
\begin{pmatrix}
e^{2i\phi} \\
e^{i\phi} \\
e^{0} \\
e^{-i\phi} \\
e^{-2i\phi}
\end{pmatrix}
\xrightarrow{\text{rotation by } \alpha}
\begin{pmatrix}
e^{2i(\phi + \alpha)} \\
e^{i(\phi + \alpha)} \\
e^{0} \\
e^{-i(\phi + \alpha)} \\
e^{-2i(\phi + \alpha)}
\end{pmatrix}
$$

$$(\text{I}) \qquad\qquad\qquad\qquad (\text{II})$$

The new matrix (II) can be written as the product of two matrices as given below.

$$
\begin{pmatrix}
e^{2i(\phi + \alpha)} \\
e^{i(\phi + \alpha)} \\
e^{0} \\
e^{-i(\phi + \alpha)} \\
e^{-2i(\phi + \alpha)}
\end{pmatrix}
=
\begin{pmatrix}
e^{2i\alpha} & 0 & 0 & 0 & 0 \\
0 & e^{i\alpha} & 0 & 0 & 0 \\
0 & 0 & e^{0} & 0 & 0 \\
0 & 0 & 0 & e^{-i\alpha} & 0 \\
0 & 0 & 0 & 0 & e^{-2i\alpha}
\end{pmatrix}
\begin{pmatrix}
e^{2i\phi} \\
e^{i\phi} \\
e^{0} \\
e^{-i\phi} \\
e^{-2i\phi}
\end{pmatrix}
$$

Thus, the matrix for rotation operation is

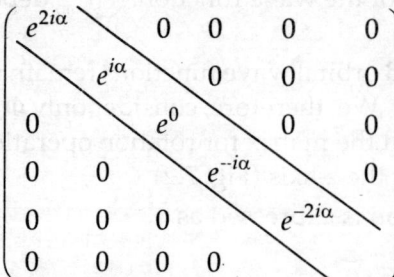

The total character for the operation is the summation of the diagonal elements

$$\chi(\alpha) = \chi(C_n) = e^{2i\alpha} + e^{i\alpha} + e^0 + e^{-i\alpha} + e^{-2i\alpha}$$
$$= e^{-2i\alpha} (e^0 + e^{-i\alpha} + 2^{i\alpha} + e^{3i\alpha} + e^{4i\alpha})$$
$$= \frac{\sin\left(l + \dfrac{1}{2}\right)\alpha}{\sin(\alpha/2)} \qquad\qquad\qquad ...(7.1)$$

where l is the orbital angular momentum quantum number, (i.e. azimuthal quantum number). $l = 2$ for d orbitals.

The above formula is independent of the point group and $\chi(\alpha)$ value depends on the value of l and hence the orbital.

In case of C_2 rotation, $\alpha = \dfrac{2\pi}{2} = \pi = 180°$

for C_3 rotation, $\alpha = \dfrac{2\pi}{3} = 120°$

for C_4 rotation, $\alpha = \dfrac{2\pi}{4} = 90°$

Substituting the values of α in cases of the different operations, the total character for the operation can be calculated.

For C_2, $\chi(\alpha) = \chi(C_2) = \dfrac{\sin\left(2 + \dfrac{1}{2}\right)\pi}{\sin\left(\dfrac{\pi}{2}\right)} = 1$

For C_3, $\chi(\alpha) = \chi(C_3) = \dfrac{\sin\left(2 + \dfrac{1}{2}\right)\dfrac{2\pi}{3}}{\sin\left\{\left(\dfrac{2\pi}{3}\right) \Big/ 2\right\}} = -1$

For C_4, $\chi(\alpha) = \chi(C_4) = \dfrac{\sin\left(2 + \dfrac{1}{2}\right)\dfrac{2\pi}{4}}{\sin\left\{\left(\dfrac{2\pi}{4}\right) \Big/ 2\right\}} = -1$

The above result may be presented in a tabular form.

Symmetry operation (C_n)	Angle of rotation (α)	Character (χ)
C_2	180°	1
C_3	120°	−1
C_4	90°	−1

Character of E

$\chi(E)$ can be determined following Eq. (7.1) by substituting $\alpha = 0$, so that

$$\chi(C_n) = \chi(E) = e^0 (e^0 + e^0 + e^0 + e^0 + e^0) = 1 \times 5 = 5$$

For identity operation, (i.e. when $\alpha = 0$) each of the diagonal elements is equal to 1 and the character is equal to $(2l + 1)$. Therefore the character

$$\chi(E) = (2l + 1) \text{ (orbital degeneracy)} \qquad \qquad ...(7.2)$$

This value is determined in the following way:

For any atomic orbital the general formula of $\chi(C_n)$ is given by

$$\chi(C_n) = \frac{\sin\left(l + \dfrac{1}{2}\right)\alpha}{\sin\left(\dfrac{\alpha}{2}\right)}$$

This value can not be determined directly since

$$\chi(C_n) = \frac{\sin\left(l + \dfrac{1}{2}\right) \times 0}{\sin(0/2)} = \frac{0}{0} \text{ (indeterminate)}$$

However, by using the l Hospital's rule, we get

$$\underset{\alpha \to 0}{\text{Lt}} \ \frac{\sin\left(l + \dfrac{1}{2}\right)\alpha}{\sin\left(\dfrac{\alpha}{2}\right)}$$

$$= \underset{\alpha \to 0}{\text{Lt}} \left[\frac{d}{d\alpha}\left\{\left(\sin l + \frac{1}{2}\right)\alpha\right\} \Big/ \frac{d}{d\alpha}\left\{\sin\left(\frac{\alpha}{2}\right)\right\} \right]$$

$$= (2l + 1) \ \underset{\alpha \to 0}{\text{Lt}} \ \frac{\cos\left(l + \dfrac{1}{2}\right) \times 0}{\cos(0/2)}$$

$$= (2l + 1) \times \frac{1}{1} = (2l + 1)$$

Thus, the total character (χ) for the reducible representation of which the d orbitals form the basis can be shown as follows:

$$\chi \text{ (character)}$$

O	E	$6C_4$	$3C_2 (= C_4^2)$	$8C_3$	$6C_2'$
Γ_d	5	–1	1	–1	1

It is to be noted that the total character for the C_4 operation obtained from the matrix consideration is the same as was obtained by considering the changes in the individual d orbitals on performing the operations. The reducible representation (Γ_d) so obtained above can be reduced to its components irreducible representations by using the standard reduction formula.

The partial form of the point group is shown below.

O	E	$6C_4$	$3C_2 (= C_4^2)$	$8C_3$	$6C_2'$
E	2	0	2	–1	0
T_2	3	–1	–1	0	1
Γ_d	5	–1	1	–1	1

$$n(E) = \frac{1}{24} [(1)(5)(2) + (6)(-1)(0) + (3)(1)(2) + 8(-1)(-1) + (6)(0)(1)] = 1$$

$$n(T_2) = \frac{1}{24} [(1)(5)(3) + (6)(-1)(-1) + (3)(1)(-1) + 8(-1)(0) + (6)(1)(1)] = 1$$

So we have

$$\Gamma_d = E + T_2$$

Lower case Mulliken symbols are used to represent reducible representations for one electron atomic orbitals. So we may write

$$\Gamma_d = e + t_2$$

This is for O point group*. On adding inversion (i) we get the irreducible representations for the O_h point group**. Thus for O_h point group we may write

$$\Gamma_d = E_g + T_{2g}$$
$$\Gamma_d = e_g + t_{2g}$$

The orbitals corresponding to these irreducible representations are

$$e_g = d_{z^2}, d_{x^2-y^2}$$
$$t_{2g} = d_{xy}, d_{yz}, d_{zx}$$

Thus, we see that on imposing an O_h field, pentadegenerate d orbitals split up into two sets, doublydegenerate, e_g ($d_{z^2}, d_{x^2-y^2}$) and triplydegenerate, t_{2g} (d_{xy}, d_{yz}, d_{zx}). Orbitals which are doubly degenerate ($d_{z^2}, d_{x^2-y^2}$) possess 1g subscript and triplydegenerate orbitals (d_{xy}, d_{yz}, d_{zx}) have 2g subscript.

We can use group theory for tetrahedral complexes (ML_4) also as done for O_h complexes. Tetrahedral complex ML_4 belongs to T_d point group. So the total character for reducible representation can be obtained by performing the symmetry operations of T_d group on the d orbitals which form the basis for the representation of the group.

T_d	E	$8C_3$	$3C_2$	S_4	σ_d
Γ_d	5	–1	1	–1	1

This is reduced to the irreducible representation using the character table of T_d point group and standard reduction formula as

$$\Gamma_d = E + T_2$$

In the tetrahedral group, there is no inversion operation (centre of symmetry) so the subscript g or u are not used in the symbols of irreducible representations. d_{xy}, d_{yz} and d_{zx} orbitals belong to T_2 representation and $d_{x^2-y^2}, d_{z^2}$ orbitals belong to E representation

---

* O is the subgroup of O_h
** Point group O_h is the direct product group, $O \times C_i$

$$e = d_{x^2-y^2}, d_{z^2}$$
$$t_2 = d_{xy}, d_{yz}, d_{zx}$$

Hence in tetrahedral field also, the five degenerate d orbitals of the free metal ion splits into two sets of degenerate orbitals (e and t_2) and in this way a tetrahedral complex is formed.

7.2 SPLITTING OF OTHER TERMS UNDER WEAK INTERACTIONS

Since s orbital has a spherical symmetry with no θ and ϕ dependent terms, the representation based on s orbital will include one as the character in all the rotational operations. Hence the terms with S belongs to A_{1g} representation.

For the p orbitals, $l = 1$. The character of the rotational operation is given by Eq. (7.1)

$$\chi(\alpha) = \chi(C_n) = \frac{\sin\left(l+\dfrac{1}{2}\right)\alpha}{\sin\left(\dfrac{\alpha}{2}\right)} \quad (\alpha \neq 0)$$

For $l = 1$, the above equation may be written as

$$\chi(\alpha) = \chi(C_n) = \frac{\sin\left(1+\dfrac{1}{2}\right)\alpha}{\sin\left(\dfrac{\alpha}{2}\right)} = \sin\left(\frac{3\alpha}{2}\right) \Big/ \sin\left(\frac{\alpha}{2}\right)$$

For C_2 rotation, $\alpha = 180°$, $\chi(C_2) = \dfrac{\sin 270°}{\sin 90°} = -1$

For C_3 rotation, $\alpha = 120°$, $\chi(C_3) = \dfrac{\sin 180°}{\sin 60°} = 0$

For C_4 rotation, $\alpha = 90°$, $\chi(C_4) = \dfrac{\sin 135°}{\sin 45°} = 1$

Since $\chi(E) = 2l + 1$, so $\chi(E) = 2.1 + 1 = 3$. So for p orbitals the reducible representations (Γ_p) is then

O	E	$6C_4$	$3C_2 (= C_4^2)$	$8C_3$	$6C_2$
Γ_p	3	1	−1	0	−1

From the character table of the group O, the above representation corresponds to the irreducible representation t_1. Hence the p term of d^2 configuration belongs to T_{1g} representation under octahedral environment.

For the f orbitals, $l = 3$ and the character of rotational operations is as follows:

C_2 rotation, $\alpha = 180°$, $\chi(C_2) = \dfrac{\sin 630°}{\sin 90°} = -1$

C_3 rotation, $\alpha = 120°$, $\chi(C_3) = \dfrac{\sin 420°}{\sin 90°} = 1$

C_4 rotation, $\alpha = 90°$, $\chi(C_4) = \dfrac{\sin 315°}{\sin 45°} = -1$

Since $\chi(E) = 2l + 1$. $\therefore \chi(E) = 2.3 + 1 = 7$

Thus the reducible representation (Γ_f) of f orbitals is

O	E	$6C_4$	$3C_2 (= C_4^2)$	$8C_3$	$6C_2$
Γ_f	7	−1	−1	1	−1

From the character of the group O, this is equivalent to $a_2 + t_1 + t_2$. Since orbitals are odd to inversion, the f orbitals split into a_{2u}, t_{1u} and t_{2u}. So the term F under octahedral symmetry will split into A_{2g}, T_{1g} and T_{2g}.

In case of g orbitals ($l = 4$), the character of rotational operation is given by

C_2 rotation, $\alpha = 180°$, $\chi(C_2) = \dfrac{\sin 810°}{\sin 90°} = 1$

C_3 rotation, $\alpha = 120°$, $\chi(C_3) = \dfrac{\sin 540°}{\sin 60°} = 0$

C_4 rotation, $\alpha = 90°$, $\chi(C_4) = \dfrac{\sin 405°}{\sin 45°} = 1$

Since $\chi(E) = 2l + 1$, so $\chi(E) = 2.4 + 1 = 9$

Hence the reducible representation is

O	E	$6C_4$	$3C_2 (= C_4^2)$	$8C_3$	$6C_2$
Γ_g	9	1	1	0	1

From the character table of the group O, this is equivalent to a_1. e, t_1 and t_2. As the g orbitals are even to inversion, the g orbitals split into a_{1g}, l_g, t_{1g} and t_{2g}. So the G terms splits into A_{1g}, E_g, T_{1g} and T_{2g} representations in octahedral environment.

The chemical environment does not interact directly with the electron spins and thus the states splitted from a particular term have the same spin multiplicity as that of the involved terms. Table 7.1 correlates terms with states along spin multiplicity and their numbers.

Table 7.1 Correlation between terms and the corresponding states

Term	No. of configurations. [Spin multiplicity $(2s + 1) \times$ dimension of the term $(2L + 1)$]	States	No. of states. [Spin multiplicity $(2s + 1) \times$ dimesnion of the state $(2L + 1)$]
^1S	$(1) \times (2.0 + 1) = 1$	$^1A_{1g}$	$(1) \times (2.0 + 1) = 1$
^3P	$(3) \times (2.1 + 1) = 9$	$^3T_{1g}$	$(3) \times (3) = 9$
^1D	$(1) \times (2.2 + 1) = 5$	$^1E_g + {}^1T_{2g}$	$(1) \times 2 + (1) \times (3) = 5$
^3F	$(3) \times (2.3 + 1) = 21$	$^3A_{2g} + {}^3T_{1g} + {}^3T_{2g}$	$(3) \times (1) + (3) \times (3) + (3) \times (3) = 21$
^1G	$(1) \times (2.4 + 1) = 9$	$^1A_{1g} + {}^1E_g + {}^1T_{1g} + {}^1T_{2g}$	$(1) \times (1) + (1) \times (2) + (1) \times (3) + (1) \times (3) = 9$

The same treatment is easily extended to d orbitals in other types of environment such as square planar on tetragonally distorted octahedral structure of point group symmetry D_{4h}.

D_4	E	$2C_4$	C_2	$2C_2'$	$2C_2''$
Γ_d	5	−1	1	1	1

This reducible representation Γ_d consists of $A_1 + B_1 + B_2 + E$ since

D_4	E	$2C_2$	$C_2 (= C_4^2)$	$2C_2'$	$2C_2''$
A_1	1	1	1	1	1
B_1	1	−1	1	1	−1
B_2	1	−1	1	−1	1
E	2	0	−2	0	0
Γ_g	5	1	−1	1	1

As the d orbitals are symmetric to inversion, the five d orbitals transform according to the irreducible representations $A_{1g} + B_{1g} + B_{2g} + E_g$ of the point group D_{4h}. Similarly, in the point group D_3, the irreducible representation for the d orbitals is obtained as follows:

D_3	E	$2C_3$	$3C_2$
Γ_d	5	−1	1

This reducible representation Γ_d consists of $A_1 + 2E$. Since

D_3	E	$2C_3$	$3C_2$
A_1	1	1	1
$2E$	4	−2	0
Γ_d	5	−1	1

The characters for the symmetry operations i and σ, like E, can not be obtained directly from the general formula Eq. (7.1), but are given by following formulae.

$$\chi(E) = 2l + 1 \text{ (orbital degeneracy)} \qquad \qquad ...(7.3)$$

$$\chi(\sigma) = \pm \sin\left(l + \frac{1}{2}\right)\pi = (-1)^l \chi(C_2) \qquad \qquad ...(7.3a)$$

$$\chi(i) = \pm (2l + 1) = \pm \chi(E) = (-1)^l \chi(E) \qquad \qquad ...(7.4)$$

where the sign + is taken when l is even and the − sign is taken when l is odd.

The reducible representation for the d orbitals in the point group C_{2v} is obtained by following Eqs (7.1)–(7.4).

$$\chi(C_2) = \frac{\sin\left(\dfrac{5\pi}{2}\right)}{\sin\left(\dfrac{\pi}{2}\right)} = 1$$

$$\chi(E) = 2l + 1 = 2 \times 2 + 1 = 5 \qquad [\because l = 2]$$
$$\chi(\sigma) = (-1)^l \, \chi(C_2) = (-1)^2 \, \chi(C_2) = 1 \times 1 = 1$$

Thus the following representation is obtained for the d orbitals under the point group C_{2v}.

C_{2v}	E	C_2	$\sigma_v(xz)$	$\sigma_v(yz)$
Γ_d	5	1	1	1

This reducible representation Γ_d is made of $2A_1 + A_2 + B_1 + B_2$. This is shown below.

C_{2v}	E	C_2	$\sigma_v(xz)$	$\sigma_v(yz)$
$2A_1$	2	2	2	2
A_2	1	1	-1	-1
B_1	1	-1	1	-1
B_2	1	-1	-1	1
Γ_d	5	1	1	1

The same treatment may be applied to electrons in other types of orbitals. The results are shown in Table 7.2 for the O_h point group. An s orbital in O_h environment, transforms into a totally symmetric representation A_{1g}. The set of p orbitals remains unsplit, transforming as t_{1u}. The same information could have been obtained directly from the O_h character table, where it is observed that (x, y, z) form a basis for the t_{1u} representation of O_h group. All orbitals with higher angular momentum quantum number (l) are split into two or more sets in O_h point group.

Point group symmetries such as D_3, D_{2d} and T_d are often encountered in transition metal compounds. In such cases we could determine the splitting of various sets of orbitals in these above mentioned point group symmetries. However, an alternative and simpler procedure of getting this information is to use the results obtained for octahedral symmetry and derive the required information from the correlation (Table 7.2).

Table 7.2 Splitting of one electron level in an octahedral environment

Type of level	l	$\chi(E)$	$\chi(C_2)$	$\chi(C_3)$	$\chi(C_4)$	Irreducible representation spanned
s	0	1	1	1	1	A_{1g}
p	1	3	-1	0	1	T_{1u}
d	2	5	1	-1	-1	$E_g + T_{2g}$
f	3	7	-1	1	-1	$A_{2u} + T_{1u} + T_{2u}$
g	4	9	1	0	1	$A_{1g} + E_g + T_{1g} + T_{2g}$
h	5	11	-1	-1	1	$E_u + 2T_{1u} + T_{2u}$
i	6	13	1	1	-1	$A_{1g} + A_{2g} + E_g + T_{1g} + 2T_{2g}$

Table 7.3 shows how the representations of group O_h are changed or decomposed into those of its subgroups when the symmetry is altered or lowered.

Table 7.3 Correlation table for group O_h

O_h	O	T_d	D_{4h}	D_{2d}	C_{4v}	C_{2v}	D_{3d}	D_3	C_{2h}
A_{1g}	A_1	A_1	A_{1g}	A_1	A_1	A_1	A_{1g}	A_1	A_g
A_{2g}	A_2	A_2	B_{1g}	B_1	B_1	A_2	A_{2g}	A_2	B_g
E_g	E	E	$A_{1g}+B_{1g}$	A_1+B_1	A_1+B_1	A_1+A_2	E_g	E	A_g+B_g
T_{1g}	T_1	T_1	$A_{2g}+E_g$	A_2+E	A_2+E	$A_2+B_1+B_2$	$A_{2g}+E_g$	A_2+E	A_g+2B_g
T_{2g}	T_2	T_2	$B_{2g}+E_g$	B_2+E	B_2+E	$A_1+B_1+B_2$	$A_{1g}+E_g$	A_1+E	$2A_g+B_g$
A_{1u}	A_1	A_2	A_{1u}	B_1	A_2	A_2	A_{1u}	A_1	A_u
A_{2u}	A_2	A_1	B_{1u}	A_1	B_2	A_1	A_{2u}	A_2	B_u
E_u	E	E	$A_{1u}+B_{1u}$	A_1+B_1	A_2+B_2	A_1+A_2	E_u	E	A_u+B_u
T_{1u}	T_1	T_2	$A_{2u}+E_u$	B_2+E	A_1+E	$A_1+B_1+B_2$	$A_{2u}+E_u$	A_2+E	A_u+2B_u
T_{2u}	T_2	T_1	$B_{2u}+E_u$	A_2+E	B_1+E	$A_2+B_1+B_2$	$A_{1u}+E_u$	A_1+E	$2A_u+B_u$

The results we have obtained so far for single electrons in various types of orbitals also apply to the behaviour of terms arising from groups of electrons. For example, just as a single d electron in a free atom has a wave function that belongs to a five fold degenerate set corresponding to the five values which m_l may take in the $\Phi(\phi)$ part of the wave function, so a D state arising from any group of electrons has a completely analogous five fold degeneracy because of the five values (+L to –L) that the quantum number M_L may take. Further, the splitting of a D term will be the same as the splitting of the set of one-electron d orbitals. This is quite understandable since the factor $\Phi(\phi)$ of the wave function for a D term is $e^{i\,M_L\phi}$ which is completely analogous to the $\Phi(\phi)$ factor $e^{i\,m_l\phi}$ for a single electron. Exactly the same relationship exists between p orbitals and P states, f orbitals and F states and so on. Thus the results given in Table 7.3 for the splitting of various one-electrons orbitals apply equally to the splitting of analogous Ressell-Saunders terms.

In Table 7.4, (where the splitting of one-electron levels in various symmetries are shown) small letters are used to represent the states for a single electron in the environments of various symmetries. Thus $s, p, d, f, ...,$ etc. represent their states in the free atom. But the capital letters are used to represent states into which the crystal field around the metal ion splits, the terms of the free ion. Thus for instance, an F state of a free ion will be split into states A_2, T_1 and T_2 when the ion is placed at the centre of a tetrahedral environment. This is shown below along with other chemical environment.

Free ion terms	States in point groups		
	O_h	T_d	D_{2h}
${}^1S$	${}^1A_{1g}$	${}^1A_1$	${}^1A_{1g}$
${}^1G$	${}^1A_{1g}$ ${}^1T_{2g}$	${}^1A_1$ ${}^1T_2$	$2{}^1A_{1g}$ ${}^1B_{2g}$
	${}^1E_g$	${}^1E$	${}^1A_{2g}$ $2{}^1E_g$
	${}^1T_{1g}$	${}^1T_1$	${}^1B_{1g}$

^3P	$^3T_{1g}$	$^3T_1$	$^3A_{2g}$
			^3E_g
^1D	^1E_g	^1E	$^1A_{1g}\ 1E_g$
	$^1T_{2g}$	$^1T_2$	$^1B_{1g}$
			$^1B_{2g}$
^3F	$^3A_{2g}$	$^3A_2$	$^3A_{2g}\ 2^3E_g$
	$^3T_{1g}$	$^3T_1$	$^3B_{1g}$
	$^3T_{2g}$	$^3T_2$	$^3B_{2g}$

It is pertinent to mention regarding the splitting of terms of the free ion in chemical environments. The chemical environment does not interact directly with the electron spins and thus all of the states into which a particular term is split have the same spin multiplicity as the parent term.

The subscripts g and u used in various states are governed by certain rules.

If the point group of the environment has no centre of symmetry (i), then no subscripts are used, as they cannot have any meaning.

For example, in an octahedral complex the metal ion is at the centre of inversion (also known as *centre of symmetry i*, and wrt this inversion centre (i) the d orbitals maintain the same sign of their wave functions (i.e. same sign on the lobes) on inversion. But a tetrahedron has no centre of inversion (i) and hence the g subscripts are dropped in a tetrahedral crystal field (Table 7.4).

Table 7.4 Splitting of one-electron levels in various symmetries

Type of level	Symmetry of Environment				
	O_h	T_d	D_{4h}	D_3	D_{2d}
s	a_{1g}	a_1	a_{1g}	a_1	a_1
p	t_{1u}	t_2	$a_{2u}+e_u$	a_2+e	b_2+e
d	e_g+t_{2g}	$e+t_2$	$a_{1g}+b_{1g}+b_{2g}+e_g$	a_1+2e	$a_1+b_1+b_2+e$
f	$a_{2u}+t_{1u}+t_{2u}$	$a_2+t_1+t_2$	$a_{2u}+b_{1u}+b_{2u}+2e_u$	a_1+2a_2+2e	$a_1+a_2+b_2+2e$
g	$a_{1g}+e_g+t_{1g}+t_{2g}$	$a_1+e+t_1+t_2$	$2a_{1g}+a_{2g}+b_{1g}+2e_g$	$2a_1+a_2+3e$	$2a_1+a_2+b_1+b_2+2e$
h	$e_u+2t_{1u}+t_{2u}$	$e+t_1+2t_2$	$a_{1u}+2a_{2u}+b_{1u}+b_{2u}$ $+3e_u$	a_1+2a_2+4e	$a_1+a_2+b_1+2b_2+3e$
i	$a_{1g}+a_{2g}+e_g+t_{1g}$ $+2t_{2g}$	$a_1+a_2+e+t_1$ $+2t_2$	$2a_{1g}+a_{2g}+2b_{1g}+2b_{2g}$ $+3e_g$	$3a_1+2a_2+4e$	$2a_1+a_2+2b_1+2b_2+3e$

When the environment does have a centre of symmetry (i), the subscripts are determined by the type of orbital, all atomic orbitals (AOs) for which quantum number l is even ($s, d, g, ...$) being centre symmetric and hence of g character, and all AOs for which l is odd ($p, f, h, ...$) being antisymmetric to inversion and thus of u character.

If the environment does not have a centre of symmetry, the g and u subscripts are inapplicable.

For point groups having a centre to which the inversion operation may be referred, the *g* and *u* characters will be determined by the nature of the one-electron wave functions of the individual electrons making up the configuration from which the term is derived.

7.3 CONSTRUCTION OF CORRELATION DIAGRAMS (OCTAHEDRAL AND TETRAHEDRAL)

Electron–electron repulsions cause a given electron configuration to be split into terms. For the d^2 configuration, electron–electron interactions come into play, giving rise not only the ground state free ion term (^3F) but also a number of excited terms ^3P, ^1G, ^1D and ^1S as well. If the separation between terms is large compared to the perturbation produced by the ligands, we have the *weak field condition*. On the other hand, if the ligand field splitting is large in comparison to the energy difference between terms, we have the *strong field* case.

Figure 7.3 shows the splitting of free ion terms of a d^2 configuration in presence of weak octahedral field (left side of a diagram). The right side of the diagram shows the effects of a strong octahedral field. The complete diagram (Fig. 7.3), often called correlation diagram, for a d^2 ion in an octahedral environment may be drawn as described in the following.

On the extreme left, the free-ion terms in the increasing order of energies are depicted, i.e.

$$E\,(^3F) < E\,(^1D) < E\,(^3P) < E\,(^1G) < E\,(^1S)$$

Immediately right to the above energies, the states into which these free-ion terms split under the influence of weak interaction in the octahedral environment are shown. The spin multiplicities of these states are the same as that of the corresponding free-ion terms.

On the extreme right, the energy levels of the three configurations t_{2g}^2, $t_{2g}^1 e_g^1$ and e_g^2 in the increasing order are shown. The energy difference between the two successive levels is Δ_0.

Immediately left to these levels, the states involved in these levels are shown which arise due to the relaxation of strong interaction between the ion and the environment. The spin multiplicity of these states are not known as the distribution of two electrons may lead to singlets or triplets.

In constructing the correlation diagram, the following two principles hold good.

1. In going from weak to strong interactions with the environment, the symmetry properties as well as their spin multiplicity are preserved. This means that there must exist a out-to-one correspondence between the states on the either sides of the correlation diagram.

2. In going from weak to strong interaction with the environment, the states of the same spin multiplicity and symmetry do not cross each other **(noncrossing rule)**.

With the above two principles, we can correlate states on either sides as described in the following.

(i) There is only one A_{2g} state on either side of the diagram $^3A_{2g}$ in the term ^3F and A_{2g} in the energy level e_g^2. We join these two states. Since the spin multiplicity is preserved, the A_{2g} state in e_g^2 level must also be triplet.

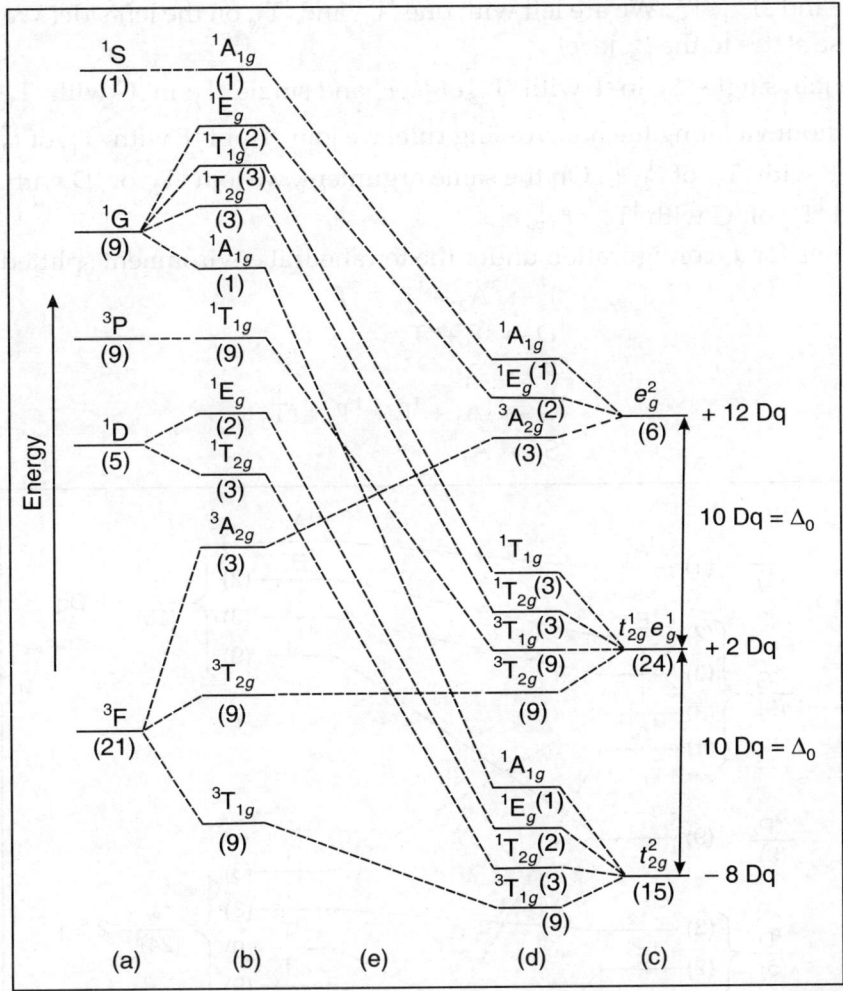

Fig. 7.3 Correlation diagram for a d^2 ion in an octahedral field (a) free ion terms (b) weak field terms (c) strong field ground and excited configurations (d) strong field terms (e) intermediate field region. The numbers in parenthese indicate how many microstates are associated with each term or configuration

(ii) There are two A_{1g} states on either side of the figure—one $^1A_{1g}$ state from ^1S term and the other $^1A_{1g}$ from ^1G term from weak field side whereas from strong field side we get $^1A_{1g}$ from e_g^2 configuration and another $^1A_{1g}$ state from t_{2g}^2 configuration. We join $^1A_{1g}$ of ^1S with the $^1A_{1g}$ of e_g^2 and $^1A_{1g}$ of ^1G with $^1A_{1g}$ of t_{2g}^2. Here noncrossing rule is not violated. It is to be noted that A_{1g} in each of e_g^2 and t_{2g}^2 is singlet.

(iii) There are two E_g states on either side of Fig. 7.3—one ^1E_g comes from ^1G term and the other ^1E_g state comes from ^1D term in low octahedral field while two E_g comes from e_g^2 and the other from t_{2g}^2 configuration. We join ^1E_g of ^1G with ^1E_g of e_g^2 and ^1E_g of ^1D with ^1E_g of t_{2g}^2. It is to be noted that E_g in each of e_g^2 and t_{2g}^2 is singlet.

(iv) There are still three T_{1g} and three T_{2g} states. On the left hand side, we have $^3T_{1g}$, $^3T_{1g}$, $^1T_{1g}$ and $^3T_{2g}$, $^1T_{2g}$, $^1T_{2g}$ states. On the right hand side, there are two T_{1g} and two T_{2g} states in $t_{2g}^1 e_g^1$ each of the two involves one singlet and one triplet, i.e. $^1T_{1g}$,

$^3T_{1g}$ and $^1T_{2g}$, $^3T_{2g}$. We are left with one $^3T_{1g}$ and $^1T_{2g}$ on the left side. We associate these states to the t^2_{2g} level.

(v) We join single $^3T_{2g}$ in ^3F with $^3T_{2g}$ of $t^1_{2g} e^1_g$ and single $^1T_{1g}$ in ^1G with $^1T_{1g}$ of $t^1_{2g} e^1_g$.

(vi) Without violating the noncrossing rule, we join $^3T_{1g}$ of ^3F with $^3T_{1g}$ of t^2_{2g} and $^3T_{1g}$ of ^3P with $^3T_{1g}$ of $t^1_{2g} e^1_g$. On the same argument, we join $^1T_{2g}$ of ^1D with $^1T_{2g}$ of t^2_{2g} and $^1T_{2g}$ of ^1G with $^1T_{2g}$ of $t^1_{2g} e^1_g$.

The terms for d^2 configuration under the tetrahedral environment splitted up as

$$^3F \rightarrow {}^3A_2 + {}^3T_1 + {}^3T_2$$
$$^1D \rightarrow {}^1E + {}^1T_2$$
$$^3P \rightarrow {}^3T_1$$
$$^1G \rightarrow {}^1A_1 + {}^1E + {}^1T_1 + {}^1T_2$$
$$^1S \rightarrow {}^1A_1$$

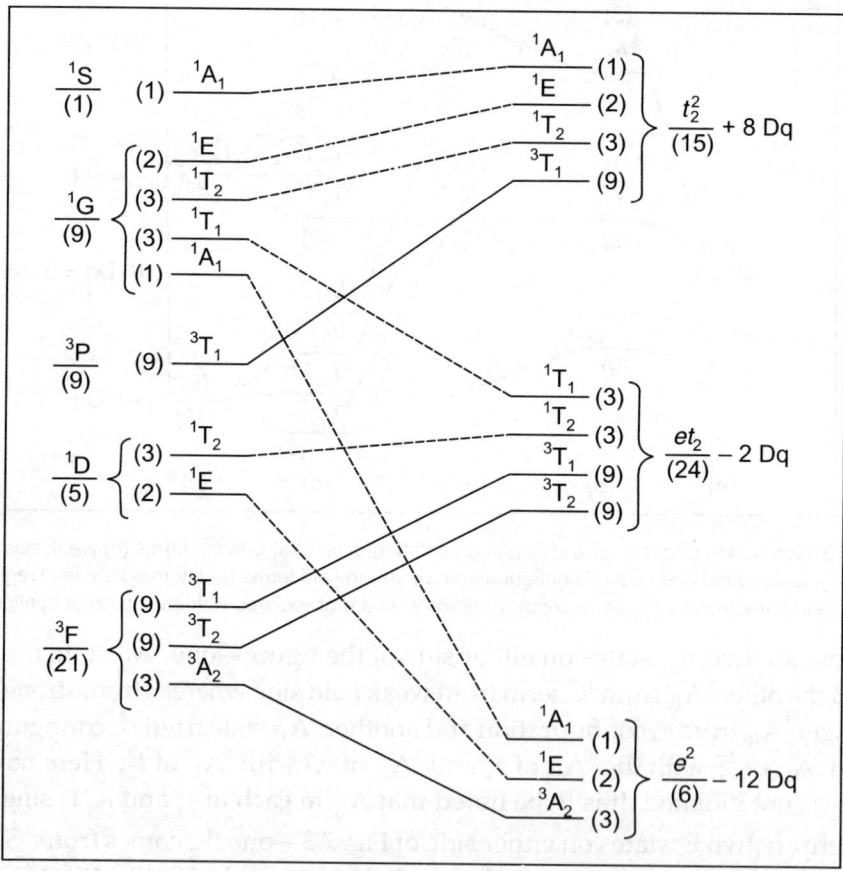

Fig. 7.4 Correlation diagram of d^2 ion in tetrahedral environment

The correlation diagram (Fig. 7.4) can be completed by joining the states on the left hand side (weak interaction) with those of right hand side (strong interaction) as described below:

1. $^3A_2$ state ^3F is joined with A_2 state in e^2. This establishes the triplet state for A_2 in e^2.

2. $^1A_1$ state in ^1S is joined with A_1 state in t_2^2 and $^1A_1$ in ^1G is joined with A_1 state in e^2. These generate single spin multiplicity in both A_1 states on the right hand side.

3. ^1E state in ^1G is joined with E state in t_2^2 and ^1E state in ^1D is joined with E state in e^2, again generating single spin multiplicity in both E state on the right side.

4. Assigning T_1 states in $e\, t_2$ $(e\, t_2 \rightarrow 2T_1 + 2T_2)$ as $^1T_1$ and $^3T_1$, we join the states $^1T_1$ in $e\, t_2$ with $^1T_1$ in ^1G and T_1 in $e\, t_2$ with $^3T_1$ in ^3F. Similarly assigning T_2 states in $e\, t_2$ as $^1T_2$ and $^3T_2$, we join the states $^1T_2$ in $e\, t_2$ with $^1T_2$ in ^1D and $^3T_2$ in $e\, t^2$ with $^3T_2$ in ^3F.

5. We are left with $^3T_1$ in P and $^1T_2$ in ^1G on the left hand side and T_1 and T_2 in t_2^2 on the right side. We join $^3T_1$ in P with T_1 in t_2^2 $(t_2^2 \rightarrow A_1 + E + T_1 + T_2$, establishing triple spin degeneracy in T_1) and $^1T_2$ in ^1G and T_2 in t_2^2 (establishing triple spin degeneray in T_2).

This completes the correlation diagram (Fig. 7.4).

7.4 CONFIGURATION OF d^2 ION IN LARGE SPLITTING

The states for very strong interaction with the O_h field are obtained from the possible configurations, t_{2g}^2, $t_{2g}^1 e_g^1$ and e_g^2 in increasing order of energy. The states corresponding to such configurations are obtained by reducing the direct product representations $T_{2g} \times T_{2g}$, $T_{2g} \times E_g$ and $E_g \times E_g$ of the O_h point group (small letter t, e, ..., will be used for orbitals and capital T, E, ..., for states).

7.4.1 Configuration t_{2g}^2

Both the electrons occupy the stable t_{2g} orbitals. The direct product of $t_{2g} \times t_{2g}$ and the irreducible representations contained in it are shown in Table 7.5.

Table 7.5 Irreducible representations contained in t_{2g}^2 configuration

O_h		E	$8C_3$	$6C_2$	$6C_4$	$3C_2 (= C_4^2)$	i	$6S_4$	$8S_6$	$3\sigma_h$	$6\sigma_d$
I	t_{2g}	3	0	1	−1	−1	3	−1	0	−1	1
II	$t_{2g} \times t_{2g}$	9	0	1	1	1	9	1	0	1	1
III	A_{1g}	1	1	1	1	1	1	1	1	1	1
	E_g	2	−1	0	0	2	2	0	−1	2	0
	T_{1g}	3	0	−1	1	−1	3	1	0	−1	−1
	T_{2g}	3	0	1	−1	−1	3	−1	0	−1	1

The entry I represents irreducible representation t_{2g}, II represents direct product and III represents irreducible representations in the direct product.

7.4.2 Configuration $t_{2g}^1 e_g^1$

Each of the orbitals t_{2g} and e_g contains one electron. The direct product $t_{2g} \times e_g$ and the irreducible representations contained in it are shown in Table 7.6.

Table 7.6 Irreducible representations contained in $t_{2g}^1 e_g^1$ configuration

O_h		E	$8C_3$	$6C_2$	$6C_4$	$3C_2 (= C_4^2)$	i	$6S_4$	$8S_6$	$3\sigma_h$	$6\sigma_d$
I	t_{2g}	3	0	1	−1	−1	3	−1	0	−1	1
	e_g	2	−1	0	0	2	2	0	−1	2	0
II	$t_{2g} \times e_g$	6	0	0	0	−2	6	0	0	−2	0
III	$\begin{cases} T_{1g} \end{cases}$	3	0	−1	1	−1	3	1	0	−1	−1
	$\begin{cases} T_{2g} \end{cases}$	3	0	0	−1	−1	3	−1	0	−1	1

Entry I represents the irreducible representations. II represents direct product and III represents irreducible representations in the direct product

7.4.3 Configuration e_g^2

Both the electrons are excited to e_g orbitals. The direct product of $e_g \times e_g$ and the irreducible representations contained in it are shown in Table 7.7.

Table 7.7 Irreducible representations contained in e_g^2 configuration

O_h		E	$8C_3$	$6C_2$	$6C_4$	$3C_2 (= C_4^2)$	i	$6S_4$	$8S_6$	$3\sigma_h$	$6\sigma_d$
I	e_g	2	−1	0	0	2	2	0	−1	2	0
II	$e_g \times e_g$	4	1	0	0	4	4	0	1	4	0
III	$\begin{cases} A_{1g} \end{cases}$	1	1	1	1	1	1	1	1	1	1
	$\begin{cases} A_{2g} \end{cases}$	1	1	−1	−1	1	1	−1	1	1	−1
	$\begin{cases} E_g \end{cases}$	2	−1	0	0	2	2	0	−1	2	0

Entry I represents the irreducible representations. II represents direct product and III represents the irreducible representations in the direct product.

From the above discussion (in a strong O_h field), we have

Electronic Configuration	Group Theoretical Terms
t_{2g}^2	$^1A_{1g} + {}^1E_g + {}^1T_{2g} + {}^3T_{1g}$
$t_{2g}^1 e_g^1$	$^1T_{1g} + {}^1T_{2g} + {}^3T_{1g} + {}^3T_{2g}$
e_g^2	$^1A_{1g} + {}^1E_g + 3A_{2g}^2$

d^3 Configuration :

The E term splitting in d^3 ion is the inverse of d^2 ion. The F term always splits in the sequence T_{1g}, T_{2g}, A_{2g} with T_{2g} in the middle. For d^3 ion, S = 3/2 and M_S = 3/2. The ground state (Fig. 7.5(a)) can only be written in one way if we have to retain a spin of 3/2. Therefore, it is an orbital singlet. The wave function can be described by the determinant $| xz^+, yz^+, xy^+ >$ which can be shown to transform as $^3A_{2g}$ under the O_h symmetry operator.

The first excited state (Fig. 7.5 (b)) is six fold orbitally degenerate becoming two or three fold degnerate terms when interelectronic repulsion is taken into account. The next excited state (Fig. 7.5 (c)) is three fold orbitally degenerate.

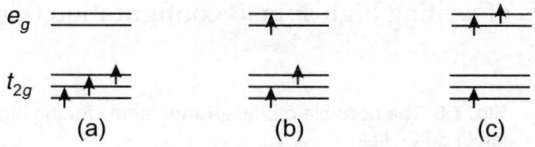

Fig. 7.5 Strong field electronic configuratrions for a d^3 ion (S = 3/2, M_S = 3/2) in an O_h field (a) orbital singlet ground state (b) Six fold orbitally degenerate first excited state (c) three fold orbitally degenerate second excited state

The free d^3 ion has spin quadrate 4_F, and 4_P states which split up in octahedral environment into ${}^4A_{2g}$, $2{}^4T_{1g}$ and ${}^4T_{2g}$ terms with ${}^4T_{2g}$ in the middle. Configuration (a) (Fig. 7.5 (a)) is orbital singlet and belongs to ${}^4A_{2g}$. The only spin-triplet term derived from e_g^2 configuration is ${}^3A_{2g}$. Therefore, the term for configuration (c) (Fig. 7.5 (c)) (i.e. $t_{2g}^1 e_g^2$) may be evaluated by the product $A_{2g} \times T_{2g} = T_{1g}$. Hence, the configuration (c) (Fig. 7.5 (c)) belongs to ${}^4T_{1g}$ and by elimination, configuration (b) (Fig. 7.5 (b)) spans ${}^4T_{2g}$ and ${}^4T_{1g}$ terms.

Terms from $t_{2g}^n e_g^m$ configurations may be obtained from the direct products paying close attention to the possible spin multiplicities.

For $t_{2g}^2 e_g^1$:

In case of d^2 ion, t_{2g}^2 spans

$$t_{2g}^2 \rightarrow t_{2g}^1 \times t_{2g}^1 \rightarrow {}^3T_{1g} + {}^1T_{2g} + {}^1E_g + {}^1A_{1g}$$

It is to be noted that the multiplication of a singlet term by e_g will give only doublet term (A × E = E), but that multiplication of a triplet term will give both doublet and quadrate terms, hence

$$t_{2g}^2 e_g^1 \equiv e_g^1 t_{2g}^2 \equiv e_g^1 (t_{2g}^1 \times t_{2g}^1)$$

Therefore $e_g^1 t_{2g}^2$ spans as $E_g \times (T_{2g} \times T_{2g})$

$= E_g \times ({}^3T_{1g} + {}^1A_{1g} + {}^1E_g + {}^1T_{2g})$

$= (E_g \times {}^3T_{1g}) + (E_g \times {}^1A_{1g}) + (E_g \times {}^1E_g) + (E_g \times {}^1T_{2g})$

$= ({}^4T_{1g} + {}^4T_{2g} + {}^2T_{1g} + {}^2T_{2g}) + {}^2E_g + (A_{1g} + A_{2g} + E_g) + ({}^4T_{1g} + {}^4T_{2g} + {}^2T_{1g} + {}^2T_{2g})$

$= {}^4T_{1g} + {}^4T_{2g} + 2({}^2T_{1g}) + 2({}^2T_{2g}) + 2({}^2E_g) + {}^2A_{1g} + {}^2A_{2g}$

By similar arguments, we shall get the following terms from $t_{2g}^1 e_g^2$ configuration

$$t_{2g}^1 e_g^2 \equiv t_{2g}^1 \times (e_g^1 \times e_g^1) = T_{2g} \times (E_g \times E_g)$$
$$= T_{2g} \times ({}^3A_{2g} + {}^1E_g) + {}^1A_{1g}$$
$$= {}^4T_{1g} + 2({}^2T_{1g}) + 2({}^2T_{2g})$$

t_{2g}^3 spans ${}^4A_{2g} + {}^2T_{1g} + {}^2T_{2g} + {}^2E_g$

d^4 ion :

The d^4 ion in an octahedral field has two possible ground states $t_{2g}^3 e_g^1$ (${}^5E_g$) or t_{2g}^4 (${}^3T_{1g}$) according to the strength of the crystal field. The high spin (S = 2) and the low spin (S = 1) configurations are shown in Figs 7.6 and 7.7.

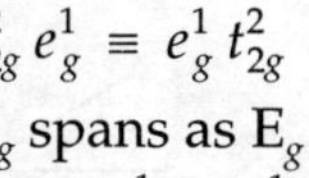

Fig. 7.6 High spin ground state of d^4 ion (S = 2, M_S = 2) in O_h field

Fig. 7.7 Low spin ground state of d^4 ion (S = 1, M_S = 1) in an O_h field

There are two ways of writing high spin d^4 configuration (Fig. 7.8). This results ^5E_g term.

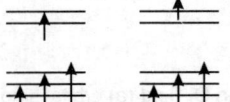

Fig. 7.8 The possible orbital arrangements for the high spin ground state of the d^4 ion in an O_h field

Alternatively, we recall that the maximum spin t_{2g}^3 configuration belongs to $^4A_{2g}$. Hence, using direct product for maximum spin terms, we get

$$t_{2g} \times t_{2g} \times t_{2g} \text{ yields } {}^4A_{2g} + ...$$

and

$$t_{2g} \times e_g \text{ yields } E_g$$

In the weak field, ^5D ground term state of d^4 spans ^5E_g and $^5T_{2g}$ and hence orbital triplet configuration (Fig. 7.9) belongs to $^5T_{2g}$. This is also obtained using direct product

$$t_{2g} \times t_{2g} \text{ yields } {}^3T_{1g} + ...$$

$$e_g \times e_g \text{ yields } {}^3A_{2g} + ...$$

and

$$T_{1g} \times A_{2g} = T_{2g}$$

In a weak field, the ground state is ^5E_g and the complex is said to be 'high-spin' or 'spin-free' with four unpaired electrons. As the crystal field strength increases there comes a value beyond which the $^3T_{1g}$ term becomes the ground state, with two unpaired electrons. Such complexes are termed *low-spin* or *spin-paired*. An incomplete diagram showing the spin quintets and spin triplets of d^4 is given in Fig. 7.9.

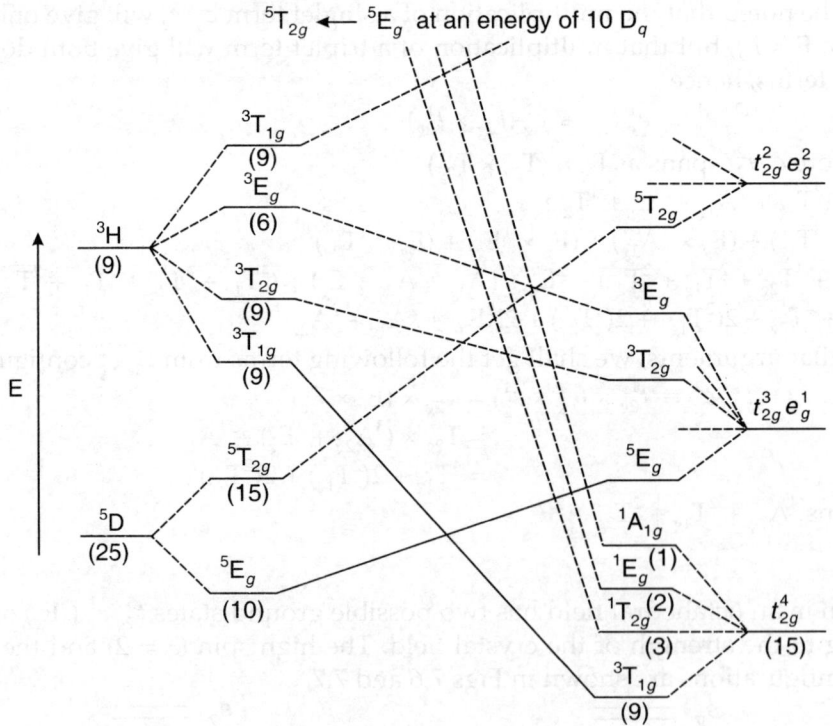

Fig. 7.9 Partial correlation diagram for d^{4*}

---

* **Courtesy:** Lever, ABP, *Inorganic Electronic Spectroscopy*, 1st Edn, Elsevier, New York (1968).

Spin-Orbit Interaction in Complexes: The Double Group

In the preceding section, we showed that for any symmetry operation corresponding to a rotation by an angle α on an orbital or state wave function having an angular momentum quantum number l or L, the character $\chi(\alpha)$ for which this forms a basis is given by Eq. (7.5).

$$\chi(\alpha) = \sin\left(L + \frac{1}{2}\right)\alpha \Big/ \sin\left(\frac{\alpha}{2}\right) (\alpha \neq 0) \qquad \qquad ...(7.5)$$

When we use this formula to derive crystal field states of a complex, we have made the assumption that spin-orbit interaction is weak and hence is ignored.

When spin-orbit interaction is significant, the quantum number that defines a state is total angular momentum J, instead of L. Quantum number J can be an integer on a half-integer. When J is an integer, we can again make use of Eq. (7.5) and replace L with J. However, when J is a half-integer, a complication arises:

$\chi(\alpha + 2\pi)$ no longer equals to $\chi(\alpha)$, as it should mathematically:

$$\chi(\alpha + 2\pi) = \sin\left[\left(J + \frac{1}{2}\right)(\alpha + 2\pi)\right]\Big/ \sin\left[\frac{(\alpha + 2\pi)}{2}\right]$$

$$= \sin\left[2\pi + \left(J + \frac{1}{2}\right)\alpha\right]\Big/ \sin\left(\pi + \frac{\alpha}{2}\right)$$

$$= \sin\left[\left(J + \frac{1}{2}\right)\alpha\right]\Big/\left(-\sin\frac{\alpha}{2}\right)$$

$$= -\chi(\alpha) \qquad \qquad ...(7.6)$$

In this event, the characters of rotation α and $(\alpha + 2\pi)$ are not identical.

$$\chi(\alpha + 2\pi) = -\chi(\alpha) = (-1)^{2J}\chi(\alpha)$$

For identity operation, i.e. when $\alpha = 0$ each of the diagonal elements of the matrix is equal to 1 and the character is equal to $(2l + 1)$.

Thus $\qquad \qquad \chi(\alpha) = 2l + 1$ (when $\alpha = 0$) $\qquad \qquad ...(7.7)$

Further by analogy with Eq. (7.2) for $\alpha = 0$,

$$\chi(\alpha) = (2J + 1) \qquad \qquad ...(7.8)$$

For $\alpha = 2\pi$

$$\chi(\alpha) = 2J + 1 \qquad \text{(when, J is an integral)}$$

and $\qquad \qquad \chi(\alpha) = -(2J + 1)$ (when, J is a half-integral) $\qquad ...(7.9)$

Of course, for any real physical system a rotation of 2π brings the system back into itself, so that characters for rotations by α and $(\alpha + 2\pi)$ ought to be identical. As a result, the mathematical function is introduced that the rotation 2π is treated as a **separate symmetry operation** labelled as R, and not as an identity operation E. Instead of the straight forward symmetry point groups met in previous chapter, we are now dealing with groups of operations which are called *double groups*, and are specified by primes, e.g. D_4', O' etc. They contain different classes of operations and different symmetry species from the simple point groups, e.g. D_{4h}, O_h to which they are related. Also the species are usually given by the symbol Γ, rather than Mullikan symbols. In Table 7.8, we reproduce the character table for the double group O′ which is used, in connection with spin-orbital coupling, instead of the O_h point group for octahedral complexes. As a result of this, we expect the spin-orbital component levels not be labelled by the straight forward Mullikan symmetry levels A_{1g}, E_g, T_{1g}, etc. of the O_h point group, but by Γ_7, Γ_8, etc. of the double group O′ (Table 7.8).

Table 7.8 Character table for the double group O′

O′	E	R	$4C_3$ $4(C_3^2 R)$	$4(C_3 R)$ $4C_3^2$	$3C_2$ $3(C_2 R)$	$3C_4$ $3(C_4^3 R)$	$3(C_4 R)$ $3C_4^3$	$6C_2'$ $6(C_2' R)$
Γ_1	1	1	1	1	1	1	1	1
Γ_2	1	1	1	1	1	−1	−1	−1
Γ_3	2	2	−1	−1	2	0	0	0
Γ_4	3	3	0	0	−1	1	1	−1
Γ_5	3	3	0	0	−1	−1	−1	1
Γ_6	2	−2	1	−1	0	$\sqrt{2}$	$-\sqrt{2}$	0
Γ_7	2	−2	1	−1	0	$-\sqrt{2}$	$\sqrt{2}$	0
Γ_8	4	−4	−1	1	0	0	0	0

Note: $C_3^2 R$ means rotation by 2π followed by C_3^2; it always gives the same character as C_3^1 and so belongs to the same class.

Before going into the spin-orbital splitting of actual ligand field levels, let us first illustrate the use of the double group O′ in predicting the species of the levels which arise from the configuration d' when we apply the octahedral ligand field after the spin–orbital coupling. Remember, however, that this is not the normal procedure. For the solitary d electron ($l = 2$, $s = 1/2$) j may take the values 5/2 or 3/2 resulting $^2D_{5/2}$ and $^2D_{3/2}$ levels have the energies shown in Fig. 7.10. Now applying Eqs (7.5) and (7.7) (with l replaced by j) for each of the operations of the group O′, we obtain the so-called representation (Γ) for $j = 5/2$ and $j = 3/2$ below.

$$\chi(\alpha) = \chi(E) = 2J + 1$$

For $J = 5/2$, $\left. \chi(E) = 2 \cdot \dfrac{5}{2} + 1 = 6 \right\}$

$J = 3/2$, $\left. \chi(E) = 2 \cdot \dfrac{3}{2} + 1 = 4 \right\}$

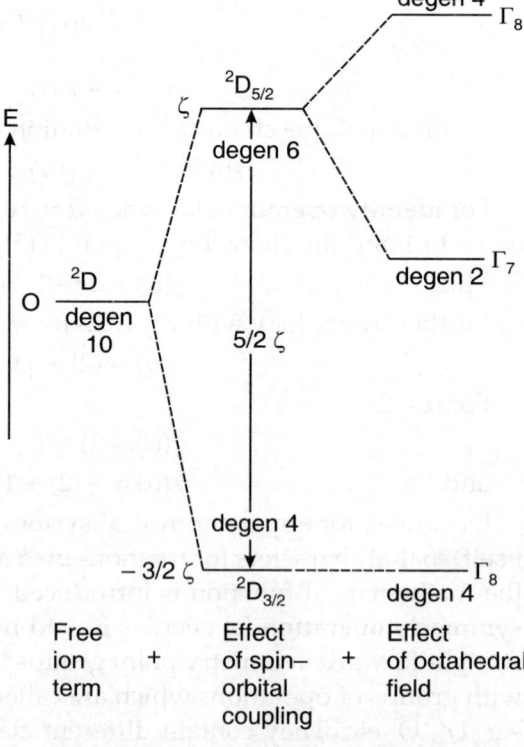

Fig. 7.10 Effect of spin–orbital coupling and an octahedral crystal field on the energy of the ^2D term for d^1.

For half-integral J values, $\chi(RC_n) = -\chi(C_n)$

For $J = 5/2$, $\chi(C_3) = \dfrac{\sin\left\{\left(\dfrac{5}{2} + \dfrac{1}{2}\right)\left(\dfrac{2\pi}{3}\right)\right\}}{\sin\left\{\dfrac{(2\pi/3)}{2}\right\}} = \dfrac{\sin\left(3 \times \dfrac{2\pi}{3}\right)}{\sin\left(\dfrac{\pi}{3}\right)} = \dfrac{\sin 2\pi}{\sin 60°} = 0$

For $J = 3/2$, $\chi(C_3) = \dfrac{\sin\left\{\left(\dfrac{3}{2} + \dfrac{1}{2}\right)\left(\dfrac{2\pi}{3}\right)\right\}}{\sin\left\{\dfrac{2\pi/3}{2}\right\}} = \dfrac{\sin\left(\dfrac{4\pi}{3}\right)}{\sin\left(\dfrac{\pi}{3}\right)} = \dfrac{-\sin\left(\dfrac{\pi}{3}\right)}{\sin\left(\dfrac{\pi}{3}\right)} = -1$

For $J = 5/2$, $\chi(C_2) = \dfrac{\sin\left\{\left(\dfrac{5}{2} + \dfrac{1}{2}\right)\left(\dfrac{2\pi}{2}\right)\right\}}{\sin\left\{\dfrac{(2\pi/2)}{2}\right\}} = \dfrac{\sin\left(3 \times \dfrac{2\pi}{3}\right)}{\sin 90°} = \dfrac{\sin 2\pi}{\sin 90°} = 0$

For $J = 3/2$, $\chi(C_2) = \dfrac{\sin\left\{\left(\dfrac{3}{2} + \dfrac{1}{2}\right)\left(\dfrac{2\pi}{2}\right)\right\}}{\sin\left\{\dfrac{2\pi/2}{2}\right\}} = \dfrac{\sin 2\pi}{\sin 90°} = 0$

For $J = 5/2$, $\chi(C_4) = \dfrac{\sin\left\{\left(\dfrac{5}{2} + \dfrac{1}{2}\right)\dfrac{2\pi}{4}\right\}}{\sin\left\{\dfrac{2\pi/4}{2}\right\}} = \dfrac{\sin\left(3 \times \dfrac{\pi}{2}\right)}{\sin\left(\dfrac{\pi}{4}\right)} = -\sqrt{2}$

For $J = 3/2$, $\chi(C_4) = \dfrac{\sin\left\{\left(\dfrac{3}{2} + \dfrac{1}{2}\right)\dfrac{\pi}{2}\right\}}{\sin\left(\dfrac{\pi}{4}\right)} = \dfrac{\sin \pi}{\sin\left(\dfrac{\pi}{4}\right)} = 0$

The reducible representation thus leads to the following:

O′	E	R	C_3	C_3R	C_2	C_4	C_4R	C_2'
$\Gamma_{J=5/2}$ or $\chi^J_{=5/2}$	6	−6	0	0	0	$-\sqrt{2}$	$\sqrt{2}$	0
$\Gamma_{J=3/2}$ or $\chi^J_{=3/2}$	4	−4	−1	1	0	0	0	0

By comparison with the O′ character table (Table 7.8), we see that

$$\Gamma_{J=3/2} = \Gamma_8$$

and $\Gamma_{J=5/2}$ is reducible, being the sum of the representations Γ_7 and Γ_8, i.e.

$$\Gamma_{J=5/2} = \Gamma_7 + \Gamma_8$$

The $^2D_{5/2}$ level is, therefore, split as shown in Fig. 7.10*. $^2D_{5/2}$ will lie the higher; this can only be done by energy calculations, not by group theory; but the higher is actually Γ_8.

When $\Delta_0 \gg \zeta$, we apply the spin-orbital coupling as a perturbation of the ligand field levels. For $^2D(d^1)$, these are of course $^2T_{2g}$ and ^2E_g. In the double group O', T_{2g} transforms on Γ_5 and E_g as Γ_3 (this may be seen by using Eqs (7.5) and (7.7) to get the representation for the five d orbitals ($l = 2$) in O'; the representation is reducible to $\Gamma_3 + \Gamma_5$. So far, we have considered only the orbital angular momentum. Now we shall consider, the spin angular momentum ($s = 1/2$) separately, using the same formulae [Eqs (7.1) and (7.2)] only replacing l by s and noting that Eq. (7.6) will hold for J replaced by S.

$$\chi(\alpha) = \frac{\sin\left(s + \frac{1}{2}\right)\alpha}{\sin\left(\frac{\alpha}{2}\right)} \quad (\alpha \neq 0) \qquad \text{...(7.10)}$$

$$\left. \begin{array}{ll} \chi(\alpha) = 2s + 1 & (\alpha = 0) \\ \chi(\alpha) = -(2s + 1) & (\alpha = 2\pi) \end{array} \right\} \qquad \text{...(7.11)}$$

This gives the O' representation:

	E	R	C_3	$C_3 R$	C_2	C_4	$C_4 R$	C_2'
$\Gamma_{s=1/2}$	2	-2	1	-1	0	$\sqrt{2}$	$\sqrt{}$	0

which is by comparison with Table 7.8, just Γ_6. Finally, the species of the spin-orbital levels arising from $^2E_g(\Gamma_3)$ and $^2T_{2g}(\Gamma_5)$ are obtained simply by taking the direct product of the characters of each separately with the characters of the spin function Γ_6.
For $^2E_g : \Gamma_3 \times \Gamma_6$ gives

	E	R	C_3	$C_3 R$	C_2	C_4	$C_4 R$	C_2'
$\Gamma_3 \times \Gamma_6$	4	-4	-1	1	0	0	0	0

which is seen from Table 7.8 to be just Γ_8.
For $^2T_{2g} : \Gamma_5 \times \Gamma_6$ gives

	E	R	C_3	$C_3 R$	C_2	C_4	$C_4 R$	C_2'
$\Gamma_5 \times \Gamma_6$	6	-6	0	0	0	$-\sqrt{2}$	$\sqrt{2}$	0

which from Table 7.8 is seen to be just the sum of representations for Γ_7 and Γ_8. Hence the $^2T_{2g}$ level splits whereas the ^2E_g level does not; and the energy diagram is that of Fig. 7.11. Once again, group theory does not tell us which of the levels is the higher—to do this we must perform the actual calculations.

Note (from column E in the character tables) that each Γ_8, has a total degeneracy of 4 and Γ_7, a total degeneracy of 2. The total degeneracy of 10 of the ^2D term is therefore preserved. Again, the $\Gamma_8 (^2T_{2g})$ level is depressed by one-half the amount the Γ_7 level is raised, to preserve the centre of gravity. It is important to note that Γ_8 occurs twice

---

* At this stage we can not predict which of Γ_7 or Γ_8 form

in the splitting scheme. The two Γ_8 levels interact with each other (having the same symmetry species), tending to be mutually 'repelled' in much the same way as we have seen configuration interaction to 'repel' the octahedral $^3T_{1g}(P)$ and $^3T_{1g}(F)$ levels of d^2. This is more pronounced, the closer the ζ value is to the splitting parameter Δ. Further more, we have also to take into account the noncrossing rule with the result that, in weak field–strong field correlation diagrams including spin–orbital coupling, no levels having identical Γ species can be allowed to cross. Combination of Figs 7.10 and 7.11 produces correlation diagram of d^1 ion (Fig. 7.12).

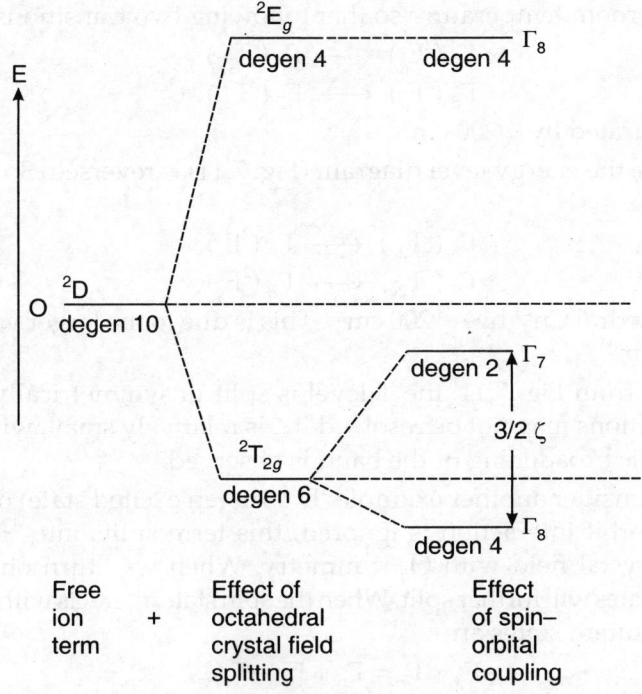

Fig. 7.11 Effect of spin-orbital coupling on the energy of the $^2T_{2g}$ and ^2E_g levels of $d^1(^2D)$ in an octahedral crystal field

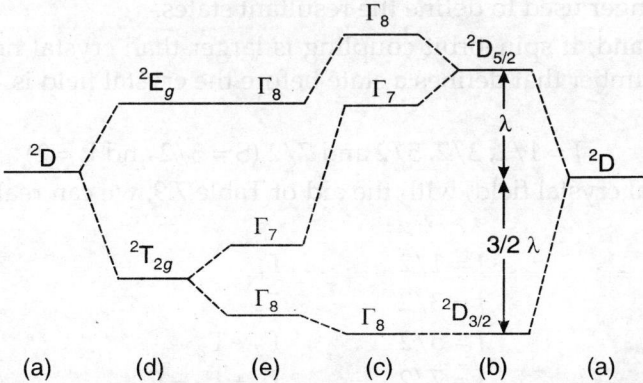

Fig. 7.12 d^1 correlation diagram relating spin–orbit terms with their crystal field and free ion parents. (a) spherically perturbed free ion (b) strong spin–orbit coupling, but Dq = 0, (c) spin–orbit coupling, i.e. λ large, and weak crystal field, i.e. Dq small, but non-zero, (d) Dq non-zero, but $\lambda = 0$, (e) both Dq and λ non-zero, but λ small (weak spin–orbit coupling). [Not to scale.]

Having achieved our aim of showing, though very qualitatively and simply, how spin–orbital coupling may work in such a way as to split some of the free ion energy levels, we can now return to the effect that this has on the spectrum. To a first approximation we have said that for d_1 there is one transition.

$$^2E_g \longleftarrow {}^2T_{2g}$$

However, spin–orbital coupling splits the ground $^2T_{2g}$ level into Γ_7 and Γ_8 separated by $3/2\,\zeta$ which amounts to about 200 cm^{-1} (ζ for $Ti^{3+} \sim 155 \text{ cm}^{-1}$) for $[Ti(H_2O)_6]^{3+}$. This is of the order of kT for T ~ 300°K so that population of both Γ_8 and Γ_7 levels of $^2T_{2g}$ state will occur at room temperature so that following two transitions will occur and

$$\Gamma_8 \, (^2E_g) \longleftarrow \Gamma_8 \, (^2T_{2g})$$
$$\Gamma_8 \, (^2E_g) \longleftarrow \Gamma_7 \, (^2T_{2g})$$

the bands are separated by ~ 200 cm^{-1}.

In d^9 complexes, the energy level diagram (Fig. 7.11) is reversed. So the two possible transitions are

$$\Gamma_8 \, (^2T_{2g}) \longleftarrow \Gamma_8 \, (^2E_g)$$
$$\Gamma_7 \, (^2T_{2g}) \longleftarrow \Gamma_8 \, (^2E_g)$$

separated in octahedral Cu^{2+} by ~ 1200 cm^{-1}. This is due to the larger value of ζ for Cu^{2+} (ζ for $Cu^{2+} = 830 \text{ cm}^{-1}$).

As can be seen from Fig. 7.11, the T level is split unsymmetrically, the individual spin–orbital transitions may not be resolved if ζ is relatively small, with the result that only an asymmetric broadening of the band is observed.

We shall now consider another example ^4D term (an excited state) of a d^5 octahedral complex. If spin–orbit interaction is ignored, this term splits into $^4\Gamma_3$ and $^4\Gamma_5$ states (Table 7.9) in a crystal field with O_h symmetry. When we "turn on" the spin–orbit interaction, these states will further split. When the spin state interacts with the orbitals parts (Γ_8 and Γ_5) the resultant states are

$$\Gamma_3 \times \Gamma_8 = \Gamma_6 + \Gamma_7 + \Gamma_8$$
$$\Gamma_5 \times \Gamma_8 = \Gamma_6 + \Gamma_7 + {}^2T_8$$

This means that $^4\Gamma_3$ splits into three states and $^4\Gamma_5$ into four. Note that spin quantum number S is no longer used to define the resultant states.

On the other hand, if spin–orbit coupling is larger than crystal field interaction, J is the quantum number that defines a state before the crystal field is "turned on". For ^4D, we have

$$J = 1/2, 3/2, 5/2 \text{ and } 7/2 \ (S = 3/2 \text{ and } L = 2)$$

In an octahedral crystal field, with the aid of Table 7.9, we can really see that these states split into

$$J = 1/2 \qquad \Gamma_6$$
$$J = 3/2 \qquad \Gamma_8$$
$$J = 5/2 \qquad \Gamma_7 + \Gamma_8$$
$$J = 7/2 \qquad \Gamma_6 + \Gamma_7 + \Gamma_8$$

The way these two sets of states are correlated is shown in Fig. 7.13. It is important to note that, whenever we are dealing with spin-orbit interaction in a system with half-integral J values, we should employ the double group.

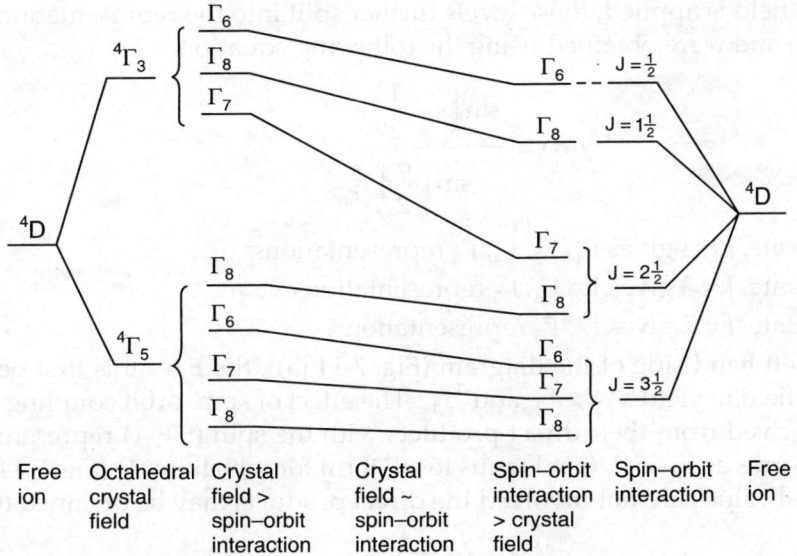

| Free ion | Octahedral crystal field | Crystal field > spin–orbit interaction | Crystal field ~ spin–orbit interaction | Spin–orbit interaction > crystal field | Spin–orbit interaction | Free ion |

Fig. 7.13 Relative effects of spin–orbit coupling and octahedral crystal field interaction on the electronic state ^{4}D.

Table 7.9 Representations under O or O′ symmetry, spanned by a state characterised by an integral or half integral j value

J	E	R	$4C_3$ $4C_3^2 R$	$4C_3^2$ $4C_3 R$	$3C_2$ $3C_2 R$	$3C_4$ $3C_4^3 R$	$3C_4^3$ $3C_4 R$	$6C_2'$ $6C_2' R$	Representations in O or O′
0	1								Γ_1
1/2	2	–2	1	–1	0	$\sqrt{2}$	$-\sqrt{2}$	0	Γ_6
1	3								Γ_4
$1\frac{1}{2}$	4	–4	–1	1	0	0	0	0	Γ_8
2	5								$\Gamma_3 + \Gamma_5$
$2\frac{1}{2}$	6	–6	0	0	0	$-\sqrt{2}$	$\sqrt{2}$	0	$\Gamma_7 + \Gamma_8$
3	7								$\Gamma_2 + \Gamma_4 + \Gamma_5$
$3\frac{1}{2}$	8	–8	1	–1	0	0	0	0	$\Gamma_6 + \Gamma_7 + \Gamma_8$
4	9								$\Gamma_1 + \Gamma_3 + \Gamma_4 + \Gamma_5$
$4\frac{1}{2}$	10	–10	–1	1	0	$\sqrt{2}$	$-\sqrt{2}$	0	$\Gamma_6 + 2\Gamma_8$
5	11								$\Gamma_3 + 2\Gamma_4 + \Gamma_5$
$5\frac{1}{2}$	12	–12	0	0	0	0	0	0	$\Gamma_6 + \Gamma_7 + 2\Gamma_8$
6	13								$\Gamma_1 + \Gamma_2 + \Gamma_3 + \Gamma_4 + 2\Gamma_5$
$6\frac{1}{2}$	14	–14	1	–1	0	$-\sqrt{2}$	$\sqrt{2}$	0	$\Gamma_6 + 2\Gamma_7 + 2\Gamma_8$

The same treatment may be applied to any ion. For example, Fig. 7.14 illustrates the spin–orbit coupling diagram for a d^8 ion in an octahedral environment (^{3}F term only). The ^{3}F term, with LS coupling splits into $^3F_4$, $^3F_3$ and $^3F_2$ (Fig. 7.14 (b)). When a weak octahedral field is applied, these levels further split into the representations shown in Fig. 7.14 (c) and were obtained using the following equation.

$$\chi(\alpha) = \frac{\sin\left(J+\frac{1}{2}\right)\alpha}{\sin\left(\frac{\alpha}{2}\right)} \qquad \ldots(7.12)$$

For $^3F_4$ state, $J = 4$ gives Γ_1, Γ_3, Γ_4, Γ_5 representations

For $^3F_3$ state, $J = 3$ gives Γ_2, Γ_4, Γ_5 representations

For $^3F_2$ state, $J = 2$ gives Γ_3, Γ_5 representations

On the left hand side of the diagram (Fig. 7.14 (a)), the F term is first perturbed by the crystal field to yield $^3A_{2g}$, $^3T_{2g}$ and $^3T_{1g}$. The effect of spin–orbit coupling upon these levels is derived from their direct products with the spin (S = 1) representation. This spin transforms as Γ_4 or Γ_1 (analogous to a P term for which L = 1) (Fig. 7.14). Since the half integral values are not involved the direct products may be obtained (Chapter 3).

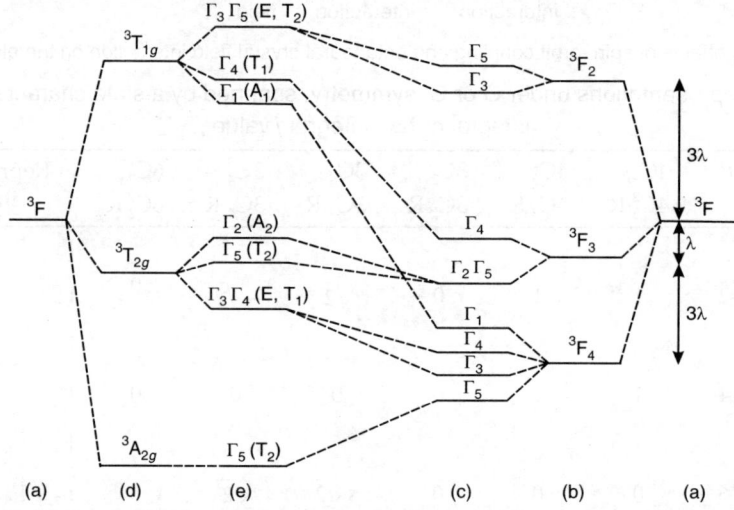

Fig. 7.14 d^8 correlation diagram relating to spin–orbit terms with their crystal field and free ion parents (^3F term only).

The multiple band observed in the spectra of many octahedral nickel(II) complexes near 13000 cm^{-1} may be derived from transitions to various spin–orbit components of the $^3T_{1g}(F)$ level. Alternatively the multiple band may arise as a consequence of the presence of a spin forbidden transition to the $^1E_g(D)$ level which gains intensity from the spin allowed band.

In fact, the appearance of spin forbidden transitions is a direct consequence of the presence of spin–orbit coupling. Formally, all transitions involving a change in spin are strictly forbidden. The rule is relaxed by the mechanism of spin–orbit coupling which can couple in terms of the same spin–orbit representation together. For example, in the case cited immediately above the spin–orbit representation of the ^1E_g level will be simply Γ_3 or E since the spin is zero. Figure 7.14 reveals that both $^3T_{1g}$ and $^3T_{2g}$ spin–orbit components contain Γ_3. Thus some singlet character will be mixed into triplet level and vice versa, by the spin-orbit mixing of the Γ_3 levels.

Since spin–orbit effects arise from coupling of spin and orbital momenta of electrons, we are concerned with the direct product representation of these two effects. As an example, we shall now work out the effects of an octahedral field and spin–orbit coupling on the ^4F free ion state of a d^7 ion. The total representation in the O point group

$$\chi_T (l = 3) = A_2 + T_1 + T_2$$

A d^7 ion in a weak octahedral field leads to a $^4T_{1g}$ ground state and $^4T_{2g}$ and $^4A_{2g}$ excited states. In the O' double group, these correspond to $T_1' (\Gamma_4)$, $T_2' (\Gamma_5)$, and $A_2' (\Gamma_2)$. Using $S = 3/2$ and substituting S for l in Eq. (7.5), we generate in the O' point group an irreducible representation of $^1G(\Gamma_8)$, i.e. one of the new irreducible representations of the double group. Now, we take the direct products of the spin and orbital parts and decompose them as before, leading to

$$\Gamma_2 \times \Gamma_8 = \Gamma_8$$
$$\Gamma_4 \times \Gamma_8 = \Gamma_6 + \Gamma_7 + 2\Gamma_8$$
$$\Gamma_5 \times \Gamma_8 = \Gamma_6 + \Gamma_7 + 2\Gamma_8$$

Spin–orbit effects do not split Γ_2, but they split Γ_4 into four states and Γ_5 into four states. We could have converted L and S to J and employed Eq. (7.12) on J values, i.e. $9/2, 7/2, 5/2$ and $3/2$ to obtain the double group representations.

J		Representations in O*
9/2	–	$\Gamma_6 + 2\Gamma_8$
7/2	–	$\Gamma_6 + \Gamma_7 + \Gamma_8$
5/2	–	$\Gamma_7 + \Gamma_8$
3/2	–	Γ_8

This procedure would have been followed if spin–orbit coupling were comparable to or greater than the crystal field. Using the approach employed above, we have assumed a large crystal field and a small spin–orbit perturbation on it, we can summarise the results with Fig. 7.15.

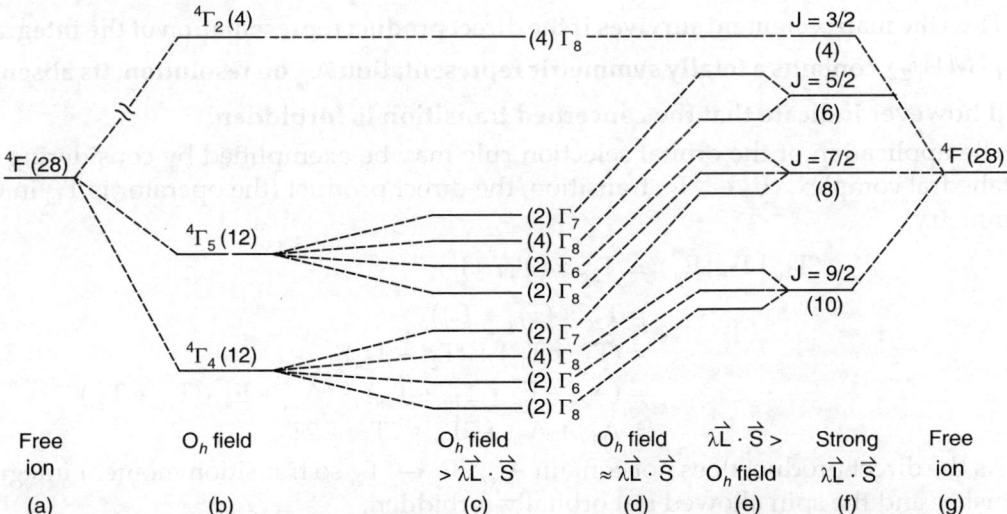

Fig. 7.15 (a) The gaseous ion (b) split by a strong O_h field (c) followed by smaller $\lambda \vec{L} \cdot \vec{S}$. On the right (g) free ion (f) spin by large spin–orbit coupling (e) followed by a weaker ligand field. Part (d) indicates the correlation of states in the intermediate region. For convenience, none of the states are shown to cross. States of different double group symmetries may cross. States of the same double group symmetry will undergo configuration interaction

7.5 SPECTRAL PROPERTIES OF TRANSITION METAL COMPLEXES

In a complex that possesses a centre of symmetry, all states arising from a d^n configuration must have g character inherent in the d orbitals. Therefore, the centrosymmetric molecules (O_h, D_{4h} symmetry) the following transitions are not feasible:

gerade ←——|——→ gerade and ungerade ←——|——→ ungerade

Such transitions are Laporte forbidden. Since the dipole moment operator is odd with respect to inversion, it can only couple an odd to an even wave function since $u \times u = g$. The selection rule:

Electric dipole transitions between states of equal parity are forbidden.

However, gerade ←——→ ungerade on vice-versa transitions are permitted. This means d-p, d-f transitions are permitted. For orbitally allowed transitions this rule is in the form

$$\Delta l = \pm 1$$

and is known as **Laperte's rule**.

From symmetry point of view, the existence on nonexistence of transitions depend upon the survival or disappearance respectively of the transition moment integral.

$$\langle \psi_{1\,(orbital)} | \widehat{M} | \psi_{2\,(orbital)} \rangle$$

If the wave functions ψ_1 and ψ_2 transform as the irreducible representations Γ_1 and Γ_2 of the group to which the molecule belongs and the operator $\widehat{M}$ transforms as Γ_M of that group, the transition moment P is zero unless the reducible representation contains the totally symmetric representation (A_{1g}) of the group. In O_h and T_d groups, the operator M transforms as follows:

Operator	O_h	T_d
M	T_{1u}	T_2

Thus the matrix element survives if the direct product representation of the integral- $\langle \psi_1 | \widehat{M} | \psi_2 \rangle$ **contains a totally symmetric representation A_{1g} on resolution. Its absence will however indicate that the concerned transition is forbidden.**

The application of the orbital selection rule may be exemplified by considering d^1 octahedral complex. $^2E_g \leftarrow {}^2T_{2g}$ transition, the direct product (the operator is T_{1u} in O_h symmetry).

$$\langle T_{2g} | \widehat{T}_{1u} | E_g \rangle = T_{2g} \times \langle \widehat{T}_{1u} \times E_g \rangle$$
$$= T_{2g} \times (T_{1u} + T_{2u})$$
$$= (T_{2g} \times T_{1u}) + (T_{2g} + T_{2u})$$
$$= (A_{2u} + E_u + T_{1u} + T_{2u}) + (A_{1u} + E_u + T_{1u} + T_{2u})$$
$$= A_{1u} + A_{2u} + 2E_u + 2T_{1u} + 2T_{2u}$$

As the direct product does not contain A_{1g}, $2E_g \leftarrow {}^2T_{2g}$ so transition moment integral vanishes and the spin allowed but orbitally forbidden.

We shall now consider $^3T_{2g} \leftarrow {}^3T_{1g}$ transition of octahedral $[V(H_2O)_6]^{3+}$ complex ion.

The direct product

$$\langle T_{1g} | \hat{T}_{1u} | T_{2g} \rangle = T_{1g} \times (T_{1u} \times T_{2g})$$
$$= T_{1g} \times (A_{2u} + E_u + T_{1u} + T_{2u})$$
$$= T_{2u} + T_{1u} + T_{2u} + A_{1u} + E_u + T_{1u} + T_{2u} + A_{2u}$$
$$+ E_u + T_{1u} + T_{2u}$$
$$= 4T_{2u} + 3T_{1u} + A_{1u} + 2E_u + A_{2u}$$

does not contain A_{1g}. So the transition moment integral vanishes and $^3T_{2g} \leftarrow {}^3T_{1g}$ transition is spin allowed but orbitally forbidden.

Similarly, for $^3A_{2g} \leftarrow {}^3T_{1g}$ transition, the operator M being the same (T_{1u}) in O_h symmetry, the direct product

$$\langle T_{1g} | \hat{T}_{1u} | T_{2g} \rangle = T_{1g} \times (T_{1u} \times A_{2g}) = T_{1g} \times T_{2u}$$
$$= A_{2u} + E_u + T_{1u} + T_{2u}$$

does neither contain A_{1g}, nor even A_{1g}. So both moment integral vanishes and $^3A_{2g} \leftarrow {}^3T_{1g}$ transition is spin allowed but orbitally forbidden by the symmetry of the electronic wave functions.

But the transition metal complexes are coloured. Light absorption does occur even though it is forbidden. This means that Laperte's selection rule is relaxed. There are two ways for the relaxation of selection rule.

1. A complex may undergo some kind of distortion resulting the change of metal-ligand bond lengths and bond angles. Thus the effective group to which the molecule belongs is not perfect O_h but some lower group which may not have a centre of symmetry. Thus the parity selection rule is relaxed by some mechanism which temporarily or permanently removes the centre of symmetry.

2. The parity selection rule may be further relaxed when the centre of symmetry of the molecule is temporarily removed by an odd vibration of the molecule.

The explanation for the appearance of the corresponding absorption band with weak intensity in terms of vibronic coupling was provided by Van Vleck. According to this mechanism, certain vibrational modes are also excited along with the electronic transition and the transition moment integral (intensity integral) should be

$$\langle (\psi_{\text{gr state}}^{\text{vib}} \ \psi_{\text{gr state}}^{\text{elec}}) | \hat{M} | (\psi_{\text{ex state}}^{\text{elec}} \ \psi_{\text{ex state}}^{\text{vib}}) \rangle$$

Vibrational ground state $\psi_{\text{gr state}}^{\text{vib}}$ is totally the symmetric representation A_{1g} and can be ignored. For $^2E_g \leftarrow {}^2T_{2g}$ transition in $[Ti(H_2O)_6]^{3+}$ complex ion, we may write

$$\langle T_{2g} | \hat{T}_{1u} | E_g \ \Gamma_{\text{ex state}}^{\text{vib}} \rangle$$
$$= T_{2g} \times T_{1u} \times E_g \times \Gamma^{\text{vib}}$$
$$= \Gamma^{\text{vib}} \times T_{2g} (T_{1u} \times E_g)$$
$$= \Gamma^{\text{vib}} \times T_{2g} (T_{1u} + T_{2u})$$
$$= \Gamma^{\text{vib}} \times (A_{2u} + E_u + T_{1u} + T_{2u} + A_{1u} + E_u + T_{1u} + T_{2u})$$
$$= \Gamma^{\text{vib}} \times (A_{1u} + A_{2u} + 2E_u + 2T_{1u} + 2T_{2u}) \qquad \dots (7.13)$$

Thus the transition moment integral does not vanish if Γ^{vib} contains any of the irreducible representations in the parenthesis of the right hand side of Eq. (7.13) the symmetries of the normal modes of an octahedral ML_6 molecule are A_{1g}, E_g, T_{2g}, $2T_{1u}$, $2T_{2u}$ (Table 7.10).

$$\Gamma^{vib} = A_{1g} + E_g + T_{2g} + 2T_{1u} + T_{2u}$$

Table 7.10 Fundamental skeletal of normal modes of vibration of some of the more commonly occurring centric and acentric molecules

Molecular skeleton	Group	Normal modes of vibrations
ML_6	O_h	$A_{1g} + E_g + T_{2g} + 2T_{1u} + T_{2u}$
ML_6	D_3	$3A_1 + 2A_2 + 5E$
ML_5	D_{3h}	$2A_1' + 3E' + 2A_2'' + E'''$
ML_5	C_{4v}	$3A_1 + 2B_1 + B_2 + 3E$
ML_4	D_{4h}	$A_{1g} + B_{1g} + B_{2g} + A_{2u} + B_{2u} + 2E_u$
ML_4	C_{4h}	$A_g + 2B_g + A_u + B_u + 2E_u$
ML_4	T_d	$A_1 + E + 2T_2$
ML_4	D_{2d}	$2A_1 + B_1 + 2B_2 + 2E$
$ML_4 L_2'$	D_{4h}	$2A_{1g} + B_{1g} + B_{2g} + E_g + 2A_{2u} + B_{2u} + 3E_u$
$ML_2 L_2'$	D_{2h}	$2A_g + B_{1g} + 2B_{1u} + 2B_{2u} + 2B_{3u}$
$ML_3 L'$	C_{3v}	$3A_1 + 3E$
$ML_2 L_2'$	C_{2v}	$4A_1 + A_2 + 2B_1 + 2B_2$

The underlined representations are also present in the parenthesis of the RHS of Eq. (7.13). Hence the direct product representation of $\Gamma^{vib} \times (A_{1u} + A_{2u} + 2E_u + {}^2T_{1u} + {}^2T_{2u})$ contains a totally symmetric component A_{1g}.

$(A_{1g} + E_g + T_{2g} + 2T_{1u} + T_{2u})(A_{1u} + A_{2u} + 2E_u + 2T_{1u} + 2T_{2u}) = A_{1g} + \dots\dots\dots$

Note that
$$T_{1u} \times T_{1u} = A_{1g} + E_g + T_{1g} + T_{2g}$$
$$T_{2u} + T_{2u} = A_{1g} + E_g + T_{1g} + T_{2g}$$

The transition moment integral (intensity integral) becomes nonzero because at least one of the two T_{1u} modes or the T_{2u} mode of vibration is excited simultaneously with the electronic excitation. Physically this means that the atomic displacement vectors in the two T_{1u} modes and the T_{2u} mode of vibration are such that they destroy the centre of symmetry of the molecule during such vibrations.

Similarly for ${}^3T_{2g} \leftarrow {}^3T_{1g}$ transition of $[V(H_2O)_6]^{3+}$ complex ion the direct product

$T_{1g} \times \Gamma^{vib}_{gr\,state} \times T_{1u} + T_{2g} \times \Gamma^{vib}_{ex\,state}$

$= T_{1g} \times T_{1u} \times T_{2g} \times \Gamma^{vib}_{ex\,state}$

$= \Gamma^{vib}_{ex\,state} \times T_{1g} \times (T_{1u} \times T_{2g})$

$= \Gamma^{vib}_{ex\,state} \times T_{1g} \times (A_{2u} + E_u + T_{1u} + T_{2u})$

$= \Gamma^{vib}_{ex\,state} \times (T_{2u} \times T_{1u} + T_{2u} + A_{1u} + E_u + T_{1u} + T_{2u} + A_{2u} + E_u + T_{1u} + T_{2u}$

$$= \Gamma^{vib}_{ex\,state} \times (A_{1u} + A_{2u} + 2E_u + 3T_{1u} + 4T_{2u}) \qquad \qquad ...(7.14)$$

$$\Gamma^{vib}_{ex\,state} = A_{1g} + E_g + 2T_{1u} + T_{2g} + T_{2u}$$

The underlined representations are also present in the parenthesis of RHS of Eq. (7.14). Hence the direct product representation of $\Gamma^{vib}_{ex\,state} \times (A_{1u} + A_{2u} + 2E_u + 3T_{1u} + 4T_{2u})$ contains a totally symmetric component A_{1g}. So the intensity integral becomes nonzero and $^3T_{2g} \leftarrow {}^3T_{1g}$ transition is allowed transition.

The possible spin allowed transitions for a d^2 ion in a tetrahedral environment are

$$^3T_2 \leftarrow {}^3A_2, \quad {}^3T_1 (F) \leftarrow {}^3A_2 \text{ and } {}^3T_1 (P) \leftarrow {}^3A_2$$

Here the dipole moment operator $\widehat{M}$ transforms to T_2: considering only the electronic (not simultaneously vibrational) transition, the transition moment integral can be expressed as below.

For
$$\begin{aligned}
{}^3T_2 \leftarrow {}^3A_2 \quad &: \quad A_2 \times T_2 \times T_2 = A_2 + E + T_1 + T_2 \\
{}^3T_1 (F) \leftarrow {}^3A_2 \quad &: \quad A_2 \times T_2 \times T_1 = A_1 + E + T_1 + T_2 \\
{}^3T_1 (P) \leftarrow {}^3A_2 \quad &: \quad A_2 \times T_2 \times T_1 = A_1 + E + T_1 + T_2
\end{aligned}$$

The totally symmetric representation A_1 is contained for the last two transitions. Hence the transition moment integrals for them do not vanish. So $^3T_1 (F) \leftarrow {}^3A_2$ and $^3T_1 (P) \leftarrow {}^3A_2$ transitions are symmetry allowed transitions and should give rise to strong absorption bands in the electronic spectrum. But the reduction result for $^3T_2 \leftarrow {}^3A_2$ transition does not contain A_1 and hence its intensity integral vanishes. Consequently this transition is not allowed without consideration of vibronic coupling. Therefore, we need to consider the vibronic transition moment integral. For $^3T_2 \leftarrow {}^3A_2$ transition

$$T_2 \times \Gamma^{vib}_{ex\,state} \times T_2 \times A_2 \times \Gamma^{vib}_{gr\,state}$$

$$= T_2 \times \Gamma^{vib}_{ex\,state} \times T_2 \times A_2 \qquad \text{(ground vibrational wave function transform as } A_1$$

$$= \Gamma^{vib} \times (T_2 \times T_2 \times A_2) \qquad \text{and can be ignored)}$$

$$= \Gamma^{vib} \times (A_2 + E + T_1 + T_2) \qquad \qquad ...(7.15)$$

For a tetrahedral molecule $\Gamma^{vib} = A_1 + E + 2T_2$

As $E \times E$ $(E_1 \times E_1 = E_2 \times E_2 = A_1 + A_2 + E_2)$ and $T_2 \times T_2$ $(T_2 \times T_2 = A_1 + E + T_1 + T_2)$ contain the totally symmetric representation A_1, the integral in Eq. (7.15) does not vanish. Hence the transition $^3T_2 \leftarrow {}^3A_2$ becomes vibronically allowed if the E or T_2 mode of vibration is excited simultaneously with the electronic transition.

8

Molecular Orbital Theory and its Applications in Inorganic Chemistry

8.1 LINEAR COMBINATION OF ATOMIC ORBITAL THEORY

Valence bond theory successfully explains the structure of the molecules but can not explain the electronic spectra of the molecule or interpret their magnetic properties quantitatively. An alternate theory, proposed by Hund and Mulliken is called the molecular orbital theory.

According to this theory when atoms come closer during the formation of molecules, atomic orbitals of same symmetry and same or close energy combine to form molecular orbitals. Molecular orbital wave functions are constructed by linear combination of atomic orbital (LCAO) wave functions. For a diatomic molecule AB, the molecular orbital wave function ψ_{MO} is obtained by linear combination of atomic orbitals of A and B atoms having same symmetry and energy.

$$\psi_{MO} = C_1\phi_1 + C_2\phi_2 \qquad \qquad ...(8.1)$$

where ϕ_1 and ϕ_2 are atomic orbital wave functions of atoms A and B and C_1 and C_2 are constants or coefficients or mixing constants. These coefficients indicate the relative contributions of each atomic orbital wave function to the LACO–MO wave function.

Schrödinger's differential equation relating to the energy of a system in three dimensions as

$$\frac{\partial^2\psi}{\partial x^2} + \frac{\partial^2 y}{\partial y^2} + \frac{\partial^2\psi}{\partial z^2} + \frac{8\pi^2 m}{h^2}(E - V)\ \psi = 0$$

Rearranging the above relation, we have

$$\left[V - \frac{h^2}{8\pi^2 m}\left(\frac{\partial^2\psi}{\partial^2 x^2} + \frac{\partial^2\psi}{\partial^2 y^2} + \frac{\partial^2\psi}{\partial^2 z^2}\right)\right]\ \psi = E\psi$$

The left hand side of the above relation can be considered as the action of an operator on ψ, which represents a molecular orbital. This operator is called the Hamiltonian operator and is represented as H. Thus

$$H\psi = E\psi \qquad \qquad ...(8.2)$$

Equation (8.2) states that, if ψ is an eigen function and an Hamiltonian operator (H) on this function will yield the same function (ψ) multiplied by a constant E called eigen value.

Multiplying both sides of Eq. (8.2) by ψ and integrating over the space, we get

$$\int \psi H\psi\, d\tau = E \int \psi\psi\, d\tau \quad [d\tau = dx\, dy\, dz]$$

$$E = \frac{\int \psi H \psi \, d\tau}{\int \psi^2 \, d\tau} \qquad ...(8.3)$$

Substituting for ψ from Eq. (8.1), we have

$$E = \frac{\int (C_1 \phi_1 + C_2 \phi_2) H(C_1 \phi_1 + C_2 \phi_2) \, d\tau}{\int (C_1 \phi_1 + C_2 \phi_2)^2 \, d\tau}$$

$$= \frac{\int (C_1 \phi_1 H C_1 \phi_1 + C_1 \phi_1 H C_2 \phi_2 + C_2 \phi_2 H C_1 \phi_1 + C_2 \phi_2 H C_2 \phi_2) \, d\tau}{\int (C_1^2 \phi_1^2 + 2C_1 C_2 \phi_1 \phi_2 + C_2^2 \phi_2^2) \, d\tau}$$

$$= \frac{C_1^2 \int \phi_1 H \phi_1 \, d\tau + 2C_1 C_2 \int \phi_1 H \phi_2 \, d\tau + C_2^2 \int \phi_2 H \phi_2 \, d\tau}{C_1^2 \int \phi_1^2 \, d\tau + 2C_1 C_2 \int \phi_1 \phi_2 \, d\tau + C_2^2 \int \phi_2^2 \, d\tau} \qquad ...(8.4)$$

Equation (8.4) can be simplified by the following substitutions for the integrals:

$$H_{ii} = \int \phi_i H \phi_i \, d\tau \quad \text{(Coulomb integrals)} = \alpha$$

$$H_{ij} = \int \phi_i H \phi_j \, d\tau \quad \text{(Resonance integrals or exchange integrals)} = \beta$$

$$S_{ij} = \int \phi_i \phi_j \, d\tau \quad \text{(Overlap integrals)}$$

$$S_{ii} = \int \phi_i \phi_i \, d\tau$$

Thus we get

$$E = \frac{C_1^2 H_{11} + 2C_1 C_2 H_{12} + C_2^2 H_{22}}{C_1^2 S_{11} + 2C_1 C_2 S_{12} + C_2^2 S_{22}}$$

$$E \left(C_1^2 S_{11} + 2 C_1 C_2 S_{12} + C_2^2 S_{22} \right) = C_1^2 H_{11} + 2 C_1 C_2 H_{12} + C_2^2 H_{22} \qquad ...(8.5)$$

To get the minimum energy of the system, we set

$$\left(\frac{\partial E}{\partial C_1} \right)_{C_2} = 0 \quad \text{and} \quad \left(\frac{\partial E}{\partial C_2} \right)_{C_1} = 0$$

Differentiating Eq. (8.5) with respect to C_1, we get

$$2C_1 S_1 E + C_1^2 S_{11} \left(\frac{\partial E}{\partial C_1} \right)_{C_2} + 2C_1 C_2 S_{12} \left(\frac{\partial E}{\partial C_1} \right)_{C_2} + 2C_2 S_{12} E + C_2^2 S_{22} \left(\frac{\partial E}{\partial C_1} \right)_{C_2}$$

$$= 2C_1 H_{11} + 2 C_2 H_{12}$$

After rearrangement, we get

$$\left(\frac{\partial E}{\partial C_1} \right)_{C_2} [C_1{}^2 S_{11} + 2 C_1 C_2 S_{12} + C_2^2 S_{22}] = 2C_1 (H_{11} - S_{11} E) + 2C_2 (H_{12} - S_{12} E)$$

$$\therefore \qquad \left(\frac{\partial E}{\partial C_1}\right)_{C_2} = \frac{2C_1(H_{11}-S_{11}E)+2C_2(H_{12}-S_{12}E)}{C_1^2 S_{11}+2C_1 C_2 S_{12}+C_2^2 S_{22}}$$

Since $\qquad \left(\dfrac{\partial E}{\partial C_1}\right)_{C_1} = 0$, we may write

$$C_1(H_{11}-S_{11}E) + C_2(H_{12}-S_{12}E) = 0 \qquad \qquad ...(8.6)$$

Similarly, on differentiating Eq. (8.5) with respect to C_2, we get

$$C_1(H_{12}-S_{12}E) + C_2(H_{22}-S_{22}E) = 0 \qquad \qquad ...(8.7)$$

Equations (8.6) and (8.7) are called 'secular equations'

$$\left.\begin{array}{l} C_1(H_{11}-S_{11}E)+C_2(H_{12}-S_{12}E) = 0 \\ C_1(H_{12}-S_{12}E)+C_2(H_{22}-S_{22}E) = 0 \end{array}\right\} \qquad ...(8.8)$$

Clearly, $C_1 = C_2 = 0$ is a solution of Eq. (8.8). But it is a meaningless solution, as ψ becomes zero, and for C_1 and C_2 to be nonvanishing, we need to have

$$\begin{vmatrix} (H_{11}-S_{11}E) & (H_{12}-S_{12}E) \\ (H_{12}-S_{12}E) & (H_{22}-S_{22}E) \end{vmatrix} = 0$$

This is known as a secular determinant, where the only unknown is E.

Since $\qquad H_{11} = \alpha,\ H_{22} = \alpha,\ H_{12} = \beta$ and $S_{11} = S_{22} = 11$

$$\left(\because S_{ii} = \int \phi_i\,\phi_i,\,d\tau = \int \phi^2\,d\tau = 1\right), \text{ we may write}$$

$$\begin{vmatrix} \alpha-E & \beta-S_{12}E \\ \beta-S_{12}E & \alpha-E \end{vmatrix} = 0$$

For a homonuclear diatomic species such as H_2^+, we can substitute $S_{12} = S$. So we have

$$\begin{vmatrix} \alpha-E & \beta-S_{12}E \\ \beta-S_{12}E & \alpha-E \end{vmatrix} = 0$$

or $\qquad (\alpha-E)^2 - (\beta-SE)^2 = 0 \quad \text{or} \quad (\alpha-E) = \pm(\beta-SE)$

Taking +ve value

$$\alpha-E = \beta-SE$$
$$\alpha-\beta = E(1-S)$$
$$E = \frac{\alpha-\beta}{1-S}$$

Taking –ve value

$$\alpha-E = -(\beta-SE)$$
$$\alpha+\beta = E(1+S)$$
$$\therefore \qquad E = \frac{\alpha+\beta}{1+S} \quad \text{or} \quad \frac{\alpha-\beta}{1-S}$$

By substituting these values of E in Eq. (8.6) we get

$$C_1 = (\alpha - SE) + C_2 (\beta - SE) = 0 \qquad [\because H_{11} = \alpha; H_{12} = \beta; S_{11}; S_{12} = S]$$

$$\therefore \qquad C_1 = \left(\frac{\beta - SE}{\alpha - SE} \right) C_2$$

When

$$E = \left(\frac{\alpha + \beta}{1 + S} \right)$$

$$\therefore \qquad C_1 = C_2$$

When

$$E = \left(\frac{\alpha - \beta}{1 - S} \right); \ C_1 = -C_2$$

Thus

$$C_2 = \pm C_1$$

For calculating C_1 we must normalise ψ $(= C_1\phi_1 + C_2\phi$ remember $C_1 = C_2)$

$$\therefore \qquad \int \psi^2 \, d\tau = C_1^2 \int \phi_1^2 \, d\tau \pm 2C_1^2 \int \phi_1\phi_2 \, d\tau + C_1^2 \int \phi_2^2 \, d\tau = 1$$

$$C_1^2 \int \phi_1^2 \, d\tau \pm 2C_1^2 \int \phi\phi_2 \, d\tau + C_1^2 \int \phi_2^2 \, d\tau = C_1^2 \pm 2\, C_1^2, S + C_1^2$$

$$= 2\, C_1^2 \pm 2\, C_1^2 S$$

$$\therefore \qquad C_1^2 (2 \pm 2S) = 1 \quad \therefore \quad C_1 = \pm \frac{1}{\sqrt{2 \pm 2S}}$$

The + sign corresponds to $C_1 = C_2$ and – sign to $C_1 = -C_2$. Thus two normalised wave function are

$$\psi_b = \frac{1}{\sqrt{2 + 2S}} (\phi_1 + \phi_2) \qquad\qquad ...(8.9)$$

$$\psi^* = \frac{1}{\sqrt{2 - 2S}} (\phi_1 - \phi_2) \qquad\qquad ...(8.10)$$

ψ_b and ψ^* stand for bonding and antibonding molecular orbitals.

ψ can be positive and negative; ψ^2 is positive and represents the electron density in the region of overlap between the atoms in a molecule.

If two atomic wave functions are orthogonal,

$$S_{12} = S = \int \phi_1\phi_2 \, d\tau = 0$$

Equations (8.9) and (8.10) then reduced to

$$\psi_b = \frac{1}{\sqrt{2}} (\phi_1 + \phi_2) \qquad\qquad ...(8.11)$$

$$\psi^* = \frac{1}{\sqrt{2}} (\phi_1 - \phi_2) \qquad\qquad ...(8.12)$$

The constants $\frac{1}{\sqrt{2}}$ in Eqs (8.11) and (8.12) for bonding and antibonding MOs is called the *normalisation constant*. When the linear combination extends over several atomic orbital wave functions such as $\phi_1, \phi_2, \phi_3, \ldots$ so that

$$\psi = C_1 \phi_1 + C_2 \phi_2 + C_3 \phi_3 + \ldots$$

The normalisation constant N is given by

$$N = \frac{1}{\sqrt{C_1^2 + C_2^2 + C_3^2 + \ldots}} \qquad \ldots(8.13)$$

The wave functions (ψ) for MOs are normalised by multiplying with a normalisation constant N such that

$$\int (N\psi)^2 \, d\tau = 1,$$

i.e. the probability of finding the electron outside the nucleus is unity.

Let us consider a diatomic molecule of atoms 1 and 2. We consider LCAO of ϕ_1 and ϕ_2.

$$\psi = C_1 \phi_1 + C_2 \phi_2$$

The normalised wave equation can be written as

$$\int (N\psi)^2 \, d\tau = \int N^2 (C_1 \phi_1 + C_2 \phi_2)^2 \, d\tau = 1$$

$$N^2 \left[\int (C_1^2 \phi_1^2 + C_2^2 \phi_2^2 + 2 C_1 C_2 \phi_1 \phi_2) \, d\tau \right] = 1$$

$$N^2 \left[C_1^2 \int \phi_1^2 \, d\tau + C_2^2 \int \phi_2^2 \, d\tau + 2 C_1 C_2 \int \phi_1 \phi_2 \, d\tau \right] = 1$$

Since each atomic wave function ϕ_1 is normalised,

$$\int \phi_1^2 \, d\tau = \int \phi_2^2 \, d\tau = 1$$

The overlap integral, $\int \phi_1 \phi_2 \, d\tau = 0$ in LCAO approximation.

$$\therefore \quad N = \frac{1}{\sqrt{C_1^2 + C_2^2}} . \text{ Therefore } N_2 [C_1^2 + C_2^2] = 1$$

In general, when a system under study is a molecule composed of n atoms as the trial function, which is to become a molecular orbital, is a linear combination of atomic orbitals (LCAO).

$$\psi = C_1 \phi_1 + C_2 \phi_2 + \ldots + C_n \phi_n$$

where $\phi_1, \phi_2 \ldots \phi_n$ are known atomic functions and the corresponding secular determinant has the dimensions $n \times n$:

$$\begin{vmatrix} H_{11} - ES_{11} & H_{12} - ES_{12} & \cdots & H_{1n} - ES_{1n} \\ H_{21} - ES_{21} & H_{22} - ES_{22} & \cdots & H_{2n} - ES_{2n} \\ \vdots & & \vdots & \\ H_{n1} - ES_{n1} & H_{n2} - ES_{n2} & \vdots & H_{nn} - ES_{nn} \end{vmatrix} = 0 \qquad \ldots(8.14)$$

There are now n roots of E. Substituting each E into the secular equation

$$\left. \begin{aligned} (H_{11} - ES_{11}) \, C_1 + (H_{12} - ES_{12}) \, C_2 + \ldots + (H_{1n} - ES_{1n}) \, C_n &= 0 \\ (H_{21} - ES_{21}) \, C_1 + (H_{22} - ES_{22}) \, C_2 + \ldots + (H_{2n} - ES_{2n}) \, C_n &= 0 \\ \vdots \qquad\qquad\qquad\qquad & \\ (H_{n1} - ES_{n1}) \, C_1 + (H_{2n} - ES_{2n}) \, C_2 + \ldots + (H_{nn} - ES_{nn}) \, C_n &= 0 \end{aligned} \right\} \qquad \ldots(8.15)$$

We get a set of n coefficients. In other words, there are all together n sets of coefficients, or n molecular orbitals.

8.2 MOLECULAR ORBITALS IN COMPLEX COMPOUNDS

In the crystal field treatment of complexes, the interaction between the central metal ion and the ligands is taken to be purely electrostatic, i.e. ionic bonding. No overlap of the metal orbitals and the ligand orbitals is considered, i.e. covalent bonding is entirely ignored. In the molecular orbital theory, the combining atomic orbitals generate a set of bonding molecular orbitals (MOs) of lower energy and an equal number of antibonding MOs of higher energy. Noninteracting atomic orbitals remain as nonbonding MOs with the energy remaining unchange.

8.2.1 σ-Bonding in Octahedral Complexes

The MO theory starts with the premise that metal orbitals and the ligand orbitals will overlap whenever symmetry permits. This means for any MO to be possible, the symmetry of the orbitals on the metal and the ligand must match. The MOs used here will be of linear combinations of atomic orbitals (LCAO) type. A coordinate system for such a complex is shown in Fig. 8.1. The representation generated by the six ligand orbtials.

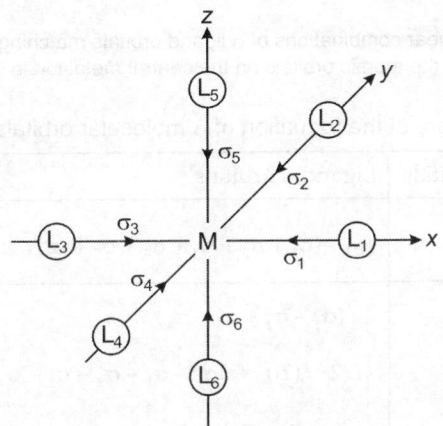

Fig. 8.1 A coordinate system for ML_6, where the ligands have only σ bond to bond with the metal ion

Q_h	E	$8C_3$	$6C_2$	$6C_4$	$3C_2$	i	$6S_4$	$8S_6$	$3\sigma_h$	$6\sigma_d$
Γ_σ	6	0	0	2	2	0	0	0	4	2

that reduces $\Gamma_\sigma = A_{1g} + E_g + T_{1u}$. The six linear combinations of ligand orbitals can be readily generated, and they match in symmetry with the suitable metal orbitals. These are summarised in Table 8.1 and illustrated in Fig. 8.2. The molecular orbital energy level diagram for an octahedral ML_6 complex with only σ bonding is shown in Fig. 8.3.

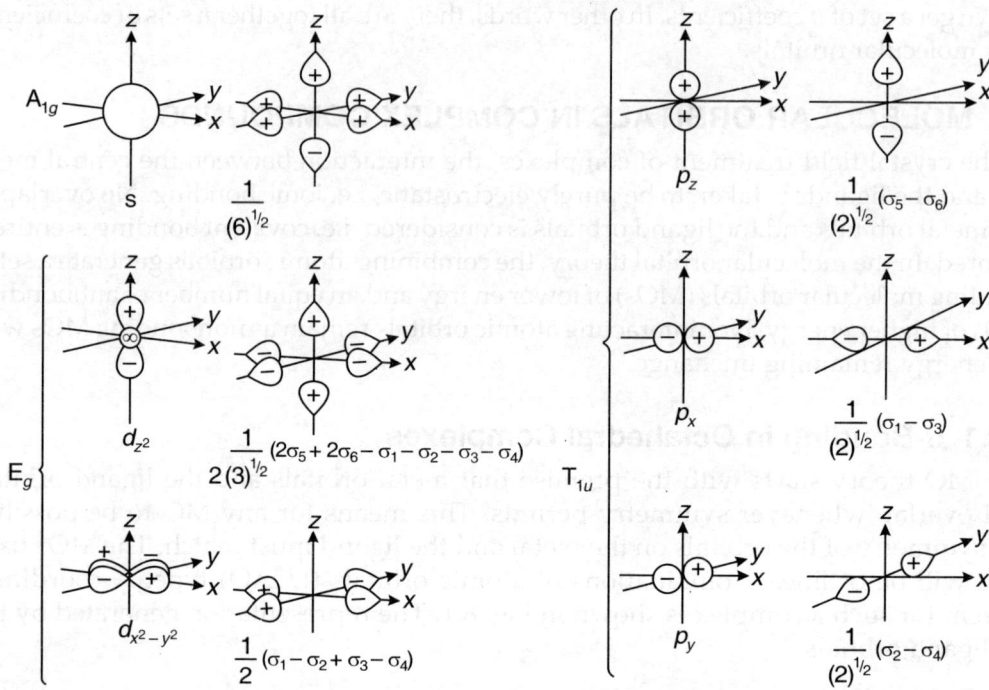

Fig. 8.2 Linear combinations of σ ligand orbitals matching symmetry
with the atomic orbitals on the central metal ion in ML_6

Table 8.1 Summary of the formation of σ molecular orbitals (Fig. 8.2) in ML_6

Symmetry	Metal orbitals	Ligand σ orbitals	Molecular orbitals
A_{1g}	s	$(6)^{-\frac{1}{2}}(\sigma_1 + \sigma_2 + \sigma_3 + \sigma_4 + \sigma_5 + \sigma_6)$	$1a_{1g}, 2_{1g}$
E_g	$\begin{cases} d_{x^2-y^2} \\ d_{z^2} \end{cases}$	$\begin{cases} \frac{1}{2}(\sigma_1 - \sigma_2 + \sigma_3 - \sigma_4) \\ (12)^{-\frac{1}{2}}(2\sigma_5 + 2\sigma_6 - \sigma_1 - \sigma_2 - \sigma_3 - \sigma_4) \\ (2)^{-\frac{1}{2}}(\sigma_1 - \sigma_3) \end{cases}$	$1e_g, 2e_g$
T_{1u}	$\begin{cases} p_x \\ p_y \\ p_z \end{cases}$	$\begin{cases} (2)^{-\frac{1}{2}}(\sigma_1 - \sigma_3) \\ (2)^{-\frac{1}{2}}(\sigma_2 - \sigma_4) \\ (2)^{-\frac{1}{2}}(\sigma_5 - \sigma_6) \end{cases}$	$1t_{1u}, 2t_{1u}$
T_{2g}	$\begin{cases} d_{xy} \\ d_{yz} \\ d_{zx} \end{cases}$	—	$1t_{2g}$

Examining the energy level diagram shown in Fig. 8.3, we see that the electron pairs from the six ligands enter into the $1\,t_{1u}$, $1\,a_{1g}$ and $1\,e_g$ orbitals, thus leaving $1\,t_{2g}$ and $2\,e_g$ orbitals to accommodate the d electrons from the metal ion. Also note that the $1\,t_{2g}$ orbitals are more stable than the $2\,e_g$ orbitals and the energy gap between them is one again called Δ_0. This is the legal agreement with the results of crystal field theory. But the two theories arrive at their results in different ways. In crystal field theory, the t_{2g} orbitals are more stable than the e_g orbitals because the former have lobes pointing between the ligands and the latter have lobes pointing at the ligands. On the other hand, in molecular orbital theory, the $1\,t_{2g}$ orbitals are more stable than the $2\,e_g$ orbitals because the former are nonbonding orbitals, while the latter are antibonding σ^* orbitals. In Fig. 8.3, we may

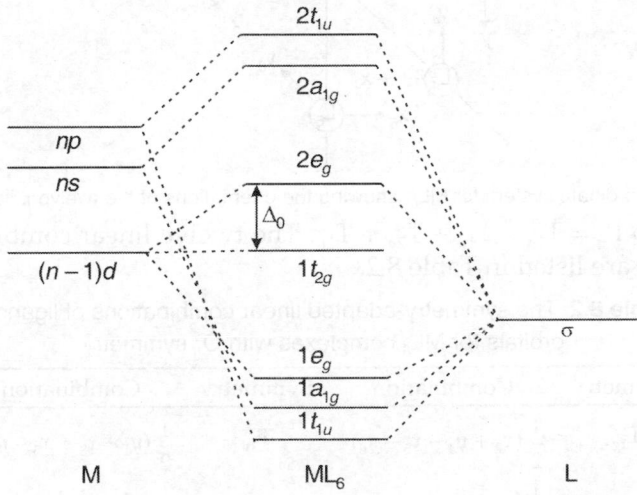

Fig. 8.3 The molecular orbital energy level diagram for an octahedral complex with only σ-bonding between the metal and the ligands

construct the MO diagrams of $[M(H_2O)_6]^{n+}$ and $[M(NH_3)6]^{n+}$ ($n = 2$ or 3). In case of $[Ti(H_2O)_6]^{3+}$, we have to accommodate twelve ligand electrons and one metal electron in the MOs and the ground state configuration of $[Ti(H_2O)_6]^{3+}$ is

$$1\,t_{1u}^6\,1a_{1g}^2\,1\,e_g^4\,1\,t_{2g}^1$$

with $S = 1/2$. Figure 8.3 is consistent with paramagnetic nature ($\mu_{eff} = 1.7$ BM) of the complex due to the presence of one unpaired electron in the t_{2g} level.

In case of $[Ni(MHs)_6]^{2+}$ ion, the filled t_{2g} orbitals of Ni^{2+} ion remain nonbonding. Two electrons of Ni^{2+} occupy antibonding $2e_g$ orbital, giving two unpaired electrons.

$$1\,t_{1u}^6\,1\,a_{1g}^2\,1\,e_g^4\,1\,t_{2g}^6\,2\,e_g^1\,2\,e_g^1$$

8.2.2 Octahedral Complexes with π-Bonding

When the ligands in ML_6 complex also have π orbitals available for interaction, the bonding picture becomes more complicated. If the ligand π orbitals are oriented in the manner depicted in Fig. 8.4, it can be readily shown that the representation generated by the twelve ligand π orbitals is

Q_h	E	$8C_3$	$6C_2$	$6C_4$	$3C_2 (= C_2^2)$	i	$6S_4$	$8S_6$	$3\sigma_h$	$6\sigma_d$
Γ_π	12	0	0	0	-4	0	0	0	0	0

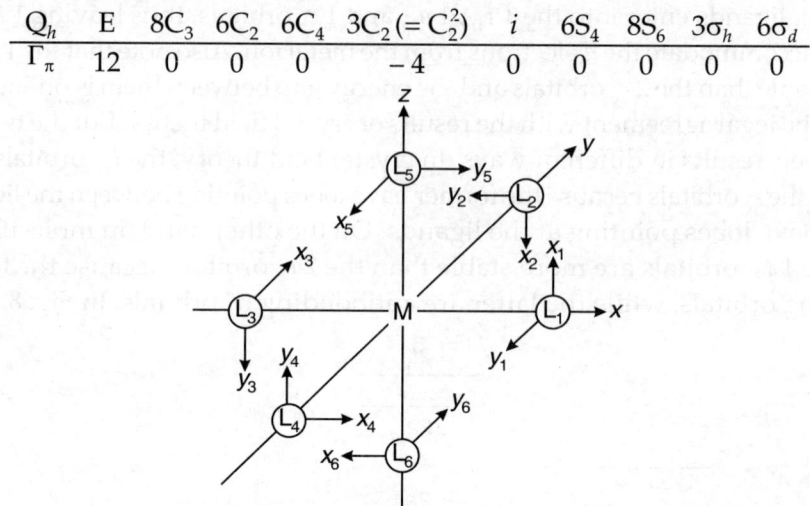

Fig. 8.4 A coordinate system for ML_6, showing the orientations of the twelve π ligand orbitals

This reduces to $\Gamma_\pi = T_{1g} + T_{1u} + T_{2g} + T_{2u}$. The twelve linear combinations with the proper symmetry are listed in Table 8.2.

Table 8.2 The symmetry-adapted linear combinations of ligand π orbitals for ML_6 complexes with O_h symmetry

Symmetry	Combination	Symmetry	Combination
T_{1g}	$\frac{1}{2}(x_2 + y_4 - x_5 - y_6)$	T_{1u}	$\frac{1}{2}(y_1 - x_3 + x_5 - y_6)$
	$\frac{1}{2}(x_1 + y_3 - y_5 - x_6)$		$\frac{1}{2}(-y_2 + x_4 + y_5 - x_6)$
	$\frac{1}{2}(-y_1 + y_2 - x_3 + x_4)$		$\frac{1}{2}(x_1 - x_2 - y_3 + y_4)$
T_{2g}	$\frac{1}{2}(x_2 + y_4 + x_5 + y_6)$	T_{1u}	$\frac{1}{2}(y_1 - x_3 - x_5 + y_6)$
	$\frac{1}{2}(x_1 + y_3 + y_5 + x_6)$		$\frac{1}{2}(-y_2 + x_4 - y_5 + x_6)$
	$\frac{1}{2}(y_1 + y_2 + x_3 + x_4)$		$\frac{1}{2}(-x_1 - x_2 + y_3 + y_4)$

Since there are no metal orbitals with either T_{1g} or T_{2u} symmetry, the combinations of ligands orbitals with these symmetry given in Table 8.2 are essentially nonbonding

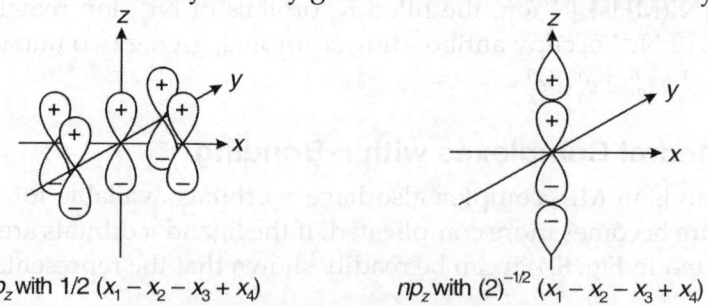

np_z with $1/2\,(x_1 - x_2 - x_3 + x_4)$ np_z with $(2)^{-1/2}\,(x_1 - x_2 - x_3 + x_4)$

Fig. 8.5 Comparison of the overlaps between the T_{1s}, σ and π ligand combinations with the same np_z orbital on the metal. Note that the σ type overlap will lead to more significant interaction between the ligands and the metal orbitals. Also, linear combinations of ligand π orbitals with T_{1u} symmetry will have less effective overlap with the metal np orbitals than the σ combinations (*see* Table 8.1) of the same symmetry. The comparison of these two types of overlap is illustrated in Fig. 8.5. In other words, the molecular orbitals consisting mostly of the ligand π orbitals with T_{1u} symmetry is essentially nonbonding or at most weakly bonding. So the most important ligand orbitals participating in π bonding are the combinations of T_{2g} symmetry. Figure 8.6 shows how these combinations overlap with the metal d orbitals.

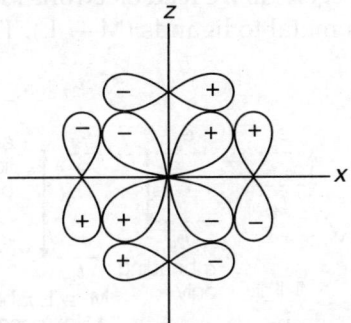

Fig. 8.6 The π-type overlap with T_{2g} symmetry between the ligand and the metal

When the ligands have filled π orbitals to bond with the metal, examples of which include F^-, Cl^-, OH^-, ..., etc. the schematic energy level diagram for such complexes is shown in Fig. 8.7. Note that the orbitals up to and including $1t_{1g}$ and $1t_{2u}$ are filled between electrons from ligands and we fill in the metal d electrons starting from $2t_{2g}$. Now the $2e_g$ orbitals remain antibonding σ^* in nature, while the $2t_{2g}$ orbital have become antibonding π^* (recalling that these orbitals called $1t_{2g}$ as shown in Fig. 8.3, are nonbonding in a complex with only σ bonding).

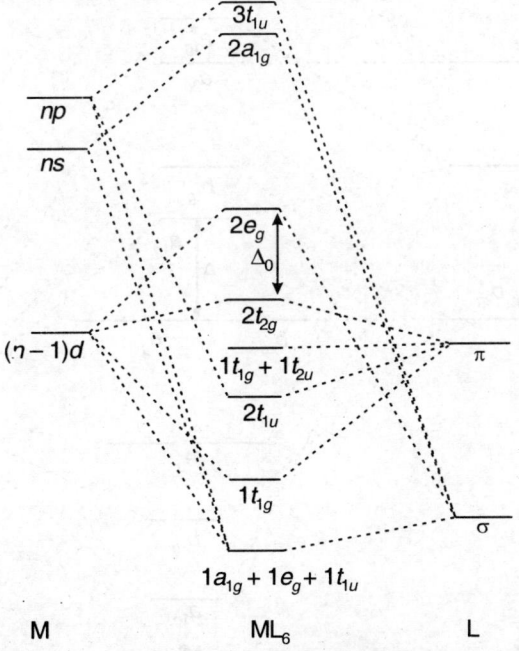

Fig. 8.7 Schematic energy level diagram for an octahedral ML_6 complex in which the ligands have filled π orbitals for bonding. Compared to complexes with only σ bonding, the present system has a smaller Δ_0 [see also Fig. 8.8(a).]

As a result, compared to the complexes with only σ bonding, the present system has a smaller Δ_0. As illustrated in Fig. 8.8(a), such a decrease in Δ_0 implies an electron flow from ligands to metal (L → M).

Conversely, when the ligands have low lying empty π orbitals for bonding, examples include CO, CN^-, PR_3, etc. the originally nonbonding t_{2g} orbitals now become π bonding in nature, while e_g orbitals remains σ* antibonding. As a result, compared to complexes with only σ bonding, Δ_0 becomes larger. As we feed electrons into these π bonding t_{2g} orbitals, electrons appear to flow from metal to ligands (M → L). This is illustrated in Fig. 8.8(b).

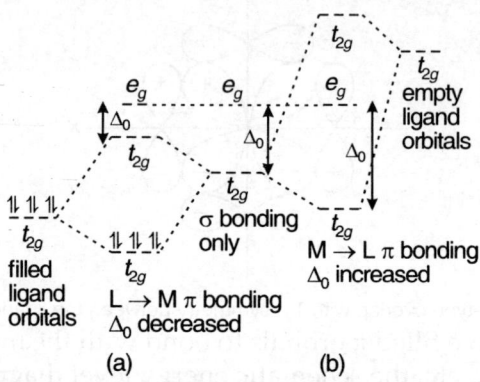

Fig. 8.8 Comparison of the effects of π bonding using (a) filled π ligand orbitals for L → M donation and (b) empty π ligand orbitals for M → L donation. Note that, compared to complexes with only σ bonding, the former leads to a smaller Δ_0 and the latter to a larger one

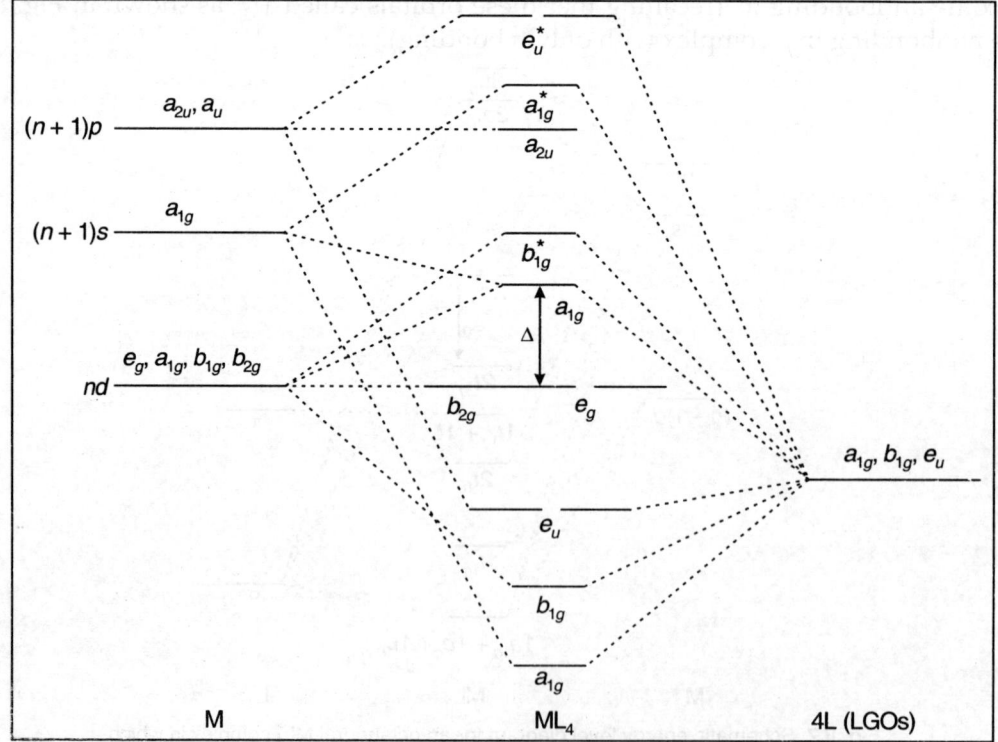

Fig. 8.9 A σ MO diagram for a square planar ML_4 complex (D_{4h} symmetry)

Thus we see that

(i) in complexes with only σ bonding, the e_g orbitals are antibonding σ^* and t_{2g} orbitals *are nonbonding* and

(ii) in complexes with ligands having filled π orbitals, the e_g orbitals remain antibonding σ^* and t_{2g} orbitals *have become antibonding* π^*, leading to a smaller Δ_0 and

(iii) in complexes with ligands having empty π^* orbitals for bonding, the e_g orbitals are still antibonding σ^* and t_{2g} *orbitals have become π bonding*, thus yielding a larger Δ_0.

In square planar environment, the metal d level is split into a_{1g} (d_{z^2}), e_g (d_{xz}, d_{yz}), b_{2g} (d_{xy}) and b_{1g} $(d_{x^2-y^2})$ orbitals. The p level loses its degeneracy and appears as a_{2u} (p_z) and e_u (p_x, p_y). The four ligands will approach to the metal atom along the x and y axes. These ligands will give rise to ligand group orbitals (LGOs) of a_{1g}, b_{1g}, and e_g symmetry. They will interact with metal orbitals of the same symmetry leading to the σ MO diagram shown in Fig. 8.9. It is to be noted that the a_{1g} LGO overlaps with both a_{1g} metal orbitals $(5d_{z^2}$ and $6s$ for platinum metal orbitals) producing three MOs of this symmetry. a_{2u}, e_g and the b_{2g} orbitals remain nonbonding as there is no matching symmetry of the orbital present in the LGOs.

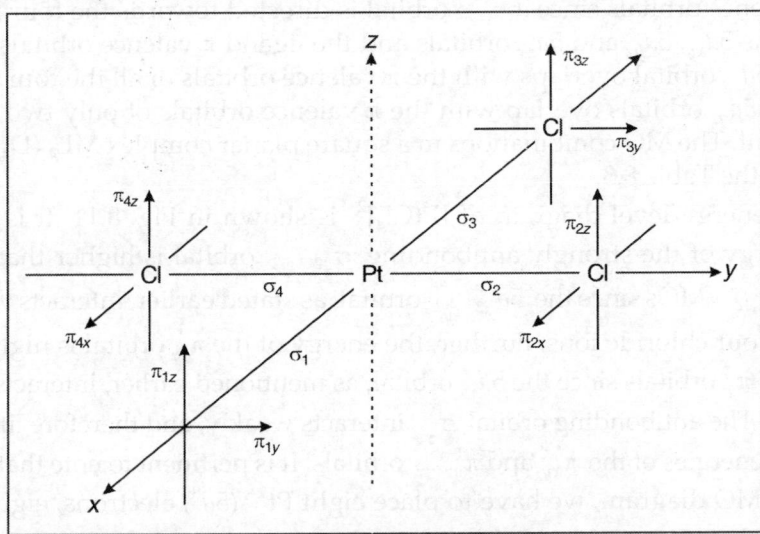

Fig. 8.10 Coordinate system of square planar $[PtCl_4]^{2-}$

Table 8.3 Molecular orbital combinations in square planar complex

Symmetry representation	Metal orbital(s)	Composite ligand σ group orbital	Composite ligand π group orbital
a_{1g}	s, d_{z^2}	$\frac{1}{2}(\sigma_1 + \sigma_2 + \sigma_3 + \sigma_4)$	
b_{1g}	$d_{x^2-y^2}$	$\frac{1}{2}(\sigma_1 - \sigma_2 + \sigma_3 - \sigma_4)$	
e_u	p_x	$\frac{1}{\sqrt{2}}(\sigma_1 - \sigma_2)$	
a_{2u}	p_y	$\frac{1}{\sqrt{2}}(\sigma_3 - \sigma_4)$	
	p_z		$\frac{1}{2}(\pi_1 + \pi_2 + \pi_3 + \pi_4)$

Contd...

Table 8.3 Molecular orbital combinations in square planar complex (Contd...)

Symmetry representation	Metal orbital(s)	Composite ligand σ group orbital	Composite ligand π group orbital
e_g	d_{xz}		$\frac{1}{\sqrt{2}}(\pi_1 - \pi_3)$
	d_{yz}		$\frac{1}{\sqrt{2}}(\pi_2 - \pi_4)$
b_{2g}	d_{xy}		$\frac{1}{2}(\pi_1 - \pi_2 + \pi_3 - \pi_4)$
a_{2g}			$\frac{1}{2}(\pi_1 + \pi_2 + \pi_3 + \pi_4)$
b_{2u}			$\frac{1}{2}(\pi_1 - \pi_2 + \pi_3 - \pi_4)$

The coordinate system of $[PtCl_4]^{2-}$, a square planar complex is shown in Fig. 8.10. The metal orbitals which can form the σ MOs are $5d_{x^2-y^2}$, $5d_{z^2}$, $6s$, $6p_x$ and $6p_y$. Of these orbitals, $5d_{z^2}$ is directed along the z-axis, and, therefore, interacts weakly with the four ligand σ valence orbitals. But the $5d_{x^2-y^2}$ orbital interacts very strongly with the four ligand σ valence orbitals since $d_{x^2-y^2}$ orbital is directed towards the four ligands. The overlap of the $5d_{xy}$, $5d_{yz}$ and $5d_{zx}$ orbitals and the ligand π valence orbitals forms the π bonds. The $5d_{xy}$ orbital overlaps with the π valence orbitals of all the four ligands. But the $5d_{xz}$ and $5d_{yz}$ orbitals overlap with the π valence orbitals of only two ligands and are equivalent. The MO combinations in a square planar complex ML_4 (D_{4h} symmetry) are as given the Table 8.3.

The MO energy level diagram of $[PtCl_4]^{2-}$ is shown in Fig. 8.11. It is to be noted that the energy of the strongly antibonding $\sigma^*_{x^2-y^2}$ orbital is higher than that of the $\sigma^*_{z^2}$, π^*_{xy}, π^*_{yz}, π^*_{zx} MOs since the $5d_{x^2-y^2}$ orbital, as stated earlier, interacts very strongly with all the four chloride ions. Further, the energy of the π^*_{xy} orbital is higher than that of the π^*_{zx} and π^*_{yz} orbitals since the $5d_{xy}$ orbital, as mentioned earlier, interacts with all the four ligands. The antibonding orbital $\sigma^*_{z^2}$ interacts weakly, and therefore, its energy lies between the energies of the π^*_{xy} and $\pi^*_{yz\,zx}$ orbitals. It is pertinent to note that $\Delta_1 > \Delta_2 \sim \Delta_3$. Since in the MO diagram, we have to place eight Pt^{2+} ($5d^8$) electrons, eight σ chloride electrons, and sixteen π chloride electrons, the ground state electronic configuration of $[PtCl_4]^{2-}$ is $(\sigma^b)^8(\pi)^{16}(\pi^*_{zx,yz})^4(\sigma^*_{z^2})^2(\sigma^*_{xy})^2$ with S = 0. This energy level diagram of $[PtCl_4]^{2-}$ (Fig. 8.11) is consistent with the diamagnetic nature of $[PtCl_4]^{2-}$ ($\mu_{eff} = 0$).

8.2.3 Tetrahedral Complex

We shall now consider and briefly discuss the MO treatment of tetrahedral ML_4 type complex. The central metal atom M of the first transition series uses the $4s$, $4p$ and $3d$ orbitals. For the donor atom L, the s and p orbitals of σ type and two pπ orbitals on each donor atom L are used in Fig. 8.11. The coordinate system for a tetrahedral complex ML_4 is shown in Fig. 8.12.

As noted earlier in a T_d complex, the lobes of the d_{xy}, d_{yz} and d_{zx} orbitals are directed towards the mid-points of the cube edge at an angle of 35°16′ wrt the ligands. The lobes of the d_{z^2} and $d_{x^2-y^2}$ are directed towards the faces of the cube and bisect the angles between

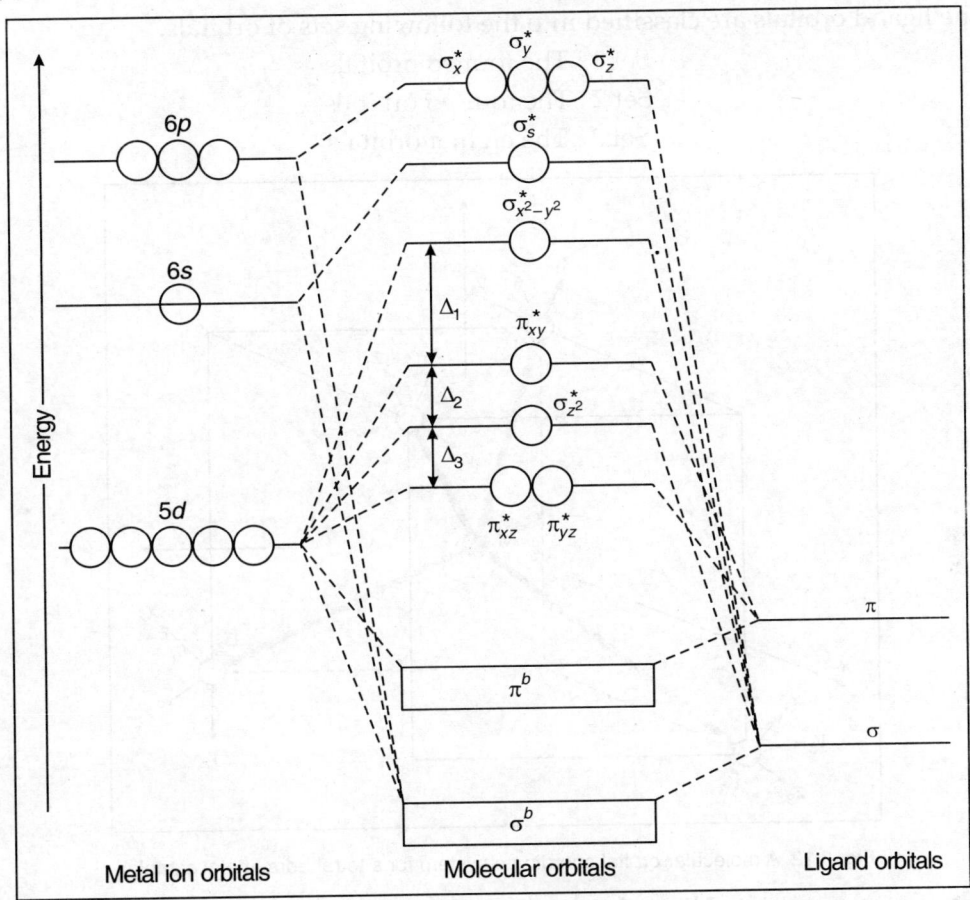

Fig. 8.11 Molecular orbital diagram of $[PtCl_4]^{2-}$

pairs of ligands making an angle of $109°28'/2 = 54°44'$ with the ligands. In case of σ bonding, one $s\sigma$ or $p\sigma$ orbital of each ligand is used for bonding to the metal. This leaves two mutually perpendicular p orbitals, each ligand which are capable of π bonding to the metal.

The metal s and p orbitals have a_1 and t_2 symmetries respectively. The five d orbitals are split into two sets: e (d_{z^2} and $d_{x^2-y^2}$) and t_2 (d_{xy}, d_{yz}, d_{zx})

$$a_1 \ : \ s$$
$$e \ : \ d_{x^2-y^2}, d_{z^2}$$
$$t_1 \ : \ none$$
$$t_2 \ : \ p_x, p_y, p_z$$
$$\qquad d_{xy}, d_{yz}, d_{zx}$$

The t_2 set of p and d orbitals of the metal could be used for σ bonding but are poor for π type of overlap because of their unfavourable orientation w.r.t. $p\pi$ orbitals of the ligand. The e type (d_{z^2} and $d_{x^2-y^2}$) orbitals can only make appreciable overlap with the $p\pi$ orbitals of the ligands.

The ligand orbitals are classified into the following sets of orbitals.

Set 1 : The four $s\sigma$ orbitals

Set 2 : The four $p\sigma$ orbitals

Set 3 : The eight π orbitals

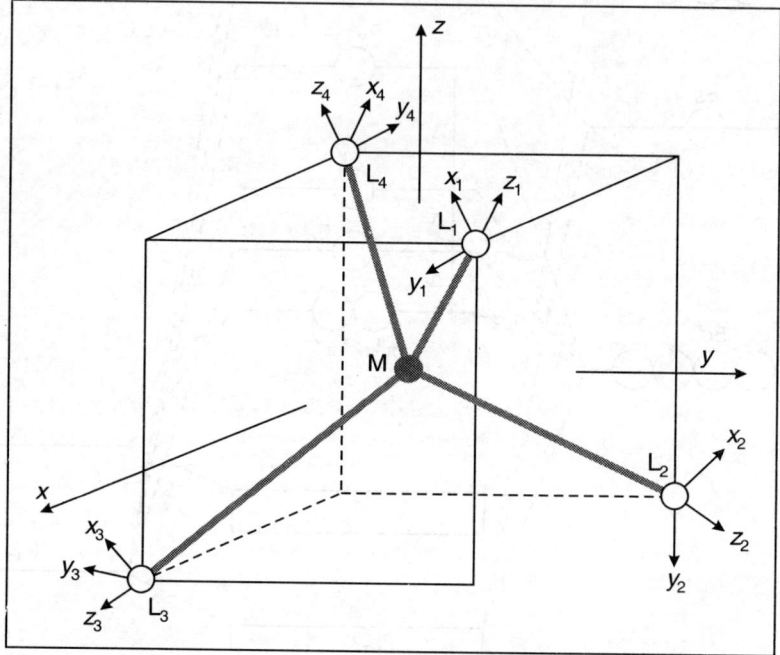

Fig. 8.12 A molecular orbital coordinate system for a tetrahedral metal complex

The symmetry adopted ligand group for σ bonding are

$$a_1 = \tfrac{1}{2}(\sigma_1 + \sigma_2 + \sigma_3 + \sigma_4)$$

where σ orbitals is a p_z or s orbital on the ligand and it matches with $4s$ atomic orbital of the metal.

The group orbitals of the ligand which matches with the $4p_x$, $4p_y$, $4p_z$ orbitals of the metal are

$$\psi\,(t_2) = \tfrac{1}{2}(\sigma_1 - \sigma_2 + \sigma_3 - \sigma_4),$$

$$\tfrac{1}{2}(\sigma_1 + \sigma_2 - \sigma_3 - \sigma_4),$$

$$\tfrac{1}{2}(\sigma_1 - \sigma_2 - \sigma_3 + \sigma_4)$$

For the metal-ligand π bonding, we may construct the following ligand group orbitals by referring to Fig. 8.12.

$$\psi\,(e) = \tfrac{1}{2}\,(p_x^1 - p_x^2 - p_x^3 + p_x^4),$$

$$\tfrac{1}{2}\,(p_y^1 - p_y^2 - p_y^3 + p_y^4)$$

This matches with the $3d_{z^2}$, $3d_{x^2-y^2}$ orbitals of the metal.

The ligand group orbital which matches with the d_{xy}, d_{yz}, d_{zx} orbitals of the metal is given below:

$$\psi\,(t_2) = \tfrac{1}{4}(p_x^1 - p_x^2 - p_x^3 - p_x^4) + \tfrac{\sqrt{3}}{4}(-p_y^1 - p_y^2 - p_y^3 + p_y^4),$$

$$\tfrac{1}{4}(p_x^1 - p_x^2 + p_x^3 - p_x^4) + \tfrac{\sqrt{3}}{4}(p_y^1 - p_y^2 + p_y^3 - p_y^4),$$

$$-\tfrac{1}{2}(p_x^1 + p_x^2 + p_x^3 + p_x^4)$$

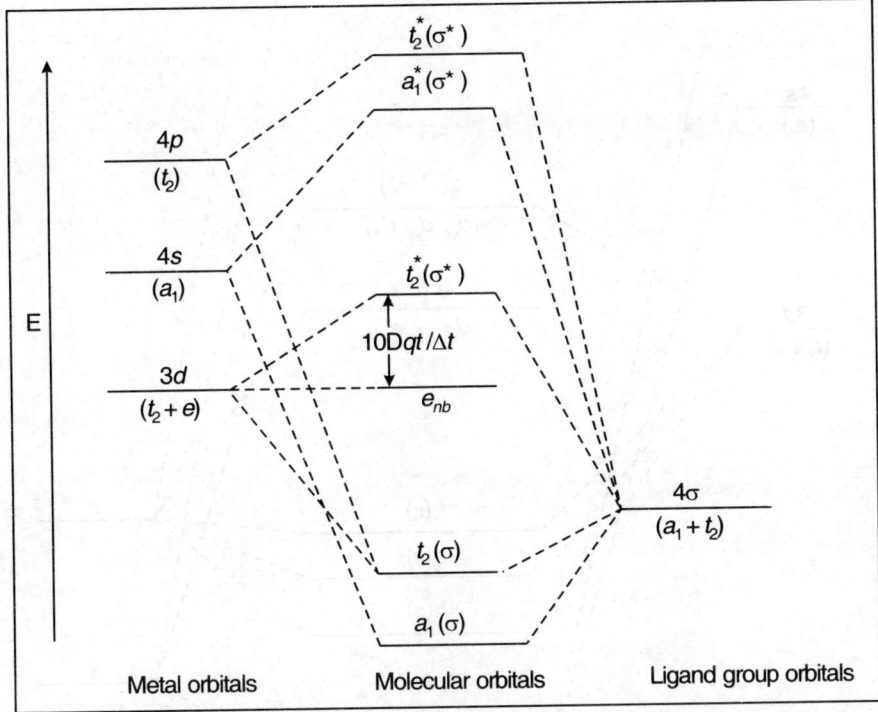

Fig. 8.13 MO energy diagram for a tetrahedral complex with metal-ligand σ bonding only

The t_2 LGOs can interact with both sets of metal t_2 orbitals (p and d) to give three sets of σ MOs — one bondings one slightly antibonding and one clearly antibonding. A qualitative MO diagram is shown in Fig. 8.13 where M – L for σ bonding alone is constructed and in Fig. 8.14 where both σ and π bonding are taken into account. It is to be noted that in contrast to the O_h case, the metal e orbitals are now nonbonding. The separation between e and the next higher t_2^* orbital is D_t (10 D_qt).

We shall consider VCl_4, a tetrahedral complex. The MO diagram of VCl_4 is shown in Fig. 8.15. Looking at this figure, we note that there are two sets of antibonding molecular orbitals, viz. $\sigma_{xy}^*, \sigma_{xz}^*, \sigma_{yz}^*$ and $\pi_{x^2-y^2}^*, \pi_{z^2}^*$ derived from the 3d valence orbitals. The energy of the former set is higher than that of the latter. Since in the MOs we have to accommodate one V^{4+} ($3d^1$) electron, eight σ chloride electrons, and sixteen π chloride electrons, the ground state electronic configuration of VCl_4 is $(\sigma^b)^8\,(\pi)^{16}\,[\pi^*(d)]^1$ with $S = \tfrac{1}{2}$.

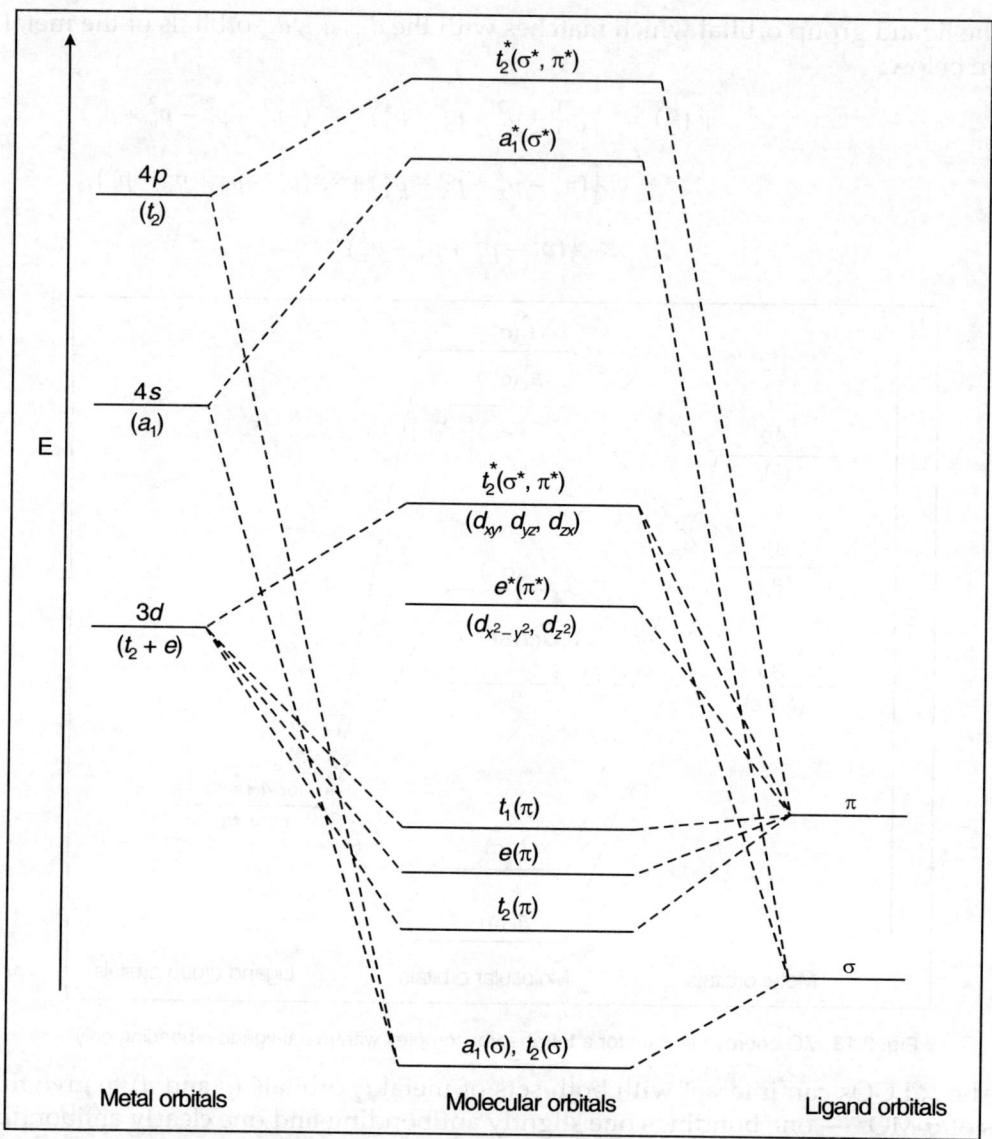

Fig. 8.14 MO energy diagram for a tetrahedral complex with σ and π bonding

The energy level diagram of the complex (Fig. 8.15) is consistent with the paramagnetic nature ($\mu_{\text{eff}} = 1.7$ BM) of the complex.

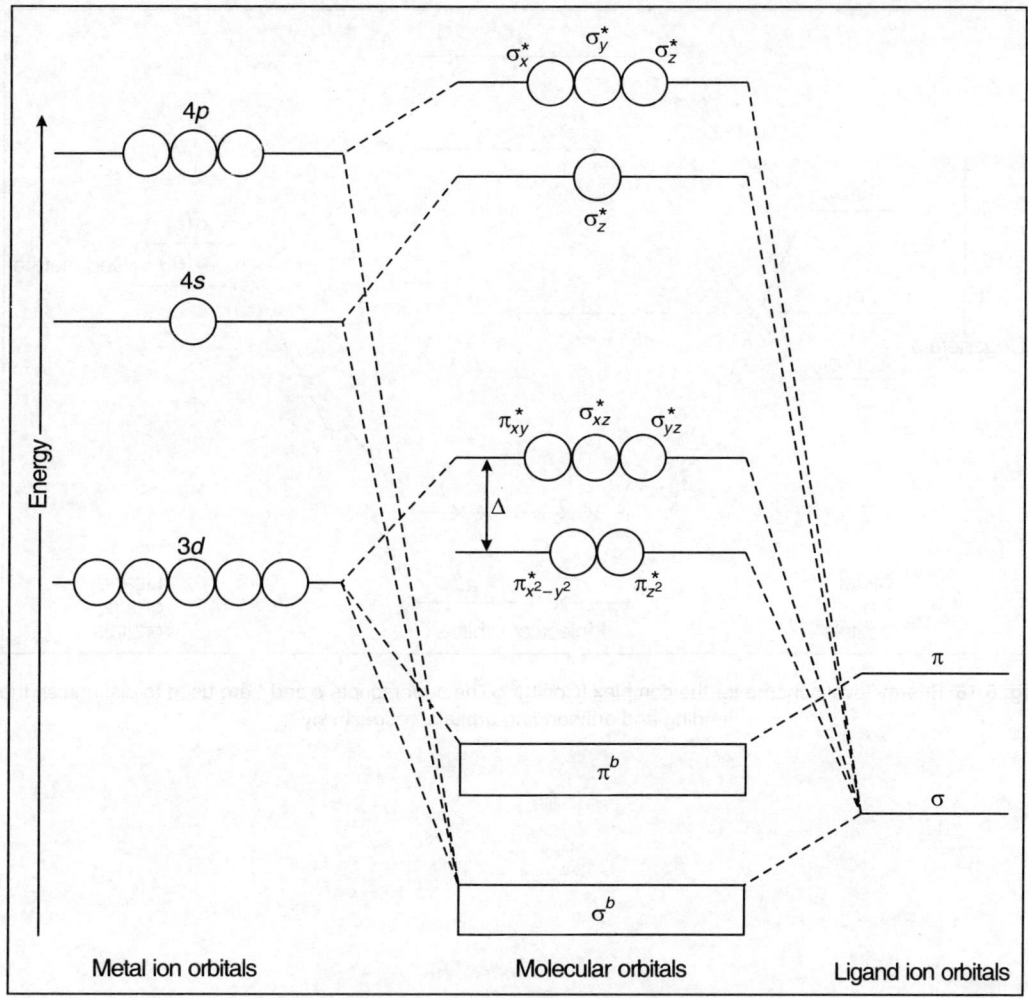

Fig. 8.15 Molecular orbital diagram of tetrahedral VCl_4
$(\sigma^b = a_1(\sigma^b), t_2(\sigma^b); \pi^b = t_2(\pi^b), e(\pi^b), t_1(\pi);$
$e(\pi^*) = \pi^*_{x^2-y^2}, \pi^*_{z^2}; t_2(\sigma^*, \pi^*) = \sigma^*_{xy}, \sigma^*_{xz}, \sigma^*_{yz};$
$a_1(\sigma^*) = \sigma^*_z; t_2(\sigma^*, \pi^*) = \sigma^*_x, \sigma^*_y, \sigma^*_z)$

For tetrahedral $[CoCl_4]^{2-}$ complex, the chloride ligands provide two electrons each for a total of eight, and the d^7 cobalt(II) ion furnishes seven, giving an overall total of fifteen. Twelve electrons will fill the six lowest energy MOs $a_1^2(b)t_2^6(b)e^4$ with the final three electrons remaining unpaired and occupying the slightly antibonding t_2^* molecular orbitals (Fig. 8.16).

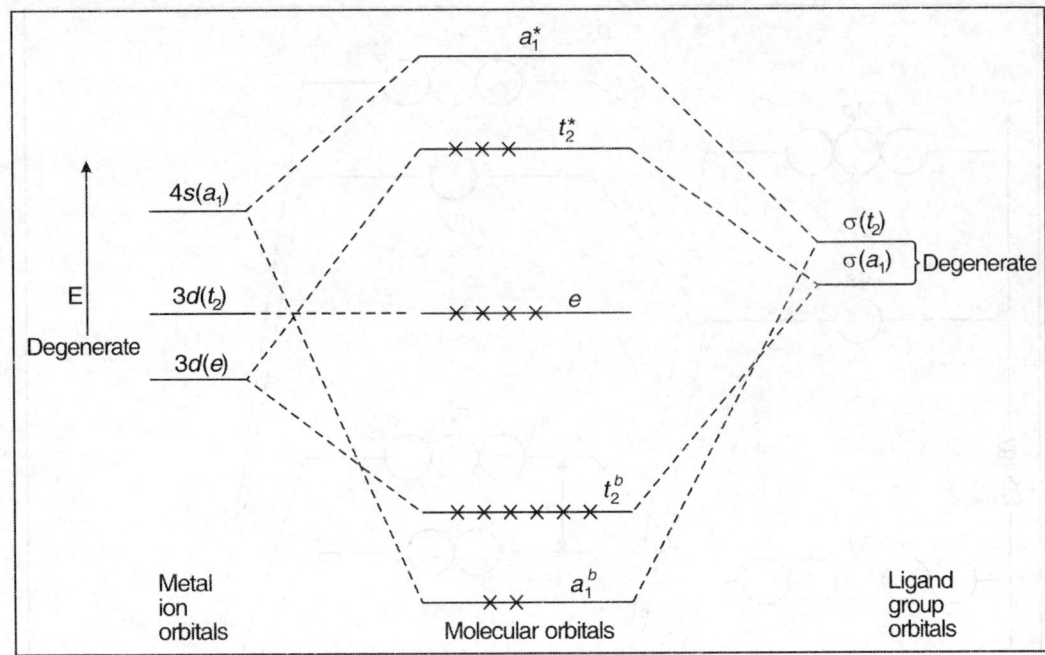

Fig. 8.16 Energy level scheme for the complex $[CoCl_4]^{2-}$. The superscripts b and $*$ are used to distinguish the bonding and antibonding orbitals, respectively

9

Molecular Orbital Theory and its Applications in Organic Chemistry

This chapter deals with the application of group theory to conjugated molecules which are planar organic molecules which are conventionally described as consisting of alternate double and single bonds. Some examples of conjugated hydrocarbons are shown in Fig. 9.1.

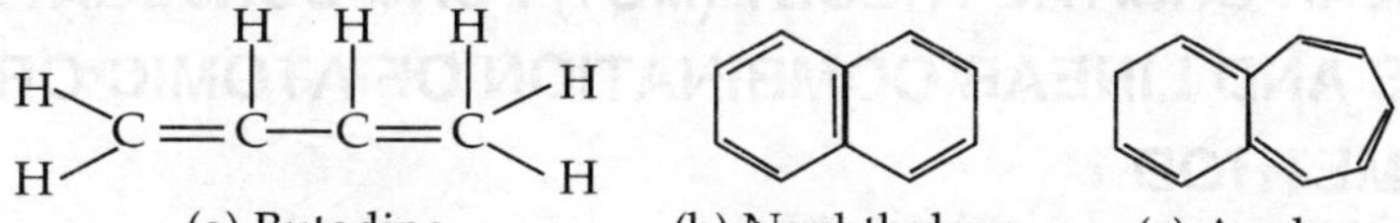

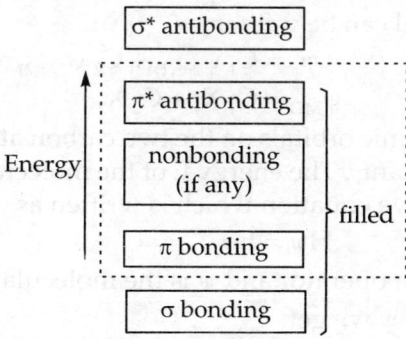

(a) Butadine   (b) Naphthalene   (c) Azulene

Fig. 9.1 Examples of conjugated hydrocarbons

The π electrons in these molecules are delocalised over all the carbon atoms which are sp^3 hybridised. A conjugated molecule possesses resonance energy or delocalisation energy, and this delocalisation energy can be obtained using molecular orbital theory.

A schematic energy level diagram for conjugated polyenes is shown in Fig. 9.2. It is obvious that the chemical and physical properties of these compounds are mostly controlled by the orbitals within the dotted lines, i.e. the π and π^* molecular orbitals, as well as the nonbonding orbitals, if there are any. Hence to study the electronic structure of these systems, we can ignore the σ_* and σ^* orbitals and concentrate on the π and π^* orbitals.

Fig. 9.2 Schematic energy level diagram for conjugated polyenes

In conjugated polyenes, each carbon atom contributes one p orbital (p_z orbital) illustrated in Fig. 9.3. These π electrons are distributed over the π molecular orbitals extended over the entire molecule.

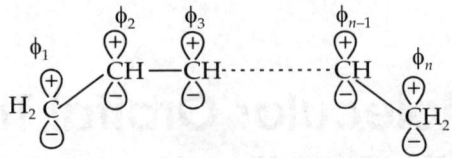

Fig. 9.3 Labeling of the π atomic orbitals in a conjugated polyene

The π molecular orbitals may be constructed by the linear combination of carbon $2p_z$ orbitals, i.e.

$$\psi\pi = C_1\,\psi_{2pz(1)} + C_2\psi_{2pz(2)} + \ldots + C_n\,\psi_{2pz}(n)$$
$$= C_1\phi_1 + C_2\phi_2 + \ldots + C_n\phi_n$$
$$= \sum_{i=1}^{n} C_i\phi_i \quad \text{or} \quad \Sigma\,C_2\,\psi_{2p_z\,(i)} \qquad \ldots(9.1)$$

9.1 MOLECULAR ORBITAL THEORY (MOT) FOR π CONJUGATED SYSTEMS AND LINEAR COMBINATION OF ATOMIC ORBITALS (LCAO) METHOD

One can construct a molecular orbital to a reasonable first approximation by to each of the atoms in a molecule, this is called the linear combination of atomic orbitals. (LCAO approximation). A linear combination is simply a sum of the respective one-electron wave function (ϕ_i) with various weight coefficients (C_i). Most commonly, only valence-shell atomic orbitals are used for forming molecular orbitals. For a diatomic molecule AB, the bonding π molecular orbital (ψ_{MO}) is obtained by linear combination of atomic orbitals of A (ϕ_A) and B (ϕ_B) atoms having same symmetry and energy

$$\psi_{MO} = C_1\,\phi_A + C_2\,\phi_B \qquad \ldots(9.2)$$

where ϕ_A and ϕ_B are atomic orbital wave functions of atoms A and B and C_1 and C_2 are the coefficients of these orbitals and are known as mixing constants.

For ethylene molecule, the linear combination of the hybridized p orbitals on the two carbon atoms of ethylene generates bending and antibonding π molecular orbitals. The bonding π molecular orbital can be written as

$$\psi_\pi = C_1\,\psi_{2pz(1)} + C_2\,\psi_{2pz(2)} \qquad \ldots(9.2a)$$

or $$\psi_{MO} = C_1\,X_1 + C_2\,X_2$$

where X_1 and X_2 are the atomic orbitals on the two carbon atoms $2p_{z(1)}$ and $2p_{z(2)}$ and C_1 and C_2 are the mixing constants. The energy E of the molecular orbital can be obtained by solving Schrödinger wave equation which is written as

$$H\psi = E\psi \qquad \ldots(9.3)$$

where H is the Hamiltonian operator and ψ is the molecular orbital.

Multiplying Eq. (9.3) by ψ we get

$$\psi\,H\,\psi = E\,\psi^2 \qquad \ldots(9.4)$$

Integration over all space then gives

$$\int\psi H\psi\,d\tau = E\int\psi^2\,d\tau \;[d\tau = dx\,dy\,dz]$$

or

$$E = \frac{\int\psi H\psi d\tau}{\int\psi d\tau} \qquad \ldots(9.5)$$

Substituting $C_1 X_1 + C_2 X_2$ for ψ_1, we get

$$E = \frac{\int (C_1 X_1 + C_2 X_2) H (C_1 X_1 + C_2 X_2) d\tau}{\int (C_1 X_1 + C_2 X_2)^2 d\tau}$$

$$= \frac{\int (C_1 X_1 H C_1 X_1) C_1 X_1 H C_2 X_2 H C_1 X_1 + C_2 H C_2 X_2) d\tau}{\int (C_1^2 X_1^2 + 2 C_1 C_2 X_1 X_2 + C_2^2 X_2^2) d\tau} \qquad …(9.6)$$

Equation (9.6) can be written in a simple form by making the following substitutions.

$$H_{11} = \int X_1 H X_1 \, d\tau$$

$$H_{12} = X_{21} = \int X_1 HX_2 \, d\tau = \int X_2 H X_1 \, d\tau$$

$$H_{22} = \int X_2 H X_2 \, d\tau$$

$$S_{11} = \int X_1^2 \, d\tau$$

$$S_{12} = \int X_1 X_2 \, d\tau$$

$$S_{22} = \int X_2^2 \, d\tau$$

Thus Eq. (9.6) for E reduces to

$$= \frac{C_1^2 H_{11} + 2C_1 C_2 H_{12} + C_2^2 H_{22}}{C_1^2 S_{11} + 2 C_1 C_2 S_{12} + C_2^2 S_{22}} \qquad …(9.7)$$

To determine the coefficients C_1 and C_2, the energy E must be minimised with respect to C_1 and C_2 independently.

When E is minimised with respect to C_1, we have

$$\frac{\sigma E}{\sigma C_1} = \frac{(2C_1 H_{11} + C_2 H_{12})(C_1^2 S_{11} + 2C_1 C_2 S_{12} + C_2^2 S_{22})}{(C_1^2 S_{11} + 2C_1 C_2 S_{12} + C_2^2 S_{22})^2}$$

$$- \frac{(2C_1 S_{11} + 2C_2 S_{12})(C_1^2 H_{11} + 2C_1 C_2 H_{12} + C_2^2 H_{22})}{(C_1^2 S_{11} + 2C_1 C_2 S_{12} + C_2^2 S_{22})^2} = 0$$

Therefore,

$$\frac{(2C_1 H_{11} + 2C_2 H_{12})(C_1^2 H_{11} + 2C_1 C_2 H_{12} + C_2^2 S_{22})}{(C_1^2 S_{11} + 2C_1 C_2 S_{12} + C_2^2 S_{22})^2}$$

$$= \frac{(2C_1 S_{11} + 2C_2 S_{12})(C_1^2 H_{11} + 2C_1 C_2 H_{12} + C_2^2 H_{22})}{(C_1^2 S_{11} + 2C_1 S_{12} C_2 + C_2^2 S_{22})^2}$$

$$\frac{(2C_1 H_{11} + 2C_2 H_{12})}{(2C_1 S_{11} + 2C_2 S_{12})} = \frac{(C_1^2 H_{11} + 2C_1 C_2 H_{12} + C_2^2 H_{22})}{(C_1^2 S_{11} + 2C_1 C_2 S_{12} + C_2^2 S_{22})} \qquad …(9.8)$$

Comparing Eq. (9.7) and (9.8), we can write

$$\frac{(2C_1 H_{11} + 2C_2 H_{12})}{(2C_1 S_{11} + 2C_2 S_{12})} = E \qquad \qquad ...(9.9)$$

or $C_1(H_{11} - E S_{11}) + C_2 (H_{12} - ES_{12}) = 0$ \qquad ...(9.10)

Equations (9.9) and (9.10) are called *secular equations*. Clearly $C_1 = C_2 = 0$ is a solution of Eqs (9.9) and (9.10). But it is a meaningless solution as ψ becomes zero. Thus, for C_1 and C_2 to be be nonvanishing, we need to have

$$\begin{vmatrix} (H_{11} - S_{11}E) & (H_{12} - S_{12} E) \\ (H_{12} - S_{12} E) & (H_{22} - S_{22}E) \end{vmatrix} = 0 \qquad \qquad ...(9.11)$$

This is known as *secular determinant*, where the only known is E.

In the general case, where

$\psi = C_1 X_1 + C_2 X_2 + C_3 X_3 + ... + C_n X_n$ the secular determinant becomes

$$\begin{vmatrix} (H_{11} - E S_{11}) & (H_{12} - E S_{12}) & ... & (H_{1n} - E S_{1n}) \\ (H_{21} - E S_{21}) & (H_{22} - E S_{22}) & ... & (H_{2n} - E S_{2n}) \\ (H_{31} - ES_{31}) & (H_{32} - E S_{23}) & ... & (H_{3n} - E S_{3n}) \\ \vdots & \vdots & ... & \vdots \\ (H_{n1} - E S_{n1}) & (H_{n2} - E S_{n2}) & ... & (H_{nn} - E S_{nn}) \end{vmatrix} = 0 \qquad ...(9.12)$$

This $n \times n$ secular determinant is solved by applying the approximations suggested by applying the approximations suggested by German Physicist E Huckel. The Huckel approximations are as follows.

1. The overlap integral, S_{ij}. It is a measure of how much the orbitals i and j overlap. S_{ij} is written as

$$S_{ij} = \int X_i X_j \, d\tau$$

where X_i and X_j are the ith and jth atomic orbitals.

$$S_{ij} = X_i X_j \, d\tau = 0; \text{ if } i \neq j$$

That is there is no overlap between orbitals on different atoms.

$S_{12} = S_{21} = S_{13} = S_{31} = S_{23} = S_{32} = 0$

If the overlap integral $S_{ij} = 0$ then X_i and X_j functions are *orthogonal*.

$$S_{ij} = \int X_i S_j \, d\tau = 1 \text{ if } i = j$$

i.e. \qquad $S_{11} = S_{22} = S_{33} = S_{44} = ... = S_{nn} = 1$

In this case X_i and X_j functions are normalised, we always take normalised atomic orbitals such that $S_{ij} = 1$. To form a bond there must be an overlap of orbital.

2. The coulomb integral, H_{ij}, is expressed as α ($\alpha < 0$)

$$H_{ij} = \int X_i H X_j \, d\tau$$

If $i = j$, this $H_{ij} = \int X_i H X_i \, d\tau = \alpha$

H_{ii} is called the coulomb integral and *refers to as the energy of an electron* in carbon p-π orbital before interaction with other orbitals.

3. The resonance integral or exchange integral H_{ij} is expressed as β ($\beta < 0$)

$H_{ij} = \int X_i H X_j \, d\tau = \beta$ if X_i and X_j are neighbouring atomic orbitals:

$H_{12} = H_{21} = H_{23} = H_{32} = \beta$

$H_{ij} = 0$ if X_i and X_j are not neighbouring atomic orbitals.

$H_{13} = H_{31} = H_{24} = H_{42} = 0$

Resonance integrals denote the energy of interaction between orbitals on atoms i and j.

The $n \times n$ secular determinant obtained using the variation method can be simplied assuming $S_{ij} = 0$, $S_{ii} = 1$, $H_{ii} = \alpha$ and $H_{ij} = \beta$; if the ith and jth orbitals are on adjacent atoms otherwise $H_{ij} = H_{ji} = 0$. Thus, the $n \times n$ secular determinant reduces to

$$\begin{vmatrix} (H_{11}-E) & H_{12} & \cdots & \\ H_{21} & (H_{12}-E) & \cdots & 0 \\ \cdots & \cdots & \cdots & 0 \\ \cdots & \cdots & \cdots & \cdots \\ \cdots & \cdots & \cdots & \cdots \\ 0 & \cdots & \cdots & \cdots \end{vmatrix} \qquad \ldots(9.13)$$

Since $S_{11} = S_{22} = S_{33} = S_{44} = 1$

$S_{12} = S_{13} = S_{14} = S_{23} = S_{24} = S_{34} = 0$

$S_{14} = S_{13} = S_{24}$ ($H_{ij} = 0$, orbitals are not on adjacent atom)

$S_{11} = S_{22} = S_{33} = S_{44} = \alpha$ (Coulomb integral)

$S_{12} = S_{23} = S_{34} = \beta$ (resonance integral)

$$\begin{vmatrix} (\alpha - E) & \beta & 0 & 0 & \cdots & 0 \\ \beta & (\alpha - E) & \beta & 0 & \cdots & 0 \\ 0 & \beta & (\alpha - E) & \beta & \cdots & \cdots \\ \cdots & \cdots & \cdots & \cdots & \cdots & \cdots \\ 0 & \cdots & \cdots & \cdots & \cdots & (\alpha - E) \end{vmatrix} = 0 \qquad \ldots(9.14)$$

A secular determinant can be solved early if a molecule contains small number of π electrons. For example, the secular determinants of ethylene and butadiene will take the form:

$$\begin{vmatrix} \alpha - E & \beta \\ \beta & \alpha - E \end{vmatrix} = 0 \qquad \ldots(9.15)$$

and

$$\begin{vmatrix} (\sigma - E) & \beta & 0 & 0 \\ \sigma & (\sigma - E) & \beta & 0 \\ 0 & \beta & (\sigma - E) & \beta \\ 0 & 0 & \beta & (\sigma - E) \end{vmatrix} = 0 \qquad \ldots(9.16)$$

After solving for E in Eq. (9.12), we can then substitute each E in the following secular equations to determine the values of coefficient C_i.

$$(H_{11} - ES_{11})C_1 + (H_{12} - ES_{12})C_2 + \quad \cdots \quad + (H_{1n} - ES_{1n})C_n = 0$$
$$(H_{12} - ES_{12})C_1 + (H_{22} - ES_{22})C_2 + \quad \cdots \quad + (H_{2n} - ES_{2n})C_n = 0$$
$$\cdots \qquad\qquad \cdots \qquad\qquad \cdots \qquad\qquad \quad \ldots(9.17)$$
$$(H_{1n} - ES_{1n})C_1 + (H_{2n} - ES_{2n})C_2 + \quad \cdots \quad (H_{nn} - ES_{nn})C_n = 0$$

Now we apply this approximation to the π system of a few polyenes.

9.1.1 Ethylene (C_2H_4)

The two atomic orbitals participating in π bonding are shown in Fig. 9.4. Labeling the carbon atoms 1 and 2, secular equations are

$$(H_{11} - E S_{11}) C_1 + (H_{12} - E S_{12}) C_2 = 0 \qquad\qquad \ldots(9.18)$$

$$(H_{12} - E S_{12}) C_1 + (H_{22} - E S_{22}) C_2 = 0 \qquad\qquad \ldots(9.19)$$

Since $\quad H_{11} = H_{22} = \alpha; S_{11} = S_{22} = 1; H_{12} = \beta; S_{12} = 0$

So the above equations reduced to

$$(\alpha - E) C_1 + \beta C_2 = 0 \qquad\qquad \ldots (9.20)$$

$$BC_1 + (\alpha - E) C_2 = 0 \qquad\qquad \ldots (9.21)$$

The secular determinant will take the form

$$\begin{vmatrix} (\alpha - E) & \beta \\ \beta & (\alpha - E) \end{vmatrix} = 0$$

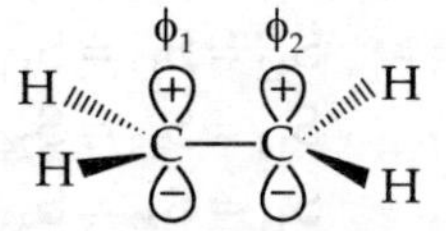

Dividing each element of the above determinant by β, we have

Fig. 9.4 The atomic orbitals for π bonding in ethylene

$$\begin{vmatrix} \left(\dfrac{\alpha - E}{\beta}\right) & 1 \\ 1 & \left(\dfrac{\alpha - E}{\beta}\right) \end{vmatrix} = 0$$

We put $\dfrac{\alpha - E}{\beta} = x$ in the above determinant and get

$$\begin{vmatrix} x & 1 \\ 1 & x \end{vmatrix} = 0$$

or $\qquad x^2 - 1 = 0 \qquad \therefore \quad x = \pm 1$

Substituting the values of x in the expression $x = \dfrac{\alpha - E}{\beta}$, we get

For $x = +1$, $\dfrac{\alpha - E}{\beta} = 1 \quad \therefore \alpha - E = \beta \therefore E = \alpha - \beta$ $\qquad\qquad \ldots(9.22)$

For $x = -1$, $\dfrac{\alpha - E}{\beta} = 1 \quad \therefore \alpha - E = \beta \therefore = \alpha - \beta$ $\qquad\qquad \ldots(9.23)$

We denote $E_1 = \alpha + \beta = E(\pi)$ ...(9.24)

$E_2 = \alpha - \beta = E(\pi*)$ ...(9.25)

Since the values of α and β are always negative ($\alpha < 0$, $\beta < 0$) so the lower energy level will be $E_1 = \alpha + \beta$ and the higher energy level will be $E_2 = \alpha - \beta$ (Fig. 9.5).

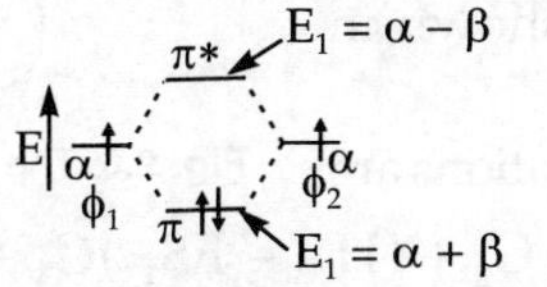

Fig. 9.5 The energy level diagram for the π molecular orbitals of ethylene

Substituting $E_1 = \alpha + \beta$ and $E_2 = \alpha - \beta$ into Eqs (9.20) and (9.21), we get

$\{\alpha - (\alpha + \beta)\} C_1 + \beta C_2 = 0 \quad \therefore C_1 = C_2$

$\beta C_1 + \{\alpha - (\alpha - \beta\} C_2 = 0 \qquad C_1 = C_2$

Applying normalising condition we have

$C_1^2 + C_2^2 = 1 \text{ or } 2C_1^2 = 1 [\because C_1 = C_2] \therefore C_1 = \dfrac{1}{\sqrt{2}}$

when $E = E_1$, we have $C_1 = C_2$ or

$$\psi_1 = \psi(\pi) = \frac{1}{\sqrt{2}} (\phi_1 + \phi_2)$$ ...(9.26)

when $E = E_2$, we have $C_1 = -C_2$, or

$$\psi_2 = \psi(\pi*) = \frac{1}{\sqrt{2}} (\phi_1 - \phi_2)$$ ...(9.27)

The two electrons in ethylene occupy the $\pi \psi(\pi)$ orbital corresponding to the energy $(\alpha + \beta)$. Hence the energy for a π bond in the Huckel model is

$$E\pi = 2 \times (\alpha + \beta) = 2\alpha + 2\beta$$ ...(9.28)

Stabilisation energy of molecules due to π-MO (π-bond)

The stabilisation energy of a of molecules due to formation of MOs is calculated in the following manner:

Energy of two electrons in π-MO (ψ_1) — energy of two electrons in P_z atomic orbitals

$$= 2(\alpha + \beta) - 2\alpha = 2\beta$$ ...(9.29)

9.1.2 Butadiene (C_4H_6)

We now proceed to use the same method to treat a higher homology of ethylene, namely butadiene. The four atomic orbitals of this molecule particularly in π-bonding are shown in Fig. 9.6.

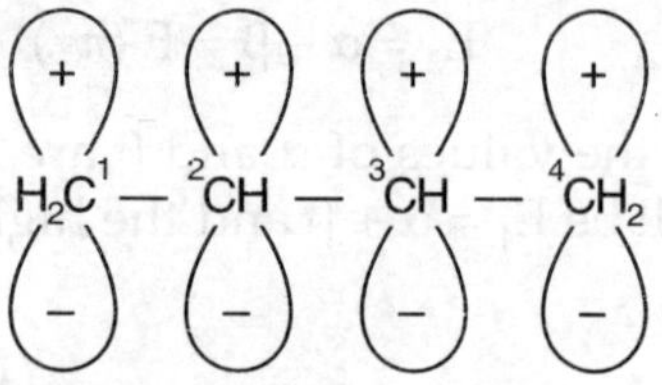

Fig. 9.6 The atomic orbitals for π bonding in butadiene

For this system the secular equations are

$$(H_{11} - E\,S_{11})\,C_1 + (H_{12} - ES_{12})\,C_2 + (H_{13} - ES_{13})C_3 + (H_{14} - ES_{14})\,C_4 = 0 \qquad ...(9.30)$$

$$(H_{21} - E\,S_{21})\,C_1 + (H_{22} - ES_{22})\,C_2 + (H_{23} - ES_{23})C_3 + (H_{24} - ES_{24})\,C_4 = 0 \qquad ...(9.31)$$

$$(H_{31} - E\,S_{31})\,C_1 + (H_{32} - ES_{32})\,C_2 + (H_{33} - ES_{33})C_3 + (H_{34} - ES_{34})\,C_4 = 0 \qquad ...(9.32)$$

$$(H_{41} - E\,S_{41})\,C_1 + (H_{42} - ES_{42})\,C_2 + (H_{43} - ES_{43})C_3 + (H_{44} - ES_{44})\,C_4 = 0 \qquad ...(9.33)$$

Since $\quad H_{11} = H_{22} = H_{33} = H_{44} = \alpha$

$$H_{12} = H_{21} = H_{23} = H_{32} = H_{34} = H_{43} = \beta$$

$$S_{11} = S_{22} = S_{33} = S_{44} = 1$$

$$S_{12} = S_{21} = S_{13} = S_{31} = S_{23} = S_{32} = 0$$

So the Eqs (9.30)–(9.33) reduce to

$$(\alpha - E)\,C_1 + \beta C_2 = 0 \qquad\qquad ...(9.34)$$

$$\beta\,C_1 + (\alpha - E)\,C_2 + \beta C_3 = 0 \qquad\qquad ...(9.35)$$

$$\beta\,C_2 + (\alpha - E)\,C_3 + \beta C_4 = 0 \qquad\qquad ...(9.36)$$

$$\beta\,C_3 + (\alpha - E)\,C_4 = 0 \qquad\qquad ...(9.37)$$

The secular determinant will be of the following form

$$\begin{vmatrix} (\alpha - E) & \beta & 0 & 0 \\ \beta & (\alpha - E) & \beta & 0 \\ 0 & \beta & (\alpha - E) & \beta \\ 0 & 0 & \beta & (\alpha - E) \end{vmatrix} = 0$$

Dividing each element of the above determinant by β we have

$$\begin{vmatrix} \left(\dfrac{\alpha - E}{\beta}\right) & 1 & 0 & 0 \\ 1 & \left(\dfrac{\alpha - E}{\beta}\right) & 1 & 0 \\ 0 & 1 & \left(\dfrac{\alpha - E}{\beta}\right) & 1 \\ 0 & 0 & 1 & \left(\dfrac{\alpha - E}{\beta}\right) \end{vmatrix} = 0$$

We put $\left(\dfrac{\alpha - E}{\beta}\right) = x$ in the above determinant and we have

$$\begin{vmatrix} x & 1 & 0 & 0 \\ 1 & x & 1 & 0 \\ 0 & 1 & x & 1 \\ 0 & 0 & 1 & x \end{vmatrix} = 0$$

Expanding this determinant, we have

$$x^4 - 3x^2 + 1 = 0$$

The four roots of the above relation are -1.618, -0.618, $+0.618$, and $+1.618$, since $\dfrac{\alpha - E}{\beta} = x$, we first put $x = -1.618$ and get $\left(\dfrac{\alpha - E_1}{\beta} = -1.618\right)$ $\therefore$ $E_1 = \alpha + 1.618\,\beta$. Since α and β have negative values, $E_1 = \alpha + 1.618\,\beta$ corresponds to the lowest energy π–MO
The other three energy values are

$$E_2 = \alpha + 0.618\,\beta$$

$$E_3 = \alpha - 0.618\,\beta$$

$$E_4 = \alpha - 0.618\,\beta$$

Since β is negative ($\beta < 0$), it follows
$E_1 < E_2 < E_3 < E_4$
The coefficients C_1, C_2, C_3, C_4, of the molecular orbitals corresponding to the energies E_1, E_2, E_3 and E_4 can be determined by substituting the above energies one by one in the secular Eqs (9.34)–(9.37). For example, if we put $E_1 = \alpha + 1.618\,\beta$ in Eqs (9.34)–(9.37), we get

From Eq. (9.34) $(\alpha - \alpha - 1.618\,\beta)\,C_1 + \beta\,C_2 = 0$ $[\because C_2 = 1.618 C_1]$ ...(9.38)

From Eq. (9.35) $\beta C_1 + (\alpha - \alpha - 1.618\,\beta)\,C_2 + \beta\,C_3 = 0$

$$C_1 - 1.618\,C_2 + C_3 = 0$$

$$C_1 - 1.618\,(1.618\,C_1) + C_3 = 0 \qquad\qquad [\because C_2 = 1.618 C_1]$$

$$C_1 - 2.618\,C_1 + C_3 = 0 \qquad\qquad [\because C_3 = 1.618 C_1] \,...(9.39)$$

$\dfrac{C_2}{C_3} = 1$ $\therefore$ $\boxed{C_2 = C_3}$ ...(9.40)

From Eq. (9.36)
$\beta\,C_2 + (\alpha - \alpha - 1.618\,\beta)\,C_3 + \beta C_4 = 0$

$$C_2 - 1.618\,C_3 + C_4 = 0$$

$$1.618\,C_1 - 1.618\,(1.618)\,C_1 + C_4 = 0$$

$$1.618\,C_1 - 2.618\,C_1 + C_4 = 0$$

$- C_1 + C_4 = 0$ $\qquad\qquad\qquad \therefore \boxed{C_4 = C_1}$ ...(9.41)

From Eq. (9.37) $\beta C_3 (\alpha - \alpha - 1.618 \beta) C_4 = 0 \quad C_3 = 1.618 C_4$ ...(9.42)

Applying normalising condition

$$C_1^2 + C_2^2 + C_3^2 + C_4^2 = 1$$

$$C_1^2 + (1.618 C_1)^2 + (1.618 C_1)^2 + C_1^2 = 10 \qquad\qquad [\because C_2 = C_3 = 1.618]$$

$$C_1^2 [1 + 2 \times 2.618 + 1] = 1 \qquad\qquad [C_1 = C_4]$$

$$C_1^2 = \frac{1}{7.236}$$

$$= 0.1382 \qquad\qquad \because C_1 = \pm 0.372 \text{...(9.43)}$$

$\therefore C_1 = C_4 = 0.372$ ...(9.44)

$\quad C_2 = C_3 = 1.618 C_1 = 1.618 \times 0.372 = 0.602$ ...(9.45)

Thus the molecular orbitals having energy E_1 is

$\psi_1 (\pi) = 0.372 \phi_1 + 0.602 \phi_2 + 0.602 \phi_3 + 0.372 \phi_4$

$\qquad = 0.372 \psi_{2pz(1)} + 0.602 \psi_{2pz(2)} + 0.602 \psi_{2pz(3)} + 0.372 \psi_{2pz(4)}$ ...(9.46)

If we put $E_2 = \alpha + 0.618 \beta$ in Eqs (9·34)–(9·37), we get

$(\alpha - \alpha - 0.618 \beta) C_1 + \beta C_2 = 0 \qquad\qquad \therefore \boxed{C_2 = 0.618 C_1}$

$\quad \beta C_1 + (\alpha - \alpha - 0.618 \beta) C_2 + \beta C_3 = 0$ ...(9.47)

$\qquad C_1 - 0.618 C_2 + C_3 = 0$

$\qquad C_1 - 0.618 \ (0.618 C_1) + C_3 = 0$

$\qquad C_1 (1 - 0.382) + C_3 = 0 \qquad\qquad \therefore \boxed{C_3 = -0.618 C_1}$ ...(9.48)

$$\frac{C_2}{C_3} = 1 \qquad\qquad \therefore \boxed{C_2 = -C_3}$$ ...(9.49)

$\quad \beta C_2 + (\alpha - \alpha - 0.618 \beta) C_3 + \beta C_4 = 0$

$\qquad C_2 - 0.618 C_3 + C_4 = 0$

$\quad 0.618 C_1 - 0.618 \ (-0.618) C_1 + C_4 = 0$

$\qquad 0.618 C_1 + 0.382 C_1 + C_4 = 0$

$\qquad\qquad C_1 + C_4 = 0 \qquad\qquad \therefore \boxed{C_1 = -C_4}$ ...(9.50)

$\quad \beta C_3 + (\alpha - \alpha - 0.618 \beta) C_4 = 0$

$\qquad\qquad C_3 = 0.618 C_4$ ...(9.51)

Applying normalising condition

$$C_1^2 + C_2^2 + C_3^2 + C_4^2 = 1$$

$$C_1^2 + (0.618\,C_1)^2 + (-0.618\,C_1)^2 + C_1^2 = 1 \qquad [\because C_1 = C_4]$$

$$C_1^2 [1 \times 2 + 0.382 \times 2] = 1 \therefore C_1^2 = \frac{1}{2.764} = 0.362$$

$$C_1 = 0.602$$

$$\therefore \qquad C_2 = 0.618\,C_1 = 0.618 \times 0.602 = 0.372$$

$$\therefore \qquad C_3 = -0.372 \text{ and } C_4 = -0.602$$

So the molecular orbitals having energy E_2 will be

$$\Psi_2\,(\pi) = 0.602\,\phi_1 + 0.372\,\phi_2 - 0.372\,\phi_3 - 0.602\,\phi_4$$

$$= 0.602\,\psi_{2pz(1)} + 0.372\,\psi_{2pz(2)} - 0.372\,\psi_{2pz(3)} - 0.602\,\psi_{2pz(4)}$$

In a similar way, the remaining two molecular orbital can be obtained

$$\Psi_3\,(\pi) = 0.602\,\phi_1 - 0.372\,\phi_2 - 0.372\,\phi_3 + 0.602\,\phi_4$$

$$= 0.602\,\psi_{2pz(1)} - 0.372\,\psi_{2pz(2)} - 0.372\,\psi_{2pz(3)} + 0.602\,\psi_{2pz(4)}$$

$$\Psi_4\,(\pi) = 0.372\,\phi_1 + 0.602\,\phi_2 + 0.602\,\phi_3 - 0.372\,\phi_4$$

$$= 0.372\,\psi_{2pz(1)} - 0.602\,\psi_{2pz(2)} + 0.602\,\psi_{2pz(3)} - 0.372\,\psi_{2pz(4)}$$

These molecular diagrams are represented schematically in Fig 9.7.

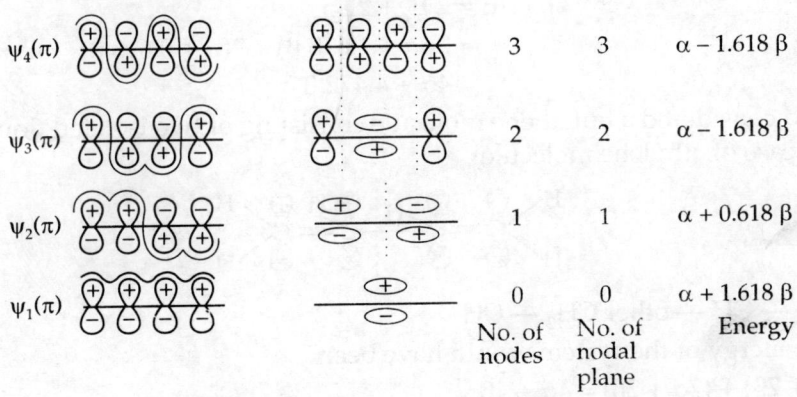

Fig. 9.7 π and π^* molecular orbitals of butadiene

All the π molecular orbitals have a nodal plane (plane along which there is no probability of finding the electron) along the internuclear axis. The wave function $\Psi_1\,(\pi)$ does not contain any node and is thus a strongly bonding orbital. $\Psi_1\,(\pi)$ has no nodal plane perpendicular to the internuclear axis.

The function $\Psi_2\,(\pi)$ has a nodal plane between second and third carbon atoms. It has a partial bonding character as it accumulates charges between atoms 1 and 2, between atoms 3 and 4. $\Psi_2\,(\pi)$ has one nodal plane.

$\Psi_3(\pi^*)$ has two nodal planes and $\Psi_4(\pi^*)$ has three nodal planes perpendicular to the internuclear axis.

This determines the order of there energies. $\Psi_1(\pi)$ and $\Psi_2(\pi)$ are lower in energy than the component atomic orbitals and hence are π bonding in nature. On the other hand, $\Psi_3(\pi)$ and $\Psi_4(\pi)$ are higher in energy then their component atomic orbitals and hence are π^* antibonding in nature (Fig. 9.8). The four electrons in the p_z orbitals of the carbon atoms are accommodated in the bonding $\Psi_1(\pi)$ and $\Psi_2(\pi)$ molecular orbitals.

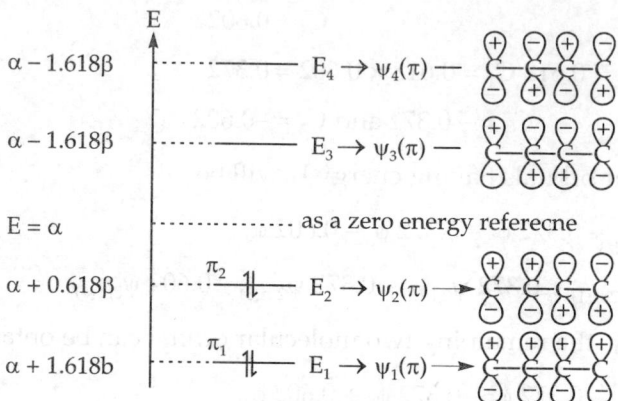

Fig. 9.8 π-MO energy levels for butadiene

Butadiene molecule contains four π-electrons and thus its ground state electronic configuration is

$$(\Psi_1)^2 (\Psi_2)^2$$

with ground state energy

$$E = 2E_1 + 2E_2$$
$$= 2(\alpha + 1.618\,\beta) + 2(\alpha + 0.618\,\beta)$$
$$= 4\alpha + 4.472\,\beta \qquad \ldots (9.52)$$

If we had considered a butadiene molecule consisting of two isolated double bonds (as seen in case of ethylene molecule)

one $CH_2 - CH -$ other $CH_2 - CH-$
the total energy of the system would have been

$$E = (2\alpha + 2\beta) + (2\alpha + 2\beta) = 4\alpha + 4\beta \qquad \ldots (9.53)$$

Hence, by allowing the four π electrons to delocalise over the entire molecular skeleton, there is a gain in stability, which is called *delocalisation energy* (DE). For butadiene

$$DE = 4\alpha + 4\alpha + 4.472\beta - (4\alpha + 4\beta)$$
$$= 0.472\,\alpha < 0 \text{ (as } \beta \text{ is a negative quantity)} \qquad \ldots (9.54)$$

Hence delocalisation of π-electrons over the entire molecule has made it more stable by an amount of 0.472β.

9.1.3 Benzene (C_6H_6)

In benzene molecule, the six $2p$ atomic orbitals taking part in the π bonding to form three π bonds (Fig. 9.9). In benzene molecule, the a bond is formed by sp^2 hybrid orbitals and each carbon atom has a p_z orbital which combine to form three π bonds. The π electrons are delocalised over the whole molecule and delocalisation energy is considered to stabilise the molecule.

The 6×6 Secular determinant in the present case is

$$
\begin{vmatrix}
\alpha - E & \beta & 0 & 0 & 0 & \beta \\
\beta & \alpha - E & \beta & 0 & 0 & 0 \\
0 & \beta & \alpha - E & \beta & 0 & 0 \\
0 & 0 & \beta & \alpha - E & \beta & 0 \\
0 & 0 & 0 & \beta & \alpha - E & \beta \\
\phi\beta & 0 & 0 & 0 & \beta & \alpha - E
\end{vmatrix} = 0
$$

Dividing each element by β and replacing $(\alpha - E)/\beta$ by x, we have

$$
\begin{vmatrix}
x & 1 & 0 & 0 & 0 & 1 \\
1 & x & 1 & 0 & 0 & 0 \\
0 & 1 & x & 1 & 0 & 0 \\
0 & 0 & 1 & x & 1 & 0 \\
0 & 0 & 0 & 1 & x & 1 \\
1 & 0 & 0 & 0 & 1 & x
\end{vmatrix} = 0
$$

The above determinant becomes

$$x^6 - 6x^4 + 9x^2 - 4 = 0$$
$$(x^2 - 4)(x^2 - 1)^2 = 0$$
$$x = \pm 2, \pm 1 \text{ and } \pm 1$$

Hence the six energy expressions are

$$E_1 = \alpha + 2\beta \qquad \left.\begin{matrix} E_4 \\ E_5 \end{matrix}\right\} \alpha - \beta$$

$$\left.\begin{matrix} E_2 \\ E_3 \end{matrix}\right\} = \alpha + \beta \qquad E_6 = \alpha - 2\beta$$

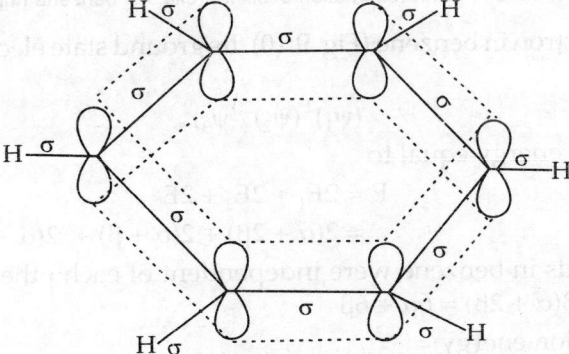

Fig. 9.9 Display of benzene molecule

The first three energy values E_1, E_2, and E_3, are lower than α, and the corresponding molecular orbitals are bonding molecular orbitals. The other three molecular orbitals as antibonding molecular orbitals. The energies E_2 and E_3 and also E_4 and E_5 constitute the degenerate levels.

The coefficients in the wave function can be determined as given below.

$$\psi_1 = \frac{1}{\sqrt{6}} \left(\psi_{2pz(1)} + \psi_{2pz(2)} + \psi_{2pz(3)} + \psi_{2pz(4)} + \psi_{2pz(5)} + \psi_{2pz(6)} \right)$$

$$\psi_2 = \frac{1}{2} \left(\psi_{2pz(2)} + \psi_{2pz(3)} - \psi_{2pz(5)} - \psi_{2pz(6)} \right)$$

$$\psi_3 = \frac{1}{2\sqrt{3}} \left(2\psi_{2pz(1)} + \psi_{2pz(2)} - \psi_{2pz(3)} - \psi_{2pz(4)} - \psi_{2pz(5)} + \psi_{2pz(6)} \right)$$

$$\psi_4 = \frac{1}{2} \left(\psi_{2pz(2)} - \psi_{2pz(3)} + \psi_{2pz(5)} - \psi_{2pz(6)} \right)$$

$$\psi_5 = \frac{1}{2\sqrt{3}} \left(2\psi_{2pz(1)} - \psi_{2pz(2)} - \psi_{2pz(3)} + \psi_{2pz(4)} - \psi_{2pz(5)} - \psi_{2pz(6)} \right)$$

$$\psi_6 = \frac{1}{\sqrt{6}} \left(\psi_{2pz(1)} - \psi_{2pz(2)} + \psi_{2pz(3)} - \psi_{2pz(4)} + \psi_{2pz(5)} - \psi_{2pz(6)} \right)$$

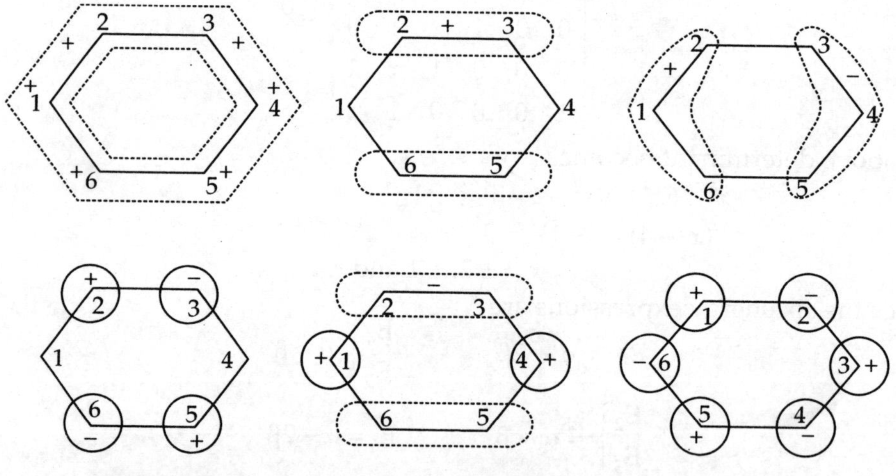

Fig. 9.10 Schematic representation of six π-orbitals of benzene (top view)

There are six π electron in benzene (Fig. 9.10). Its ground state electronic configuration is

$$\left(\psi_1 \right)^2 \left(\psi_2 \right)^2 \left(\psi_3 \right)^2$$

with ground state energy equal to

$$E = 2E_1 + 2E_2 + 2E_2$$
$$= 2(\alpha + 2\beta) + 2(\alpha + \beta) + 2(\alpha + \beta) = 6\alpha + 8\beta$$

If the three π bonds in benzene were independent of each other, i.e. localised they would have energy $3(\alpha + 2\beta) = 6\alpha + 6\beta$

Hence delocalisation energy =

$$6\alpha + 8\beta - (6\alpha + 6\beta) = 2\beta$$

9.1.4 Some Useful Quantities

Electron Densities: Any molecular orbital may be expressed as

$$\psi_j = \sum_{r=1}^{n} C_{jr}\, \psi_{2p_z(r)}$$

where the subscript j refers to the jth molecule orbital and the subscript r refers to the rth atom of the molecule. The quantity C_{jr}^2 gives the electron density on the rth atom due to the jth molecular orbital. The total π electron density on atom r may be calculated from

$$q_r = \Sigma nj\, C_{jr}^2$$

where rj represents the number of electrons occupying the jth molecular orbital. The summation is taken over all the MOs. The sum of all the qs equals the total number of electrons in system.

Let us consider ethylene molecule. The two electrons have occupied π_u – MO of ethylene. Therefore, ED_1 (electron density at carbon 1) and ED_2 (electron density at carbon 2) may be calculated in the following way.

no. of electrons × square of the coefficient (C_2^{11})

$$ED_1 = q_1 = 2 \times \frac{1}{2} = 1 \text{ [note that } C_{11} = \frac{1}{\sqrt{2}} \text{ for ethylene]}$$

$$ED_2 = q_2 = 2 \times \frac{1}{2} = 1 \text{ [note that } C_{12} = \frac{1}{\sqrt{2}} \text{ for ethylene]}$$

∴ Total electron density at carbon 1 and carbon 2 comes out to be 1. This indicates that each carbon atom gets a contribution of one electron out of the two π electrons.

Butadiene

$$q_1 = 2C_{11}^2 + 2C_{21}^2 = 2\,(0.372)^2 + 2(0.602)^2 = 1.00$$

$$q_2 = 2C_{12}^2 + 2C_{22}^2 = 2\,(0.602)^2 + 2(0.372)^2 = 1.00$$

Similarly $q_3 = q_4 = 1.00$

Here π electron charge is unity at each carbon atom

Butadiene cation

$$q_1 = 2C_{11}^2 + C_{21}^2 = 2(0.372)^2 + (0.602)^2 = 0.640$$

$$q_2 = 2C_{12}^2 + C_{22}^2 = 2(0.602)^2 + (0.372)^2 = 0.860$$

$$q_3 = q_2$$

$$q_4 = q_1$$

π-Bond order

The π-bond order between atoms r and s is defined as

$$P_{rs} = \Sigma r_j\, C_{jr}\, C_{js} = 2\Sigma\, C_{jr}\, C_{js}$$

where rj = no of electrons or jth MOs

For butadiene, we have

$$P_{12} = 2C_{11} C_{12} + 2C_{21} C_{22}$$

$$= 2(0.372)(0.602) + 2(0.602)(0.372)$$

$$= 0.894$$

$$P_{23} = 2C_{12} C_{13} + 2C_{22} C_{23}$$

$$= 2(0.602)(0.602) + 2(0.372)(-0.372)$$

$$= 0.447$$

Thus $C_1 - C_2$ bond has more double bond character than $C_2 - C_3$ bond. Consequently the bond length $C_1 - C_2$ is smaller than that of $C_2 - C_3$.

Free Valence Index

Free valence index is defined by Coulson as

$$F_\mu = n_{max} - n_r$$

where n_r is the sum of the π electron bond orders of all the bonds coming from atom μ and n_{max} is the maximum value of such bond orders, usually taken as $\sqrt{3}$.

For butadiene $\overset{1}{C}H_2 = \overset{2}{C}H - \overset{3}{C}CH = \overset{4}{C}H_2$

$$F_1 = \sqrt{3} - P_{12} = 1.732 - 0.894 = 0.838$$

$$F_8 = \sqrt{3} - P_{34} = 1.732 - 0.894 = 0.838$$

$$F_2 (= F_3) = \sqrt{3} - P_{23} = 1.732 - 0.894 - 0.447 = 0.391$$

Thus the terminal atom are more reactive than the central atoms.

9.2 SYMMETRY RULES FOR CHEMICAL REACTIONS: PREDICTING THE COURSE OF A REACTION BY CONSIDERING THE SYMMETRY OF THE WAVE FUNCTION

Pericyclic reactions include the Diels–Alder reactions, dimerisation of olefinic compounds, electrocyclic reactions. Electrocyclic reactions are those reactions where the bonding of the two terminal carbon atoms of a polyene occurs to from a ring or the reverse of it to form a polyene.

Diels–Alder reaction

(Butadiene)(ethylene) Cyclohexene   (Cyclobutene)

Electrocyclic reaction

(Cyclobutene)

Some of these reactions proceed thermally, others require light, but they really proceed with a defined stereochemistry. If a compound undergoes both thermally and photochemically induced reactions, the products often have different reactions, the products often have different stereo chemistry. The thermal and photochemical cyclisation of a butadiene bearing different substituents at its terminal carbon atoms is represented by Eqs. (9.55) and (9.56) respectively.

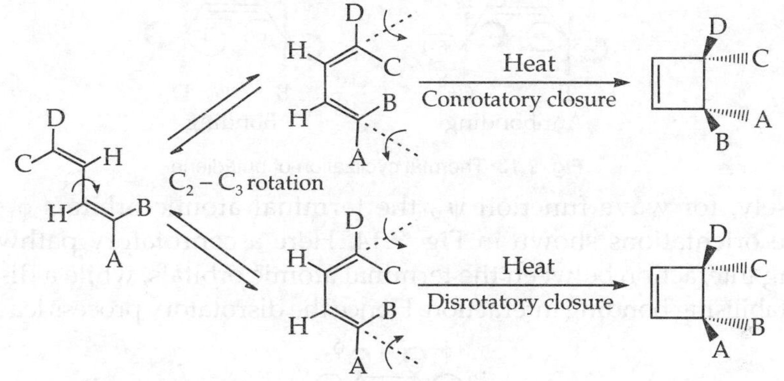

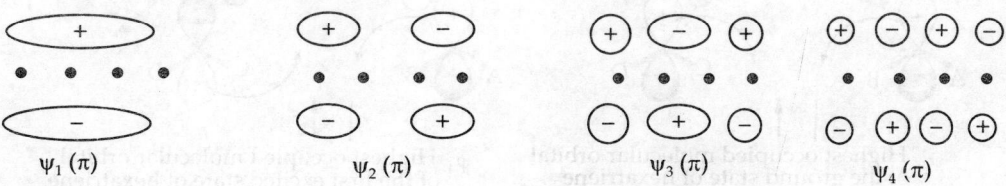

...(9.55)

...(9.56)

The former process is called *conrotatory* where the two π orbitals can rotate in the same direction. The later process is known as *disrotatory* where the two π orbitals can rotate in the opposite direction (attention is paid to the nature of the overlap between the orbitals which combine to form a σ-bond during the cyclisation process).

The atomic orbitals taking part in the π-bonding of this molecule is shown in Fig. 9.11.

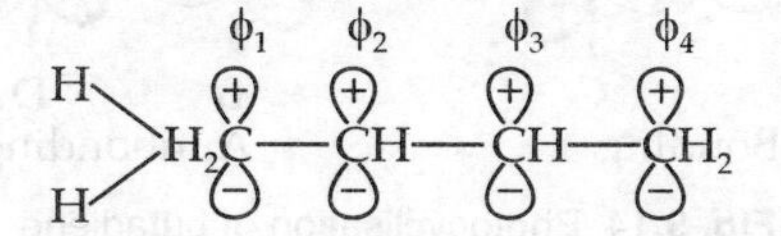

$$\phi_1 \quad \phi_2 \quad \phi_3 \quad \phi_4$$

$$H_2C-CH-CH-CH_2$$

Fig. 9.11 The atomic orbitals for π-bonding in butadiene

The π and π* molecular orbitals of butadiene and their corresponding wave functions are shown pictorially in Fig. 9.12.

$$\psi_1 (\pi) \qquad \psi_2 (\pi) \qquad \psi_3 (\pi) \qquad \psi_4 (\pi)$$

Fig. 9.12 π and π* molecular orbitals of butadiene

Attention is paid to the nature of the overlap between the orbitals which combine to form a σ bond during the cyclisation process. For wave function ψ_2, the terminal atomic orbitals ϕ_1 and ϕ_4 have the relative orientations as shown in Fig. 9.13.

It is evident that a conrotatory process leads to a bonding interaction between ϕ_1 and ϕ_4, while a disrotatory process leads to an antibonding interaction between ϕ_1 and ϕ_4, clearly the conrotatory process prevails in this case.

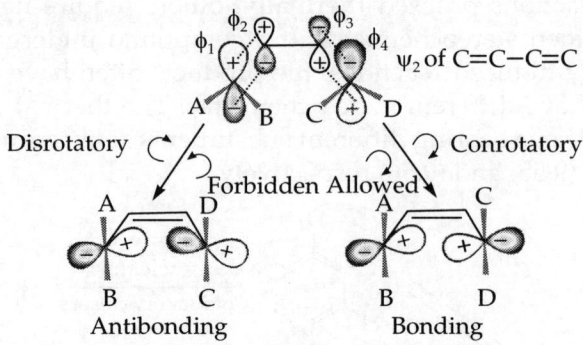

Fig. 9.13 Thermal cyclization of butadiene

Conversely, for wave function ψ_3, the terminal atomic orbitals ϕ_1 and ϕ_4 have the relative orientations shown in Fig. 9.14. Here a conrotatory pathway yields an antibonding interaction between the terminal atomic orbitals, while a disrotatory step leads to a stabilising bonding interaction. Hence the disrotatory process leads in this case.

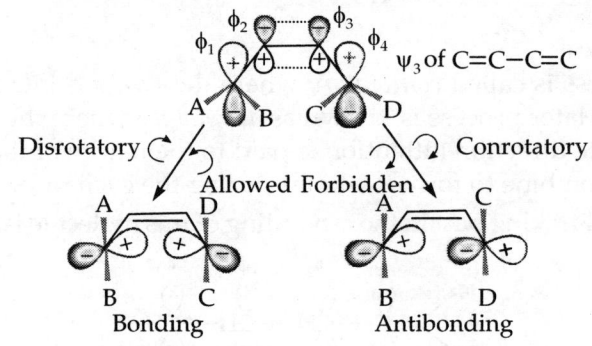

Fig. 9.14 Photocyclisation of butadiene

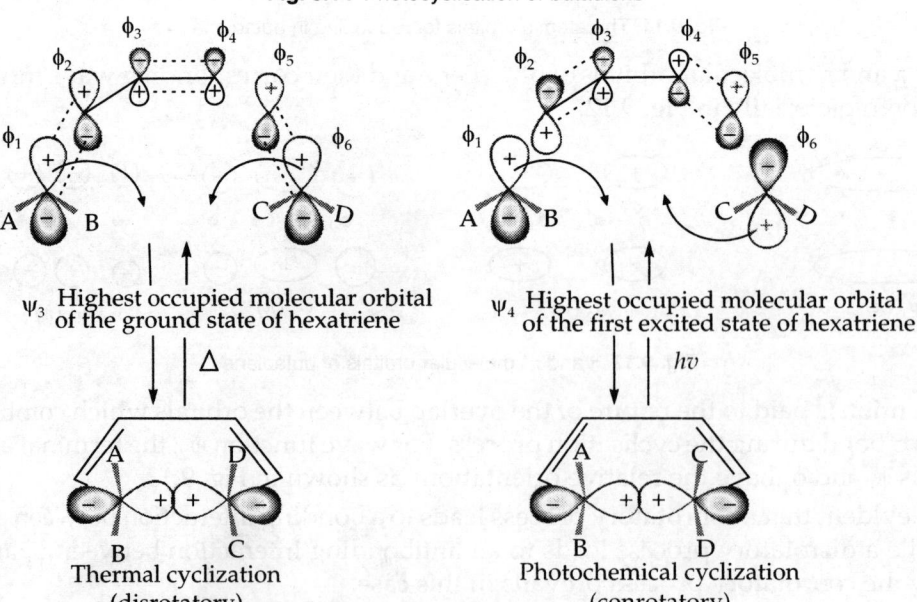

Fig. 9.15 Thermal and photochemical cyclisation reactions of hexatriene

For the thermal and photochemical cyclisation of hexatriene ($_1CH_2 = _2CH - _3CH = _4CH - _5CH = _6CH_2$), the controlling HOMO's are ψ_3 and ψ_4 respectively. As shown in Fig. 9.15, the allowed pathway for the thermal reaction is disrotatory. On the other hand, the allowed pathway for the photochemical reaction is conrotatory. These results are just the opposite of those found for the cyclisation reaction of butadiene (Figs 9.13 and 9.14)

RB Woodward and R Hoffman proposed theories to explain the above mentioned results as well as those for other reactions. This is known as *conservation of orbital symmetry* or Woodward–Hoffman rules. The symmetry rules are summed up as order:

1. During the entire process of transformation there should be a conservation of one symmetry element.
2. The orbital symmetries with respect to this conserved symmetry element remain unchanged (conserved) in the transformation. An orbital which is symmetric(s) with respect to the conserved symmetry element *remains so, all through, until it merges* (correlates) *with a symmetry orbital of the product*.
   Similarly, an antisymmetric one (A) merges with an antisymmetric orbital.
3. An orbital symmetry correlation diagram can thus be drawn up in which the new crossing rule will apply.
   These three symmetry rules are supplemented by a couple of energy rules:
   (a) In thermal reaction, a ground state configuration of the reactant must correlate with ground state configuration of the product in the orbital symmetry correction diagram.
   (b) In photochemical reaction, the first excited state configuration of the reactant should merge into the first exited state of the product.

It should be noted that both butadiene and cyclobutane belong to C_{2v} point group and possess all the symmetry elements of the group. During the conversion of butadiene to cyclobutene or vice-versa ($\Large(\!\rightleftharpoons\!\square)$), one or other symmetry element is preserved throughout the entire reaction course depending up on the nature of the process, viz. conrotation or disrotation (Fig. 9.16).

In conrotation, it is the C_2 symmetry axis which is conserved in all the conformations starting from the initial cyclobutene to the end material butadiene. Similarly in disrotation it is the σ_v which is conserved in the stages from the starting material to the end product.

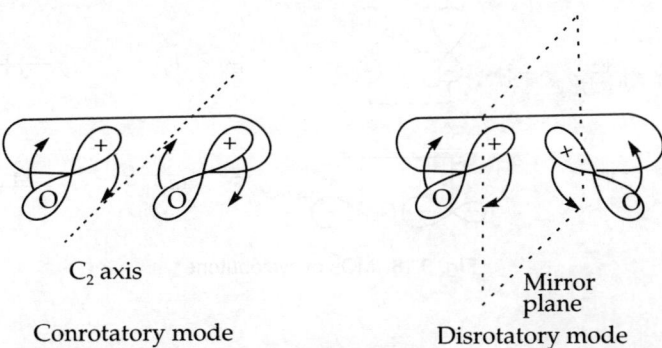

C$_2$ axis

Conrotatory mode

Mirror plane

Disrotatory mode

Fig. 9.16 The symmetry of the conrotatory and disotatory processes

In order to study the conservation of orbital symmetry, we have to consider other MOs also rather than HOMO discussed earlier. The MOs of butadiene (Fig. 9.17) and cyclobutene (Fig. 9.18) can be correlated by their symmetry type as shown in Figs 9.19 and 9.20.

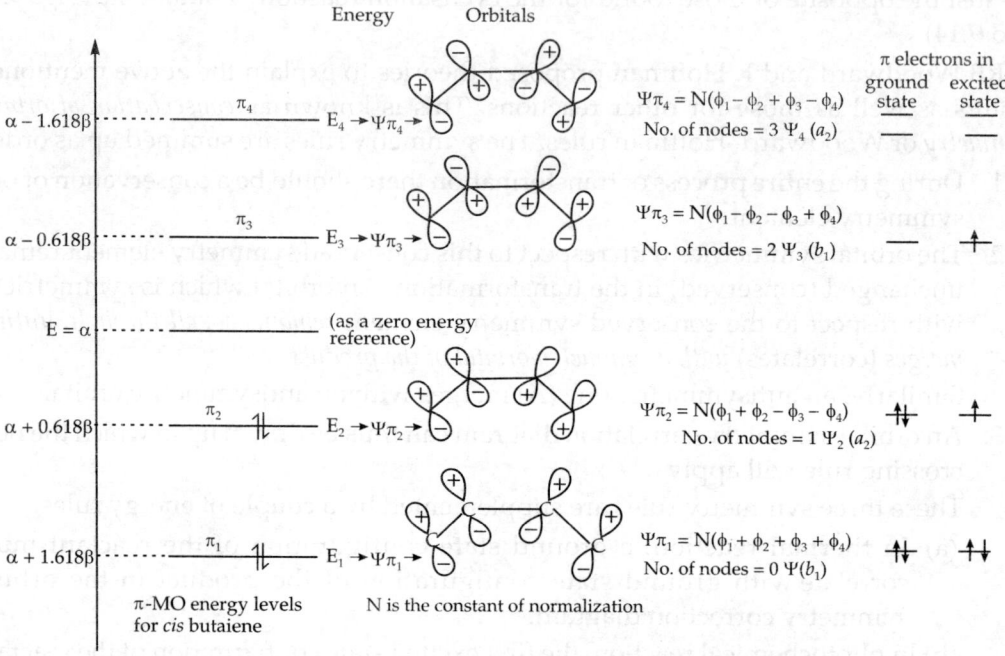

Fig. 9.17 Relative π-MO energy levels, diagram for *cis*-butadiene

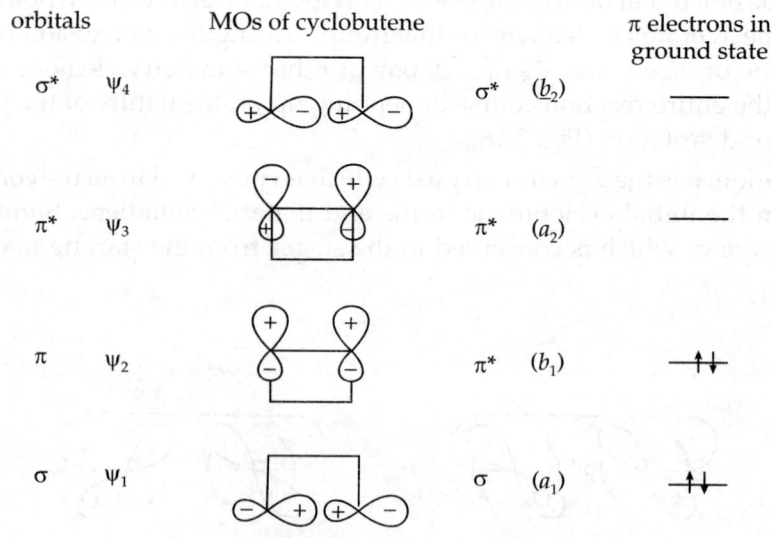

Fig. 9.18. MOs of cyclobutene

MOs

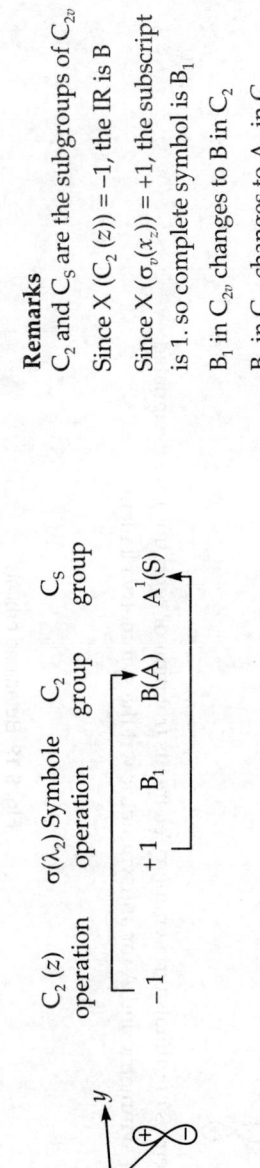

1. ψ_1 of butadiene

$\xrightarrow{C_2(z) \text{ operation}}$

2. ψ_2

$\xrightarrow{C_2(z) \text{ operation}}$

$C_2(z)$ operation	$\sigma(\lambda_2)$ operation	Symbole	C_2 group	C_S group
-1	$+1$	B_1	$B(A)$	$A^1(S)$

Remarks

C_2 and C_S are the subgroups of C_{2v}

Since $X(C_2(z)) = -1$, the IR is B

Since $X(\sigma_v(x_z)) = +1$, the subscript is 1. so complete symbol is B_1

B_1 in C_{2v} changes to B in C_2

B_1 in C_{2v} changes to A_1 in C_s

Model Shows that

For $C_2(z)$ rotation since of lobes change 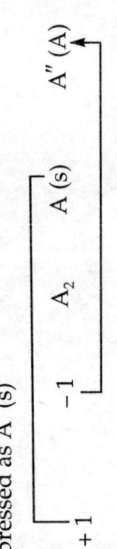 so (-1) comes

For $\sigma_v(xz)$ operation since of lobes remain same so $(+1)$ comes

Since the character of $C_2(z)$ operation is -1 (i.e. $X(C_2(z)) = -1$)

so B is antisymmetric (A) in C_2 group. It is written

As $X(\sigma_v(xz)) = +1$, So A^1 is symmetric(s) in Cs group. This is expressed as A^1(s)

$+1$	-1	A_2	A (s)	A'' (A)

Since the character of $C_2(z)$ is $+1$, so A is symmetric(s). So the symbol will be A(s). The character of $\sigma_v(xz)$ is -1 so A'' will be known unsymmetric (A) and represented as A''(s)

Remarks

Since $X(C_2(z)) = +1$, the IR to A

since $X(\sigma_v(xz)) = -1$, the subscript is 2. The complete symbol is A_2

A_2 in C_{2v} changes to A in C_2 and A'' in C_S

Fig. 9.19 Butadiene orbitals (Contd...)

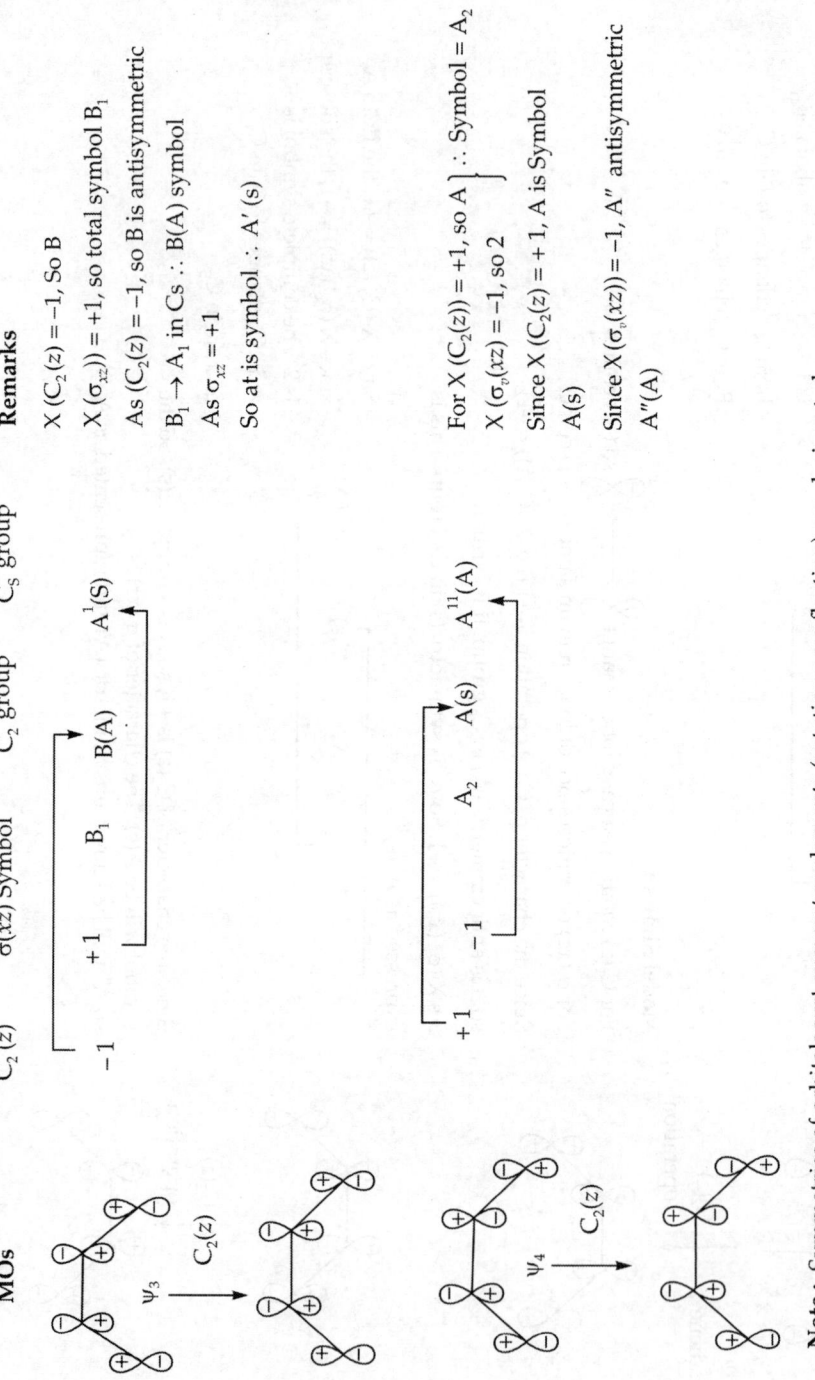

MOs	C$_2$(z)	σ(xz) Symbol	C$_2$ group	C$_S$ group	Remarks

3.

$\psi_3 \xrightarrow{C_2(z)}$

-1 $\quad$ $+1$ $\quad$ B$_1$ $\quad$ B(A) $\quad$ A^1(S)

X (C$_2$(z)) = –1, So B

X (σ$_{xz}$)) = +1, so total symbol B$_1$

As (C$_2$(z)) = –1, so B is antisymmetric

B$_1 \rightarrow$ A$_1$ in Cs ∴ B(A) symbol

As σ$_{xz}$ = +1

So at is symbol ∴ A' (s)

4.

$\psi_4 \xrightarrow{C_2(z)}$

$+1$ $\quad$ -1 $\quad$ A$_2$ $\quad$ A(s) $\quad$ A^{11}(A)

For X (C$_2$(z)) = +1, so A $\Big\}$ ∴ Symbol = A$_2$

X (σ$_v$(xz) = –1, so 2

Since X (C$_2$(z) = +1, A is Symbol

A(s)

Since X (σ$_v$(xz)) = –1, A'' antisymmetric

A''(A)

Fig. 9.19 Butadiene orbitals

Note : Symmetries of orbitals wrt symmetry elements (rotation or reflection) are designated as S for symmetric and A for antisymmetric wrt the symmetry element.

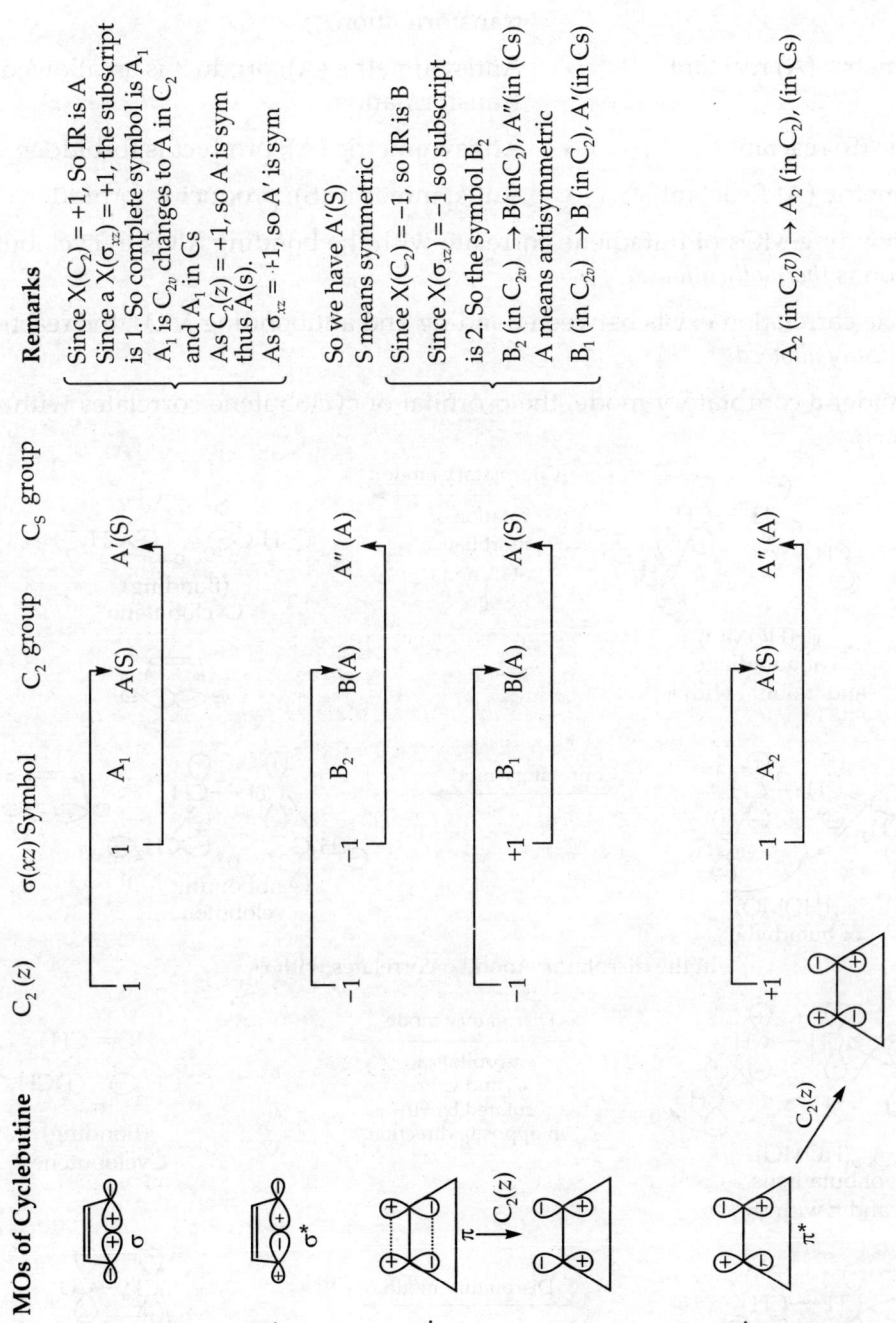

Fig. 9.20 Cyclobutene orbitals

The selection rules derived by Woodward–Hoffman rules are

Symmetric (S) reactant	→	Symmetric (S) loose line; product is an allowed transformation
Antisymmetric (A) reactant	→	Antisymmetric (A); product is an allowed transformation.
Symmetric (S) reactant	→	Antisymmetric (A); product is forbidden
Antisymmetric (A) Reactant	→	Antisymmetric (S); product is forbidden

If the bonding MOs of butadiene correlate with the bonding MOs in cyclobutene, the reaction is *thermally allowed.*

But if the correlation exists between bonding and antibonding MOs, the reaction is *photochemically allowed.*

Thus under a conrotatory mode, the σ-orbital of cyclobutene correlates with ψ_2 of butadiene.

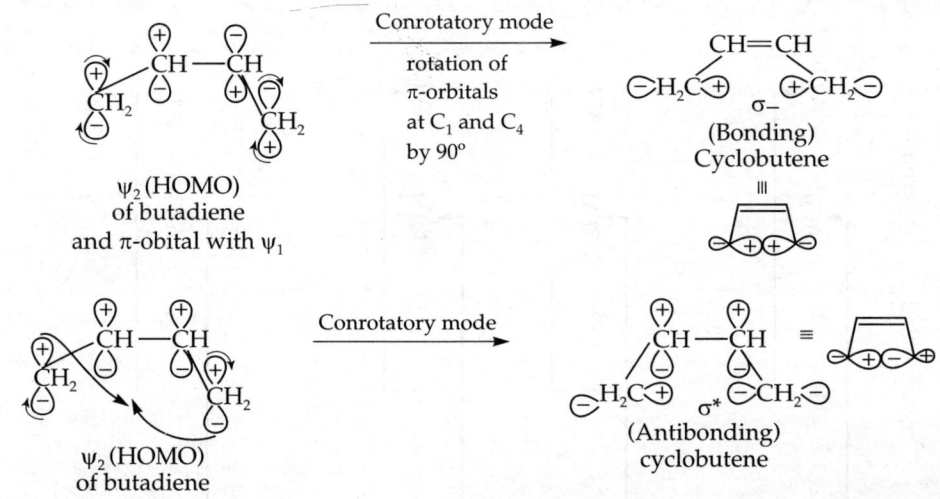

In the disrotatory mode σ correlates with ψ_1

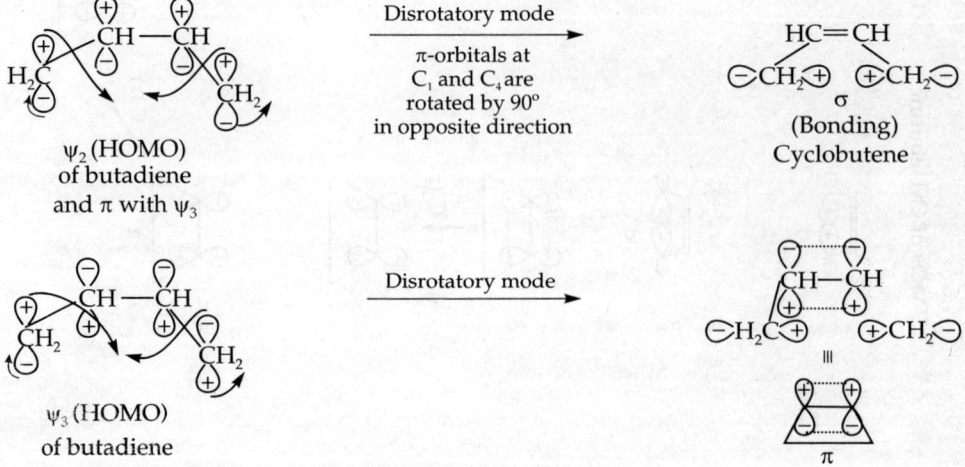

Table 9.1 Shows the orbital correlation between butadiene and cyclobutane orbitals

Table 9.1 Symmetry properties of butadiene and cyclobutane orbitals under conrotatory and disrotatory modes

Cylcobutene	C_2 (conrotatory mode)	Cs (disrotatory mode)
σ (Bonding)	a_1	a_1'
σ* (Antibonding)	b_2	a_2''
π (Bonding)	b_1	a_2'
π* (Antibonding)	a_2	a_1''
Butadiene		
ψ_1 (Bonding)	b_1	a_1'
ψ_2 (Bonding)	a_1	a_1''
ψ_3 (Antibonding)	b_2	a_2'
ψ_4 (Antibonding)	a_1	a_2''

The ground state configuration of cyclobutene is $\sigma^2\pi^2$. Hence the symmetry configuration in the conrotatory mode is $a_1^2 b_1^2$. which is totally symmetric, A $((a_1^2(b_1^2)) = (A_1 \times A_1)(B_1 \times B_1) = A_1 \times A_1 = A_1)$. This will correlate with the ground state of butadiene $\psi_1^2 \psi_2^2 = (b_1^2)(a_1^2)$ which is also totally symmetric, A $(b_1^2)(a_1^2) = (B_1 \times B_1)(A_1 \times A_1) = A_1 \times A_1 = A_1)$.

The first excited state of cyclobutene is $\sigma^2 \pi\pi^* (\equiv (a_1^2)(b_1)(a_2) = (A_1 \times A_2 = B_2 = B_2)$ for a conrotatory mode. This correlates with the excited $\psi_1 \psi_2^2 \psi_4 (\equiv (b_1 (a_1)^2)(a_1) = B_1 \times (A_1 \times A_1) \times A_1 = B_1 \times A_1 \times A_1 = B_1 \times A_1 = B_1)$ state of butadiene. Notice that this is *not* the lowest excited state of butadiene which would be $\psi_1^2 \psi_2 \psi_3 (\equiv (b_1)^2 (a_1)(b_2) = (B_1 \times B_1) (A_1 \times B_2 = A_1 \times A_1 \times B_2 = A_1 \times B_2 = B_2)$. Figure 9.21 is a correction diagram for cyclobutane–butadiene under the the conrotatory mode.

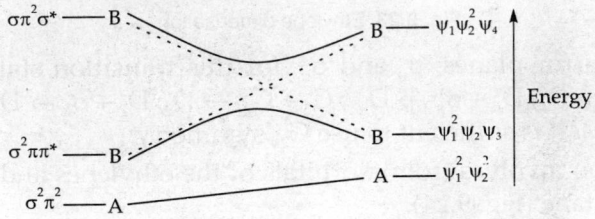

$\sigma\pi^2\sigma^*$ —— B

$\sigma^2\pi\pi^*$ —— B

$\sigma^2\pi^2$ —— A

B —— $\psi_1\psi_2^2 \psi_4$

Energy

B —— $\psi_1^2 \psi_2 \psi_3$

A —— $\psi_1^2 \psi_2^2$

Fig. 9.21 Correlation diagram for cyclobutane–butadiene under conrotatory mode

The excited state of cyclobutane $\sigma^2\pi\pi^*$ correlates with the upper excited state of butadiene $\psi_1 \psi_2^2 \psi_4$ but because of the noncrossing rule, the actual correctations are given by the full curved lines.

We can construct an exactly analogous diagram for the disrotatory mode, as in Fig. 9.22. Here we have the ground state $\sigma^2 \pi^2$ ($\equiv (a_1')^2 (a_2')^2 = (A_1' \times A_1')(A_2' \times A_2') = A_1' \times A_1' = A_1' = A_1'$ for a disrotatory mode of the cyclobutane corresponding with $\psi_1^2 \psi_3^2$ ($\equiv (a_1')^2 (a_2')^2 = (A_1' \times A_1')(A_2' \times A_2') = A_1' \times A_1' = A_1'$) and the first excited state $\alpha^2 \pi\pi^*$ ($\equiv (a_1')^2 (a_2')(a'') = (A_1' \times A_1')(A_2' \times A_1'') = A_1' \times A_2'' = A_2''$) corresponding to $\psi_1^2 \psi_2 \psi_3$ ($\equiv (a_1')^2 (a_2'')(a_2') = (A_1' \times A_1')(A_2'' \times A_2') = A_1' \times A_1'' = A_1''$ and again the noncrossing rule applies.

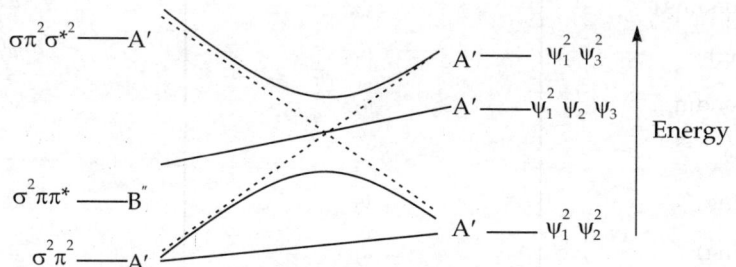

Fig. 9.22 Correlation diagram for cyclobutane–butadiene under disrotatory mode

We shall now consider pericyclic reactions involving two separate molecules like the Diels–Alder reaction. The simplest case would be ethylene dimerisation.

we assume that the two ethylene molecules approach each other in parallel planes, one above the other (Fig. 9.23).

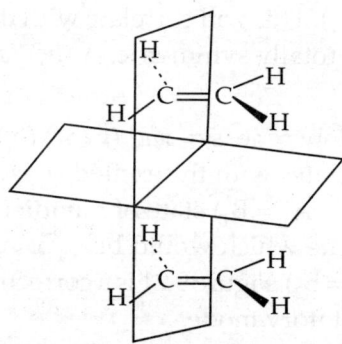

Fig. 9.23 Ethylene dimerisation

There are two mirror planes, σ_v and σ_v', for this transition state which is of D_{2h} symmetry ($C_n + C_2 \rightarrow D_n$; $D_n + \sigma_h \rightarrow D_{nh}$; $C_2 + C_2' \rightarrow D_2$; $D_2 + \sigma_h \rightarrow D_{2h}$; $D_{2h} \rightarrow C_2 + 2C_2' + \sigma_h$; $C_{2v} \rightarrow C_2 + 2\sigma_v$). It is sufficient to use C_{2v} symmetry.

The actual reaction involves from π-orbitals of the ethylenes and four σ-orbitals of the product cyclobutane (Fig. 9.24).

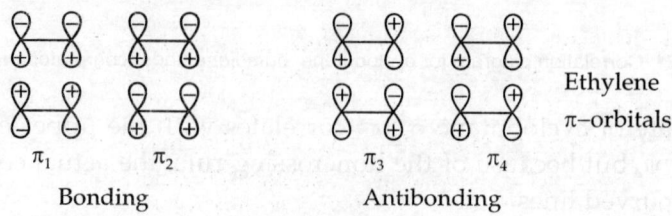

Fig. 9.24 (Contd...)

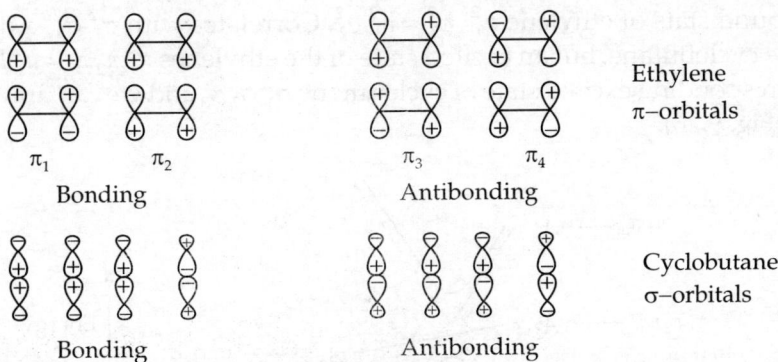

Fig. 9.24

In Table 9.2, we designate the symmetries of these orbitals under the operations of the C_{2v} point group (Fig. 9.25).

Table 9.2 The symmetries of the orbitals involved in the demeristation of ethylene

Ethylene π-orbital	C_2	σ_v	Symmetry symbol	Cyclobutene σ-orbital
π_1	+1	+1	a_1	σ_1
π_2	−1	−1	b_2	σ_3
π_3	−1	+1	b_1	σ_2
π_4	+1	−1	a_2	σ_4

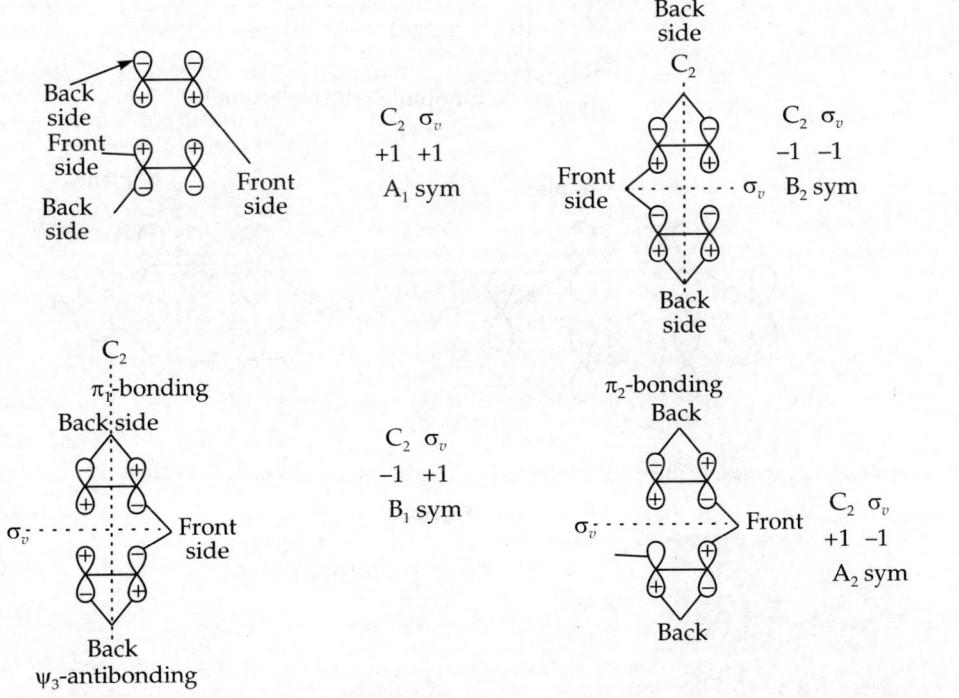

Fig. 9.25

The ground stats of ethylene $\pi_1^2 \ \pi_2^2 = a_1^2 \ b_2^2$. Correlates with $\sigma_1^2 \ \sigma_3^2$, a doubly excited state of the cyclobutane, but an excited state of the ethylenes $\pi_1^2 \ \pi_2 \ \pi_3 = a_1^2 \ b_2 \ b_1$ correlates with a corresponding excited state of cyclobutane $\sigma_1^2 \ \sigma_2 \ \sigma_3$ and we can draw a correlation diagram (Fig. 9.26).

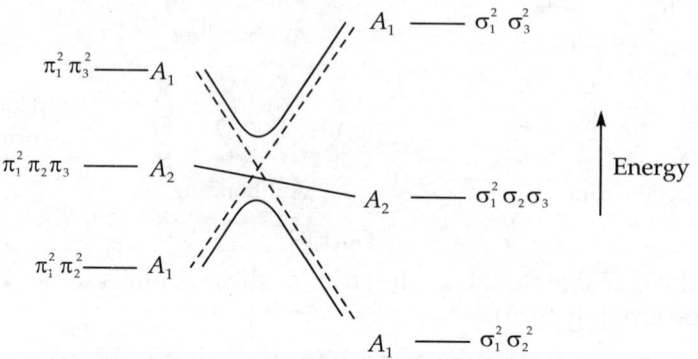

Fig. 9.26 Correlation diagram for ethylene dimerisation

From the above discussion, it would appear that ethylene dimerisation (or dissociation of cyclobutane) was favoured for a photochemical process.

We can now consider the reaction between butadiene (diene contains $4\pi e^-$) and ethylene (dienophile contains $2\pi e^-$) in which we have a σ–plane (Fig. 9.27). The symmetry is Cs. The symmetries of the orbitals involved in the cyclo-addition of ethylene and butadiene is shown in the Table 9.3.

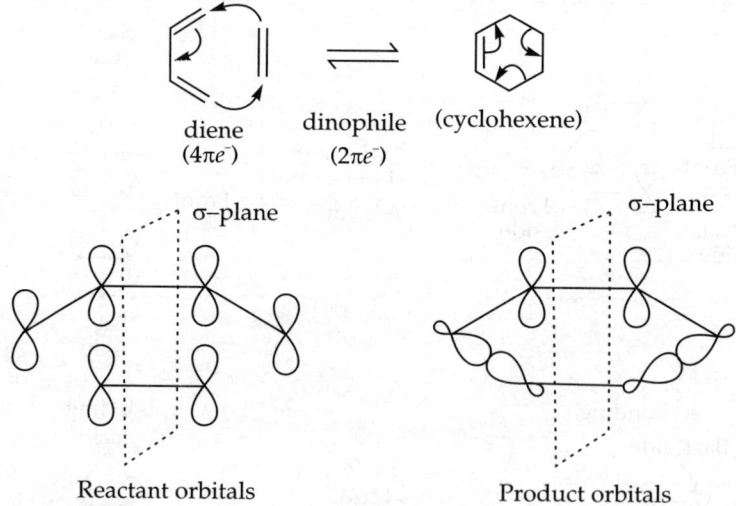

Fig. 9.27

Table 9.3 The symmetries of the orbitals involved in cycloaddition of ethylene and butadiene

Reactant orbitals		Symmetry symbol	Product orbitals	
Butadiene	ψ_1	a'	σ_1	Cyclohexene
	ψ_2	a''	σ_2	
	ψ_3	a'	σ_3^*	
	ψ_4	a''	σ_4^*	
Ethylene	π_e	a'	π_c	
	π_e^*	a''	π_c^*	

The ground states of the reactants $\psi_1^2\,\psi_2^2\,\pi_e^2$ ($\equiv (a')^2\,(a'')^2\,(a')^2 = (A' \times A')\,(A'' \times A'')$ $(A' \times A') = A' \times A' = A'$) correlates with the ground state of the products $\sigma_1^2\,\sigma_2^2\,\pi_e^2$ ($\equiv (a')^2\,(a'')^2\,(a')^2 = A'$) so that this reaction should proceed thermally.

PROBLEMS AND SOLUTIONS

Problem 1. Using the Huckel approximations, show that for the allyl radical ($\overset{\bullet}{\underset{1}{C}H_2} - \underset{2}{C}H = \underset{3}{C}H_2$), the wave functions and the corresponding energies are given by

$$\psi_1 = \frac{1}{2}\,\psi_{2p_z(1)} + \frac{1}{\sqrt{2}}\,\psi_{2p_z(2)} + \frac{1}{2}\,\psi_{2p_z(3)};\ E_1 = \alpha + \sqrt{2}\,\beta$$

$$\psi_2 = \frac{1}{\sqrt{2}}\,\psi_{2p_z(1)} - \frac{1}{\sqrt{2}}\,\psi_{2p_z(2)};\ E_2 = \alpha$$

$$\psi_3 = \frac{1}{2}\,\psi_{2p_z(1)} - \frac{1}{\sqrt{2}}\,\psi_{2p_z(2)} + \frac{1}{2}\,\psi_{2p_z(3)};\ E_3 = \alpha - \sqrt{2}\,\beta$$

Solution

The secular determinant for allyl radical is given by ($\overset{\bullet}{\underset{1}{C}H_2} - \underset{2}{C}H = \underset{3}{C}H_2$)

$$\begin{vmatrix} \alpha - E & \beta & 0 \\ \beta & \alpha - E & \beta \\ 0 & \beta & \alpha - E \end{vmatrix} = 0$$

Dividing each element of the above determinant by β and then setting $\dfrac{\alpha - E}{\beta} = x$, we get

$$\begin{vmatrix} x & 1 & 0 \\ 1 & x & 1 \\ 0 & 1 & x \end{vmatrix} = 0,\ \text{and on expanding gives } x^3 - 2x = 0,$$

$x = -\sqrt{2},\,0,\,\sqrt{2}$

The corresponding energy values are

$$E_1 = \pi \div \sqrt{2}\,\beta$$
$$E_2 = \alpha$$
$$E_3 = \alpha - \sqrt{2}\,\beta$$

So the secular equations are

$$C_1 x + C_2 = 0$$
$$C + C_2 x + C_3 = 0$$
$$C_2 + C_3 x = 0$$

For $x = -\sqrt{2}$, we shall have

$$-\sqrt{2}\ C_1 + C_2 = 0 \qquad \qquad \dots(1)$$
$$C_1 - \sqrt{2}\ C_2 + C_3 = 0 \qquad \qquad \dots(2)$$
$$C_2 - \sqrt{2}\ C_3 + C_3 = 0 \qquad \qquad \dots(3)$$

From Eq. (1) $\qquad\qquad\qquad\qquad C_1 = \dfrac{C_2}{\sqrt{2}}$

From Eq. (3) $\qquad\qquad\qquad\qquad C_3 = \dfrac{C_2}{\sqrt{2}}$

Using normalisation criterion

$$C_1^2 + C_2^2 + C_3^2 = 1$$

$$\left(\frac{C_2}{\sqrt{2}}\right)^2 + C_2^2 + \left(\frac{C_2}{\sqrt{2}}\right)^2 = 1 \qquad \therefore C_2 = \pm\left(\frac{1}{\sqrt{2}}\right);$$

So the first wave function is $\quad C_1 = \dfrac{C_2}{\sqrt{2}} = \pm\dfrac{1}{\sqrt{2}}\cdot\dfrac{1}{2} = \dfrac{1}{2}; C_3 = \pm\dfrac{1}{\sqrt{2}}\cdot\dfrac{1}{\sqrt{2}} = \pm\dfrac{1}{2}$

$$\psi_1 = \frac{1}{2}\ \psi_{2p_z(1)} + \left(\frac{1}{\sqrt{2}}\right)\psi_{2p_z(2)} + \frac{1}{2}\ \psi_{2p_z(3)}$$

Working out similarly with $x = 0$ and $x = \sqrt{2}$ we get

$$\psi_2 = \left(\frac{1}{\sqrt{2}}\right)\psi_{2p_z(1)} - \left(\frac{1}{\sqrt{2}}\right)\psi_{2p_z(3)}$$

$$\psi_3 = \frac{1}{2}\ \psi_{2p_z(1)} - \left(\frac{1}{\sqrt{2}}\right)\psi_{2p_z(2)} + \frac{1}{2}\ \psi_{2p_z(3)}$$

STUDY QUESTIONS

1. Construct the orbital and state (configuration) correlation diagrams for conrotatory and disrotatory ring opening of cyclohexadiene to *cis*-hexadiene. Discuss its thermal photochemical allowedness.

2. Choose the correct option :

   The reaction is

   (a) photochemically allowed by symmetry
   (b) thermally allowed by symmetry
   (c) not symmetry, allowed under any condition
   (d) requires a catalyst

3. Indicate, by Δ or $h\nu$, whether the following reactions are thermally or photo-chemically allowed.

4. Calculate the electron densities and bond orders for (i) allyl radical (ii) allyl cation, and (iii) allyl anion.

5. Define the following statements with their utilities.
   (i) Delocalisation energy
   (ii) Electron density on an atom of a molecule
   (iii) Bond order (iv) Free valence

6. Describe the Huckel approximation and show how these help simplifying the LCAO-MO framework of π-electrons of conjugated hydrocarbons.

7. Derive the ground state electronic configuration and ground state energy for each of the following molecules: (i) ethylene (ii) butadiene and (ii) benzene.

8. Work out the molecular orbitals of allyl radical ($CH_2\!\!=\!\!CH\!\!-\!\!\overset{\bullet}{C}H_2$, point group C_{2v}). Find out their energies and compare with those of cyclopropenyl.

   Ans : $\psi_1\,(A_2) = \dfrac{1}{\sqrt{2}}\ (\phi_1 - \phi_3);\ E_1 = \alpha$

   $\psi_2\,(B_2) = \left(\dfrac{1}{2}\right)(\phi_1 + \sqrt{2}\ \phi_2 + \phi_3);\ E_2 = \alpha + \sqrt{2}\ \beta$

   $\psi_2\,(B_2) = \left(\dfrac{1}{2}\right)(\phi_1 - \sqrt{2}\ \phi_2 + \phi_3);\ E_2 = \alpha - \sqrt{2}\ \beta.$

9. Work out π molecular orbitals of cyclopropenyl and their energies.

   Ans : $\psi_1\,(A) = \left(\dfrac{1}{\sqrt{3}}\right)\ (\phi_1 + \phi_2 + \phi_3);\ E_1 = \alpha$

   $\psi_2\,(E) = \left(\dfrac{1}{\sqrt{6}}\right)(2\phi_1 - \phi_2 - \phi_3);\ E_2 = \alpha - \beta$

   $\psi_3\,(E) = \dfrac{1}{\sqrt{2}}\ (\phi_2 - \phi_2);\ E_3 = \alpha - \beta.$

10. Work out the π molecular orbitals and their energies for the molecule $C_4\,H_4$.

    Ans : $\psi\,(A) = \dfrac{1}{2}\ (\phi_1 + \phi_2 + \phi_3 + \phi_4);\ E = \alpha + 2\beta$

    $\psi\,(Ea) = \left(\dfrac{1}{\sqrt{2}}\right)\ (\phi_1 - \phi_3);\ \psi\,(Ea) = \left(\dfrac{1}{\sqrt{2}}\right)\ (\phi_2 - \phi_4);\ E = \alpha$

    $\psi\,(B) = \dfrac{1}{2}\ (\phi_2 - \phi_2 + \phi_3 - \phi_4);\ E = \alpha - 2\beta.$

10

Applications of Group Theory to Electronic Spectroscopy— UV and Visible Spectroscopy

Spectroscopy is defined as the interaction of electromagnetic radiation with matter. An important application of group theory is to predict the probability of transitions, i.e. allowed transitions and forbidden transitions, number of bonds in the spectrum and their intensities. In this chapter, we shall see how symmetry principles continue to be applied in interpreting the electronic transitions in transition metal complexes. The complex metal ion may be centrosymmetric (O_h, D_{4h}, etc. symmetry) or noncentro-symmetric one (T_d, D_3, etc. symmetry) depending on the possession or the lack of an inversion centre, respectively. Coupling between electronic and vibrational motion plays an important role in electronic spectroscopy. The group theoretical aspects of this coupling are discussed and presented to obtain valuable information about the molecular structure both in the ground state as well as in excited state, to derive selection rules for spectroscopy transitions, interpret various bands, lines, etc. in spectra, describe the types of orbitals used in bonding and study a number of additional molecular properties.

10.1 LAW OF LIGHT ABSORPTION

If a beam of monochromatic light of intensity I_0 passes through a solution of path length t (measured in cm) containing solute of molar concentration C, thus the transmitted light of intensity I is related to the incident light intensity I_0 by

$$E = \log_{10}\left(\frac{I_0}{I}\right) = \in Ct \qquad \qquad ...(10.1)$$

or
$$\frac{I_0}{I} = 10^{\in Ct}$$

or   Transmittance $(T) = \dfrac{I_0}{I} = 10^{-\in Ct}$ $\qquad \qquad ...(10.2)$

where $\in$ is known as *molar absorptivity* or *molar extinction coefficient* (in units of litre·mol^{-1}cm^{-1} or cm^2·mol^{-1}) E is the optical density or *absorbance* and T is the *transmittance*.

10.2 BAND INTENSITIES

When the electronic absorption spectra are recorded on gaseous substances, they often contain sharp and hence symmetric bands. In such cases, it is easy to measure the peak height $\in_{max}$. The magnitude of $\in_{max}$ is a direct measure of the probability of the electronic transition and hence the intensity of the band (Fig. 10.1).

But when the spectra are recorded on solutions or solids, the bands observed are 'broad' and 'unsymmetric'. In such cases, the peak height ($\in_{max}$) is not an accurate measure of the intensity of the band but rather the 'area' under the band. Therefore, the area of the entire band can be obtained by integrating, i.e.

$$I = \int_{v_1}^{v_2} \in dv \qquad \qquad ...(10.3)$$

where I is the integrated intensity, v_1 and v_2 are the starting and ending frequencies of the band.

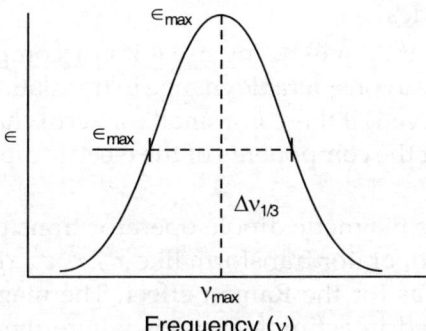

Fig. 10.1 Typical absorption curve defining $\in_{max}$ and $\Delta v_{1/3}$

The integrated intensity I is expressed as *oscillator strength* (f) is proportional to integrated intensity (I).

$$f \propto I$$

$$f = 0.102\left(\frac{mc^2}{N\pi e^2}\right)\int_{v_1}^{v_2} \in dv \qquad \left(\begin{array}{l} m = \text{mass of the electron} \\ e = \text{charge of the electron} \\ N = \text{Avogadro's number} \\ c = \text{velocity of light} \end{array}\right)$$

$$= 4.315 \times 10^{-9} \int_{v_1}^{v_2} \in dv$$

$$\cong 4.6 \times 10^{-9} \in_{max} \Delta v_{1/2} \qquad \qquad ...(10.4)$$

where $\Delta v_{1/2}$ is the half width of the band, i.e. width of the band or peak at $\in_{max/2}$.

The magnitude of f can be estimated from the Eq. (10.4) and its magnitude depends upon the extent to which the absorption band is allowed or forbidden and is related to the dipole structure D.

$$f = \left(\frac{8\pi^2 mc}{3h}\right) GvD = 1.096 \times 10^{11} \, GvD \qquad \qquad ...(10.5)$$

G is a *number* referring to the degeneracies of the states concerned, v is the wave number of the transition and h is the Planck's constant.

Dipole structure D is related to dipole moment operator M

$$D^{1/2} = \int \psi_a \widehat{M} \psi_b \, d\tau = P$$

$$P = <\psi_a |M| \psi_b> \qquad \qquad ...(10.6)$$

where ψ_a and ψ_b are the total wave functions of the states between which the transition is occurring. [Eq. (10.6)] is known as the *transition moment P*.

Now $\qquad \qquad P^2 = D$

So the square of the transition moment P^2 is called the dipole strength D of the transition.

10.3 ELECTRONIC DIPOLE SELECTION RULES

Transitions may occur through electric dipole, magnetic dipole or electric quadrupole mechanisms. As far as transition metal (d block) complexes are concerned, their spectra usually arise through electric dipole transitions; magnetic and electric quadrupole transitions have a very much lower probability of occurance (i.e. give rise to very low intensity absorption). Magnetic dipole transitions are observed in f block elements, i.e. in lanthanides and actinides.

An electric dipole transition will be intense if it is accompanied by a large change in dipole moment, i.e. there is a considerable change in translation during the transition, and will be weak (or not observed) if there is a small (or zero) change in the dipole moment. Group theory reveals that the components of the electric dipole operator transform like translations.

The components of the magnetic dipole operator transform like rotations; those of the quadrupole moment operator transform like $x^2, y^2, z^2, xy, yz, zx$ and therefore have the same selection rules as for the Raman effect. The magnetic quadrupole allowed transitions can be detected in actinide systems where they can be quite strong. The intensities of electric dipole, magnetic dipole and electric quadrupole transitions are of the order of $6.5 \times 10^{-36}, 8.7 \times 10^{-41}$ and 4×10^{-42} CGS units respectively.

The molar extinction coefficient ($\in$) cannot be readily handled theoretically. It is more convenient to talk of the *oscillator strength* (f) of a transition.

M [Eq. (10.6)] is the dipole moment operator with components M_x, M_y and M_z, such that, for electric dipole allowed transitions.

$$M_x = -e\sum_i x_i, \qquad \qquad ...(10.7)$$

i.e. the intensity of absorption band depends on the relative orientation of the electric vector of the incident light, and the symmetry axes of the molecule. The term is summed over all electrons i of charge $-e$. From Eq. (10.5) we have

$$f = (1.096 \times 10^{11} \times G)\, v \cdot D$$
$$= \text{constant} \cdot v \cdot D$$
$$= \text{constant} \cdot v \langle \psi_a \mid \widehat{M} \mid \psi_b \rangle^2 \qquad ...(10.8)$$
$$[\because D = \langle \psi_a \mid \widehat{M} \mid \psi_b \rangle^2]$$

Since transition moment vector **M** for a molecule can be resolved with components M_x, M_y and M_z, so Eq. (10.8) may be written as

$$f = \text{constant} \cdot v\, [\langle \psi_a \mid \widehat{M}_x \mid \psi_b \rangle^2 + \langle \psi_a \mid \widehat{M}_y \mid \psi_b \rangle^2 + \langle \psi_a \mid \widehat{M}_z \mid \psi_b \rangle^2]$$
$$...(10.9)$$

Each transition moment component, e.g. $\langle \psi_a \mid \widehat{M}_x \mid \psi_b \rangle$, is a number and hence it should remain unchanged on performing all the operations of the group to which the molecule belongs. Therefore,

$$\chi(R)\, (\langle \psi_a \mid \widehat{M}_x \mid \psi_b \rangle) = +1\, (\langle \psi_a \mid \widehat{M}_x \mid \psi_b \rangle) \qquad ...(10.10)$$

Same is true for the total transition moment integral

$$\chi(R)\, (\langle \psi_a \mid \widehat{M} \mid \psi_b \rangle) = +1\, (\langle \psi_a \mid \widehat{M} \mid \psi_b \rangle) \qquad ...(10.11)$$

Such a situation of integral invariance under all operation can occur only if

(a) the integrand transforms as the totally symmetric representation A (A stands for a totally symmetric irreducible representation) of the group concerned, or

(b) if the integral is zero.

Thus to be nonzero, the direct product of the representations of the initial and final wave functions with the particular dipole moment operator component, must contain A_{1g} (or equivalent totally symmetric representation).

Since each dipole moment operator component (M_x, M_y, M_z) transforms as translations, it has odd (u) parity. In order to get a g representation in the total direct product it is necessary that the direct product of the representations of the initial and final wave functions should be odd, since $u \times u = g$. This can occur only if one of the wave functions is odd and the other even.

$$\psi_a \times \psi_b = g \times u = u \qquad (\psi_a\text{-ground state wave function};$$
$$\psi_b\text{-excited state wave functions})$$

$$\langle \psi_a | \widehat{M} | \psi_b \rangle = g \times u \times u = u \times u \times g = g$$

The transition will be allowed if the triple-product is of g-type (totally symmetric). This is the basis for Laporte, or parity selection rule.

p orbitals have u character,

d orbitals have g character,

f orbitals have u character.

$\widehat{M}$ (dipole moment operator) has u character. Thus for d–d transition, we may write

$$\langle g | \hat{u} | g \rangle \rightarrow g \times u \times g \rightarrow g \times u \rightarrow u \qquad \text{(forbidden transition)}$$

Similarly for p–p and f–f transitions

$$\langle u | \hat{u} | u \rangle \rightarrow u \times \hat{u} \times u \rightarrow u \times g \rightarrow u \qquad \text{(forbidden transition)},$$

But for d–p transition and d–f transition

$$\langle g | \hat{u} | u \rangle \rightarrow \langle g \times g \rangle \rightarrow g \qquad \text{(allowed transition)}$$

This selection rule is variously called as **Laporte, parity** or **orbital symmetry selection rule**. This rule can be stated as **electric–dipole transition between states of same parity are forbidden.**

This means that p–p, d–d, f–f transitions are forbidden, but s–p, p–d or d–f transitions would be allowed. This rule may be expressed in the form:

$$\Delta l = \pm 1 \qquad \qquad ...(10.12)$$

and is known as **Laporte's rule** after its discover.

Since dipole moment operator (M) is a one-electron operator, i.e. it operates on one electron at a time, an electric dipole transition can only involve one electron excitation, hence: *Transitions involving excitation of two or more electrons are forbidden.*

Let us consider ethylene molecule σ_b and π_b molecular orbitals (MOs) are filled with pair of electrons and there are vacant π^* and σ^* MOs. The symmetries of the MOs are as follows:

$$\sigma_b - g \qquad \qquad \sigma^* - u$$
$$\pi_b - u \qquad \qquad \pi^* - g$$

Hence $\sigma_{(g)} \rightarrow \sigma^*_{(u)}$ and $\pi_{(u)} \rightarrow \sigma^*_{(g)}$ are only allowed transitions. The $\pi_{(u)} \rightarrow \pi^*_{(g)}$ transitions occurs at 171 nm (near UV region) and $\sigma_{(g)}$ $\sigma^*_{(u)}$ transition occurs in the far-UV region (Fig. 10.2).

Thus the π electrons are responsible for energy absorptions occurring in near UV region.

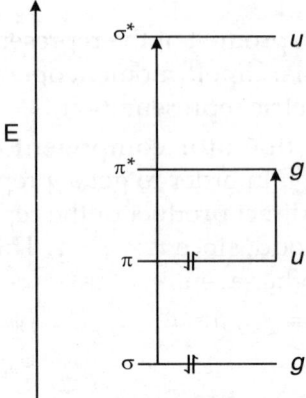

Fig. 10.2 Relative energies of ethylene molecular orbitals

As the number of conjugated double bonds go on increasing in the molecule, the separation between π (bonding) and π^* (antibonding) goes on decreasing in the visible region giving the colour to the compound. In this way long conjugated systems become coloured.

$$\sigma_{(g)} \rightarrow \pi_{(g)}^* \text{ and } \pi_{(u)} \rightarrow \sigma_{(u)}^*,$$

are forbidden transitions and should not take place. However, forbidden transitions are observed in many compounds with low intensity (weak bands). Let us see how such forbidden transitions may occur.

It is known that the excitations of electrons from the ground state to the excited state is followed by changes in the vibrational and rotational levels in the molecule. The electrons may go from the ground vibrational level in the ground electronic state to higher vibrational level in excited electronic states (Fig. 10.3).

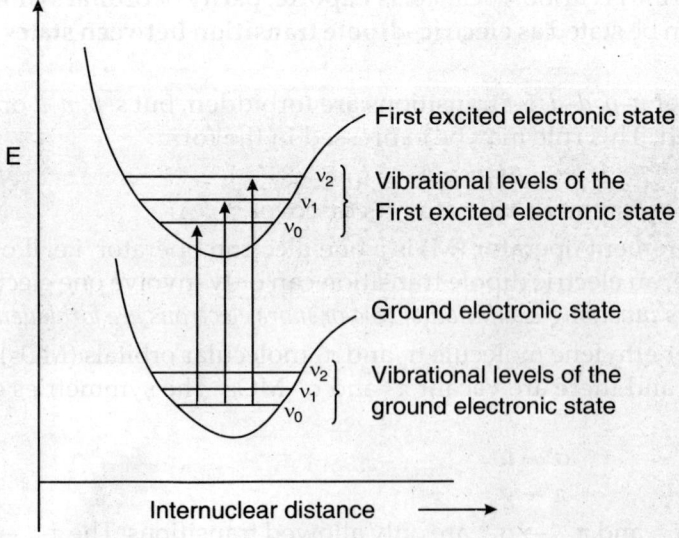

Fig. 10.3 Morse curves for the ground and excited states of a diatomic molecule

The possible transitions are $v_0 \rightarrow v_0'$

$v_0 \rightarrow v_1'$

$v_0 \rightarrow v_2'$

$v_0 \rightarrow v_3'$, etc.

These transitions correspond to different amounts of energies. Thus due to vibrational changes involved, there is splitting or broadening of the electronic transition bands.

So the total wave function ψ can be factorised into rotational ψ^R, vibrational ψ^V and electronic ψ^E components, such that

$$\psi = \psi^R \cdot \psi^V \cdot \psi^E \qquad ...(10.13)$$

This is usually known as the **Born–Oppenheimer** approximation. Electronic transitions take place in such a short time interval ($\sim 10^{-15}$ s.) that the nuclei may be assumed to remain stationary during the transition (**Franck–Condon transition**).

At the instant of excitation, the excited state has the same nuclear geometry as the ground state, and may be highly excited, vibrationally. Relaxation to the ground vibration of the excited electronic state occurs very rapidly (picoseconds). The equilibrium nuclear configuration of the excited state, may or may not be the same as that of the ground state.

Since our point of interest at this moment is on the electronic properties, so the rotational and vibrational parts of the wave function may be ignored and we shall consider only the electronic component of the wave function. Let ψ_a be the electronic component of the wave function. If ψ_a consists of an orbital(o) component (ψ_a^o), and a spin (s) component (ψ_a^s) then

$$\psi_a = \psi_a^o \cdot \psi_a^s \qquad ...(10.14)$$

Since dipole moment operator $(\widehat{M})$, does not couple electronic spins, they must be 'integrated out' of the transition moment expression [(Eq. (10.8)] to give

$$\sqrt{f} = \text{constant} \langle \psi_a^o \, | \, M \, | \, \psi_b^o \rangle \langle \psi_a^s \, | \, \psi_b^s \rangle \qquad ...(10.15)$$

Since the second term $\langle \psi_a^s \, | \, \psi_b^s \rangle$ is the product of $+1/2$ and $-1/2$ spins, it is always odd and zero, i.e. the spins are orthogonal. Thus the second integral is zero if the spin wave functions are ψ_a and ψ_b differ. Hence **transitions between wave functions of different spin are forbidden.** More simply:

$$\Delta s = 0 \qquad ...(10.16)$$

is the spin selection rule for electronic transitions.

10.4 DERIVATION OF SOME SELECTION RULES

(1) For molecules with a centre of symmetry, allowed transitions are $g \rightarrow u$ or $u \rightarrow g$ (g and u refer to gerade and ungerade, which are German for even and odd, respectively). The d and s orbitals are g, and p orbitals are u. All wave functions in a molecule with a centre of symmetry (i) are g or u. **All components of the vector $\widehat{M}$ in a point group containing an inversion centre (i) are necessarily ungerade.**

$$\Gamma_{\psi g} \times \Gamma_{operator} \times \Gamma_x = \Gamma$$
$$u \times u \times u = u \quad \text{forbidden}$$
$$u \times u \times g = g \quad \text{allowed}$$

$$g \times u \times g = u \quad \text{forbidden}$$
$$g \times u \times u = g \quad \text{allowed}$$

This leads to the selection rule that $g \to u$ and $u \to g$ are allowed, but $g \to g$ and $u \to u$ are forbidden. Therefore, $d \to d$ transitions in transition metal complexes with a centre of symmetry (i) are forbidden. Values of $\in$ for the d–d transitions in $[Ni(H_2O)_6]^{2+}$ are ~ 20.

(2) Transitions between states of different multiplicity are forbidden.

(3) Transitions in molecules without a centre of symmetry depend upon the symmetries of the initial and final states. If the direct product of these and any one of $\widehat{M}_x$, $\widehat{M}_y$ or $\widehat{M}_z$ is A_{1g} or A_1 the transition is allowed. If all integrals are odd, the transition is forbidden. Thus for allowed transition

$$\langle \Psi_{\text{ground state}} \mid \widehat{M} \mid \Psi_{\text{excited state}} \cdots \to A_{1g} \text{ or } A_1$$

Let us discuss two types of compounds, to work out the number of allowed electronic transitions.

(i) Aldehydes in which a chromophoric group is responsible for absorption and

(ii) Aromatic compounds (benzene) where π orbitals are delocalised over the whole molecule.

10.4.1 Spectrum of Formaldehyde Molecule (HCHO)

The formaldehyde molecule (or any symmetric ketone) belongs to C_{2v} point group. A molecular orbital description of the valence electrons in this molecule is:

$$(\sigma_{CO})^2 (\sigma_{CO})^2 n^2 (\pi_{CO}^*)^0 (\sigma_{CO}^*)^0$$

The relative energies of these orbitals are indicated in Fig. 10.4. The transitions (1), (2), (3), and (4) are referred to as $n \to \pi^*$, $\pi \to \sigma^*$, $\pi \to \pi^*$ and $\sigma \to \sigma^*$, respectively. The shapes of the MOs are also shown in Fig. 10.4.

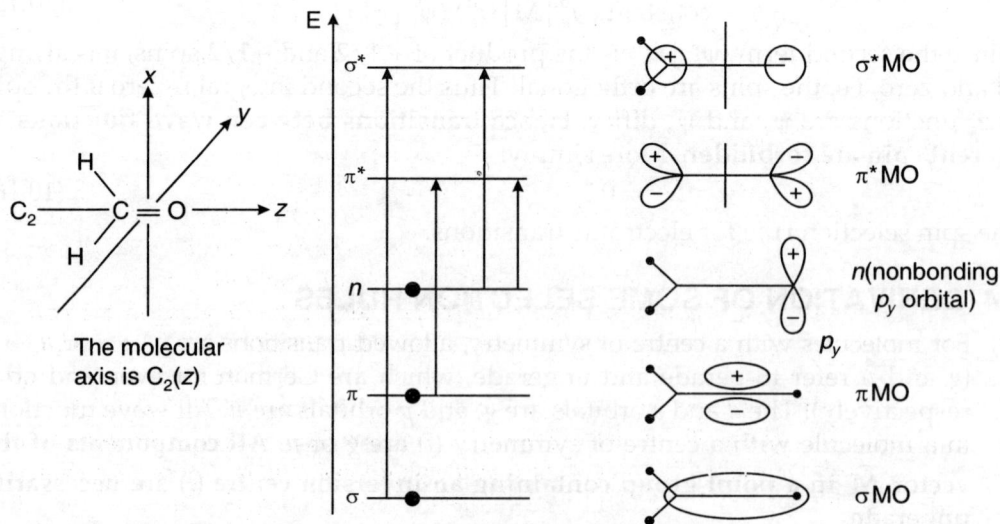

Fig. 10.4 Shapes of the molecular orbitals of formaldehyde and relative energies of the carbonyl molecular orbitals in formaldehyde

The molecular axis C_2 is z-axis, x-axis is $\perp$ to z-axis and in the plane of the paper and y-axis is $\perp$ to the plane of the paper. By convention, the yz plane is selected to contain the four atoms of formaldehyde.

If we perform the operations of C_{2v} point group over the molecular orbitals and assign characters based on how they undergo change, we can find the irreducible representations to which the molecular orbitals belong (if it remains unchanged, the character is +1, if it changes to opposite, the character is –1). The C_{2v} character table is duplicated here in Table 10.1. The MO transformation can be observed by performing $E, C_2, \sigma_V (xz), \sigma_V (yz)$ operations as shown in Table 10.1.

Table 10.1 Character table for the C_{2v} point group*

MOs	E	C_2	$\sigma_V (xz)$	$\sigma_V (yz)$		Corresponding irreducible representation of C_{2v}
σ	+ 1	+ 1	+ 1	+ 1	z	A_1
π	+ 1	– 1	+ 1	– 1	x, R_{xy}	B_1
n	+ 1	– 1	– 1	+ 1	y, R_z	B_2
π^*	+ 1	– 1	+ 1	– 1	x, R_y	B_1
σ^*	+ 1	+ 1	+ 1	+ 1	z	A_1

MOs	E	C_2	$\sigma_V (xz)$	$\sigma_V (yz)$		Corresponding irreducible representation of C_{2v}
σ	+ 1	+ 1	+ 1	+ 1	z	A_1
σ^*	+ 1	+ 1	+ 1	+ 1	z	A_1
π	+ 1	– 1	+ 1	– 1	x, R_y	B_1
π^*	+ 1	– 1	+ 1	– 1	x, R_y	B_1
n	+ 1	– 1	– 1	+ 1	y, R_z	B_2

The symmetry operations $E, C_2, \sigma_V(xz), \sigma_V (yz)$ performed on the π orbital produce the result + 1, – 1, + 1, – 1. This result is identical to that listed for the irreducible representation B_1 in the Table 10.1. The orbital is said to belong to (or to transform as) the symmetry species b_1, the lower case letter being employed for an orbitals and the upper case letters being reserved to describe the symmetry of the entire ground or excited state. Similarly, if the other orbitals of formaldehyde such as n, π^*, σ and σ^* are subjected to the above symmetry operations, it can be shown that those orbitals belong to the irreducible representations $b_2, b_1, a_1,$ and a_1, respectively. Note that a representation does not change sign with rotation about the n-fold axis, but b does.

The symmetry species of a state is the product of the symmetry species of each of the odd electron orbitals. In case of $n \rightarrow \pi^*$ transition, the ground state has both the electrons in the nonbonding orbitals $(n^{1t})\pi^{**}$. Each electron has b_2 symmetry. The symmetry of the spectral state is $b_2 \times b_2 = A_1$ ($b \times b = A, 2 \times 2 = 1 \therefore b_2 \times b_2 = A_1$).

In the excited state, one electron is transferred to the π^* MO orbital $(n^1 \pi^{*1})$. Thus one unpaired electrons unpaired electron is in the n orbital with b_2 symmetry and the other one in the π^* orbital with b_1 symmetry.

The direct product is given by:

$$b_1 \times b_2 = \frac{\begin{array}{cccc} E & C_2 & \sigma_v(xz) & \sigma_v'(yz) \\ (1)(1) & (-1)(-1) & (1)(-1) & (-1)(1) \end{array}}{}$$

$$\text{result} = \quad +1 \qquad +1 \qquad -1 \qquad -1 \quad = A_2$$

The resulting irreducible representation is A_2. The excited state from this transition is thus described as A_2 and the transition as $A_1 \rightarrow A_2$. The spin multiplicity is usually included, so the complete designation becomes $^1A_1 \rightarrow {}^1A_2 \ (n, \pi^*)$.

For an $n \rightarrow \pi^*$ transition, the ground state is A_1 and the excited state is A_2. The character table indicates that no dipole moment component has symmetry A_2, i.e. irreducible representation A_2 does not correspond to any of the vectors in C_{2v} character table and hence this $n \rightarrow \pi^*$ transition is not allowed.

If the state symmetry species (e.g. A_2 above) is not known, the general term symbol Γ is employed.

$$n \xrightarrow{\quad\times\quad} \pi^* \text{ (transition is not allowed electronically)}$$

Now we shall consider $\pi \rightarrow \pi^*$ transition. The ground state configuration is written as

$$\psi_{\text{ground}} = (\pi)^2$$

and the excited state as

$$\psi_{\text{excited}} = (\pi)'\,(\pi^*)'$$

Representation for ground state wave function will be

$$\Gamma_{\text{ground}} = b_1 \times b_1 = A_1$$
$$\Gamma_{\text{excited}} = b_1 \times b_1 = A_1$$

Thus the direct product representation of ψ_{ground} and ψ_{excited} can be expressed as

$$\Gamma_{\text{ground}} \times \Gamma_{\text{excited}} = A_1 \times A_1 = A_1$$

So both the ground and excited states are A_1 and the transition can be expressed as

$$^1A_1 \longrightarrow {}^1A_1$$

The superscripts show that it is a singlet to singlet transition. The product of the ground and excited state is A_1. This corresponds to the Z vector (Table 10.1). This transition will bring change in the dipole moment in the z direction.

As $M_z = A_1$, the total integral equals to

$$\Gamma_{\text{ground}} \cdot M_z \cdot \Gamma_{\text{excited}}$$
$$= A_1 \cdot A_1 \cdot A_1 = A_1$$

This is an allowed transition

For the $\sigma \rightarrow \sigma^*$ transition, i.e. $(\sigma)^2 \, (\sigma^*)^0 \rightarrow (\sigma)^1 \, (\sigma^*)^1$, the ground state and excited state have both symmetry $A_1 \ (a_1 \cdot a_1)$ because both σ and σ^* have a_1 symmetry. Thus both ground and excited states are A_1 and hence the symmetry of the transition is also A_1 $(A_1 \cdot A_1)$. Thus will also bring change in the dipole moment in the z direction and is an allowed transition. Similarly $n \rightarrow \sigma^*$ is an allowed transition.

10.4.2 Spectrum of Benzene Molecule

The MOs of benzene have a_{1u}, e_{1g}, e_{2u} and b_{1g} symmetry. The ground and excited states of benzene is shown in Fig. 10.5.

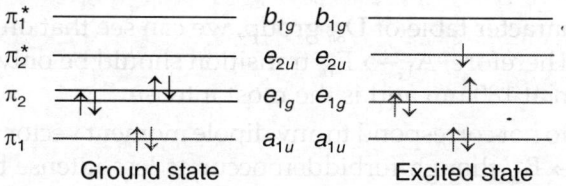

Ground state   Excited state

Fig. 10.5 The ground state and first excited state of benzene

In the ground state of the molecule, out of six electrons, two are filled in the a_{1u} and four are filled in the doubly degenerate e_{1g} state. *An electron present in an orbital (a_{1u} or e_{1g}) has the symmetry of the orbital itself.*

If there is a pair of electrons, the symmetry of this electronic arrangement is obtained by squaring the characters a_{1u} or e_{1g}.   [Electrons are paired in

$$\Gamma_{\text{ground state}} = (a_{1u})^2 (e_{1g})^2 (e_{1g})^2 \qquad a_{1u}, e_{1g}, e_{1g}$$
$$= A_{1g} \times A_{1g} \times A_{1g} \qquad \text{orbitals. So they}$$
$$= A_{1g} \times A_{1g} [A_{1g} \times A_{1g} = A_{1g}] \quad \text{correspond to}$$
$$= A_{1g} \qquad A_{1g} \text{ symmetry}]$$

Thus, since the electrons are filled in pairs, the symmetry of the ground state of benzene is A_{1g}.

In the first excited state one of the electrons from e_{1g} molecular orbital is excited to e_{2u} MO (Fig. 10.5). Since both the MOs are doubly degenerate, the excited state will correspond to different symmetries. This can be worked out by finding the product of the characters of the representations e_{1g} and e_{2u} occupied by the two unpaired electrons. Note that other electrons are paired, and belong to A_{1g} symmetry. This will not affect the total symmetry and hence **the symmetry of the state depends upon the unpaired electrons only**.

Referring to the character table of D_{6h} point group of the benzene molecule, we get the following result.

D_{2h}	E	$2C_6$	$2C_3$	C_2	$3C_2'$	$3C_2''$	σ_h	$3\sigma_v$	$3\sigma_d$	$2S_6$	$2S_3$	i
$\chi(E_{1g})\cdot\chi(E_{2u})$	2×2	1×-1	-1×-1	-2×2	0×0	0×0	-2×-2	0×0	0×0	-1×1	1×1	2×-2
	$= 4$	$= -1$	$= 1$	$= -4$	$= 0$	$= 0$	$= 4$	$= 0$	$= 0$	$= -1$	$= 1$	$= -4$

By using the reduction formula the reducible representation can be reduced to the following irreducible representations.

$$\Gamma = B_{1u} + B_{2u} + E_{1u}$$

Thus the lowest energy excited state of the benzene molecule has two nondegenerate states B_{1u} and B_{2u} and a doubly degenerate state E_{1u}. Therefore the possible transitions are

$$A_{1g} \longrightarrow B_{1u}$$
$$A_{1g} \longrightarrow B_{2u}$$
$$A_{1g} \longrightarrow E_{1u}$$

In order to know whether the transition is allowed or not, Laporte rule is used, i.e.

$$\langle \psi_{\text{ground state}} \mid \widehat{M} \mid \psi_{\text{excited state}} \rangle \text{ must be } A_{1g}$$

The ground state is A_{1g}, then $\langle | \widehat{M} | \psi_{\text{ex. state}} \rangle$ should be A_{1g}. This will be possible only if the representation of the excited state corresponds to the representation of one of the dipole moment vector $x_1 y$ or z.

Now, from the character table of D_{6h} group, we can see that only E_{1u} representation belongs to x and y. Therefore, $A_{1g} \rightarrow E_{1u}$ transition should be only allowed. This occur in benzene spectrum at 180 nm and is the most intense.

Since B_{1u} and B_{2u} do not correspond to any dipole moment vector (x, y, z), the transition $A_{1g} \rightarrow B_{1u}$ and $A_{1g} \rightarrow B_{2u}$ though forbidden occur as less intense bands at 210 nm and 250 nm, respectively. They become partially allowed due to vibronic coupling. The 210 nm band also steals intensity from the neighbouring high intensity band at 180 nm.

A band with very low intensity is deserved at 330 nm. This band has been assigned to the spin forbidden transition $^1A_{1g} \rightarrow {}^3B_{1u}$ singlet – triplet forbidden transitions (singlet)
(singlet) (triplet)
(triplet) may take place when spin – orbit coupling is present and the singlet state may have the same total angular momentum as the triplet state and the two states interact to give rise such transition. But such transitions have low intensity as observed in case of formaldehyde and benzene molecules.

10.5 ELECTRONIC SPECTRA OF METAL COMPLEXES

In a free gaseous metal ion the d orbitals are degenerate. So there will be no spectra from d–d transitions. In the metal complex the electrical field created by the ligands causes splitting of the d orbitals. In an O_h field the pentadentate d orbitals get split up into t_{2g} and e_g sets. Rearrangement of electrons takes place in such a way that the energy is least. In case of d^1 metal ion, the arrangement is $t_{2g}^1 e_g^0$ in an octahedral field. On being excited, the electron in the t_{2g} orbitals absorbs energy equal to the crystal field splitting (Δ_0) and moves to the e_g orbitals. Since the value of Δ_0 is low, the absorption takes place in the visible region and the transition metal complexes are coloured.

The absorption spectra of the octahedral complexes show that the molar absorbance of such d–d transition bands are low. This is because of the selection rules.

In case of octahedral complexes the d–d transitions $(t_{2g} \rightarrow e_g)$ are g–g transitions. This should not cause any change in the dipole moment, i.e. the transition moment (P), $\langle \psi_a | M | \psi_b \rangle = 0$. Thus for octahedral complexes with centre of symmetry (i) d–d transitions should be Laporte forbidden.

But transition metal complexes are coloured. Light absorption does occur even though it is forbidden. This means that Laporte's selection rule is relaxed by the following ways.

1. A complex may undergo some kind of distortion resulting the change of metal–ligand bond length and bond angles. Thus the effective group to which the molecule belongs is not perfect O_h but some lower group which may not have a centre of symmetry (i). Thus the parity selection rule is relaxed by some mechanism which temporarily or permanently removes the centre of symmetry (i).

2. The parity selection rule may be further relaxed when the centre of symmetry (i) of the molecule is temporarily removed by an odd vibration of the molecule.

3. Due to the coupling of the electronic and vibrational wave functions, i.e. due to vibronic coupling, the d–d transitions become partially allowed in octahedral complexes resulting low intensity bands.

4. In the complex, there may be some mixing of d and p orbitals and thus $t_{2g} \rightarrow e_g$ transitions are not purely d–d type transitions but dp–dp type transitions.

5. The intensity of the bands of some complexes is much higher than can be expected from the above three reasons. This can be explained by considering that the metal d orbitals overlap with the ligand orbitals and thus the pure d orbital character is lost. So the transition is not purely d–d type but dp–dp type.

6. The number of unpaired spins (i.e. spin multiplicity) should not change during d–d transition. In $[Mn(H_2O)_6]^{2+}$ (d^5 configuration) the ground state ($^6A_{1g}$) has spin multiplicity six but the excited state has spin multiplicity four ($^4T_{1g}$). Thus the electronic transition in $[Mn(H_2O_6]^*$ is doubly forbidden (Laporte and spin) and hence the intensity of the band is very low. This is the reason for the very light pink colour of manganese (II) salts.

7. Simultaneous excitation of more than one electron does not take place. However, low intensity bands, corresponding to two electron transitions, are observed in complex compounds.

The spectrum of titanium (II) compounds ($3d^2$), $TiCl_6^{2-}$, has peaks at 7600 (9.1) and 14500 (29.4) cm^{-1}, respectively. The lower energy band is presumably the $^3T_{2g} \leftarrow {}^3T_{1g}$ transition. Two more transitions $^3A_{2g} \leftarrow {}^3T_{1g}$ and $^3T_{1g}$ (P) $\leftarrow {}^3T_{1g}$ are expected but only one more band below 31000 cm^{-1} is observed. Since neither of these transitions are expected to have energies in excess of 31000 cm^{-1}, the band at 14500 cm^{-1} may contain both transitions. In the strong field limit, the $^3A_{2g} \leftarrow {}^3T_{1g}$ transition is forbidden, since it is a two-electron transition ($t_{2g}^2 \rightarrow e_g^2$). Thus it may be weak and be observed by the $3T_{1g}$ (P) $\leftarrow {}^3T_{1g}$ transition.

The occurance of two: electron transitions result from the mixing of either.

(i) the ground state with another configuration or

(ii) the mixing of excited state with another configuration, such that the actual transition involves only a one electron excitation.

In this case the ground state configuration t_{2g}^2 ($3T_{1g}$) is mixed with excited $t_{2g} e_g$ ($^3T_{1g}$ (P)) configuration, and the actual observed transition to e_g^2 occurs from $L_{2g} e_g$ part of the ground state. The intensity of such transitions is a function of the degree of mixing of the configurations.

Thus to be non-zero value, the direct product of the representations of the initial (A_2) and final curve functions (T_1) with particular dipole moment operator component (T_2) must contain totally symmetric representation A_1.

Since A_1 representation is present, it follows that $^4T_1 \leftarrow {}^4A_2$ transition is allowed and it is purely an electronic transition.

For the $T_2 \leftarrow A_2$ transition, the direct product $A_2 \times T_2 \times T_2$ (here $\widehat{M}$ components transform as T_2) reduces to $A_2 + E + T_1 + T_2$, i.e.

$$A_2 \times T_2 \times T_2 = A_2 \times (T_2 \times T_2)$$
$$= A_2 \times (A_1 + E + T_1 + T_2)$$
$$= A_2 + E + T_2 + T_1$$
$$= A_2 + E + T_1 + T_2$$

Since there is no A_1 component, the $^4T_2 \leftarrow {}^4A_2$ transition is forbidden. Mixing p-character into the wave functions will not help for this type of transition and as a result the $^4T_2 \leftarrow {}^4A_2$ transition is still forbidden. Accordingly, the ϵ for $^4T_1 \leftarrow {}^4A_2$ transition is 10 times greater than that of $^4T_2 \leftarrow {}^4A_2$. However, the latter gains most of its intensity by vibronic coupling.

10.5.1 Molecules without Inversion Centres (acentric or noncentro-symmetric complexes)

When a complex lacks a centre of symmetry (i) then the parity selection rule forbidding d–d transition is relaxed to some extent and d–d transition becomes possible since neither ψ_a nor ψ_b is of pure gerade nature (transition moment $P = \langle \psi_a | M | \psi_b \rangle$). If two different electronic states of d^n configuration of a metal contain different amounts of p character, a transition from one orbital to the other will be to a certain extent a $d \rightarrow p$ or $p \rightarrow d$ transition, which is highly allowed even in the free atom since the d orbitals are even to inversion and p orbitals are odd.

The triply degenerate t_2 orbitals of a tetrahedral complex do not contain only the pure d_{xy}, d_{yz}, d_{zx} orbitals. It is a set consisting of mixed character of d and p orbitals. Such dp mixing augments the magnitude of the transition probability integral.

Let us consider tetrahedral complex ion $[CoCl_4]^{2-}$. The ground state belongs to A_2 and two excited states are T_1 and T_2. In the group T_d, $\widehat{M}$ components (M_x, M_y, M_z) transform as T_2.

For the $T_1 \leftarrow A_2$ transition, the direct product

$$A_2 \times T_2 \times T_1 = A_2 \times (T_2 \times T_1)$$
$$= A_2 \times (A_2 + E + T_1 + T_2)$$
$$= A_1 + E + T_2 + T_1$$
$$= A_1 + E + T_1 + T_2$$

Since totally symmetric representation A_1 is contained in the reduction result for the $T_1 \leftarrow A_2$ transition, hence transition moment integral does not vanish. Hence it is an allowed transition.

In the group C_{4V}, the components M_x, M_y of the dipole moment operator ($\widehat{M}$) transform as E, and M_z transforms as A_1. For a transition $\psi_a - \psi_b$ to be orbitally allowed, the product $\Gamma_a \cdot \Gamma_b$ must contain E and/or A_1. Under such circumstances the triple direct product $\langle \psi_a | M | \psi_b \rangle$ must contain A_1.

Transition $\psi_b \leftarrow \psi_a$	Direct product $\Gamma_a \times \Gamma_b$	Orbitally allowed or forbidden	Polarisation
$A_2 \leftarrow A_1$	$A_1 \times A = A_2$	Forbidden	
$E \leftarrow A_2$	$A_2 \times E = E$	Allowed	x, y
			x, y
$B_2 \leftarrow B_1$	$B_1 \times B_2 = A_2$	Forbidden	
$B_2 \leftarrow B_2$	$B_2 \times B_2 = A_1$	Allowed	z
$E \leftarrow E$	$E \times E = A_1 + A_2 + B_1 + B_2$	Allowed	z

Note: $A \times A = A$; $2 \times 1 = 2$; $E \times A = E$; $B \times B = A$

$E \times E = A_1 + A_2 + B_1 + B_2$ in C_{4V} symmetry

Tetrachloro cuprate ion, $[CuCl_4]^{2-}$, a tetrahedral complex has a $^2T_2$ ground state and is expected to be Jahn–Teller distorted. The molecule possesses D_{2d} symmetry.

The energy level scheme is shown in Fig. 10.6.

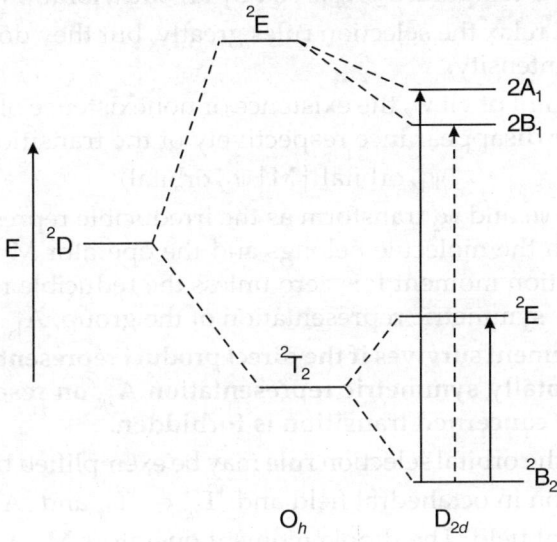

Fig. 10.6 Energy level scheme for the D_{2d} $CuCl_4^{2-}$ ion

The dipole moment operators transform as B_2 ($\parallel$) + E ($\perp$) and the selection rules are:

Transition	Direct product of the wave electronic functions	Direct product with dipole moment operators		Polari-sation
		B_2	E	
$^2E \leftarrow {}^2B_2$	E × B = E	E × B_2 = E	E × E = A_1^+ $A_2 + B_1 + B_2$	$\perp$
$^2B_1 \leftarrow {}^2B_2$	$B_1 \times B_2 = A_2$	$A_2 \times B_2 = B_1$	$A_2 \times E = E$	Forbidden
$^2A_1 \leftarrow {}^2B_2$	$A_1 \times B_2 = B_2$	$B_2 \times B_2 = A_1$	$B_2 \times E = E$	$\parallel$

10.5.2 Molecules with Inversion Centres (centrosymmetric complexes)

In a complex that possesses a centre of symmetry (i), all states arising from a d^n configuration must have g character inherent in the d orbitals. Therefore, for the centrosymmetric molecules (O_h, D_{4h} symmetry), the following transitions are not feasible.

$$\text{gerade} \longleftrightarrow\!\!\!\!| \longrightarrow \text{gerade} \quad \text{and} \quad \text{ungerade} \longleftrightarrow\!\!\!\!| \longrightarrow \text{ungerade}$$

Such transitions are Laporte forbidden. However, gerade $\longleftrightarrow$ ungerade and ungerade $\longleftrightarrow$ gerade transitions are allowed.

But transition metal complexes are coloured. Adsorption does occur, even though it is forbidden and adsorption intensities in centrosymmetric transition metal complexes are small, and smaller than similar adsorption in acentric transition metal complexes.

There are two common mechanisms by which the parity selection rule is relaxed.

1. An octahedral complex may undergo some kind of distortion resulting the removal of centre of symmetry (i) permanently or temporarily. Thus the effective group to which the molecule belongs is no longer O_h but lower group which may not have a centre of symmetry (i).

2. The parity selection rule may be further relaxed when the centre of symmetry (i) of the molecule is temporarily reserved by an odd vibration of the molecule.

Such effects do not relax the selection rules greatly, but they do provide a pathway for some absorption intensity.

From symmetry point of view, the existence or nonexistence of transitions depend upon the survived or disappearance respectively of the transition moment integral

$$\langle \psi_1, \text{orbital} \,|\, M \,|\, \psi_2, \text{orbital} \rangle$$

If the wave functions ψ_1 and ψ_2 transform as the irreducible representations Γ_1 and Γ_2 of the group to which the molecule belongs and the operator M transforms as Γ_M of that group, the transition moment P is zero unless the reducible representation $\Gamma_1 \Gamma_M \Gamma_2$ contains the totally symmetric representation of the group, A_{1g}.

Thus the matrix element survives if the direct product representation of the integral $\psi_1 M \psi_2$ contains a totally symmetric representation A_{1g} on resolution. Its absence will indicate that the concerned transition is forbidden.

The application of the orbital selection rule may be exemplified by considering $^2E_g \leftarrow {}^2T_{2g}$ transition for d^1 ion in octahedral field and $^3T_{2g} \leftarrow {}^3T_{1g}$ and $^3A_{2g} \leftarrow {}^3T_{1g}$ transitions for d^2 ion in octahedral field. The dipole moment operators M_x, M_y and M_z transform as T_{1u}, i.e. M transforms as T_{1u} in octahedral symmetry. For $^2E_g \leftarrow {}^2T_{2g}$ transition in d^1 ion, the direct product

$$
\begin{aligned}
T_{2g} \times T_{1u} \times E_g &= T_{2g} \times (T_{1u} \times E_g) \\
&= T_{2g} \times (T_{1u} + T_{2u}) \ [\text{In } O_h, E_g \times T_{1u} = T_{1u} + T_{2u}] \\
&= (T_{2g} \times T_{1u}) + (T_{2g} \times T_{2u}) \\
&= (A_{2u} + E_u + T_{1u} + T_{2u}) + (A_{1u} + E_u + T_{1u} + T_{2u}] \\
&= A_{1u} + A_{2u} + 2E_u + 2T_{1u} + 2T_{2u}
\end{aligned}
$$

As the direct product does not contain A_{1g}, so $^2E_g \leftarrow 2T_{2g}$ transition is forbidden. In case of d^2 ion, the direct product

$$
\begin{aligned}
T_{1g} \times T_{1u} \times T_{2g} &= T_{1g} \times (T_{1u} \times T_{2g}) \\
&= T_{1g} \times (A_{2u} + E_u + T_{1u} + T_{2u}) \\
&= T_{2u} + T_{1u} + T_{2u} + A_{1u} + E_u + T_{1u} + T_{2u} + A_{2u} + E_u + T_{1u} + T_{2u} \\
&= 4T_{2u} + 3T_{1u} + A_{1u} + 2E_u + A_{2u}
\end{aligned}
$$

$T_{1g} \times T_{1u} \times T_{2g}$ product also does not contain A_{1g}, so $^3T_{2g} \leftarrow {}^3T_{1g}$ transition is also not allowed.

Similarly the direct product

$$
\begin{aligned}
T_{1g} \times T_{1u} \times A_{2g} &= T_{1g} \times (T_{1u} \times A_{2g}) \\
&= T_{1g} \times T_{2u} \\
&= A_{2u} + E_u + T_{1u} + T_{2u}
\end{aligned}
\qquad
\begin{bmatrix}
\because T \times A = T \\
1 \times 2 = 2 \\
u \times g = u
\end{bmatrix}
$$

does neither contain A_{1g} nor even A_1. So $^3A_{2g} \leftarrow {}^3T_{1g}$ transition is also forbidden.

10.6 VIBRONIC COUPLING IN CENTROSYMMETRIC MOLECULES

The total wave function (ψ) for a molecule can be written as the product of an electronic wave function (ψ^E), a vibrational wave function (ψ^V) and rotational wave function (ψ^R)

$$\psi = \psi^E \cdot \psi^V \cdot \psi^R$$

The role of rotational wave function (ψ^R) is very minor and can be ignored. The different normal modes of vibration of the atoms present in the metal complex compound temporarily remove the centre of symmetry (i) of the molecule and in this way change the gerade nature of the d orbital. Thus the vibrationally excited molecule is no longer restricted by the parity rule and can undergo an electronic transition. In practice, it is assumed that the molecule is electronically excited with simultaneous excitation of an appropriate odd vibration. Such a mechanism is known as **vibronic coupling** (**VIBR**ational electr**ONIC**), Eq. (10.15) can now be rewritten, ignoring spin as:

$$\sqrt{} = \text{constant} \cdot \langle \psi_a^V \, \psi_a^0 \,|\, M \,|\, \psi_b^V \, \psi_b^0 \rangle \qquad \qquad ...(10.17)$$

and the direct product derivation must now include the direct products of the ground and excited vibrational contributions.

Normally in absorption, we deal with molecules in their ground vibrational states, with $V'' = 0$ states, which are always totally symmetric.

The ground vibrational state, $V'' = 0$ of any vibration, whatever its symmetry, is totally symmetric. **Hence ψ_{av} can be ignored if it is a vibrational ground state ($V'' = 0$), since it will not contribute to the direct product.** For the integral in Eq. (10.17) to be nonzero, the direct product of the initial and final electronic wave functions with the component of the dipole moment operator **must span the representation of an odd vibrations of the molecule.** For a ML_n molecule (L is a polyatomic ligand), the vibration of the metal ligand framework may couple to the electronic motion.

The relaxation in parity selection rule through vibronic coupling mechanism can be illustrated through an example of ML_6 (O_h), a centrosymmetric molecule. The symmetries of all the normal modes are $A_{1g} + E_g + T_{2g} + 2T_{1u} + T_{2u}$. This result suggests that the odd (u) vibrations of the ML_6 skeleton which temporarily remove the centre of inversion (i) are T_{1u} and T_{2u} type. Only the vibronic couplings involving such vibrations can allow the transitions by partially relaxing the party rule. Let us consider some of the transitions given in the following table for an illustrative account.

Transition $\psi_b \leftarrow \psi_a$	Direct product $\Gamma_j = \Gamma_{(\psi_b)} \Gamma_{(\psi_a)}$	Direct product of Γ_j with T_{1u} (dipole moment operator)	Remarks (Nature of relaxation)
1. $A_{2g} \leftarrow A_{2g}$	$A_{2g} \times A_{2g} = A_{1g}$	$A_{1g} \times T_{1u} = T_{1u}$	Vibronically allowed
2. $E_g \leftarrow E_g$	$E_g \times E_g$ $= A_{1g} + A_{2g} + E_g$	$(A_{1g} + A_{2g} + E_g) \times T_{1u}$ $= T_{1u} + T_{2u} + T_{1u} + T_{2u}$ $= 2T_{1u} + 2T_{2u}$	Vibronically allowed
3. $A_{2u} \leftarrow A_{2g}$	$A_{2u} \times A_{2g} = A_{1u}$	$A_{1u} \times T_{1u} = T_{1g}$	Forbidden
4. $A_{1g} \leftarrow A_{2g}$	$A_{1g} \times A_{2g} = A_{2g}$	$A_{2g} \times T_{1u} = T_{2u}$	Vibronically allowed
5. $T_{2g} \leftarrow A_{2g}$	$T_{2g} \times A_{2g} = T_{1g}$	$T_{1g} \times T_{1u}$ $= A_{1u} + E_u + T_{1u} + T_{2u}$	Vibronically allowed
6. $T_{2g} \leftarrow A_{2u}$	$T_{2g} \times A_{2u} = T_{1u}$	$T_{1u} \times T_{1u}$ $= A_{1g} + E_g + T_{1g} + T_{2g}$	Orbitally allowed
7. $T_{2u} \leftarrow A_{1g}$	$T_{2u} \times A_{1g} = T_{2u}$	$T_{2u} \times T_{1u}$ $= A_{2g} + E_g + T_{1g} + T_{2g}$	Forbidden

Note:

1. If the direct product of the third column ($\Gamma_j \cdot T_{1u}$) contains A_{1g} than the transition is orbitally allowed and expected to have high intensity ($\in$, 10^3–10^4 l mol^{-1}).

2. If the direct product of the third column does not contain A_{1g} but does contain one of the odd (u) vibrations of the framework, in this case T_{1u} or T_{2u} or both (T_{1u} and T_{2u}) then the transition will be vibronically allowed. Transitions 1, 2, 4, 5 are all vibronically allowed transitions and have low intensity ($\in$, $1 - 100\, l \cdot cm^{-1} \cdot mol^{-1}$).

3. Transitions 3 and 7 are both orbitally and vibronically forbidden and could not appear.

Some examples are discussed here wherein the electronically forbidden transitions become partially allowed through vibronic interactions.

[Ti(H₂O)₆]³⁺

The ground state ψ_a transforms as T_{2g} and the excited state ψ_b as E_g. For $^2E_g \leftarrow {}^2T_{2g}$ transition, we may write

$$\langle \psi_b^V \, \psi_{E_{2g}} \mid T_{1u} \mid \psi_a^V \, \psi_{2T_{2g}} \rangle$$

when an ion be in its ground vibrational state, then ψ_a^V is always symmetric, i.e. of A_{1g} symmetric. It can be ignored as it does not contribute to the direct product.

For $\quad \langle \psi_b^V \, \psi_{Eg} \mid T_{1u} \mid \psi_{T_{2g}} \rangle$ the direct product

$$\Gamma_{\psi_b^V} \times E_g \times (T_{1u} \times T_{2g}) = \Gamma_{\psi_b^V} \times E_g \times (A_{1u} + E_u + T_{1u} + T_{2u})$$
$$= \Gamma_{\psi_b^V} \times (E_u + A_{1u} + A_{2u} + E_u + T_{1u} + T_{2u} + T_{1u} + T_{2u})$$
$$= \Gamma_{\psi_b^V} \times (A_{1u} + A_{2u} + 2E_u + 2T_{1u} + 2T_{2u})$$

The symmetries of the normal modes of an octahedral ML₆ molecule are A_{1g}, E_g, T_{2g}, $2T_{1u}$, T_{2u} (Table 7.8).

$\Gamma_{\psi_b^V}$ **corresponds to the normal modes of T_{1u} and T_{2u} and can vibronically couple leading to non-zero value** of electric dipole transition integral.

This permits a d–d transition band. This is shown below:

$$(T_{1u} + T_{2u})\,[A_{1u} + A_{2u} + 2E_u + 2T_{1u} + 2T_{2u}] = A_{1g} + A_{1g} + \dots \qquad [\because T_{1u} \times T_{1u} = A_{1g} + \dots$$
$$T_{2u} \times T_{2u} = A_{1g} + \dots]$$

[V(H₂O)₆]³⁺

For the first transition $^3T_{2g} \leftarrow {}^3T_{1g}$, we may write

$$\langle \psi_b^V \, \psi_{3T_{2g}} \mid T_{1u} \mid \psi_{3T_{1g}} \rangle$$

Thus the representations become

$$\Gamma_{\psi_{bV}} \times T_{2g} \times (T_{1u} \times T_{1g}) = \Gamma_{\psi_b^V} \times T_{2g} \times (A_{1u} + E_u + T_{1u} + T_{2u})$$
$$= \Gamma_{\psi_b^V} \times (T_{2u} + T_{1u} + T_{2u} + A_{2u} + E_u + T_{1u} + T_{2u}$$
$$+ A_{1u} + E_u + T_{1u} + T_{2u})$$
$$= \Gamma_{\psi_b^V} \times (A_{1u} + A_{2u} + 2E_u + 3T_{1u} + 4T_{2u})$$

The symmetries of the normal modes of an octahedral ML₆ molecule are A_{1g}, E_g, $2T_{1u}$, T_{2g} and T_{2u} (Table 7.8). Now $\Gamma_{\psi_b^V}$ **corresponds to the normal modes of T_{1u} and T_{2u} symmetries, can vibronically couple leading to nonzero value of electric dipole transition integral.** This permits a d–d transition band.

$[Co(NH_3)_6]^{3+}$

The ground state ψ_a transform as $^1A_{1g}$ and the two excited states with the same spin (S=0) belong to T_{1g} and T_{2g}. T_{1u} is the dipole moment operator. Thus, for the $^1A_{1g} \rightarrow {}^1T_{1g}$ transition the direct product representation of ψ_b T_{1u} ψ_a is given by

$$\Gamma\,[\psi_b\,M\,\psi_a] = T_{1g} \times (T_{1u} \times A_{1g})$$
$$= T_{1g} \times (T_{1u})$$
$$= T_{1g} \times T_{1u}$$

This can be reduced to $A_{1u} + E_u + T_{1u} + T_{2u}$

Thus, if there are any normal vibrations whose first excited states belong to any of these representations, there will be nonvanishing intensity integrals. The symmetries of normal modes of an octahedral ML_6 molecule are A_{1g}, E_g, T_{2g}, $2T_{1u}$, $2T_{2u}$.

Thus, while the pure electronic transition $^1A_{1g} \rightarrow {}^1T_{1g} / {}^1T_{1g}$ is not allowed, all the transitions in which there is simultaneous excitation of a vibration of T_{1u} or T_{2u} symmetry are allowed.

Similarly, for an $^1A_{1g} \rightarrow {}^1T_{2g}$ transition

$$\Gamma\,[\psi_b\,M\,\psi_a] = T_{2g} \times (T_{1u} \times A_{1g})$$
$$= T_{2g} \times T_{1u}$$
$$= A_{2u} + E_u + T_{1u} + T_{2u}$$

Thus the $^1A_{1g} \rightarrow {}^1T_{2g}$ transition can also occur so long as there is simultaneous excitation of a T_{1u} or T_{2u} vibration.

10.7 VIBRONIC POLARISATION

Some transition metal complexes in their crystalline states can or can not absorb polarised light depending upon the orientation of the crystal with respect to plane of polarisation. The phenomenon is called **dichroism**. For an O_h complex, the direction of the electric vector of light along x, y and z axes is the same. This is possible as the axes are interchangeable by the symmetry operations of the molecule. However, in case of less symmetrical complexes where x, y and z do not belong to the same representation, we encounter the phenomenon of polarisation.

Let the sample is a single crystal in which all of the molecules have the same orientation relative to the crystallographic axes. We orient the crystal in such a way that the direction of the electric vector of the light will correspond to x, y or z direction in a coordinate for the molecule. It is then possible that some transition may occur for only one or two of these orientations but not for all three.

Let us consider $trans$-$[Co(en)_2Cl_2]Cl$ (Fig. 10.7) with a set of coordinate axes.

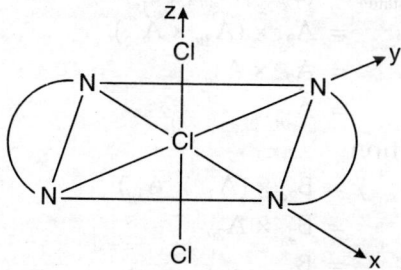

Fig. 10.7 A sketch of the $trans$-$[Co(en)_2Cl_2]^+$ ion showing a set of reference axes

The symmetry of this complex is D_{4h} and has a centre of symmetry (i). So d–d transition will occur if there is vibronic coupling. The ground state of the spin paired cobalt (III) ion is $^1A_{1g}$ both in O_h and D_{4h}. The first two singlet excited states in O_h symmetry are $^1T_{1g}$ and $^1T_{2g}$. These two $^1T_{1g}$ and $^1T_{2g}$ states split into $(^1A_{2g} + \,^1E_g)$ and $(^1B_{2g} + \,^1E_g)$ in D_{4h} respectively (Fig. 10.8).

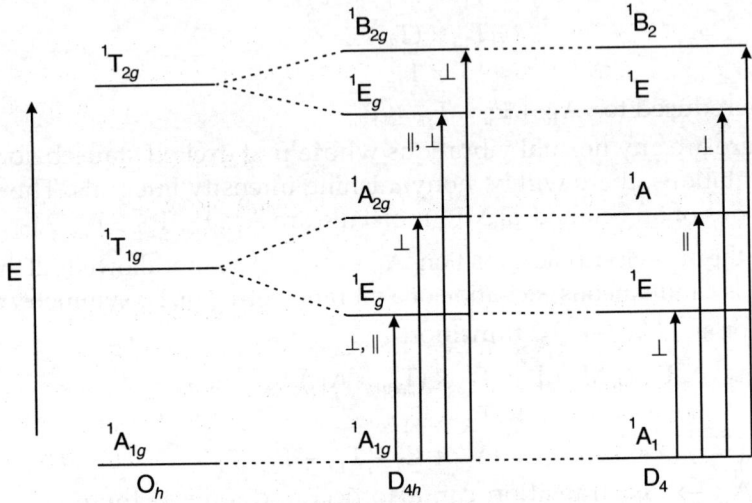

Fig. 10.8 The energy levels and transition polarisations in *trans*-[Co(en)$_2$Cl$_2$]$^+$ in O_h, D_{4h} and D_4 groups

Symmetry	Ground state	First excited state	Second excited state	Possible transitions
O_h	$^1A_{1g}$	$^1T_{1g}$	$^1T_{2g}$	$^1A_{1g} \rightarrow \,^1T_{1g}$ $^1A_{1g} \rightarrow \,^1T_{2g}$
D_{4h}	$^1A_{1g}$	$^1A_{2g} + \,^1E_g$	$^1B_{2g} + \,^1E_g$	$^1A_{1g} \rightarrow \,^1A_{2g}$ $^1A_{1g} \rightarrow \,^1B_{2g}$ $^1A_{1g} \rightarrow \,^1E_g$ $^1A_{1g} \rightarrow 1E_g$

The dipole moment operator T_{1u} in O_h symmetry transforms as A_{2u} (z) + E_u (x, y) in D_{4h} symmetry. For the representations of the purely electronic dipole moment integral, we have:

For $^1A_{2g} \leftarrow \,^1A_{1g}$ transition

$$\langle \psi_{\text{excited state}} \,|\, Z \,|\, \psi_{\text{ground state}} \rangle = \langle \psi_{A_{2g}} \,|\, A_{2u} \,|\, \psi_{A_{1g}} \rangle$$
$$= A_{2g} \times (A_{2u} \times A_{1g})$$
$$= A_{2g} \times A_{2u}$$
$$= A_{1u}$$

For $^1B_{2g} \leftarrow \,^1A_{1g}$ transition

$$\langle \psi_{B_{2g}} \,|\, A_{2u} \,|\, \psi_{A_{1g}} \rangle = B_{2g} \times (A_{2u} \times A_{1g})$$
$$= B_{2g} \times A_{2u}$$
$$= B_{1u}$$

For $\quad {}^1E_g \leftarrow {}^1A_{1g}$ transition

$$\langle \psi_{E_g} \mid A_{2u} \mid \psi_{A_{1g}} \rangle = E_g \times (A_{2u} \times A_{1g})$$
$$= E_g \times A_{2u}$$
$$= E_u$$

For $\quad {}^1A_{2g} \leftarrow {}^1A_{1g}$ transition

$$\langle \psi_{\text{excited state}} \mid (x, y) \mid \psi_{\text{ground state}} \rangle$$
$$= \langle \psi_{A_{2g}} \mid E_u \mid \psi_{A_{1g}} \rangle$$
$$= A_{2g} \times (E_u \times A_{1g})$$
$$= A_{2g} \times E_u$$
$$= E_u$$

For $\quad {}^1B_{2g} \leftarrow {}^1A_{1g}$ transition

$$\langle \psi_{B_{2g}} \mid E_u \mid \psi_{A_{1g}} \rangle = B_{2g} \times (E_u \times A_{1g})$$
$$= B_{2g} \times E_u$$
$$= E_u$$

For $\quad {}^1E_g \leftarrow {}^1A_{1g}$ transition

$$\langle \psi_{E_g} \mid E_u \mid \psi_{A_{1g}} \rangle = E_g \times (E_u \times A_{1g})$$
$$= E_g \times E_u$$
$$= A_{1u} + A_{2u} + B_{1u} + B_{2u}$$

In D_{4h} symmetry of $[ML_4 X_2]$ type, complex in normal modes of vibrations are A_{2u}, B_{2u} and E_u. The selection rule for vibronic transition is that the direct product representation of the product of the initial representation ($\psi_{\text{ground state}}$), final wave function ($\psi_{\text{excited state}}$) and the dipole moment operator (z, (x, y) component) must include an irreducible representation (IR) which is spanned by one or the other normal mode of vibrations of the species.

We thus tabulate the ingredients of the selection rule (Table 10.2).

Table 10.2 Polarised absorption by *trans*-[Co(en)$_2$Cl$_2$]Cl crystals

Transitions	Direct product of electronic wave functions	Symmetry of dipole moment operator		Final direct product with dipole movement operator	Symmetry of normal modes of vibrations	Polarised absorption, i.e. polarisation with vibronic coupling
		z component	(x, y) component			
${}^1A_{2g} \leftarrow {}^1A_{1g}$	$A_{2g} \times A_{1g} = A_{2g}$	A_{2u}	E_u	$A_{2g} \times A_{2u}$ $= A_{1u}$	A_{2u}, B_{2u} and E_u	Forbidden Allowed, i.e. (x, y) polarised, i.e. $\perp$
${}^1B_{2g} \leftarrow {}^1A_{1g}$	$B_{2g} \times A_{1g} = B_{2g}$	A_{2u}	E_u	$B_{2g} \times A_{2u} = B_{1u}$ $B_{2g} \times E_u = E_u +$		Forbidden Allowed, i.e. (x, y) polarised, i.e. $\perp$
${}^1E_g \leftarrow {}^1A_{1g}$	$E_g \times A_{1g} = E_g$	A_{2u}	E_u	$E_g \times A_{2u} = E_u$ $E_g \times E_u = A_{1u} +$ $A_{2u} +$ $B_{1u} +$ B_{2u}		Allowed, z polarised, i.e. $\parallel$ Allowed, (x, y) polarised, i.e. $\perp$

Fig. 10.9 shows the polarised absorption spectra of trans $[Co(en)_2Cl_2]Cl.HCl.2H_2O$

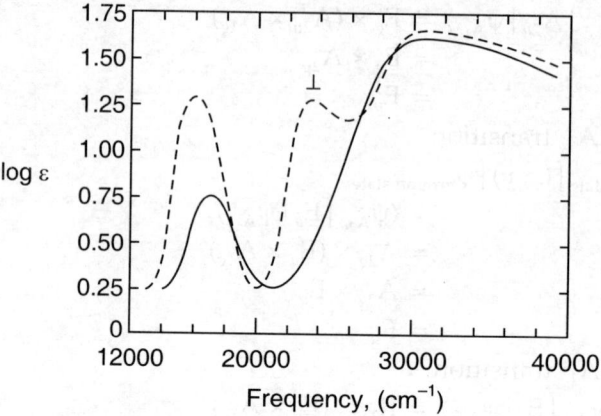

Fig. 10.9 The polarised absorption spectrum of a crystal of the complex *trans*-Co(en)$_2$Cl$_2$]$^+$ ion. The full line shows the spectrum with light polarised parallel or nearly parallel to the Cl–Co–Cl axis and the dashed line shows the spectrum with light polarised ⊥ to the Cl–Co–Cl axis

The lowest energy band appears at ~ 16000 cm^{-1} in both polarisations. This implies that the transition involves the ^1E_g component of the $^1T_{1g}$ level (refer to Fig. 10.8).

$$^1E_g \leftarrow {}^1A_{1g}$$

The second band at ~ 23000 cm^{-1} appears only in perpendicular polarisation and is evidently the A_{2g} component of $^1T_{1g}$.

The high energy absorption appears in both polarisation and is therefore attributed to the E_g component of $^1T_{2g}$.

The B_{2g} component of $^1T_{2g}$ is not observed but may be close to the E_g transition that it is not resolved.

[Co(NH$_3$)$_4$Cl$_2$]Cl

This compound of D_{4h} symmetry possesses $^1A_{1g}$ as its ground state and $^1A_{2g}$ and ^1E_g as two of its upper states emerging from $^1T_{1g}$ state of the O_h complex (Fig. 10.10).

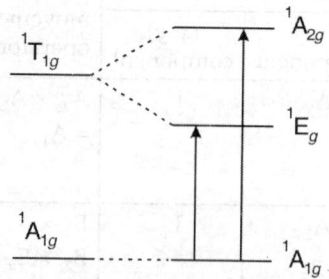

Fig. 10. 10 The energy levels of [Co(NH$_3$)$_4$Cl$_2$]Cl in O_h and D_{4h} groups

For any vibronic *d–d* transitions, there must be some interaction between vibrational wave function with the electronic wave functions, both in the ground state and upper excited states.

For $^1A_{2g} \leftarrow {}^1A_{1g}$ and $^1E_g \leftarrow {}^1A_{1g}$ transitions, we consider the following two transition moment integrals, viz.

$$\langle \psi_{A_{2g}} \, \psi_{ex}^V \,|\, Z \,|\, \psi_{A_{1g}} \, \psi_{gr}^V \rangle$$

$$\langle \psi_{E_g} \, \psi_{ex}^V \,|\, E_u \,|_{(x,\,y)} \, \psi_{A_{1g}} \, \psi_{gr}^V \rangle$$

It is to be noted that in D_{4h} point group z, x and y in the electric dipole moments transform as A_{2u} and E_u (with x and y as bases). Furthermore, the ground state vibrational wave function ψ_{gr}^V is completely symmetrical, i.e. its belongs to A_{1g}.

Exposure to z polarised radiation:

The direct product of the symmetries of the components in the integrands are

$$\Gamma_{\psi_{ex}^V} \times A_{2g} \times A_{2u} \times (A_{1g} \times A_{1g})$$
$$= \Gamma_{\psi_{ex}^V} \times A_{2g} \times (A_{2u} \times A_{1g})$$
$$= \Gamma_{\psi_{ex}^V} \times A_{2g} \times A_{2u}$$
$$= \Gamma_{\psi_{ex}^V} \times A_{1u}$$

In D_{4h} complex, the possible normal modes of vibrations are $2A_{1g}$, B_{1g}, E_g, $2A_{2u}$, B_{2u}, $3E_u$. Since no normal modes of A_{1u} exists, so z polarisation transition $^1A_{2g} \leftarrow {}^1A_{1g}$ is not feasible (possible).

For $\qquad \Gamma_{\psi_{ex}^V} \times E_g \times A_{2u} \times (A_{1g} \times A_{1g})$
$$= \Gamma_{\psi_{ex}^V} \times E_g \times (A_{2u} \times A_{1g})$$
$$= \Gamma_{\psi_{ex}^V} \times E_g \times A_{2u}$$
$$= \Gamma_{\psi_{ex}^V} \times E_u \qquad [E_g \times A_{2u} = E_u]$$

Since their exists a normal mode with E_u symmetry

$\therefore \qquad\qquad \Gamma_{\psi_{ex}^V} \times E_u = E_u \times E_u = A_{1g} + A_{2g} + B_{1g} + B_{2g}$

Therefore, the z polarised transition $^1E_g \leftarrow {}^1A_{1g}$ is feasible (possible)

Exposure to *xy* plane polarised radiations:

$$\Gamma_{\psi_{ex}^V} \times A_{2g} \times E_u \times (A_{1g} \times A_{1g})$$
$$= \Gamma_{\psi_{ex}^V} \times A_{2g} \times (E_u \times A_{1g})$$
$$= \Gamma_{\psi_{ex}^V} \times A_{2g} \times E_u$$
$$= \Gamma_{\psi_{ex}^V} \times E_u$$

Vibronic function involving in normal mode E_u leads to a final product

$$E_u \times E_u = A_{1g} + A_{2g} + B_{1g} + B_{2g}$$

So xy polarised radiation will cause a transition $^1A_{2g} \leftarrow {}^1A_{1g}$

For $\qquad \Gamma_{\psi_{ex}^V} \times E_g \times E_u \times (A_{1g} \times A_{1g})$
$$= \Gamma_{\psi_{ex}^V} \times E_g \times (E_u \times A_{1g})$$
$$= \Gamma_{\psi_{ex}^V} \times (E_g \times E_u)$$
$$= \Gamma_{\psi_{ex}^V} \times (A_{1u} + A_{2u} + B_{1u} + B_{2u})$$

$$[\text{In } D_{4h} \text{ symmetry } E_g \times E_u = A_{1u} + A_{2u} + B_{1u} + B_{2u}]$$

Vibronic function involving the normal mode A_{2u} (or B_{1u}) leads to a final result containing completely symmetric component (A_{1g}) [In D_{4h} symmetry of $ML_4 X_2$ type complex the normal modes of vibrations are A_{2u}, B_{2u} and E_u]

$\psi_1^{orbital}$ transforms as irreducible representation as Γ_1

$\psi_2^{orbital}$ transforms as irreducible representation as Γ_2

$$\left\{\begin{array}{l} A_{2u} \times A_{1u} = A_{2g} \\ A_{2u} \times A_{2u} = A_{1g} \quad \text{(Completely symmetric components, } A_{1g}) \\ A_{2u} \times B_{1u} = B_{2g} \\ A_{2u} \times B_{2u} = B_{1g} \end{array}\right.$$

$$B_{1u} \times A_{1u} = B_{1g}$$
$$B_{1u} \times A_{2u} = B_{2g}$$
$$B_{1u} \times B_{1u} = A_{1g} \quad \text{(Completely symmetric)}$$
$$B_{1u} \times B_{2u} = A_{2g}$$

So $^1E_g \leftarrow {}^1A_{1g}$ transition is possible.

Thus the feasible transitions are (vibronically polarised).

Transition	Polarised
$^1A_{2g} \leftarrow {}^1A_{1g}$	$(x, y$ polarised$)$
$^1E_g \leftarrow {}^1A_{1g}$	$(z$ polarised$)$
$^1E_g \leftarrow {}^1A_{1g}$	$(x, y$ polarised$)$

10.8 POLARISATION OF ELECTRONICALLY ALLOWED TRANSITIONS

Just as with vibronically allowed transitions, in symmetry groups, in which all Cartesian axes are not equivalent (noncubic groups), it is observed that transitions will be allowed only for certain orientations of the electric vector of the incident light. One class of compounds in which this phenomenon has been studied, both theoretically and experimentally, consists of [M(acac)$_3$] (acac H = acetylacetone) and [M(ox)$_3$]$^{3-}$ (oxH$_2$ = oxalic acid) (M(III) = Ti, V, Cr, Mn, Fe, Co) (Fig. 10.11).

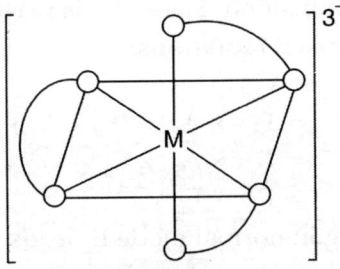

Fig. 10.11 Tris (oxalato) metal (III) ion, (O—O = oxalate ion)

If the metal ion is only influenced by the O atoms of the ligands than vibronic selection rule should be strictly followed. But if the metal ion is influenced by the molecule as a whole, electronic selection rule should be applied since the compound possesses D_3 symmetry and lacks centre of symmetry.

We choose tris (oxalato) chromium (III) ion, [Cr(ox)$_3$]$^{3-}$, where pure electronic selection rule is applied. The symmetry of [Cr(ox)$_3$]$^{3-}$ ion is reduced from O_h to D_3. The corresponding electronic states are obtained from correlation table shown below:

O_h	D_3
A_{2g}	A_2 (ground state)
T_{1g}	$A_2 + E$
T_{2g}	$A_1 + E$

The polarised absorption spectra of crystals of Na Mg Al $(Ox)_3.H_2O$ containing Cr^{3+} at 77°K is given in Fig. 10.12. The dipole moment operator transforms as $A_2(\parallel) + E(\perp)$ in the D_3 group. The $\parallel$ and $\perp$ signs refer to as the electric vector being respectively $\parallel$ and $\perp$ to the trigonal axis (C_3). The notation $\sigma(\perp)$ and $\pi(\parallel)$ may also be used. The spectra obtained with the electric vector $\parallel$ and $\perp$ to the trigonal axis are called π and σ spectra respectively.

The irreducible representations in the D_3 point group comprise A_1, A_2 and E (Fig. 10.12). Since the centre of inversion is lacking, the $3d$ orbitals can mix with odd parity atomic and molecular orbitals (Fig. 10.13).

Thus the transition will be Laporte allowed. The selection rules are given in Table 10.3.

For $A_1 \leftarrow A_2$ transition, $\langle \psi_e' | M_z | \psi_e \rangle$ becomes

$A_1 \times (A_2 \times A_2)$ ($\because$ dipole moment operator $M_z = A_2$]

$= A_1 \times A_1$ ($\because A_2 \times A_2 = A_1$; $2 \times 2 = 1$; $A_1 \times A_1 = A_1$]

$= A_1$

For $A_2 \leftarrow A_2$ transition, $\langle \psi_e' | M_z | \psi_e \rangle$ becomes

$A_2 \times (A_2 \times A_2) = A_2 \times A_1 = A_2$

and for $E \leftarrow A_2$, transition, $\langle \psi_e' | M_z | \psi_e \rangle$ becomes

$= E \times (A_2 \times A_2) = E \times A_1 = E$

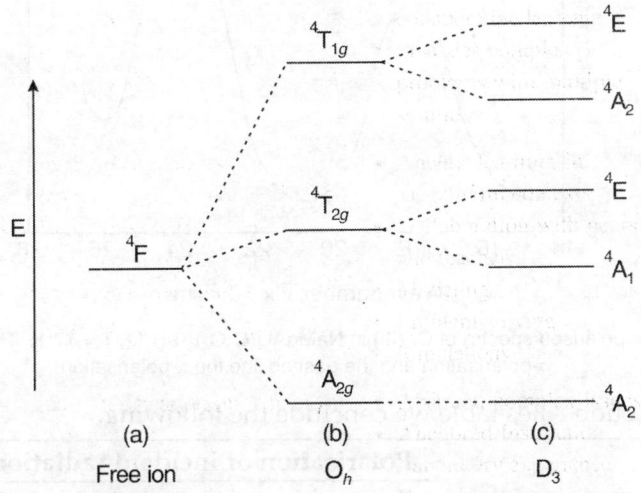

(a)   (b)   (c)

Free ion O_h D_3

Fig. 10.12 Energy level diagram for the Cr^{3+} ion: (a) spherically perturbed free ion (b) Octahedral environment (c) trigonal D_3 environment

Table 10.3 Electric dipole selection rules for D_3 symmetry

D_3	A_1	A_2	E
A_1		$\parallel$	$\perp$
A_2	$\parallel$		$\perp$
E	$\perp$	$\perp$	$\parallel$ and $\perp$

For $\langle \psi_e' \mid M_x, M_y \mid \psi_e \rangle$ (M_x, M_y are dipole moment components) the transition $A_1 \leftarrow A_2$:

$= A_1 \times (E \times A_2)$ Here the dipole moment operator is E

$= A_1 \times E$ ($\because E \times A = E$)

$= E$

For $A_2 \leftarrow A_2$ transition,

$A_2 \times (E \times A_2)$ The same dipole moment operator E

$= A_2 \times E$

$= E$

and for $E \leftarrow A_2$ transition,

$E \times (E \times A_2)$

$= E \times E$

$= A_1 + A_2 + E$

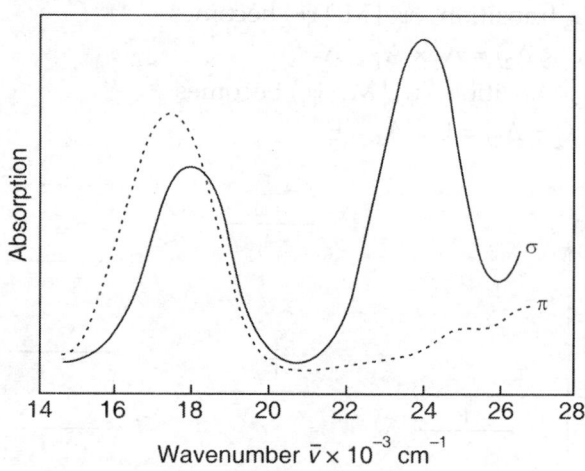

Fig. 10.13 The polarised spectra of Cr (III) in NaMgAl [(C$_2$O$_4$)$_3$·9H$_2$O, T = 77°K. The solid line is the σ-polarisation and the dashed line the π-polarisation

From the selection rules table we conclude the following:

	Polarisation of incident radiation	
Transitions	**z**	**(x, y)**
$A_1 \leftarrow A_2$	Allowed	Forbidden
$A_2 \leftarrow A_2$	Forbidden	Forbidden
$E \leftarrow A_2$	Forbidden	Allowed

[V (C$_2$O$_4$)$_3$]$^{3-}$: The solution spectrum of the [V(C$_2$O$_4$)$_3$]$^{3-}$ ion shows two bands one at 16450 ($\epsilon = 16$) and the other at 23500 cm^{-1} ($\epsilon = 12$) respectively. These bands are due to $^3T_{2g} \leftarrow {}^3T_{1g}$ and $^3A_{2g} \leftarrow {}^3T_{1g}$ transitions respectively (Fig. 10.12(b)). There is no splitting attributable to the lower symmetry of the complex. Since V(III) is linked by three bidentate ligands (C$_2$O$_4^{2-}$) so the compound belongs to D$_3$ point group. The energy level scheme for the V(III) ion in O$_h$ and D$_3$ environments as shown Fig. 10.12.

The crystal spectra studies shows a splitting of 150 cm^{-1} of the $^3T_{2g}$ (in O_h) level. The lower energy is assigned to $^3A_1 \leftarrow {}^3A_2$ ($\parallel$) and the higher to $^3E \leftarrow {}^3A_2(\perp)$ (polarisation ratio 1.3) (Figs 10.14 and 10.15).

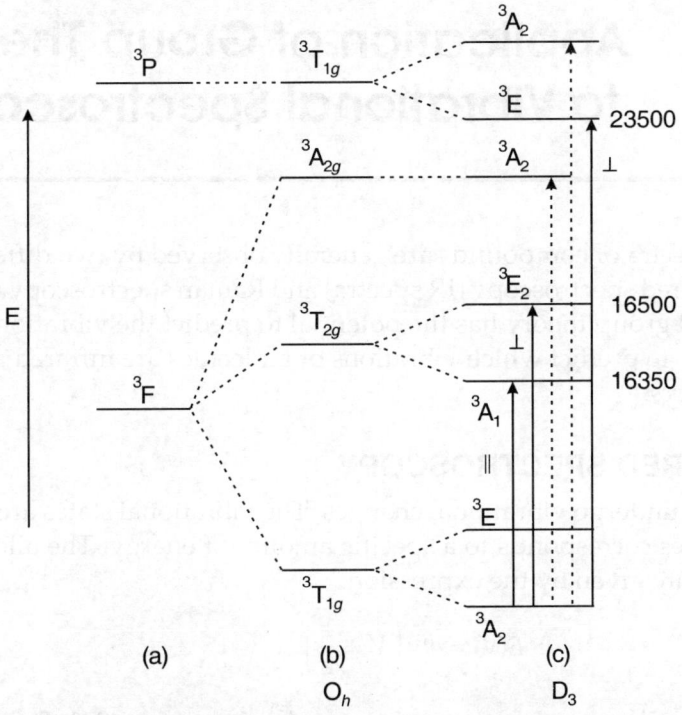

Fig. 10.14 Energy level scheme for the V^{2+} ion (a) spherically perturbed free ion (b) octahedral environment (c) trigonal D_3 environment

A weak band occurs in the crystal spectra at 23500 cm^{-1}. Transition to $^3A_2$ level is forbidden. Therefore, it is likely $^3E \leftarrow {}^3A_2$ transition, ^3E is the component of $^3T_{1g}$(P). Thus the earlier solution spectrum band assignment at 23500 cm^{-1} for $3A_{2g} \leftarrow {}^3T_{1g}$ transition is probably not correct.

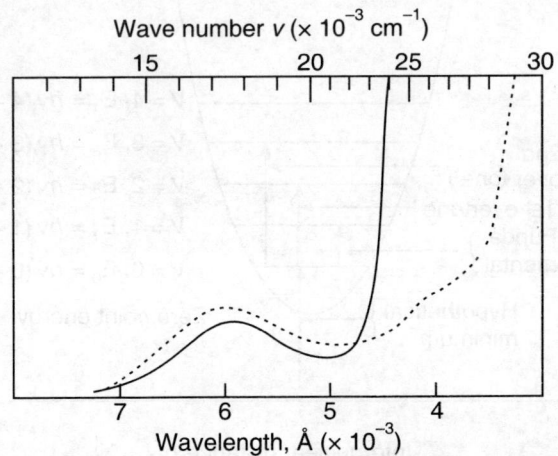

Fig. 10.15 The polarised absorption spectra of crystals of NaMgAl$(C_2O_4)_3 \cdot 9H_2O$ containing V^{3+} at 77 K

11

Application of Group Theory to Vibrational Spectroscopy

Vibrational spectra of compounds are generally observed by two different techniques. These are infrared spectroscopy (IR spectra) and Raman spectroscopy (Raman spectra). Symmetry and group theory has the potential to predict the vibrational spectrum of a compound, i.e. to predict which vibrations of molecules are infrared active and which are *Raman actives*.

11.1 INFRARED SPECTROSCOPY

Molecules can undergo vibrational changes. The vibrational states are quantised. Each one of the states corresponds to a specific amount of energy. The allowed vibrational energy levels are given by the expression:

$$E(v) = hv \left(V + \frac{1}{2} \right) \qquad \qquad ...(11.1)$$

where V is the vibrational quantum number having values 0, 1, 2, 3; hv is the energy difference between the ground and the first vibration level.

For a diatomic molecule, the atoms are constantly vibrating. The molecule absorbs a quantum of energy to attain higher vibrational levels where the two atoms stretch and compress to a greater extent than the lower vibrational level. The vibrational levels can be represented by the diagram shown in Fig. 11.1.

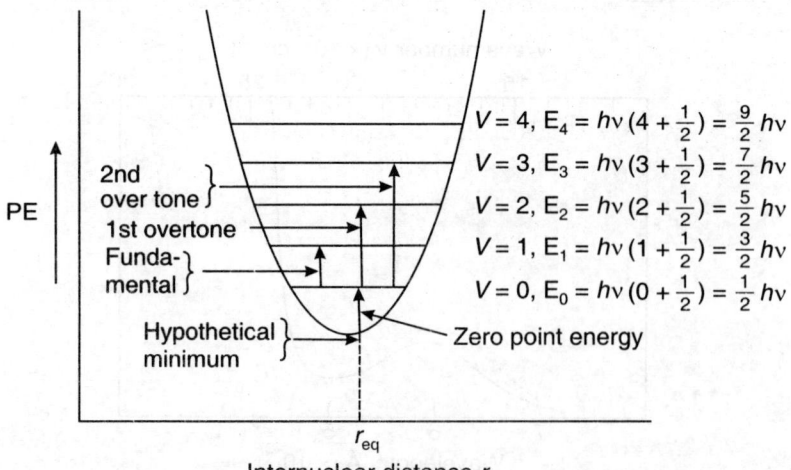

Fig. 11.1 Potential energy *vs* internuclear distance *r* (Morse curve)

The vibrational states are quantised and the minimum energy of the state is called *vibrational ground state*, i.e. when $V = 0$, $E = hv\left(V + \dfrac{1}{2}\right) = hv\left(0 + \dfrac{1}{2}\right) = \dfrac{1}{2}hv$. Thus the vibrational energy is not zero. The energy difference between the hypothetical minimum and the actual minimum energy (when $V = 0$) is called the *zero point energy* of the molecule.

The vibrational frequency of the harmonic oscillator (a linear diatomic molecule, A – B) in any vibrational level is given by

$$v = \frac{1}{2\pi}\sqrt{\frac{k}{\mu}} \qquad \qquad ...(11.2)$$

or

$$\bar{v} = \frac{1}{2\pi c}\sqrt{\frac{k}{\mu}} \qquad \qquad ...(11.3)$$

where μ is the reduced mass = $\left(\dfrac{m_A \times m_B}{m_A + m_B}\right)$

where m_A and m_B are the masses of atoms A and B respectively.

So the position of an absorption bond due to a particular transition between vibrational states (levels) can be determined using Eq. (11.3) which indicates that frequency (v) of vibration is directly proportional to the free constant k (strength of the bond) and inversely proportional to the masses of atoms bonded to one another but independent to vibrational quantum number V.

$$v \propto k \quad \text{and} \quad v \propto \frac{1}{\mu}$$

So vibrational frequency v is the same in all states (levels) though energy changes with the change of vibrational quantum number V.

The energy separation between two vibrational levels corresponds to the normal IR (range from 4000 cm^{-1} to 250 cm^{-1}) or for IR (range from 250 cm^{-1} to 50 cm^{-1}). The frequency of an IR band of a molecule depends on the bond strength (k). The order of stretching frequency is

triple bond > double bond > single bond

11.2 INFRARED SELECTION RULES

Absorption takes place in the IR region in accordance with the following selection rules.

1. For the IR absorption to take place, the necessary condition is that there should be change in the magnitude or direction of the dipole moment (μ).

   This creates an oscillating dipole moment which *interacts* with the electrical component of the IR radiation and absorption takes place.

   $$I = \langle \psi_{V'} \, \hat{\mu} \, \psi_{V''} \rangle$$

   where I is the intensity of the vibrational bond resulting from the transfer of the molecule from the ground vibrational level $\psi_{V'}$ to higher level $\psi_{V''}$.

For homodiatomic molecules, there is no change in the dipole moment following stretching and hence this does not show any IR absorption. It is said to be IR inactive. In case of heterodiatomic molecule, there is change in dipole moment. So the molecule is IR active.

2. For a heterodiatomic molecule, transition from one vibrational level to another occurs with the condition of

$$\Delta V = \pm 1$$

Since the molecules exists in the ground level V_0, $V_0 \rightarrow V_1$ transitions are most common. The intensity of such bands is more and the bond frequency is called fundamental frequency.

Forbidden transitions $V_0 \rightarrow V_2$ and $V_0 \rightarrow V_3$ involving $\Delta V = 2$ or 3, do take place though with a lower intensity. They are called first and second *overtones*.

11.3 NORMAL MODES OF VIBRATIONS[*] (FUNDAMENTAL VIBRATIONS) IN POLYATOMIC MOLECULE

Let us consider a molecule containing N atoms. If we specify the positions of each atom of the molecule in space with three coordinates x, y and z, then there will be $3N$ number of coordinates for the molecule. It is to be noted that the number of degrees of freedom possessed by a molecule = number of coordinates required to specify the position of all the atoms of the molecule completely.

So for a molecule with N atoms, the number of degrees of freedom = $3N$

A polyatomic molecule has translational, rotational and vibrational motions. Total degrees of freedom $3N$ are conveniently chosen as follows.

Translational Motion

In translational motion; the molecule is free to move from one place to another in three dimensional space. Since three coordinates (x, y, z) are required to describe the position of the centre of mass of the molecule, three degrees of freedom are utilised in describing the translational motion of the molecule.

Hence remaining degrees of freedom will be $(3N - 3)$.

Rotational Motion

A linear molecule can rotate about two mutually perpendicular axes whereas a non-linear molecule can rotate about three axes. Thus, two degrees of freedom are utilised in describing the rotational motion of a linear molecule and three for those of a nonlinear molecule.

Vibrational Motion

Vibrational degrees of freedom can be obtained by subtracting the translational and rotational degrees of freedom from the total degrees of freedom.

Vibrational degrees of freedom = Total degrees of freedom – translational degree of freedom – rotational degree of freedom

---

[*] $3N - 6$ (or $3N - 5$ for linear molecules) degrees of freedom is due to vibrational motion. These vibrations are called normal modes (stretching and bending modes).

For a linear molecule, vibrational degrees of freedom $= 3N - (3 + 2) = 3N - 5$.

Problem: How many fundamental vibrations are possible for H_2O and CO_2?

Solution: H_2O is a bent molecule. So the number of fundamental vibrations (normal modes of vibrations) will be $3N - 6 = 3 \times 3 - 6 = 3$.

CO_2 is a linear molecule. So the number of fundamental vibrations for CO_2 will be $= 3N - 5 = 3 \times 3 - 5 = 4$.

These vibrations are called 'normal modes' which are divided into stretching and bending modes.

Thus vibrations in a molecule are of two types:

(i) Bond stretching and (ii) Bending

The possible vibrations in case of bent and linear molecules are shown below:

Bent Molecule

1. **Symmetrical stretching:**

   Both sides stretch or compress together. Two figures (Fig. 11.2) represent one mode of vibration.

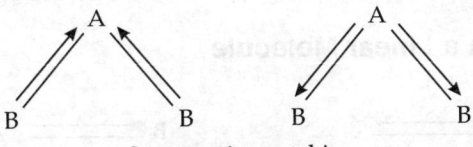

Symmetric stretching, ν_s

Fig. 11.2

2. **Asymmetric stretching:**

   When one bond is stretching, the other is compressing. Two figures (Fig. 11.3) represent one mode of vibration.

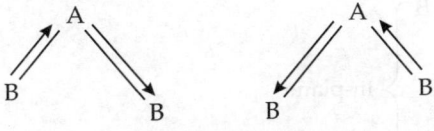

Asymmetric stretching, ν_{as}

Fig. 11.3

3. **Bending in plane:**

   The molecule undergoes bending vibration but all the three atoms are maintained in the same plane. This can be of two types: scissoring and rocking. (Fig. 11.4).

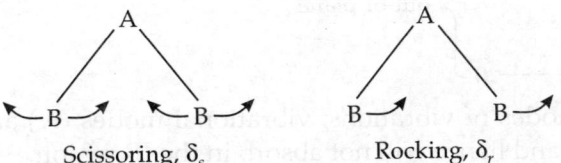

Scissoring, δ_s Rocking, δ_r

Fig. 11.4

4. **Bending out of plane:**

In this mode of bending, the atoms do not remain in the same plane but move out of plane. This may also be of two types, (+) sign shows that the atom is going above the plane of the molecule and (–) sign shows that the atom is going below the plane of the molecule (Fig. 11.5).

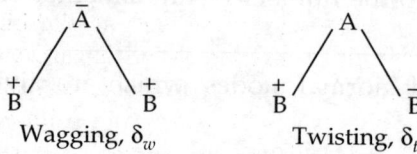

Wagging, δ_w Twisting, δ_t

Fig. 11.5

Thus out of six vibrational modes (symmetrical and asymmetrical stretching, scissoring, rocking, wagging and twisting), the rocking, wagging and twisting are part of rotation around x, y and z axes respectively, and will not absorb in the IR region. Hence the possible vibrations are symmetrical and asymmetrical stretchings and symmetrical bending in plane (scissoring). So the total number of fundamental vibrations for XY_2 molecule is $3N – 6 = 3 \times 3 – 6 = 3$

Bending vibrations are of lower energy than the stretching vibrations.

Possible Vibrations in a Linear Molecule

Stretching modes

(i)   B ⟶ A ⟵ B   B ⟶ A ⟶ B

Symmetrical stretching

(ii)   B ⟶ A ⟶ B   B ⟶ A ⟶ B

Asymmetrical stretching

Bending modes

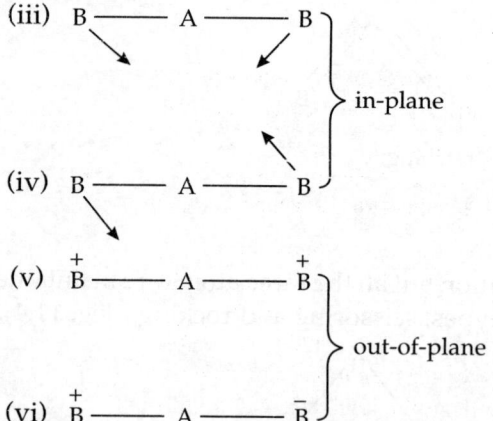

(iii)   B —— A —— B ⎫
⎬ in-plane
(iv)   B —— A —— B ⎭

(v) $\overset{+}{B}$ —— A —— $\overset{+}{B}$ ⎫
⎬ out-of-plane
(vi) $\overset{+}{B}$ —— A —— $\overset{-}{B}$ ⎭

Here out of six modes of vibrations, vibrational modes (iv) and (vi) are rotations around y and z axes and hence will not absorb in the IR region.

The mode (v) (out of plane vibration) which was partly rotation around x-axis in case of bent molecules is not a part of rotation in case of linear molecules. This is because rotation around x axis (molecular axis) is not possible.

Therefore in this case, there is an additional vibration and hence the possible total number of vibrations are 4.

$3N - 5 = 3 \times 3 - 5 = 4$.

Table 11.1 describes the normal modes of vibrations of di-, tri- and tetra-atomic molecules.

11.4 SYMMETRY BASED SELECTION RULES OF INFRARED

When a molecule is irradiated by infrared radiation, it undergoes a vibrational transition from vibrational ground state ($V = 0$ level) to vibrational first excited state ($V = 1$ level) and the absorption of radiation occurs in the infrared region of the electromagnetic spectrum. This transition is due to the **interaction between the incident radiation and the molecular dipole moment**.

Let the initial and final vibrational wave function be represented by ψ_i and ψ_f respectively. The transition moment integral (M) can be expressed as

$$M = \langle \psi_i \, \mu_k \, \psi_f \rangle$$

where μ_k is the dipole moment of the molecule. μ_k can be resolved into its three Cartesian components. So the above integral can be expressed as:

$$\left. \begin{aligned} M_x &= \langle \psi_i \, \mu_x \, \psi_f \rangle \\ M_y &= \langle \psi_i \, \mu_y \, \psi_f \rangle \\ M_z &= \langle \psi_i \, \mu_z \, \psi_f \rangle \end{aligned} \right\} \qquad \qquad ...(11.4)$$

$$\therefore \qquad M = \langle \psi_i \, \mu_x \, \psi_f \rangle + \langle \psi_i \, \mu_y \, \psi_f \rangle + \langle \psi_i \, \mu_z \, \psi_f \rangle \qquad ...(11.4a)$$

Now the selection rule can be stated in the following way.

1. For a given transition if $M = 0$, then the probability of that transition occuring is also zero. In that case the transition is forbidden in the infrared region.

2. For a given transition if $M \neq 0$, then the probability of that transition occuring is also non-zero. In that case the transition is allowed and hence occurs in the infrared region.

   If any one of the integrals is non-zero [Eq. (11.4)], the mode is said to be allowed. This means that for a given mode to be active or allowed all three integrals [Eq. (11.4)], need not to be nonzeros. Consider the following points

   (i) The wave function ψ_i that represents vibrational ground state ($V = 0$) of the molecule belongs to a totally symmetric irreducible representation (IR) in any point group.

   (ii) The dipole moment components μ_x, μ_y and μ_z transform as the IRs to which x, y and z functions belong to in any character table.

   For example in C_{2v} point group μ_x, μ_y and μ_z transform as B_1, B_2 and A_2.

   In D_{4n} point group μ_x and μ_y transform as E_4 and μ_z, belong to A_{2u}.

   (iii) The wave function ψ_f (vibrational excited state, $V = 1$) is the same as the normal mode.

Let us consider the direct product of $\Gamma(\psi_i) \times \Gamma(\mu_k) \times \Gamma(\psi_f)$ for every normal mode and to see whether the product is equal to or contains a totally symmetric IR (A_k).

Table 11.1 Normal modes of vibrations of di-, tri- and tetra-atomic molecules

Molecule	No of atoms N	Total degrees of freedom $3N$	No. of modes of vibrations		Bond stretching $(N-1)$	No. of motions of	
			For linear molecule $3N-5$	For bent molecule $3N-6$		Bending linear molecule $2N-4$	Bending nonlinear molecule $(2N-5)$
1. Diatomic molecules H_2, O_2, HCl	2	6	1	—	1	0	—
2. Nonlinear triatomic molecule SO_2, H_2O, etc.	3	9	—	3	2	—	1
3. Linear triatomic CO_2, N_2O molecules	3	9	4	—	2	2	—
4. Linear tetra-atomic molecules C_2H_2	4	12	7	—	3	4	—

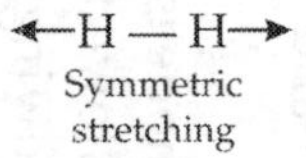

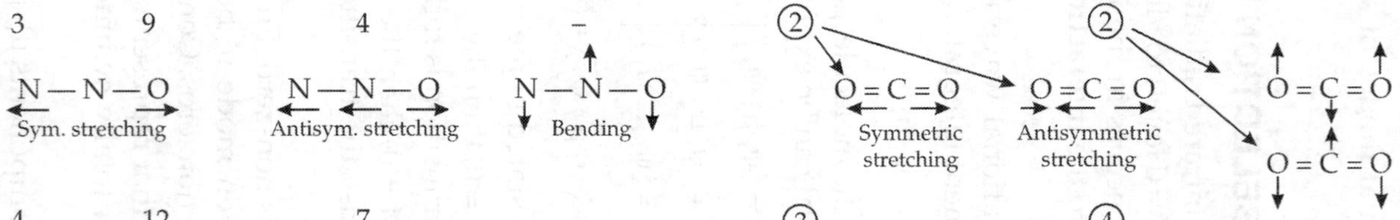

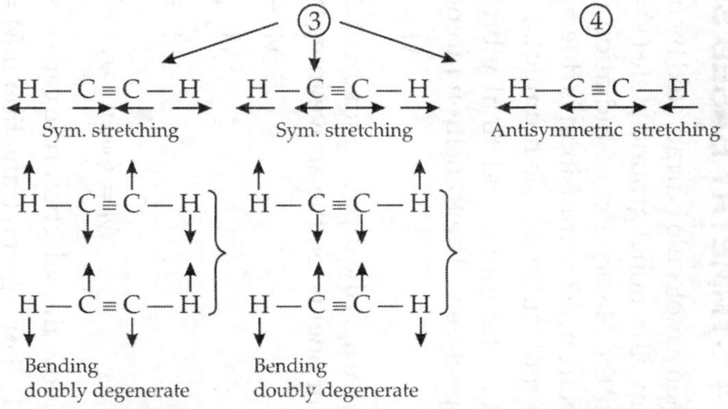

H — H Symmetric stretching

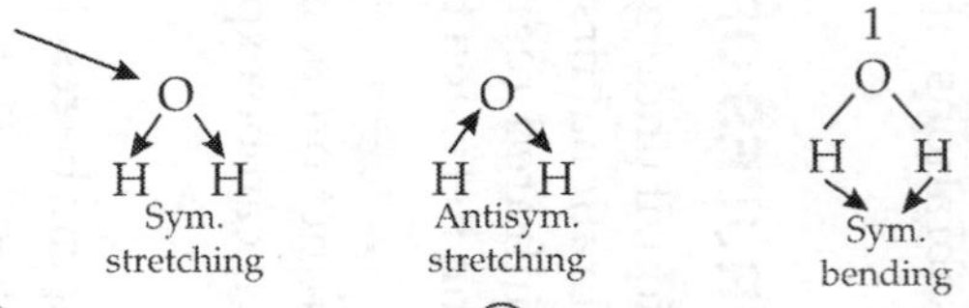

N — N — O Sym. stretching

N — N — O Antisym. stretching

N — N — O Bending

O, H H Sym. stretching

O, H H Antisym. stretching

O, H H Sym. bending

O = C = O Symmetric stretching

O = C = O Antisymmetric stretching

O = C = O

H — C ≡ C — H Sym. stretching

H — C ≡ C — H Antisymmetric stretching

H — C ≡ C — H Bending doubly degenerate

If $\Gamma(\psi_i) \times \Gamma(\mu_k) \times \Gamma(\psi_f)$ contains A_1 then the transition will be allowed and transition moment integral $M \neq 0$. If $\Gamma(\psi_i) \times \Gamma(\mu_k) \times \Gamma(\psi_f)$ does not contain A_1 then the transition will be forbidden and transition moment integral $M = 0$.

Since $\Gamma(\psi_i)$ always belongs to a totally symmetric irreducible representation (A_1), the direct product $\Gamma(\psi_i) \times \Gamma(\mu_k) \times \Gamma(\psi_f)$ is reduced to two of the function, i.e. $\Gamma(\mu_k), \Gamma(\psi_f)$. This is due to the fact that multiplication of any function by a totally symmetric irreducible representation (A_1) leaves the function unaltered. Therefore, symmetry based selection rules can be rewritten as

$$\Gamma(\psi_i)\, \Gamma(\mu_k)\, \Gamma(\psi_f)$$
$$= A\, \Gamma(\mu_k)\, \Gamma(\psi_f) = \Gamma(\mu_k)\, \Gamma(\psi_f)$$

If the product of $\Gamma(\mu_k)\, \Gamma(\psi_f)$ contains A_1 then the transition is allowed, but if the product of $\Gamma(\mu_k)\, \Gamma(\psi_f)$ does not contain A_1 then the transition is forbidden.

Further direct product of $\Gamma(\mu_k)\, \Gamma(\psi_f)$ contains A_1 if $\Gamma(\mu_k) = \Gamma(\psi_f)$ and does not contain A_1 if $\Gamma(\mu_k) \neq \Gamma(\psi_f)$.

11.5 APPLICATIONS OF THE SYMMETRY BASED SELECTION RULES FOR DETERMINING THE INFRARED ACTIVITY OF SOME SELECTED MODES

11.5.1 Bent XY$_2$ Type Molecule with C$_{2v}$ Symmetry (H$_2$O)

Water molecule contains 3 atoms. Therefore, the number of modes of vibration $= 3N - 6 = 3 \times 3 - 6 = 3$. Water molecule belongs to C_{2v} group with one operation in each class ($E, C_2, \sigma_v(xz), \sigma_v(yz)$). The total characters of the operation give the total representation

C_{2v}	E	C_2	$\sigma_v(xz)$	$\sigma_v(yz)$
No. of unshifted atom's contribution to the total	3	1	3	1
character per unshifted atom	3	−1	1	1
Total character of the representation Γ_{total} or Γ_{3N}.	9	−1	3	1

The total character Γ_{3N} can be reduced to the irreducible representation by using the reduction formula.

$$\Gamma_{3IV} = 3A_1 + 1A_2 + 3B_1 + 2B_2$$

$\Gamma_{translational\ motion}$ $\qquad = A_1 + B_1 + B_2$

$\Gamma_{rotational\ motion}$ $\qquad = A_2 + B_1 + B_2$

$\therefore \Gamma_{vibrational\ motion}$ $\qquad = \Gamma_{3N} - (\Gamma_{trans} + \Gamma_{rot})$

$\qquad\qquad\qquad\quad = 2A_1 + B_1$

Of the three modes of vibrations, two belongs to A_1 irreducible representation and one belongs to B_1 irreducible representation (symmetry species). Water being a bent molecule has one symmetrical, one asymmetrical stretching and one symmetrical bending stretch.

		E	C_2	$\sigma_v(xz)$	$\sigma_v(yz)$	Vibrational mode
v/1	Sym. stretching	1	1	1	1	A_1 (v_1)
v/3	Asymmetrical stretching	1	−1	1	−1	B_1 (v_3)
v/2	Sym. bending (in plane)	1	1	1	1	A_1 (v_2)

The next step is to find out which of these vibrations will be IR active. For the transition to be IR active, they should be associated with the change in the dipole moment, i.e. $\langle \psi_i \, \mu_k \, \psi_f \rangle$ should be non-zero. In case of polyatomic molecules the change in dipole moment takes place along x, y and z directions. Hence, the intensity of the IR band depends on three integrals [Eq. (11.4)]:

$$M_x = \langle \psi_i \, \mu_x \, \psi_f \rangle, \, M_y = \langle \psi_i \, \mu_y \, \psi_f \rangle \text{ and}$$

$$M_z = \langle \psi_i \, \mu_z \, \mu_f \rangle$$

If one of them is non-zero, the vibration will be IR active.

For water molecule, the vibrational ground state ψ_i should be A_1 symmetry. The dipole moment components μ_x, μ_y and μ_z transform as $B_1^{(x)}$, $B_2^{(y)}$ and $A_1^{(z)}$ and ψ_f has the symmetry A_1. The direct products for A_1 modes can be worked out as below.

$$\text{Integrand} \equiv A_1 \times \begin{pmatrix} B_1(x) \\ B_2(y) \\ A_1(z) \end{pmatrix} \times A_1$$

$$\langle \psi_i \, \mu \, \psi_f \rangle \equiv \psi_i \qquad \mu \qquad \psi_f$$

$$= \begin{pmatrix} A_1 \times B_1 \times A_1 \\ A_1 \times B_2 \times A_1 \\ A_1 \times A_1 \times A_1 \end{pmatrix}$$

$$= \begin{pmatrix} B_1 \times A_1 \\ B_2 \times A_1 \\ A_1 \times A_1 \end{pmatrix} = \begin{pmatrix} B_1 \\ B_2 \\ A_1 \end{pmatrix}$$

Hence, the direct product representation contains the totally symmetric irreducible representation A_1. Therefore, v_1 vibrational mode will be IR active. Similarly for v_3 (asymmetric stretching) we have

$$\langle \psi_i \mid \mu \mid \psi_f \rangle = A_1 \times \begin{pmatrix} B_1(x) \\ B_2(y) \\ A_1(z) \end{pmatrix} \times B_1$$

$$= \begin{pmatrix} A_1 \times B_1 \times B_1 \\ A_1 \times B_2 \times B_1 \\ A_1 \times A_1 \times B_1 \end{pmatrix} = \begin{pmatrix} B_1 \times B_1 \\ B_2 \times B_1 \\ A_1 \times B_1 \end{pmatrix} = \begin{pmatrix} A_1 \\ A_2 \\ B_1 \end{pmatrix}$$

As the direct product contains the totally symmetric irreducible representation (A_1), therefore v_3 vibrational mode will be IR active. Thus, all the three normal modes of vibration for H_2O are infrared active. We can therefore expect three absorption bends in the IR spectrum of H_2O.

Free H_2O molecule has three normal vibration frequencies observed at $v_1 = 3642$ cm^{-1} (A_1), $v_2 = 1595$ cm^{-1} (A_1) and $v_3 = 3756$ cm^{-1} (B_1) respectively.

Note that $v_{asy\ stretching} > v_{sym\ stretching} > v_{bending\ stretching}$

11.5.2 Square Planar XY$_4$ Type Molecule of D$_{4h}$ Symmetry (Centrosymmetric)

For a square planar complex XY_4

$$\Gamma_{vib} = A_{1g} + B_{1g} + B_{2g} + A_{2u} + B_{2u} + 2E_u \ (3N - 6 = 9)$$

The vibrational ground state belongs to the totally symmetric representation A_{1g}.

The dipole moment components μ_x and μ_y belong to the doubly degenerate irreducible representation E_u and μ_z belongs to A_{2u}.

The symmetries of the vibrational excited modes should be chosen from the normal modes of A_{1g}, B_{1g}, B_{2g}, A_{2u}, B_{2u} and E_u. On working out the direct products for each of these modes, we get the following result (Table 11.2).

Table 11.2 The direct products of $\Gamma(\mu_k)$ and $\Gamma(\psi_f)$ for possible infrared transition assignments in XY$_4$ molecule. The symmetry of the ground state, $\Gamma(\psi_i) = A_{1g}$

$\Gamma(\mu_k)$	$\Gamma(\psi_f)$	$\Gamma(\mu_k) \times \Gamma(\psi_f)$	Transition
A_{2u}	A_{1g}	$A_{2u} \times A_{1g} = A_{2u}$	Forbidden
	B_{1g}	$A_{2u} \times B_{1g} = B_{2u}$	Forbidden
	B_{2g}	$A_{2u} \times B_{2g} = B_{1u}$	Forbidden
	A_{2u}	$A_{2u} \times A_{2u} = \boxed{A_{1g}}$	*Allowed*
	B_{2u}	$A_{2u} \times B_{2u} = B_{1g}$	Forbidden
	E_u	$A_{2u} \times E_u = E_g$	Forbidden

Contd...

$\Gamma(\mu_k)$	$\Gamma(\psi_f)$	$\Gamma(\mu_k) \times \Gamma(\psi_f)$	Transition
E_u	A_{1g}	$E_u \times A_{1g} = E_u$	Forbidden
	B_{1g}	$E_u \times B_{1g} = E_u$	Forbidden
	B_{2g}	$E_u \times B_{2g} = E_u$	Forbidden
	A_{2u}	$E_u \times A_{2u} = E_g$	Forbidden
	B_{2u}	$E_u \times B_{2u} = E_g$	Forbidden
	E_u	$E_u \times E_u = \boxed{A_{1g}} + A_{2g} + E_g$	*Allowed*

Note: 1.   If any component of $\Gamma(\mu_k)$ (μ_x, μ_y, μ_z) has the same symmetry as the symmetry of $\Gamma(\psi_f)$, the product $\Gamma(\mu_k) \times \Gamma(\psi_f)$ will be totally symmetric and the integral will have nonzero value.

       2.   Generally all u type modes should be found to be infrared active while g type modes being inactive.

11.6 POLARISABILITY

When a molecule is placed in a static electric field, the nuclei are attracted towards the negative pole and the electrons toward the positive pole, inducing a dipole moment in the molecule. If μ_i represents the induced dipole moment and E the electric field, then the induced moment will be proportional to the applied electric field

$$\mu_i \propto E$$
$$\mu_i = \alpha \, E \qquad \qquad ...(11.5)$$

The proportionality constant α is called the '*polarisability*' and is defined as the electric dipole induced in the molecular system by a unit electric field.

The magnitude of this polarisability depends upon the orientation of the bonds in the molecule with respect to the electric field direction, i.e. it is anisotropy.

For atoms, which posses a spherical symmetry, the polarisability is same in all directions (isotropic), so α can be expressed by a single scalar quality, whereas for molecules with less spherical symmetry, the polarisability will not be same in all the directions (anisotropic) and then α can be best described by a tensor (a square matrix). Dipole moment is a vector, it will have three components then $\mu_{ind}^{(x)}$, $\mu_{ind}^{(y)}$ $\mu_{ind}^{(z)}$. The induced dipole moments are related to the polarisability tensor by the following equations:

$$\mu_{ind}^{(x)} = \alpha_{xx} \, E_x + \alpha_{xy} \, E_y + \alpha_{xz} \, E_z$$
$$\mu_{ind}^{(y)} = \alpha_{yx} \, E_x + \alpha_{yy} \, E_y + \alpha_{yz} \, E_z \qquad \qquad ...(11.6)$$
$$\mu_{ind}^{(z)} = \alpha_{zx} \, E_x + \alpha_{zy} \, E_y + \alpha_{zz} \, E_z$$

Equations (11.6) can be written in matrix form as

$$\begin{pmatrix} \mu_{ind}^{(x)} \\ \mu_{ind}^{(y)} \\ \mu_{ind}^{(z)} \end{pmatrix} = \begin{pmatrix} \alpha_{xx} & \alpha_{xy} & \alpha_{xz} \\ \alpha_{yx} & \alpha_{yy} & \alpha_{yz} \\ \alpha_{zx} & \alpha_{zy} & \alpha_{zz} \end{pmatrix} \begin{pmatrix} E_x \\ E_y \\ E_z \end{pmatrix}$$

Polarisability tensor

where
$$\alpha_{xx} = \left(\frac{\partial \mu_x}{\partial E_x}\right)$$

$$\alpha_{xy} = \left(\frac{\partial \mu_x}{\partial E_y}\right)$$

and
$$\alpha_{xz} = \left(\frac{\partial \mu_x}{\partial E_z}\right)$$

Since polarisability tensor is symmetric, i.e. $\alpha_{ij} = \alpha_{ji}$, only six of the nine components are distinct: α_{xx}, α_{yy}, α_{zz}, α_{xy}, α_{yz} and α_{zx}.

11.7 RAMAN SPECTROSCOPY

Raman spectroscopy is concerned with vibrational and rotational transitions, and in this respect it is similar to infrared spectroscopy.

Since the selection rules are different, the information obtained from the Raman spectrum often complements that obtained from an infrared study and provides valuable structural information.

When a beam of monochromatic light is passed through a transparent medium, a portion of light is scattered in all directions. The frequency of the scattered rays is the same as that of the incident beam. This is known as *Rayleigh scattering*. But a careful spectroscopic examination of the scattered rays reveals some faint lines other than those due to Rayleigh scattering. That is, a small portion of the rays are scattered with a different frequency. This phenomenon is called *Raman scattering* (Fig. 11.6).

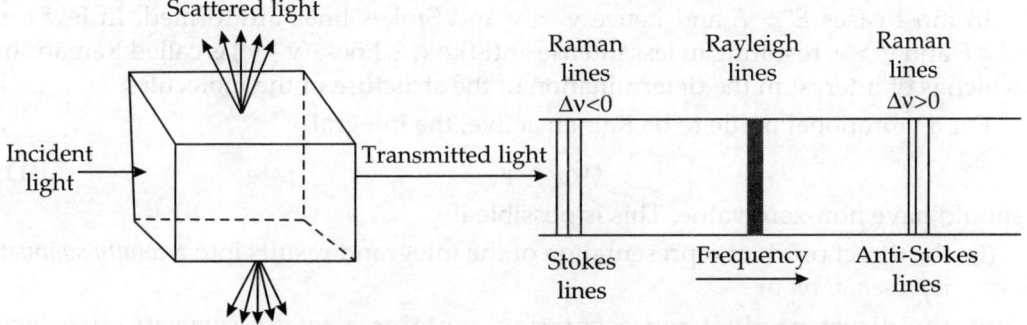

Fig. 11.6 Schematic representation of a Raman spectrum

According to Raman when the incident radiation passes through the medium, it strikes against the molecules, induces an electronic polarisation in it and get scattered. There are two types of collisions.

11.7.1 Elastic Collision

There is no transfer of energy between the incident radiation and the molecules. The scattered rays have the same energy as the incident rays, This is called *Rayleigh scattering*.

11.7.2 Inelastic Collision

The incident radiation may transfer some energy to the molecule and thus raises it to a higher electronic, vibrational, rotational or translational level. The scattered rays will have lower energy level than the incident radiation and form the *Stoke's lines*.

Conversely, the molecule may initially be in higher energy state and may transfer some energy to the incident radiation to come down to lower energy level. Thus, the scattered rays will have higher energy than the incident radiation and will form *anti-Stoke's lines*. The probability of the molecule being in the higher energy state is less and hence the intensity of the anti-Stoke's lines is less than the Stoke's lines.

Let us consider a molecule with energy E and $KE = \dfrac{1}{2}mv^2$, where m is the mass of the molecule and v its initial velocity. Let the incident radiation with frequency ν strike the molecule and undergo an inelastic collision. As a result, the energy of the molecule changes to E' and its KE changes to $\dfrac{1}{2}mv'^2$, where v' is the changed velocity of the molecule. The frequency of scattered radiation is also changed to ν'. Since the total energy has to be conserved, the following equation holds good.

$$E + \frac{1}{2}mv^2 + h\nu = E' + \frac{1}{2}mv'^2 + h\nu'$$

The change in the KE of the molecule is not significant and hence $\dfrac{1}{2}mv^2 = \dfrac{1}{2}mv'^2$ is considered. Thus, we have

$$E + h\nu = E' + h\nu'$$

or
$$E' - E = h(\nu - \nu')$$

In most cases $E' > E$ and hence $\nu' < \nu$ and Stokes lines are formed. In few cases $E' < E$ and $\nu' > \nu$, resulting in less intense anti-Stoke's lines. $(\nu - \nu')$ is called Raman shift, which is of interest in the determination of the structure of the molecules.

For a vibrational mode to be Raman active, the integral

$$\langle \psi_{V'} \, \hat{\alpha} \, \psi_{V''} \rangle \qquad \qquad ...(11.7)$$

should have non-zero value. This is possible if

  (i) the direct product representation of the integrand results into a *totally symmetric representation* or
  (ii) the direct product representation contains a *totally symmetric irreducible representation*.

The intensity of a Raman line (I) *as a result of the transfer of the molecule* from the lower vibrational level v' to higher vibrational level v'' can be determined by Eq. (11.7).

$$I = \langle \psi_{V'} \, \hat{\alpha} \, \psi_{V''} \rangle$$

Six equations can be written by putting six different values of α (α_{xx}, α_{yy}, α_{zz}, α_{xy}, α_{yz} and α_{zx}). If one of the integrals is non-zero, the vibration is Raman active.

For a vibration to be Raman active, the integral $\langle \psi_{V'} \, \hat{\alpha} \, \psi_{V''} \rangle$ should be non-zero. For this, the integral should not change on performing any operation, i.e. should be of A_1

symmetry. We know that vibrational ground level has A_1 symmetry. Hence $\langle \psi_{V'} \, \hat{\alpha} \, \psi_{V''} \rangle$ to be A_1, $\hat{\alpha} \, \psi_{V''}$ should also be A_1. This means that the higher vibrational level $\psi_{V''}$ should have the same symmetry as α i.e. α_{xx}, α_{yy}, α_{zz}, α_{xy}, α_{xz}, α_{yz}. Thus for a vibration to be Raman active, it should belong to the same species as one of the components of the polarisability.

These polarisability components correspond to the same irreducible representation as the square of the component x, y, z (x^2, y^2, z^2) or their binary products (xy, yz, zx).

11.8 SELECTION RULE FOR VIBRATIONAL RAMAN SPECTRA

In order for a vibration to be Raman active, there must be a change in polarisability (α) during the vibration. Mathematically, we may write the above statement as

$$\left(\frac{\partial \alpha_{ij}}{\partial r} \right)_{re} \neq 0$$

For a vibration to be Raman active, the derivative of one of the components of the polarisability tensor (α_{ij}) with respect to normal coordinate (r) should be non zero.

If the direct product $\Gamma\,(\psi_{V'}^{ex}) \times \Gamma(\hat{\alpha}) \times \Gamma\,(\psi_V)$ contains a totally symmetric irreducible representation of the point group, then the transition $v \rightarrow v^*$ is Raman active.

11.8.1 Examples of Molecules with Variety of Symmetry (Bent Molecule with C_{2v} Symmetry)

As discussed in case of infrared, here also the normal modes $= 2A_1 + B_1$ and the ground state ψ_i always belongs to A_1 symmetry. The polarisability components α_{ij} transform as shown below (from character table C_{2v}).

C_{2v}	α_{ij}
A_1	x^2, y^2, z^3
A_2	xy
B_1	xz
B_2	yz

The direct products for A_1 and B_1 modes can now be worked out as shown in Table 11.3.

Table 11.3 The direct products of $\Gamma(\alpha_{ij})$ and $\Gamma(\psi_f)$ for possible Raman activity assignments in H_2O (C_{2v}) molecule. The symmetry of the ground state, $\Gamma(\psi_i) = A_1$

$\Gamma(\alpha_{ij})$	$\Gamma(\psi_f)$	$\Gamma(\alpha_{ij}) \times \Gamma(\psi_f)$	Transition
A_1	$\begin{cases} A_1 \\ B_1 \end{cases}$	$A_1 \times A_1 = A_1$ $A_1 \times B_1 = B_1$	Allowed Forbidden
A_2	$\begin{cases} A_1 \\ B_1 \end{cases}$	$A_2 \times A_1 = A_2$ $A_2 \times B_1 = B_2$	Forbidden Forbidden

Contd...

$\Gamma(\alpha_{ij})$	$\Gamma(\psi_f)$	$\Gamma(\alpha_{ij}) \times \Gamma(\psi_f)$	Transition
B_1	$\begin{cases} A_1 \\ B_1 \end{cases}$	$B_1 \times A_1 = B_1$ $B_1 \times B_1 = A_1$	Forbidden Allowed
B_2	$\begin{cases} A_1 \\ B_1 \end{cases}$	$B_2 \times A_1 = B_2$ $B_2 \times B_1 = A_2$	Forbidden Forbidden

The above exercise concludes that all the three modes ($2A_1$ and B_1) are Raman active, in addition to being IR active.

Square Planar XY$_4$ Molecule (D$_{4h}$ Symmetry)

The normal modes of vibration for XY_4 (D_{4h}) molecule

$$\Gamma_{vib} = A_{1g} + B_{1g} + B_{2g} + A_{2u} + B2_u + 2E_4$$

We shall now determine which of these modes are Raman active. The vibrational ground state $\Gamma(\psi_i)$ is totally symmetric A_{1g}. The symmetries of the polarisability components, α_{ij}, are shown in the table.

D_{4h}	α_{ij}
A_{1g}	$x^2 + y^2, z^2$
B_{1g}	$x^2 - y^2$
B_{2g}	xy
E_g	yz, zx

The symmetries of the vibrational excited modes should be chosen from the worked out normal modes of A_{1g}, B_{1g}, B_{2g}, A_{2u}, B_{2u} and E_u. On working out the direct products for each of these modes we get the following type of result (Table 11.4).

Table 11.4 The direct products of $\Gamma(\alpha_{ij})$ and $\Gamma(\psi_f)$ for possible Raman activity assignments in XY$_4$ (D$_{4h}$) molecule. The symmetry of the ground state, $\Gamma(\psi_i) = A_{1g}$

$\Gamma(\alpha_{ij})$	$\Gamma(\psi_f)$/normal mode	$\Gamma(\alpha_{ij}) \times \Gamma(\psi_f)$	Raman selection rule
$\begin{cases} A_{1g} \\ B_{1g} \\ B_{2g} \\ E_g \end{cases}$	A_{1g}	$A_{1g} \times A_{1g} = A_{1g}$ $B_{1g} \times A_{1g} = B_{1g}$ $B_{2g} \times A_{1g} = B_{2g}$ $E_g \times A_{1g} = E_g$	Allowed Forbidden Forbidden Forbidden
$\begin{cases} A_{1g} \\ B_{1g} \\ B_{2g} \\ E_g \end{cases}$	B_{1g}	$A_{1g} \times B_{1g} = B_{1g}$ $B_{1g} \times B_{1g} = A_{1g}$ $B_{2g} \times B_{1g} = A_{2g}$ $E_g \times B_{1g} = E_g$	Forbidden Allowed Forbidden Forbidden
$\begin{cases} A_{1g} \\ B_{1g} \\ B_{2g} \\ E_g \end{cases}$	B_{2g}	$A_{1g} \times B_{2g} = B_{2g}$ $B_{1g} \times B_{2g} = A_{2g}$ $B_{2g} \times B_{2g} = A_{1g}$ $E_g \times B_{2g} = E_g$	Forbidden Forbidden Allowed Forbidden

Contd...

$\Gamma(\alpha_{ij})$	$\Gamma(\psi_f)$/normal mode	$\Gamma(\alpha_{ij}) \times \Gamma(\psi_f)$	Raman selection rule
A_{1g}	A_{2u}	$A_{1g} \times A_{2u} = A_{2u}$	Forbidden
B_{1g}		$B_{1g} \times A_{2u} = B_{2u}$	Forbidden
B_{2g}		$B_{2g} \times A_{2u} = B_{1u}$	Forbidden
E_g		$E_g \times A_{2u} = E_u$	Forbidden
A_{1g}	B_{2u}	$A_{1g} \times B_{2u} = B_{2u}$	Forbidden
B_{1g}		$B_{1g} \times B_{2u} = A_{2u}$	Forbidden
B_{2g}		$B_{2g} \times B_{2u} = A_{1u}$	Forbidden
E_g		$E_g \times B_{2u} = E_u$	Forbidden
A_{1g}	E_u	$A_{1g} \times E_u = E_u$	Forbidden
B_{1g}		$B_{1g} \times E_u = E_u$	Forbidden
B_{2g}		$B_{2g} \times E_u = E_u$	Forbidden
E_g		$E_g \times E_u = A_{1u} + A_{2u} + E_u$	Forbidden

Note that only g type modes have been found to be Raman active, while u modes being inactive.

11.9 USE OF SYMMETRY CONSIDERATIONS TO DETERMINE THE NUMBER OF ACTIVE INFRARED AND RAMAN LINES

A molecular vibration is infrared active (has an infrared absorption), if the dipole moment (μ) of the molecule changes during the vibration and a vibration will be Raman active if the polarisability of the molecule (α) changes during the vibration. Mathematically, we may write

$$\left(\frac{\partial \mu}{\partial r}\right)_{re} \neq 0, \text{ and } \left(\frac{\partial \alpha_{ij}}{\partial r}\right)_{re} \neq 0$$

For IR active vibration, the derivative of one of the components of dipole moment (μ) with respect to the normal coordinate r (internuclear distance, r) should be non-zero. For a vibration to be Raman active, the derivative of one of the components of polarisability tensor (α_{ij}) with respect to r should be non-zero.

The three components of μ_i Eq. (11.5) along x, y, z axes are μ_x, μ_y, μ_z and the six components of α_{ij} are α_{xx}, α_{yy}, α_{zz}, α_{xy}, α_{yz}, α_{zx}. Hence, on the basis of symmetry species of the wave functions of vibrational states involved in transition and the symmetry of dipole or polarisability operator, we arrive at the following selection rules for IR and Raman spectra. If the direct product $\Gamma(\psi^{V'}_{\text{excited state}}) \times \Gamma(\mu) \times \Gamma(\psi^{V}_{\text{ground state}})$ contains the totally symmetric irreducible representation of the point group, their the transition $V \rightarrow V'$ is IR active and, if the direct product $\Gamma(\psi^{\hat{V}'}_{\text{excited state}}) \times \Gamma(\alpha) \times \Gamma(\psi^{V}_{\text{ground state}})$ contains the totally symmetric irreducible representation of the point group, their the transition $V \rightarrow V'$ is Raman active.

The interpretation of the IR and Raman spectra is greatly simplified by taking into account the symmetry of the molecule in question. In this section, we introduce the mechanism involved.

We shall first consider a homodiatomic molecule X_2 ($X_2 = H_2, N_2, O_2, F_2, Cl_2$, etc.). As the two atoms (X) in X_2 molecule are identical, the stretching vibration causes no change in dipole moment either before or during vibration. So the vibrations are IR inactive. But if there is increase or decrease in size of the molecule during the symmetrical stretching, the polarisability (α) of the molecule changes and hence the vibration of the molecule changes and in that case the vibration becomes Raman active. In heterodiatomic molecule such as HX, CO, NO, etc. the stretching vibration causes changes in both the dipole moment and polarisability of the molecule. In that case, the molecule will be both IR and Raman active.

H₂O:

Water molecule contains three atoms. So the number of modes of vibration = $3N - 6$ = $3 \times 3 - 6 = 3$. The molecule belongs to C_{2v} point group with one operation in each class (E, C_2, $\sigma_v(xz)$, $\sigma'_v(yz)$). We shall examine the number of atoms in the molecule which are not shifted by E, C_2, $\sigma_v(xz)$ and $\sigma'_v(yz)$ operations.

- All the three atoms of water molecule are not moved by the identity (E) operation.
- The oxygen is the only atom not shifted by C_2 operation.
- The $\sigma_v(xz)$ operation results, the three atoms being unchanged.

The oxygen atom is not moved by reflection in the $\sigma'_v(yz)$ plane, but the two hydrogen atoms move. In summary

Symmetry operation:	E	C_2	$\sigma_v(xz)$	$\sigma'_v(yz)$
No. of unshifted atom:	3	1	3	1

Table 11.5 represent a testing of the contribution to the total character of each symmetry operation by each unshifted atom in the molecule. Each of the number of the unshifted atoms determined is to be multiplied by the number in Table 11.5, corresponding to be particular symmetry operation, to give the total character for that operation.

Table 11.5 Contribution to the total character per the unshifted atom

Operation	Contribution per atom	Operation	Contribution per atom
E	3	C_3^1, C_2^3	0
C_2	−1	C_4^1, C_4^3	1
σ	1	C_6^1, C_6^5	2
i	−3	S_3^1, S_3^2	−2
		S_4^1, S_4^3	−1
		S_6^1, S_6^5	0

The total characters of the operation give the total representation. This procedure is shown in Table 11.6.

Table 11.6 Calculation of the total representation for the C_{2v} structure of H_2O

C_{2v}	E	C_2	$\sigma_v(xz)$	$\sigma_v(yz)$
No. of unshifted atoms contribution to the total character per unshifted atom from Table 11.5	3 3	1 −1	3 1	1 1

Contd...

C_{2v}	E	C_2	$\sigma_v(xz)$	$\sigma_v(yz)$
Total character of the representation of which, the vibration form the basis. Γ_{total} or Γ_{3IV}	9	−1	3	1

*Γ_{3N} = No. of unshifted atom × contribution to the total character per unshifted atom.

The total representation (9, −1, 3, 1) is some combination of the characters listed for the species A_1, A_2, B_1 and B_2 in the character Table 11.7. These species are referred to as irreducible representations (e.g. +1, +1, −1, −1 is the irreducible representation for the symmetry species A_2).

Table 11.7 Irreducible representation in the C_{2v} character table

C_{2v}	E	C_2	$\sigma_v(xz)$	$\sigma'_v(yz)$		
A_1	1	1	1	1	z	x^2, y^2, z^2
A_2	1	1	−1	−1	R_z	xy
B_1	1	−1	1	−1	x, R_y	xz
B_2	1	−1	−1	1	y, R_x	yz

Our next task is to figure out what combination of the irreducible representation should be taken to give (9, −1, 3, 1) representation. The number of times (n) of an irreducible representation (Γ_i) occurs in a reducible representation is determined by using standard reduction formula

$$n(\Gamma_i) = \frac{1}{h}\left[\sum C(R)\,\chi_{RR}(R)\,\chi_{IR}(R)\right]$$

where $\chi_{RR}(R)$ is the character of the reducible representation under operation R, $\chi_{IR}(R)$ the corresponding irreducible representation character, C(R) the number of operations in the class to which R belongs to and h is the order of the group.

$$n(A_1) = \frac{1}{4}\left[(1)(9)(1) + (1)(-1)(1) + (1)(3)(1) + (1)(1)(1)\right] = 3$$
$$\text{for E} \qquad \text{for } C_2 \qquad \text{for } \sigma_v(xz) \quad \text{for } \sigma'_v(yz)$$

$$n(A_2) = \frac{1}{4}\left[(1)(9)(1) + (1)(-1)(1) + (1)(3)(-1) + (1)(1)(-1)\right] = 1$$

$$n(B_1) = \frac{1}{4}\left[(1)(9)(1) + (1)(-1)(-1) + (1)(3)(1) + (1)(1)(-1)\right] = 3$$

$$n(B_2) = -\left[(1)(9)(1) + (1)(-1)(-1) + (1)(3)(-1) + (1)(1)(1)\right] = 2$$

where n refers to the total number of times a particular species (A_1, A_2, B_1, B_2) contributes to the total representation.

The reducible representation for all motions (translation, rotation and vibration) of water molecule is, therefore, reduced to $3A_1 + 1A_2 + 3B_1 + 2B_2$

∴ $\Gamma_{3N} = 3A_1 + 1A_2 + 3B_1 + 2B_2$ (a total of 9 representation)

From the character table of C_{2v}, we have

Γ_{trans} (for right in the = $A_1 + B_1 + B_2$ (translation is motion along the x, y, z directions, character table) so it transforms in the same way as the three axes – a total of 3 representations)

$\Gamma_{rot} = A_2 + B_1 + B_2$ (rotation in three directions R_x, R_y, R_z – a total of 3 representations)

Hence, $\Gamma_{vib.} = \Gamma_{3N} - (\Gamma_{trans} + \Gamma_{rot})$

$= 3A_1 + A_2 + 3B_1 + 2B_2 - (A_1 + B_1 + B_2 + A_2 + B_1 + B_2)$

$= 2A_1 + B_1$

The three vibrational modes are shown in Table 11.8.

Table 11.8 Symmetry of molecular modes of water

All motions	Translation (x, y, z)	Rotation (R_x, R_y, R_z)	Vibration (Remaining modes)
$3A_1$	A_1		$2A_1$
A_2		A_2	
$3B_1$	B_1	B_1	B_1
$2B_2$	B_2	B_2	

Two of the modes are totally symmetric (A_1) and do not change the symmetry of the molecule, but one is antisymmetric to C_2 rotation and to reflection, perpendicular to the plane of the molecule (B_1). These models are illustrated as symmetric stretch, symmetric bend and antisymmetric stretch (Table 11.9)

Table 11.9 The vibrational modes of water

A_1	H O H	Symmetric stretch: Change in dipole moment; more distance between positive hydrogen and negative oxygen: IR active
B_1	H O H	Antisymmetric stretch: Change in dipole moment; change in distance between positive hydrogens and negative oxygen: IR active.
A_1	H O H	Symmetric bend: Change in dipole moment; angle between H—O vectors changes: IR active

In group theory terms, a vibrational mode is active in the infrared if it corresponds to an irreducible representation and has the same symmetry as the cartesian coordinate $x, y,$ or z, because a vibrational motion that shifts the centre of charge of the molecule in any of the $x, y,$ or z directions results in a change in dipole moment. Otherwise, the vibrational mode is not infrared active.

For a vibration to be Raman active, there must be a change in the polarisable tensor. The components of the polarisability tensor transform as the quadratic functions of x, y and z. Therefore, in the character tables we are looking for $x^2, y^2, z^2, xy, yz, zx$ or their combinations such as $x^2 - y^2$. Since A_1 vibration corresponds to x^2, y^2, z^2 and B_1 vibration corresponds to the binary product xz, all the three ($2A_1 + B_1$) vibrations bring change in polarisability and are Raman active. The infrared and Raman activity of H_2O fundamentals can be summarised below:

Mode	Infrared	Raman	Raman
$2A_1$	+	+	Mixed (a combination of O—H bond stretch and $\widehat{HOH}$ angle deformation)
B_1	+	+	Pure

The three vibrations of H_2O molecule absorb at 3652 cm^{-1} ($v_1(A_1)$), 1545 cm^{-1} ($v_2(A_1)$) and 3756 cm^{-1} ($v_3(B_1)$) respectively.

SO$_2$:

Sulphur dioxide belongs to C_{2v} point group and is predicted to have their normal modes from the $3N - 6$ rule ($3 \times 3 - 6 = 3$) spectral data (Table 11.10) show the presence of m their three bands. The three bands at 1361, 1151 and 519 cm^{-1} are the *fundamentals* and are referred to as the v_3 (or v_{as}), v_1 (or v_s) and v_2 (or δ) bands (Fig. 11.7). The symmetrical stretching vibrations (v_s or v_1) are of higher energy their banding vibrations (δ or v). Asymmetric stretching (v_{as} or v_3) are of higher the their symmetrical vibration. Hence the order of energies is $v_{as} > v_s > \delta$ (or $v_3 > v_1 > v_2$). The terms v_1, v_2, v_3, etc. are used for symmetrical vibrations in order of energies and their asymmetrical vibration in order of energies. In cases of molecules where bending out of plane is IR active, the term π is used to represent the vibration.

Table 11.10 Infrared spectrum of SO$_2$

v (cm^{-1})	Assignment	v (cm^{-1})	Assignment
519	v_2 (δ)	606	$v_1 - v_2$
1151	v_1 (v_s)	1871	$v_2 + v_3$
1361	v_3 (v_{as})	2305	$2v_1$
		2499	$v_1 + v_3$

Bands are also observed at 606 cm^{-1}, 1871 cm^{-1}, 2305 cm^{-1} and 2499 cm^{-1} corresponding to ($v_1 - v_2$), ($v_2 + v_3$), $2v_1$ and ($v_1 + v_3$) respectively, difference bands ($v_1 - v_2$) involve a transition from the state v_2 to which the molecule had already been excited to v_1. Combination bands ($v_2 + v_3$) or ($v_1 + v_2$) involve simultaneous excitation of both vibrational modes. The overtone of v_1 occurs at about $2v_1$ or 2305 cm^{-1} ($2v_1$ is the first overtone of the stretching vibration).

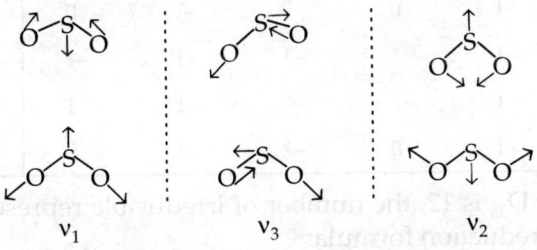

Fig. 11.7 The three fundamental vibrations for SO$_2$

The occurance of overtone and the combination bands is also governed by some symmetry requirements. The irreducible representation to which the overtone belongs can be obtained by squaring the characters of the fundamental vibration. For the overtone to occur, it must belong to irreducible representation corresponding to the vectors x, y, or z. For example, in SO_2, v_1 below to A_1. Hence, the character of the overtone $2v_1$ is $A_1 \times A_1 = A_1$ and will be IR active.

Similarly for a combination band to the IR active, the product of the characters of the two individual vibrations should belong to an irreducible representation corresponding to one of the vectors. Thus, the $(v_2 + v_3)$ and $(v_1 + v_3)$ both have characters $A_1 \cdot B_1 = B_1$ and hence are IR active.

In some cases, fundamental vibration may not be IR active but the overtone or combination bands may be IR active, as per above symmetry combinations.

As a third of the use of character tables in the analysis of IR and Raman spectra, we choose BCl_3 with D_{3h} symmetry. This compound has four atoms. The total number of degrees of freedom $= 3N = 3 \times 4 = 12$ of which translational modes $= 3$ and rotational modes $= 3$. So the number of modes of vibrations $= 3N - 6 = 3 \times 4 - 6 = 6$ of which bond stretching is $N - 1 = 4 - 1 = 3$ (since there are 3 bonds) and the remaining three $(6 - 3 = 3)$ will be bending modes. Table 11.11 shows the derivative of Γ_{total} (or Γ_{3N}) for BCl_3 molecule.

Table 11.11 Derivation of Γ_{total} for BCl_3

D_{3h}	E	$2C_3$	$3C_2$	σ_h	$2S_3$	$3\sigma_v$
Unshifted atoms	4	1	2	4	1	2
Contribution to the total character per unshifted atom (see Table 11.5)	3	0	−1	1	−2	1
Γ_{total} (or Γ_{3N}^{*})	12	0	−2	4	−2	2

$^{*}\Gamma_{3N}$ = No. of unshifted atom × contribution to the total character per unshifted atom.

The irreducible representation of D_{3h} point is shown in Table 11.12.

Table 11.12 The character table of the D_{3h} point group

D_{3h}	E	$2C_3$	$3C_2$	σ_h	$2S_3$	$3\sigma_v$		
A_1'	1	1	1	1	1	1		$x^2 + y^2, z^2$
A_2'	1	1	−1	1	1	−1	R_z	
E'	2	−1	0	2	−1	0	(x, y)	$x^2 - y^2, xy$
A_1''	1	1	1	−1	−1	−1		
A_2''	1	1	−1	−1	−1	1	z	
E''	2	−1	0	−2	1	0	(R_x, R_y)	xz, yz

Since order (h) of D_{3h} is 12, the number of irreducible representations are obtained following standard reduction formula:

$$n(\Gamma_i) = \frac{1}{h} \left[\sum C(R)\, \chi_{RR}(R)\, \chi_{IR}(R) \right]$$

$$n(A_1') = \frac{1}{12} [(1)(12)(1) + (2)(0)(1) + (3)(-2)(1) + (1)(4)(1)$$
$$+ (2)(-2)(1) + (3)(2)(1)] = 1$$

$$n(A_2') = \frac{1}{12} [(1)(12)(1) + (2)(0)(1) + (3)(-2)(-1) + (1)(4)(1)$$
$$+ (2)(-2)(1) + (3)(2)(-1)] = 1$$

$$n(E') = \frac{1}{12} [(1)(12)(2) + (2)(0)(-1) + (3)(-2)(0) + (1)(4)(2)$$
$$+ (2)(-2)(-1) + (3)(2)(0)] = 3$$

$$n(A_1'') = \frac{1}{12} [(1)(12)(1) + (2)(0)(1) + (3)(-2)(1) + (1)(4)(-1)$$
$$+ (2)(-2)(-1) + (3)(2)(-1)] = 0$$

$$n(A_2'') = \frac{1}{12} [(1)(12)(1) + (2)(0)(1) + (3)(-2)(-1) + (1)(4)(-1)$$
$$+ (2)(-2)(-1) + (3)(2)(1)] = 2$$

$$n(E'') = \frac{1}{12} [(1)(12)(2) + (2)(0)(-1) + (3)(-2)(0) + (1)(4)(-2)$$
$$+ (2)(-2)(1) + (3)(2)(0)] = 1$$

$\therefore \quad \Gamma_{total}$ (or Γ_{3N}) $= A_1' + A_2' + 3E' + 2A_2'' + E''$ (a total of 12 representations)

From the character table of D_{3h} point group, we have

$\qquad \Gamma_{trans} = E' + A_2''$ and $\Gamma_{rot} = A_2' + E''$

$\therefore \qquad \Gamma_{vib} = \Gamma_{total} - (\Gamma_{trans} + \Gamma_{rot}) = A_1' + 2E' + A''_2$ (a total of 6 representations)

Thus the six fundamental vibrations of BCl_3 transform A_2'' and $2E'$. Each E' representation describes two vibrations modes of equal energy. Thus the $2E'$ notation represents four different vibrations, two of one in and other two of another. A_1' mode is Raman active, the A_2'' is IR active, and the E' modes are both Raman and IR active.

NH₃:

Ammonia (point group C_{3v}) has a total twelve ($= 3N$, $N = 4$) degrees of freedom of which six ($= 3N - 6 = 3 \times 4 - 6 = 6$) belongs to vibrational motions. The reducible representation of NH_3 is given below.

C_{3v}	E	$2C_3$	$3\sigma_v$
Unshifted atoms	4	1	2
Contribution to the total character per unshifted atom (vide Table 11.5)	3	0	1
Γ_{total} (or Γ_{3N})	12	0	2

The irreducible representations of C_{3v} point group is shown in Table 11.13.

Table 11.13 Character table of the C_{3v} point group

C_{2v}	E	$2C_3$	$3\sigma_v$		
A_1	1	1	1	z	$(x^2 + y^2, z^2)$
A_2	1	1	-1	R_z	
E	2	-1	0	$(x, y), (R_x, R_y)$	$(x^2 - y^2, xy), (zx, yz)$

$$n(A_1) = \frac{1}{6} [(1)\,(12)\,(1) + (2)\,(0)\,(1) + (3)\,(2)\,(1)] = 3$$

$$n(A_2) = \frac{1}{6} [(1)\,(12)\,(1) + (2)\,(0)\,(1) + (3)\,(2)\,(-1)] = 1$$

$$n(E) = \frac{1}{6} [(1)\,(12)\,(2) + (2)\,(0)\,(-1) + (3)\,(2)\,(0)] = 4$$

Hence $\qquad \Gamma_{3N}$ or $(\Gamma_{total}) = 3A_1 + A_2 + 4E$

From the character table of C_{3v} point group we have

$$\Gamma_{trans} = A_1 + E \text{ and } \Gamma_{rot} = A_2 + E$$

$\therefore \qquad \Gamma_{vib} = \Gamma_{total} - (\Gamma_{trans} + \Gamma_{rot})$

$$= 2A_1 + 2E$$

The activities of these fundamentals are as follows:

Infrared active : A_1, E

Raman active : A_1, E

So all the fundamentals will appear in the both the infrared and Raman spectra.

CH_4:

Methane (point group T_d) is a non-linear molecule and contains 5 atoms. So it has a total degrees of freedom = $3 \times 5 = 15$ out of which translational mode = 3, rotational mode = 3. The remaining $15 - 6 = 9$ belongs to vibrational modes. The reducible representation is as follows:

T_d	E	$8C_3$	$3C_2$	$6C_4$	$6\sigma_d$
No. of unshifted atoms	5	2 (C-atom) + (1 H atoms)	1 (C-atom)	1 (C-atom)	2 (C-atom) + (2 H atom)
Contribution to the total character per unshifted atom (vide Table 11.5)	3	0	-1	-1	1
Γ_{total} or Γ_{3N}	15	0	-1	-1	3

The irreducible representations are obtained for the character table of T_d (Table 11.14).

Table 11.14 Character table of the T_d point group

T_d	E	$8C_3$	$3C_2$	$6S_4$	$6\sigma_d$		
A_1	1	1	1	1	1		$x^2 + y^2 + z^2$
A_2	1	1	1	-1	-1		
E	2	-1	2	0	0		$2z^2 - x^2 - y^2$
T_1	3	0	-1	1	-1	(R_x, R_y, R_z)	
T_2	3	0	-1	-1	1	(x, y, z)	(xy, xz, yz)

Now order $(h) = 24$

$$n(A_1) = \frac{1}{24} \left[(1)(15)(1) + (8)(0)(1) + (3)(-1)(1) + (6)(-1)(1) + (6)(3)(1) \right]$$

$$n(A_2) = \frac{1}{24} \left[(1)(15)(1) + (8)(0)(1) + (3)(-1)(1) + (6)(-1)(-1) + (6)(3)(-1) \right] = 0$$

$$n(E) = \frac{1}{24} \left[(1)(15)(2) + (8)(0)(-1) + (3)(-1)(2) + (6)(-1)(0) + (6)(3)(0) \right] = 1$$

$$n(T_1) = \frac{1}{24} \left[(1)(15)(3) + (8)(0)(0) + (3)(-1)(-1) + (6)(-1)(1) + (6)(3)(-1) \right] = 1$$

$$n(T_2) = \frac{1}{24} \left[(1)(15)(3) + (8)(0)(0) + (3)(-1)(-1) + (6)(-1)(-1) + (6)(3)(1) \right] = 3$$

Hence $\Gamma_{3N} = A_1 + E + T_1 + 3T_2$ (a total of 15 representations)

From the character table of T_d (Table 11.14), we have

$\Gamma_{trans} = T_2$ (a total of 3 representations)

$\Gamma_{rot} = T_1$ (a total of 3 representations)

Hence the resultant vibrations are

$$\Gamma_{vib} = \Gamma_{3N} - (\Gamma_{trans} + \Gamma_{rot})$$
$$= A_1 + E + T_1 + 3T_2 - T_2 - T_1$$
$$= A_1 + E + 2T_2 \text{ (a total of 9 representations)}$$

The character table also shows that the activities of the fundamentals are as follows:

Raman only   : A_1 and E

IR and Raman : T_2

CH₃D:

The CH_3D molecule belongs to C_{3v} point group, the reducible representation based on three vectors on each atoms is

C_{3v}	E	$2C_3$	$3\sigma_v$
Γ_{total} or Γ_{3N}	15	0	3

The resolution of this representation gives

$$\Gamma_{3N} = 4A_1 + A_2 + 5E$$

From the character table (Table 11.13), we find that

$$\Gamma_{trans} = A_1 + E, \Gamma_{rot} = A_2 + E$$

Hence $\quad \Gamma_{vib} = \Gamma_{3N} - (\Gamma_{trans} + \Gamma_{rot})$

$$= 4A_1 + A_2 + 5E - A_1 - E - A_2 - E$$

$$= 3A_1 + 3E$$

The activities of these fundamentals are shown below:

Infrared active : A_1, E

Raman active : A_1, E

SF_6:

This molecule is non-linear and belongs to the point group O_h. It has $3 \times 7 = 21$ degrees of freedom of which translational mode = 3 and rotational mode = 3. The remaining $21 - 6 = 15$ belongs to vibrational modes. The reducible representation is as follows:

O_h	E	$8C_3$	$6C_2$	$6C_4$	$3C_2 (= C_4^2)$	i	$6S_4$	$8S_6$	$3\sigma_h$	$6\sigma_d$
No. of unshifted atoms	7	1	1	3	3	1	1	1	5	3
Contribution to the total character per unshifted atom (vide Table 11.5)	3	0	−1	1	−1	−3	−1	0	1	1
Γ_{total} or Γ_{3N}	2	1	−1	3	−3	−3	−1	0	5	3

The Γ_{3N} representation can be reduced by using standard reduction formula and we get

$$\Gamma_{3N} = A_{1g} + E_g + T_{1g} + 3T_{1u} + T_{2g} + T_{2u}$$

$$= (1) + (2) + (3) + (3 \times 3 = 9) \, (3) \, (3) = 21 \text{ total modes}$$

From the character table of O_h point group, we can identify the translational and rotational representations as:

$$\Gamma_{trans} = \begin{Bmatrix} x \\ y \\ z \end{Bmatrix} \to T_{1u} \text{ and } \Gamma_{rot} = \begin{Bmatrix} R_x \\ R_y \\ R_z \end{Bmatrix} \to T_{1g}$$

Thus the resultant vibrational modes are

$$\Gamma_{vib} = \Gamma_{3N} - (\Gamma_{trans} + \Gamma_{rot})$$

$$= A_{1g} + E_g + T_{1g} + 3T_{1u} + T_{2g} + T_{2u} - T_{1u} - T_{1g}$$

$$= A_{1g} + E_g + 2T_{1u} + T_{2u} + T_{2g}$$

$$= (1) + (2) + (2 \times 3) + (3) + (3) = 15 \text{ vibrational modes}$$

The activities of these fundamentals are as follows:

Infrared active : $2T_{1u}$

Raman active : A_{1g}, E_g, T_{2g}

Inactive : T_{2u}

trans-N_2F_2:

The planar non-linear *trans*-N_2F_2 is four atomic molecule. It has twelve degrees of freedom ($3N = 3 \times 4 = 12$) of which 3 are translations motions, 3 are rotational motions and the remaining $12 - 3 - 3 = 6$ belongs to vibrational motion. The molecule belongs to C_{2h} point group. The reducible representation of this molecule is shown below:

C_{2h}	E	C_3	C_3	σ_h
No. of unshifted atoms	4	0	0	4
Contribution to the total character per unshifted atom (vide Table 11.5)	3	−1	−3	1
Γ_{total} or Γ_{3N}	12	0	0	4

The irreducible representations are obtained from the character table of C_{2h} (Table 11.15) point group.

Table 11.15 The character table of the C_{2h} point group

C_{2h}	E	C_2	i	σ_h		
A_g	1	1	1	1	R_z	x^2, y^2, z^2, xy
B_g	1	−1	1	−1	R_x, R_y	xz, yz
A_u	1	1	−1	−1	z	
B_u	1	−1	−1	1	x, y	

Since order (h) of C_{2h} is 4, the number of irreducible representations are obtained following standard reduction formula:

$$n(A_g) = \frac{1}{4} \ [(1) \ (12) \ (1) + (1) \ (0) \ (1) + (1) \ (0) \ (1) + (1) \ (4) \ (1)] = 4$$

$$n(B_g) = \frac{1}{4} \ [(1) \ (12) \ (1) + (1) \ (0) \ (−1) + (1) \ (0) \ (1) + (1) \ (4) \ (−1)] = 2$$

$$n(A_u) = \frac{1}{4} \ [(1) \ (12) \ (1) + (1) \ (0) \ (1) + (1) \ (0) \ (−1) + (1) \ (4) \ (−1)] = 2$$

$$n(B_u) = \frac{1}{4} \ [(1) \ (12) \ (1) + (1) \ (0) \ (−1) + (1) \ (0) \ (−1) + (1) \ (4) \ (1)] = 4$$

Hence $\qquad\qquad \Gamma_{3N}$ or $\Gamma_{total} = 4A_g + 2B_g + 2A_u + 4B_u$

From the character table (Table 11.15) of C_{2h}, we have

$$\Gamma_{trans} = A_u + B_u + B_u$$
$$\Gamma_{rot} = A_g + B_g + B_g$$

Hence,

$$\Gamma_{vib} = \Gamma_{total} - (\Gamma_{trans} + \Gamma_{rot})$$
$$= 3A_g + A_u + 2B_u$$

From the right columns of the character table (Table 11.15), we have

IR active   : A_u, B_u

Raman active : A_g

CO_2:

In a linear molecule ($D_{\alpha h}$ point group), there should be $3N - 5 (= 3 \times 3 - 5) = 4$ vibrations possible. There are v_s, v_{as} and bending vibrations in plane and out of plane.

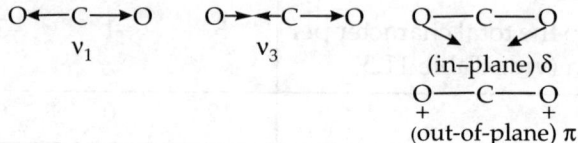

Out of four vibrations, v_s does not bring charge in the dipole moment and should be IR inactive. The IR active vibrations are v_{as}, δ and π. Though δ and π are at different vibration modes, but for a linear molecule they are equivalent and hence are degenerate, i.e. involve same energy change.

The IR bands occuring at 2349 cm⁻¹ and 667 cm⁻¹ have been assigned to v_{as} and δ or π. The band at 1340 cm⁻¹ corresponding to v_s does not occur because it is IR inactive. However, two intense bands are observed at 1388 cm⁻¹ and 1286 cm⁻¹. This splitting is due to a phenomenon known as *Fermi resonance*.

Therefore, there are two vibrations very close in energy and are of same symmetry, they combine by a quantum mechanical resonance resulting in two new vibrations— one having energy higher than the combining vibrations and another having energy lower than the combining vibrations. In case of CO_2 molecule, v_1 is expected at 1340 cm⁻¹ but is IR inactive. The first overtone $2v_2$ ($2 \times 667 = 1334$ cm⁻¹) occurs very near to the fundamental v_1. Hence, they undergo Fermi resonance resulting in two bands at 1385 cm⁻¹ and 1286 cm⁻¹.

In the absence of the resonance, $2v_2$ would have occured as a weak overtone. But due to two vibration interactions, the fundamental vibration v_1 becomes allowed since both vibrations consist partly of the fundamental vibration, the intensity of the fundamental is distributed between the two bands. Thus two intense bands are observed at 1388 cm⁻¹ and 1286 cm⁻¹.

Square Planar AB_4 Molecule

Square planar AB_4 molecule belongs to D_{4h} point group. The possible vibrations are $(3 \times 5 - 6) = 9$ and they are A_{1g}, B_{1g}, A_{2u}, B_{2g}, B_{2u} and $2E_u$ (4 vibrations) species. On refer- ing to D_{4h} character table, it can be seen that vibrations belonging to A_{2u} and E_u are IR active and A_{1g}, B_{1g}, B_{2g} are Raman active.

If the molecular structure of the unknown compound is not known, in that case the molecule is assigned a particular structure and the possible number of vibrations are theoretically calculated. The number of vibrations are compared with the actual number of bands observed in the IR spectrum of the compound. If there is an agreement, the assigned structure may be correct, otherwise an alternate structure can be opted.

Let us consider SF_4 molecule. The possible structure are given below:

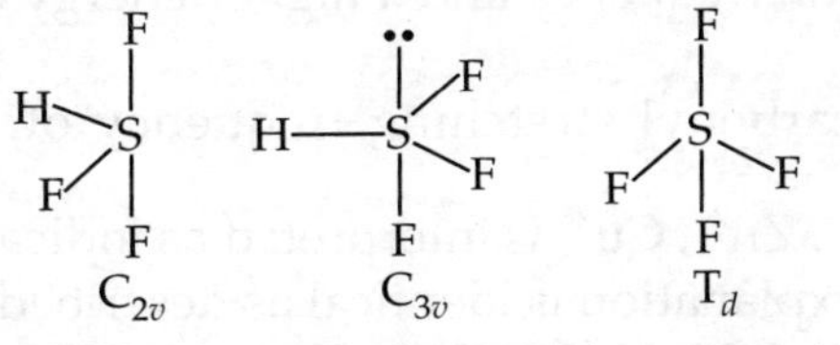

$$\text{(I) } C_{2v} \qquad \text{(II) } C_{3v} \qquad \text{(III) } T_d$$

Applying group theory, it can be shown that SF_4 molecule with C_{2v} symmetry there should be eight IR active vibrations ($4A_1$, $2B_1$ and $2B_2$); with C_{3v} symmetry there should be six IR active vibrations ($2A_1$ and $2E$), and with T_d symmetry there should be six IR active vibrations belonging to two T_2 species and should give rise to two fundamental bands in IR spectrum. The actual spectrum shows five IR fundamental bands. This rules out T_d structure. It may be having C_{2v} or C_{3v} symmetry. Some of the fundamental bands are of lower intensity and hence remain hidden under overtone and combination bands. This is a problem in the assignment of structures on the basis of IR spectra. The overtones and combination bands have to be separated from the fundamentals. On doing this, it is observed that SF_4 belongs to C_{2v} point group and hence has structure (I).

11.10 CHANGE IN THE SPECTRA OF DONOR MOLECULES UPON COORDINATION

The IR spectrum of N, N-dimethylacetamide in CCl_4 solvent has an absorption band at 1662 cm^{-1} compared to acetone (1715 cm^{-1}). Such low frequency of C=O group in N,N-dimethylacetamide is attributed to a resonance interaction with the lone pair on the nitrogen (Fig. 11.8).

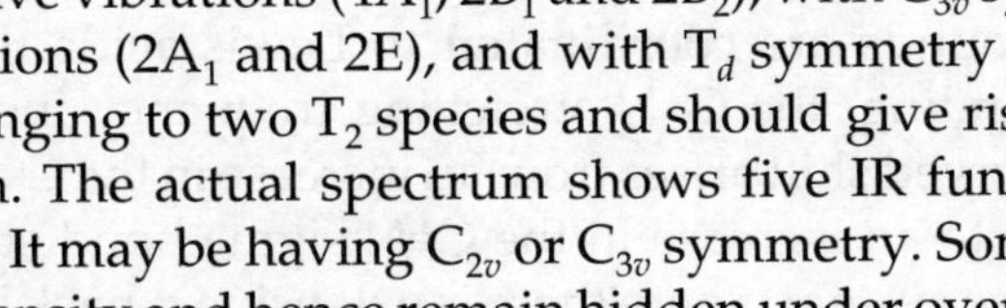

Fig. 11.8 Resonance structrues for N,N-dimethylacetamide

Upon coordination to a Lewis acid, a decrease in the frequency of this band is observed. This decrease has been attributed to the effect of oxygen coordination to the Lewis acid (X).

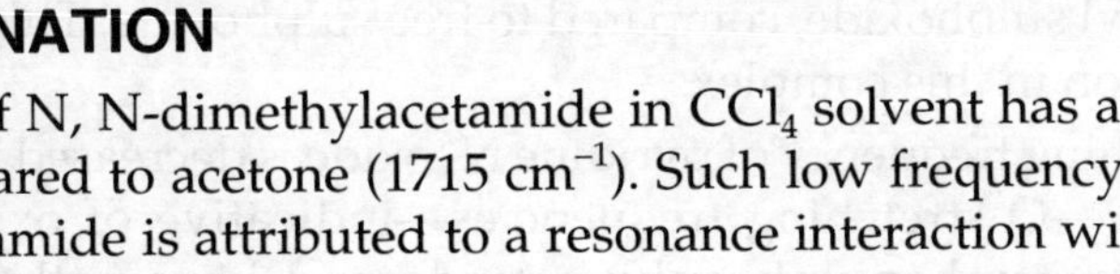

The most important effect in this case involves decreasing the carbonyl force constant by draining π-electron density out of the carbonyl group. This causes the observed decrease in the carbonyl frequency and indicates oxygen coordination. The absence of any absorption in the carbonyl region on the high frequency side of the uncomplexed carbonyl band is further supported for oxygen coordination. If there were nitrogen coordination in the complexes, the nitrogen lone pair would be involved, resulting in a decreased C—N vibration frequency and a higher-energy carbonyl absorption.

The decrease in the carbonyl stretching frequency of urea $\left(O=C\begin{smallmatrix}\diagup NH_2 \\ \diagdown NH_2\end{smallmatrix}\right)$ upon complexation to Fe^{3+}, Cr^{3+}, Zn^{2+}, Cu^{2+} is interpreted as indicating oxygen coordination in these complexes. The explanation is identical as described for the amides. Nitrogen coordinates are observed in $Pd(NH_2CONH_2)_2 Cl_2$ and $Pt(NH_2CONH_2)_2 Cl_2$ compounds. The IR spectra show the expected increase of C—O stretching frequency as well as a decrease in the C—N frequency.

When nitrile compounds are coordinated to Lewis acids that are not generally involved in p-back bonding, the C≡N stretching frequency increases. The principal contribution to the observed shift arises from an increase in the C≡N force constant. This increase is mainly due to an increase in the C≡N sigma bond strength from nitrogen rehybridisation. In those system where there is extensive π-back bending from the acid into the π^* orbitals of the nitrile group, the decreased π bond energy accounts for the decreased frequency.

A decrease in the S—O stretching frequency, indicative of oxygen coordination, is observed when dimethyl sulphoxide or tetramethylene sulphoxide is complexed to metal ion, iodine and phenol. Interestingly S—O stretching increases in the palladium complex of dimethyl sulphoxide, compared to free sulphoxide. This is an indication of sulphur coordination in this complex.

The N—O stretching frequency of pyridine N–oxide is decreased upon coordination. Decreases in the P—O stretching frequencies—indicative of oxygen coordination are observed when triphenylphosphine oxide and hexamethylphosphoramide, $OP[N(CH_3)_2]_3$, are coordinated to metal ions, phenol or iodine.

11.11 INFRARED SPECTRA OF COMPLEX COMPOUNDS

A complex compound is formed by coordination of ligand molecules (or ions) (nL^-) to the metal ion (M^{n+}).

$$M^{n+} + nL^- \longrightarrow ML_n$$

when the complex compound is formed, certain changes in the properties of the ligand is noted. These are as follows:

11.11.1 Polarisation of Bonds

Due to coordination of the ligand to the metal ion, the electron density moves from ligand to metal ion causing the polarisation of the bonds of the ligand atom. For example in $[M(NH_3)_m]^{n+}$ complex, the coordination of nitrogen atom with the metal ion causes polarisation of the N—H bond and therefore, the vibrational band (v_{N-H}) occurs at lower frequencies ~ 3100 cm^{-1} as compared to free ammonia (~ 3600 cm^{-1}). Such lowering of the frequency of the band has been correlated with the strength of the metal-ligand bond strength. The greater the metal-ligand bond strength, the greater is the lowering of v_{N-H}.

11.11.2 Symmetry Lowering

The symmetry of a ligand is changed upon complexation. For example, the symmetry change accompanying the bending of small molecules (e.g. N_2, O_2 and H_2) to transition metal ions has a dramatic influence on the infrared spectra of these materials. When azide ion is added to $[Ru(NH_3)_5(H_2O)]^{3+}$, a product containing coordinated dinitrogen, N_2, is obtained. The stretching vibration of free N_2 is infrared inactive but Raman active with a band at 2331 cm^{-1}. In $[Ru(NH_3)_5(N_2)]^{2+}$, a sharp intense line appears at 2130 cm^{-1} in the infrared spectrum that is attributed to the N—N stretching vibration. In $IrN_2Cl\,[P(C_6H_5)_3]_2$ and $HCoN_2\,[P(C_6H_5)_3]_3$ the same vibrations gives rise to bands at 2095 cm^{-1} and 2088 cm^{-1}, respectively.

Trans-$[IrCl(CO)(PPh_3)_2]$ reacts with dioxygen to give $[Ir(Cl)(CO)(O_2)(PPh_3)_2]$. The complex $[IrCl(CO)(PPh_3)_2]$ binds O_2 in such a way that the O—O bond axis is perpendicular to line drawn from iridium to the centre of the O—O bond.

An IR bond assigned to the O—O stretching vibration in the complex occurs at 857 cm^{-1}. This corresponds to an infrared inactive stretch in free O_2, which does have a Raman band at 1555 cm^{-1}.

When H_2 is added to $Ir(CO)(Cl)(PPh_3)_2$, two new sharp absorption bands arise in the infrared at 2190 and 2100 cm^{-1}. The infrared bands at 2160 and 2101 cm^{-1} in $Ir(H_2)(CO)$ $(PPh_3)_2$ are shifted to 1620 and 1548 cm^{-1} in the deuterated analogue. The agreement of these shifts with those predicted from the reduced mass indicates that there vibrations are mainly metal-hydrogen modes.

The infrared spectra of coordination compounds may be analysed by subgrouping the vibrations into ligand vibrations and metal-ligand (skeletal) vibrations. The interaction of the vibrations of one ligand with those of another ligand in the complex is considered to be negligible. Thus, the infrared spectra of metal complexes are usually interpreted by taking only a (1:1) metal to ligand complex. The skeletal vibrations comprising of metal-ligands modes are dependent upon the stereochemistry of the metal complex while the ligand vibrations are insensitive to the stereochemistry of the complex. The classic examples of aqua and ammine complexes are considered first for illustration before taking up more complex molecules.

11.11.3 Aqua Complexes

The infrared spectra of aqua complexes, $[M(H_2O)_6]^{n+}$, can be interpreted by considering a (1:1) metal to H_2O complex, $M–OH_2$, since interligand interactions can be neglected. The complex $M–OH_2$ belongs to the same point group (C_{2v}) as that of free aqua molecule (C_{2v}).

Free H_2O molecule has three normal vibration frequencies observed at $v_1 = 3642$ (A_1), $v_2 = 1595$ (A_1) and $v_3 = 3756$ (B_2) cm^{-1} respectively.

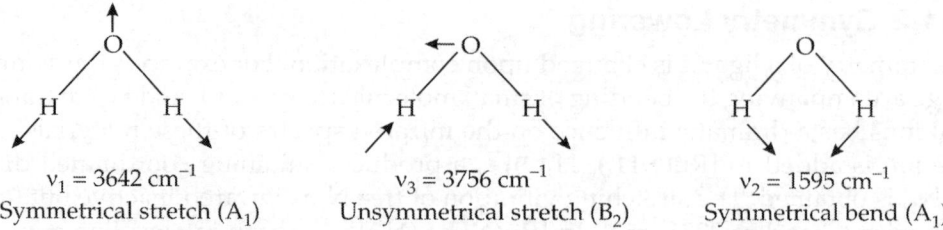

$\nu_1 = 3642 \text{ cm}^{-1}$
Symmetrical stretch (A_1)

$\nu_3 = 3756 \text{ cm}^{-1}$
Unsymmetrical stretch (B_2)

$\nu_2 = 1595 \text{ cm}^{-1}$
Symmetrical bend (A_1)

Coordinated H_2O molecule exhibits the following new modes of vibrations, in addition to the fundamentals.

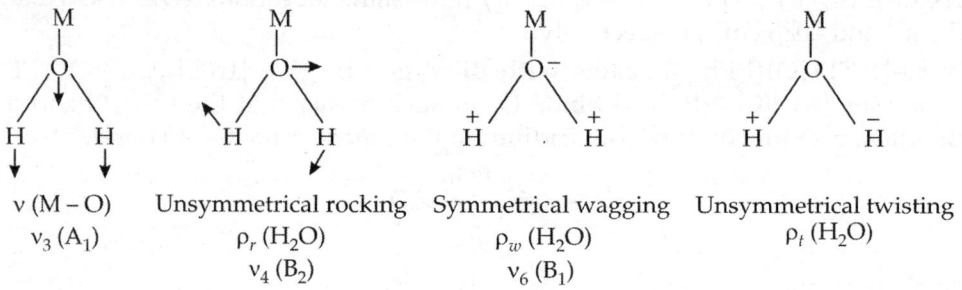

ν (M – O)
ν_3 (A_1)

Unsymmetrical rocking
ρ_r (H_2O)
ν_4 (B_2)

Symmetrical wagging
ρ_w (H_2O)
ν_6 (B_1)

Unsymmetrical twisting
ρ_t (H_2O)

Generally, the ligand stretching bands are shifted to lower frequencies upon complex formation, for example, the (O—H) bond of free H_2O occurs at 3642 cm^{-1} and is shifted to 3250 cm^{-1} in [Cr(H_2O)$_6$]Cl$_3$. The lowering of the band has been mainly assigned due to the reduction in bond order of O—H of electrons from oxygen atom of aqua molecule to the metal ion on coordination weakens O—H bond to some extent (i.e. bond order is reduced). Therefore, the force constant (K) value decreases and energy (E) also decreases, $\Delta E \propto \sqrt{\dfrac{K}{\mu}}$. The bending mode of the free H_2O molecule increases by 20 to

50 cm^{-1} on coordination of H_2O to a metal. There are atleast three new bands (unsymmetrical rocking ρ_r, symmetrical wagging ρ_w and unsymmetrical twisting ρ_t), which are activated due to coordination. The unsymmetrical rocking ρ_r and metal-oxygen stretching ν_3 (A_1) modes will become IR active if M—O bond is sufficiently covalent.

The M—OH$_2$ rocking mode (band) has been identified in the region of 700–850 cm^{-1} as a band of medium intensity. The M—OH$_2$ wagging mode is observed in the 600–700 cm^{-1} region (medium band intensity). The metal-oxygen stretching frequency, ν(M—O), has been identified in the 350–450 and 450–500 cm^{-1} region for the divalent and trivalent [M(H_2O)$_6$]$^{n+}$ complexes of the first transition series and it is of weak to medium intensity. The infrared frequencies for a few aqua complexes are shown in Table 11.16.

Table 11.16 Infrared frequencies in cm^{-1} for some aqua complexes

Complex	Stretching band $\nu(H_2O)$	Bending mode $\delta(H_2O)$	Rocking mode $\rho_r(H_2O)$	Wagging mode ρ_w	Twisting mode ρ_t
H_2O	3750, 3642	1600	–	–	–
[Ni(H_2O)$_6$]SiF$_6$	3350, 3200	1640	755	645	405
[Cr(H_2O)$_6$]Cl$_3$	3300, 3250	1650	800	541	490

Contd...

Complex	Stretching band $v(H_2O)$	Bending mode $\delta(H_2O)$	Rocking mode $\rho_r(H_2O)$	Wagging mode ρ_w	Twisting mode ρ_t
$[Cu(H_2O)_4]SO_4 \cdot H_2O$	3200, 3150	1640	887, 855	535	440
$[Zn(H_2O)_6]SO_4$	3450, 3350	1650	621	541, 555	358

11.11.4 Amine Complexes

Free NH_3 molecule (C_{3v} point group) retains its symmetry in (1 : 1) metal-NH_3 complexes. On coordination of ammonia to a metal ion, the N—H bond is weakened (i.e. reduction in bond order of N—H of complexation) and the NH_3 stretching frequencies are shifted to lower frequency. For example, the v(N—H) of free NH_3 occurs at 333 cm^{-1} and is shifted to 3240 cm^{-1} in $[Co(NH_3)_6]Cl_3$. The stronger the M—N bonding, the weaker is the N—H bond and lower are the v(N—H) stretching frequencies.

As a result of coordination, the NH_3 deformation vibrations of free NH_3 are shifted to higher frequencies. The symmetric bending of free NH_3 is most sensitive to coordination while degenerate mode is least affected. Systematic variations in v(N—H) stretching, NH_3 bending ($\delta(NH_3)$) and rocking ($\rho_r(NH_3)$) frequencies have been shown in Table 11.17.

Table 11.17 Vibrational frequencies in some ammine complexes (cm^{-1})

Complex	Stretching band $v(NH_3)$	Bending mode $v(NH_3)$	Rocking mode $v(\rho_r(NH_3))$	v(M—N)
NH_3	3335, 3419	950	–	–
$[Co(NH_3)_6]Cl_3$	3240, 3160	1329	831	477
$[Cr(NH_3)_6]Cl_3$	3260, 3180	1308	746	473
$[Zn(NH_3)_6]Cl_2$	3350, 3220	1145	685	300
$[Co(NH_3)_6]Cl_2$	3330, 3250	1163	654	324

Two new bands are observed in the M—NH_3 (1 : 1) complexes. A band due to NH_3 rocking ($\rho_r(NH_3)$) is observed near 650–850 cm^{-1} and is very sensitive to the nature of the metal ion, its oxidation state and the anion present. The metal-nitrogen (M—N) stretching frequency of the MN_6 skeleton has been assigned to medium intensity bands in the region 300–500 cm^{-1} for divalent and trivalent ammine complexes of the first transition series.

11.11.5 Nitro and Nitrito Complexes

The nitrite ion (NO_2^-) in the complex may be present in three forms.
  (i) Anionic nitrite $[ML_n] (NO_2)_m$
  (ii) Coordination of NO_2^- to metal ion through N atom, such complexes are called nitro complexes.

$$M—N \overset{O}{\underset{O}{\diagdown}}$$

Nitro complex
C_{2v}

(iii) NO_2^- coordinated through one O atom. Such complexes are known as nitrito complexes

$$M—O\diagdown_N\diagup O$$

Out of the three possible situations which one is correct can be solved by considering the IR spectra of the complexes. We shall consider one by one all (the three cases).

Anionic NO_2 group comprises C_{2v} point group. It shows three IR bands at 1250 cm^{-1} (v_1), 850 cm^{-1} (v_2) and 1335 cm^{-1} (v_3) (v_1 and v_3 are symmetric and asymmetric stretching frequencies and v_2 is the bending frequency). These bands are shifted to 1300–1370 and 1370–1490 cm^{-1}, respectively in metal nitro complexes. The shift in the symmetric stretching frequency is usually higher in nitro complexes. The NO_2 bending in metal nitro complexes is found at 800–850 cm^{-1}. A new band at ~ 625 cm^{-1} corresponding to the out of plane bending of NO_2^- (wagging). This was a part of the rotation in the free ion but becomes IR active on coordination of NO_2^- with the metal ion.

$$Co—N\diagup^{O}_{\diagdown O} \qquad\qquad Co—O—N\diagup^{\diagup}_{\diagdown} O$$

Nitro isomer (yellow)   Nitro isomer (red)

$\left.\begin{array}{l}1430\text{ cm}^{-1}\text{ (unsymmetric)}\\1315\text{ cm}^{-1}\text{ (symmetric)}\end{array}\right\}\Delta v = 115\text{ cm}^{-1}$ $\left.\begin{array}{l}1460\text{ cm}^{-1}\text{ (unsymmetric)}\\1065\text{ cm}^{-1}\text{ (symmetric)}\end{array}\right\}\Delta v = 400\text{ cm}^{-1}$

600 cm^{-1} (rocking)

If the coordination of NO_2 is through the oxygen atom then nitrito complex is formed the symmetry of NO_2^- is lowered to C_s. The two v(N—O) vibrations occur at ~ 1468 cm^{-1} and ~ 1065 cm^{-1}, v_2 band occurs at 825 cm^{-1}.

Thus, in nitro complexes, both N—O bonds are equivalent and therefore v_{as} and v_s have a separation of ~ 100 cm^{-1}.

$$M—N\diagup^{O}_{\diagdown O} \qquad\qquad M—O\diagdown_N\diagup O$$

Nitro complex   Nitrito complex
C_{2v} C_s

In nitrito complex, there is one N—O bond and another N=O and hence their stretching frequencies, $v_{N=O}$ (1468 cm^{-1}) and $v_{N—O}$ (1015 cm^{-1}) have a bigger separation. Further, the band due to the wagging mode in nitro complexes disappears in nitrito complexes.

11.11.6 Nitrate and Carbonate Complexes

Free NO_3^- ion belongs to a high symmetry point group D_{3h}. The ion has four normal vibrations (normal modes) of A_1', A_2'' and $2E'$ of the four fundamentals only, v_2 (A_2''), v_3 (E') and v_4 (E') are IR active and v_1 (A_1') is infrared inactive.

Point Group	v_1	v_2	v_3	v_1
D_{3h}	A_1' (R)	A_2'' (IR)	E' (R, IR)	E' (R, IR)

On coordination of a nitrate ion to a metal ion, the local symmetry of the nitrate ion is lowered from D_{3h} to C_s, when NO_3^- ion acts as a monodentate ligand and from D_{3h} to C_{2v} where the nitrate ion acts either as a bidentate ligand or as a bridging ligand (Fig. 11.9).

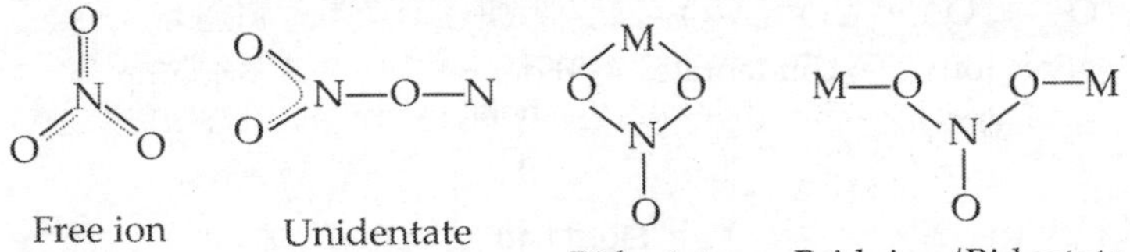

Free ion	Unidentate	Bidentate chelate	Bridging/Bidentate
D_{3h}	C_s	C_{2v}	C_{2v}

Fig. 11.9 Mode of bonding of NO_3^- ion to a metal

Thus the local symmetry at the nitrate ion is reduced to C_s and C_{2v}, depending upon the mode of coordination. In each of these cases, there are changes in the spectral activity as well as the degeneracy. Thus $A_1'(v_1)$ which is inactive in infrared, becomes active in all lower symmetries. A_2'' mode changes its symmetry and E' modes lift their degeneracies as the symmetry is lowered each of this giving rise to two IR bands.

In D_{3h} symmetry A_1', A_2'', E' are changed to A', A'', $2A'$ in C_s symmetry and A_1, B_1, A_1, B_2 in C_{2v} symmetry. The splitting of v_3 is very much larger for C_{2v} cases.

Point group	v_1	v_2	v_3		v_1	
D_{3h}	A_1' (R)	A_2'' (IR)	E'		E'	
C_{2v}	A_1 (R, IR)	B_1 (R, IR)	A_1 (R, IR)	B_2 (R, IR)	A_1 (R, IR)	B_2 (R, IR)
C_s	A' (R, IR)	A'' (R, IR)	$2A'$ (R, IR)		$2A'$ (R, IR)	

It must be noted that NO_3^- ion in both unidentate and bidentate coordination exhibits two separated bands each for v_3 and v_4. The separation is larger in bidentate mode than in monodentate. For example in $[Ni(en)_2(NO_3)_2]$, with unidentate nitrate, three bands are observed at 1420 cm^{-1}, 1305 cm^{-1} and 1008 cm^{-1} but in $[Ni(en)_2NO_3]ClO_4$ with bidentate chelating nitrate, the bands are at 147 cm^{-1}, 1290 cm^{-1} and 1025 cm^{-1}. The first two bands have a greater difference ($\Delta v = 1476 - 1290 = 186$ cm^{-1}) in case of bidentate nitrate than in case of unidentate nitrate ($\Delta v = 1420 - 1305 = 115$ cm^{-1}). However, it is difficult to distinguish between chelated and bridged nitrate by the help of IR spectra.

Carbonate (CO_3^{2-}) can be present in three different forms in complexes

(i)   In free ionic form with D_{3h} symmetry

(ii) As monodentate ligand with coordination from one oxygen atom with Cs symmetry.

(iii) As bidentate ligand coordinating with the metal ion through two oxygen atoms, the symmetry being reduced to C_{2v}.

(iv) The nitrate can also be in the bridged form coordinated to two metal ions. The symmetry is C_{2v} in this case also.

The possible modes of coordination are shown in Fig. 11.10.

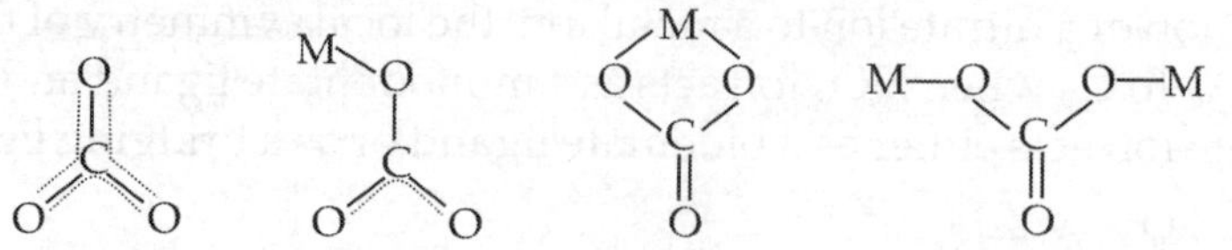

Free ion   Unidentate   Bidentate   Bridging
D_{3h} C_s chelate C_{2v}
                C_{2v}

Fig. 11.10

The local symmetry at the carbonate ion is reduced as shown above to C_s and C_{2v} depending upon the mode of coordination. In each of these cases, there are changes in the spectral activity as well as degeneracy. Thus $A'_1(\nu_1)$ which was inactive in IR, becomes active in all lower symmetries. A''_2 mode changes its symmetry (B_1 in C_{2v} and A'' in C_s) and E' modes left their degeneracies as the symmetry is lowered (E' in D_{3h} changes to A_1 and B_2 in C_{2v} and $2A'$ in C_s). Each of this giving rise to two IR bands.

It is important to note here that CO_3^{2-} ion in both unidentate and bidentate coordination exhibits two separate bands each for ν_3 and ν_4. Thus IR spectrum of $[Co(NH_3)_5CO_3]$ Br shows two separated bands at 1450 and 1370 cm^{-1} ($\Delta\nu = 80$ cm^{-1}), whereas that of $[Co(NH_3)_4CO_3]Cl$ exhibits them at 1592 cm^{-1} and 1255 cm^{-1} ($\Delta\nu = 337$ cm^{-1}). Thus it can be concluded that the smaller separation ($\Delta\nu = 80$ cm^{-1}) corresponds to unidentate coordination and the larger separation ($\Delta\nu \sim 340$ cm^{-1}) to the bidentate coordination. The bridging bidentate may give a similar spectrum to that of bidentate coordination, but distinction between these two can be made from the metal-oxygen stretching modes in the far infrared region.

11.11.7 Complexes Containing Sulphate and Perchlorate

The sulphate (T_d) ion has four normal modes of vibrations and only two of them (ν_3 and ν_4) are IR active. Upon coordination, the local symmetry of SO_4^{2-} changes to coordinated through one of the oxygen atoms as a monodentate ligand with C_{3v} symmetry. Coordinated through two oxygen atoms to the metal ion as a bidentate chelating ligand with C_{2v} symmetry. Sulphate ion can also bridge two metal ions by coordinating through one oxygen atom to both. Symmetry is C_{2v} in this case also.

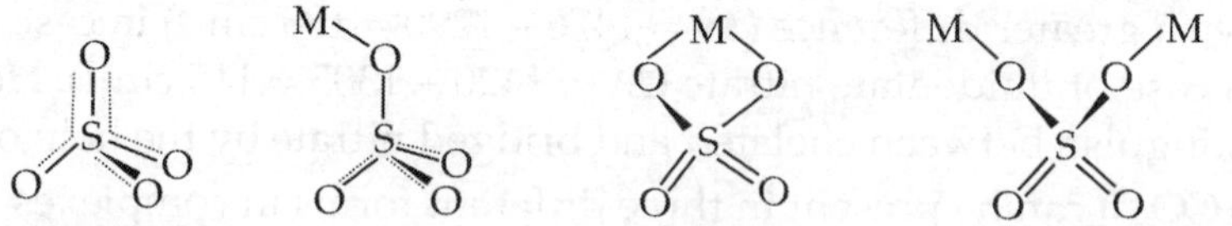

Free ion   Unidentate   Bidentate   Bridged bidentate
(T_d) complex   complex   complex
      (C_{3v}) (C_{2v}) (C_{2v})

Upon coordination to metal ion, there are changes in the spectral activity of ν_1 and ν_2 (which become active in both the Raman and IR) and the splitting of ν_3 and ν_4 bands into two ($A_1 + E$) when sulphate acts as unidentate ligand and three ($A_1 + B_1 + B_2$) components when it acts as bidented ligand.

Frequency	Free ion T_d	Unidentate ligand, C_{3v}	Bidentate ligand, C_{2v}	Bridging ligand, C_{2v}
ν_1	A_1 (R)	A_1 (R, IR)	A_1 (R, IR)	A_1 (R, IR)
ν_2	E (R)	E (R, IR)	A_1 (R, IR), A_2 (R)	A_1 (R, IR), A_2 (R)
ν_3	T_2 (R, IR)	(A_1, E) (R, IR)	(A_1, B_1, B_2) (R, IR)	(A_1, B_1, B_2) (R, IR)
ν_4	T_2 (R, IR)	(A_1, E) (R, IR)	(A_1, B_1, B_2) (R, IR)	(A_1, B_1, B_2) (R, IR)

Simple sulphates show two bands (ν_3 and ν_4) in the infrared at 1104 and 613 cm^{-1}. The other two bands, being active in Raman are found at 983 and 450 cm^{-1} respectively. Two bands are observed in $[Co(NH_3)_6]_2 (SO_4)_3$, where sulphate is ionic, they correspond to ν_3 and ν_4 vibrations.

On coordination as a monodentate ligand as in $[Co(NH_3)_5SO_4]$ Br, the symmetry of SO_4 is reduced to C_{3v}. ν_1 and ν_2 bands become IR active and occur with medium intensity at 970 cm^{-1} and 438 cm^{-1} respectively. There is also splitting of ν_3 (~ 1040 cm^{-1} and ~ 1125 cm^{-1}) and ν_4 bands into two (645 cm^{-1} and 604 cm^{-1}).

The chelating and bridging sulphate have C_{2v} symmetry. Due to further lowering of symmetry, ν_1 and ν_2 appear with medium intensity and ν_3 and ν_4 split up into three

$$\left[(NH_3)_4Co \begin{array}{c} NH_2 \\ \diagdown SO_4 \diagup \end{array} Co(NH_3)_4 \right] (NO_3)_3$$

bands each. For example, in complex show three split up bands of ν_3 vibration at 1050, 1170 and 11905 cm^{-1}. In $[Co(en)_2SO_4]$ Br, sulphate is chelated to cobalt(III) ion, the three split up bands of ν_3 vibration occur at 1211, 1176 and 1075 cm^{-1}. Thus, bridged and chelated sulphate can not be distinguished by the number of IR spectral bands. The only distinction is that ν_3 bands in chelated sulphate occur at higher energy than in case of bridged sulphate. The infrared frequencies of the free (ionic) sulphate group and its complexes are compiled in Table 11.18.

Table 11.18 Infrared frequencies of some sulphate complexes (cm^{-1})

Compound	Local symmetry of SO_4	ν_1	ν_2	ν_3	ν_4
Na_2SO_4	T_d	–	–	1104	613
$[Co(NH_3)_5SO_4]$ Br	C_{3v}	970	438	1032–1044 br	645
				1117–1143 br	604
$[Co(en)_2SO_4]$ Br	C_{2v}			1075, 1176,	
				1211	
$\left[(NH_3)_4Co \begin{array}{c} NH_2 \\ \diagdown SO_4 \diagup \end{array} Co(NH_3)_4 \right]^{3+}$	C_{2v}	995	462	1050–1060 br	641
				1105 s	610
				1170 s	571

The perchlorate ion, ClO_4^-, is weakly coordinating. Therefore, it is present in complexes as anion or as an unidentate or rarely as bidentate ligand. The tetrahedral symmetry of the free ion ClO_4^- is reduced to C_{3v} on unidentate coordination of the ion to a metal. The symmetry is further lowered to C_{2v} if ClO_4^- ion binds to a metal as a bidentate, either as

| Free ion (T_d) | Unidentate complex (C_{3v}) | Bidentate complex (C_{2v}) | Bridging bidentate (C_{2v}) |

achelate or as a bridging ligand. The infrared frequencies of free perchlorate ion and unidentately and bidentately (chelate) coordinated perchlorate ions are given in Table 11.19 as representative of such complexes. The complexes with anionic perchlorate exhibit only one stretching at 1170 cm^{-1} due to v_{as} (v_3). A less intense band due to v_s (v_1) is observed at 935 cm^{-1}. In complexes with unidentate perchlorate [Ni(CH$_3$CN)$_4$(ClO$_4$)$_2$], v_1 vibration becomes IR active and is observed as an intense band at 972 cm^{-1} v_3 band is also split up into two 1135 cm^{-1} and 1025 cm^{-1}.

Table 11.19 Infrared frequencies of some perchlorate complexes (cm^{-1})

Compound	v_1	v_2	v_3	v_4
ClO_4^- (T$_d$)	–	–	1100	620
Cu(ClO$_4$)$_2$. 3H$_2$O (C$_{3v}$)	920	480	1158	605
			1030	620, 648
Mn(ClO$_4$)$_2$. 2H$_2$O (C$_{2v}$)	945	453	1210	613
		467	1138, 1038	635, 666

In the complex [Ni(CH$_3$CN)$_2$ (ClO$_4$)]$^{-1}$ with bidentate perchlorate, v_1 is observed at 920 cm^{-1} and v_3 is split up into three bands at 1195 cm^{-1}, 1106 cm^{-1} and 1100 cm^{-1} due to C$_{2v}$ symmetry of bidentate perchlorate.

Therefore, after considering the number of bands, splitting of the bands if any in the IR spectrum of the complex one may conclude that the ligand is present as free ionic, monodentate or bidentate form in the complex. For example, in the IR spectrum of [Co(NH$_3$)$_6$]$_2$ (SO$_4$)$_3$. 5H$_2$O, v_1 = 973 cm^{-1} is observed with low intensity and v_3 and v_4 do not split. This clearly indicates that SO$_4^{2-}$ is present in this compound with T$_d$ symmetry and so SO$_4^{2-}$ is not coordinated to Co(III) ion.

In [Co(NH$_3$)$_5$ SO$_4$] Br, v_1 and v_2 are observed and v_3 and v_4 split up into two bands each, indicating C$_{3v}$ symmetry of sulphate and therefore it will be monodentate in this complex. When the splitting of v_3 and v_4 each into three bands occurs it indicates C$_{2v}$ symmetry of sulphate. Sulphate from T$_d$ symmetry attains C$_{2v}$ symmetry when it acts as either as bidentate or as bridging group.

11.11.8 Chelated and Bridged Ethylenediamine

In general, the coordination of ethylene diamine (en) with metal ion causes lowering of N—H frequency ($v_{N—H}$). Ethylene diamine can exist in different conformations as shown in Fig. 11.11. The order of the symmetry of the ligand is C$_{2v}$ > C$_{2h}$ > C$_2$.

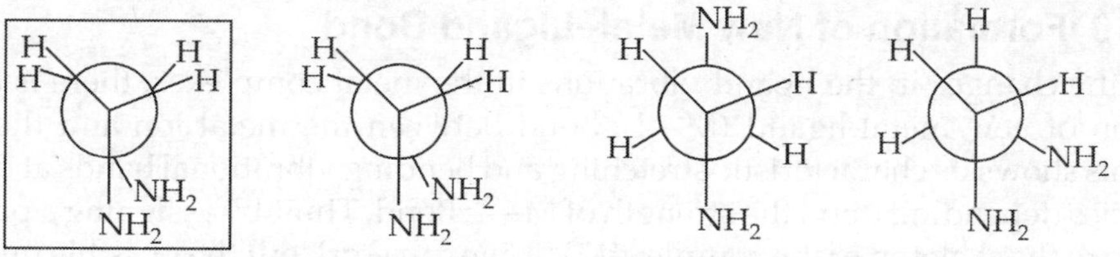

Eclipsed (*cis*)   Staggered (*trans*)   Gauche

(C_{2v}) (C_{2h}) (C_2)

Fig. 11.11

The eclipsed form of ethylene diamine has C_{2v} symmetry the staggered form has C_{2h} and the Gauche has C_2 symmetry. The staggered (C_{2h}) and Gauche (C_2) form will exhibit more IR active bands than the eclipsed (C_{2v}) form due to lower symmetry.

When ethylene diamine (en) is in the bidentate chelated form, it acquires gauche conformation and when in the bridged form it acquires staggered (*trans-*) conformation. Therefore in the IR spectrum of two complexes containing ethylene diamine as ligand, the complex which shows lesser number of bands may contain ethylene diamine in staggered form (C_{2h}). In that case en will function as a bidentate bridging ligand in that complex. But if the complex shows relatively more number of IR bands corresponding to ligand (en) frequencies, it may indicate that en is present as bidentate chelated form $(C_2$ symmetry). As the symmetry is lowered, the number of IR bands increases.

11.11.9 Oxalate Complexes

Free oxalate ion $C_2O_4^{2-}$ has D_{2h} symmetry (Fig. 11.12). When it is coordinated to a metal ion as bidentate ligand, its symmetry is reduced from D_{2h} to C_{2v}. Hence, the IR inactive vibrations in the free oxalate became IR active in the coordinated ion and hence the number of IR band in the oxalate complexes is more. In other words, in the oxalate ion all the four C—O are equivalent, whereas on coordination two COs coordinated to metal ion are C—O and the two free are C=O. Thus, there will be two different stretching vibrations corresponding to v_{C-O} and $v_{C=O}$.

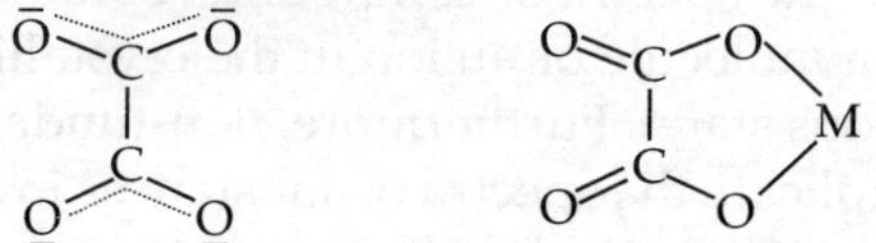

Free oxalate ion   Bindentate complex

D_{2h} C_{2v}

Fig. 11.12

Such an interpretation is empirical because it is based on the group frequency concept. In case of chelate rings, coupling between various ligand vibrations and the new M—O bending vibration takes place and a quantitative interpretation of IR spectra is possible only in terms of normal coordinate analysis. However, on the basis of IR spectral bands of different oxalate complexes, it can be generalised that as v_{M-O} increases, v_{C-O} decreases, whereas $v_{C=O}$ and v_{C-C} increases.

11.11.10 Formation of New Metal–Ligand Bond

Besides the changes in the ligand vibrations in the metal complexes, there is also the formation of new metal-ligand (M—L) bond between the metal ion and the ligand atom. This shows its characteristic stretching and bending vibrational bands at different frequencies depending upon the strength of M—L Bond. Thus, by assigning a particular structure to the skeleton of the complex $[ML_n]$, we can work out the possible number of IR active bands for the assigned point symmetry by using the concept of group theory. If the actual number of IR bands observed agree with the theoretically expected ones, the assigned structure is correct and we can arrive at the geometry of the complex.

For example, in a complex $[Co(NH_3)_6] Cl_3$ the skeleton CoN_6 has O_h symmetry from the chracter table of O_h point group, we can identify that only vibrations of two T_{1u} species are IR active. Therefore, we expect two IR bands for Co—N vibrations. The IR spectrum of $[Co(NH_3)_6] Cl_3$ shows three bands at 49 cm^{-1}, 477 cm^{-1} and 499 cm^{-1}. Occurance of three band may be due to the reduction of symmetry of $[Co(NH_3)_6]^{3+}$ in the crystalline state.

Now we shall consider $[Pt(NH_3)_4]Cl_2$. The expected structure is square planar. The symmetry of $[PtN_4]$ chromophore is D_{4h}. From the character table of D_{4h} symmetry, it can be shown that two IR active bands corresponding to A_{2u} and E_u vibrations should be observed. Actually, the IR spectrum of the complex shows a band at 510 cm^{-1} corresponding to ν_{Pt-N} vibrations.

For tetrahedral structure, as in $[CONH_3)_4]^{2+}$, we expect two IR bands for T_d symmetry corresponding to ν_3 and ν_4 vibrations. Actually two bands corresponding to ν_3 and ν_4 vibrations are observed at 430 cm^{-1} and 195 cm^{-1}.

11.12 GEOMETRICAL ISOMERISM

A number of investigations have been made to distinguish geometrical isomers by their infrared spectra. It has been found (1) that the isomer of lower symmetry generally exhibits more bands than that of higher symmetry and (2) that the *cis*- and *trans*- isomers absorb radiation of different frequencies for some ligand vibrations. Rule (1) must be applied with caution to the spectra of compounds obtained in the crystalline state, because the symmetry of a molecule or an ion in the crystalline state may be different from that in the free (gaseous state). Furthermore, non-fundamental vibrations involving lattice mode may complicate the spectra of substances in the crystalline state. Most investigations have been limited to the high frequency region where the vibrations due to the ligands appear. Since the essential difference between *cis*- and *trans*- structures is in the special arrangement of the coordinate bonds, the spectral of low frequency region, which involves the coordinate bond stretching and bending bands, often show more marked differences than do those of the high frequency region.

Square planar complexes of the type $[M(NH_3) 2X_2]$ where M = Pt, Pd and X = halogen) has two isomers as shown below.

trans-[Pd(NH$_3$)$_2$ Cl$_2$]
D_{2h}

cis-[Pd(NH$_3$)$_2$ Cl$_2$]
C_{2v}

Vibrational spectra of $[M(NH_3)_2 X_2]$ type complexes reveal that the *cis-* compound should exhibit two bands for M—X as well as M—N stretching (i.e. asymmetric and symmetric) modes while the *trans-* isomer should give only the asymmetric M—X and M—N stretching vibrations in the infrared. Thus the M—N and M—X stretching bands each split into two in the *cis-* isomer. In most cases, a second (M—N) vibration for the *cis-* compound was not observed. The infrared (IR) and Raman spectral frequencies for $[Pd(NH_3) 2Cl_2]$ and $[Pt(NH_3)_2 Cl_2]$ are given in Table 11.20.

Table 11.20 Infrared and Raman frequencies (cm^{-1}) of $[Pd(NH_3)_2 Cl_2]$ and $[Pt(NH_3)_2 Cl_2]$ complexes

Complex/Symmetry		$v_{M—N}$	$v_{M—Cl}$	Bending modes, cm^{-1}
trans-$[Pd(NH_3)_2 Cl_2]$	IR	496	333	245, 162, 137
(D_{2h})	R	494	295	222
cis-$[Pd(NH_3)_2 Cl_2]$	IR	495, 476	327, 306	245, 218, 160, 135
(C_{2v})				
trans-$[Pt(NH_3)_2 Cl_2]$	IR	506	330	
(D_{2h})				
cis-$[Pt(NH_3)_2 Cl_2]$	IR	510	324, 317	
(C_{2v})				

Octahedral complexes of the type $[Co(NH_3)_4 Cl_2]^+$ capable of showing two geometrical isomers: *cis*-$[Co(NH_3)_4 Cl_2]^+$ and *trans*-$[Co(NH_3)_4 Cl_2]^+$ with different symmetry as shown in Fig. 11.13.

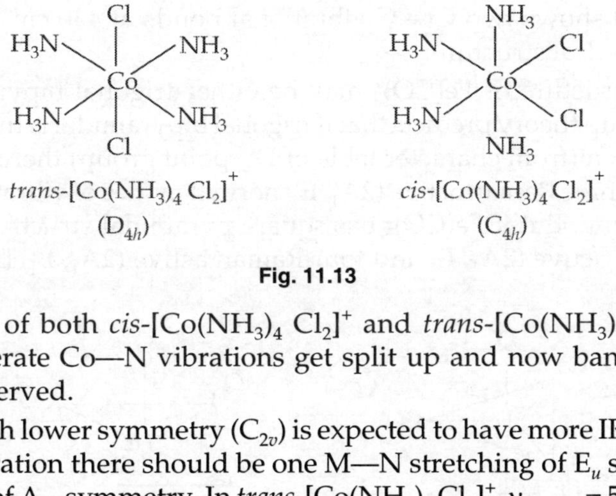

trans-$[Co(NH_3)_4 Cl_2]^+$ *cis*-$[Co(NH_3)_4 Cl_2]^+$
(D_{4h}) (C_{4h})

Fig. 11.13

The symmetry of both *cis*-$[Co(NH_3)_4 Cl_2]^+$ and *trans*-$[Co(NH_3)_4 Cl_2]^+$ is reduced and hence degenerate Co—N vibrations get split up and now bands due to Co—Cl vibrations are observed.

The *cis* form with lower symmetry (C_{2v}) is expected to have more IR active vibrations. From D_{4h} consideration there should be one M—N stretching of E_u symmetry and one M—Cl stretching of A_{2u} symmetry. In *trans*-$[Co(NH_3)_4 Cl_2]^+$, $v_{Co—N} = 501$ cm^{-1} and $v_{Co—Cl}$ $= 353$ cm^{-1} are observed. In case of cis-$[Co(NH_3)_4 Cl_2]^+$ (C_{2v}), there should be four Co—N stretching modes, two of A_1, one of B_1 and one of B_2 symmetry and also two Co—Cl stretching modes of A_1 and B_1 symmetry. Thus, the two isomers can be distinguished by the help of IR spectra.

11.13 METAL CARBONYLS

Free CO absorbs at 2141 cm^{-1}. In metal complexes the stretching frequency of free CO at 2141 cm^{-1} is shifted to lower wave numbers depending upon the mode of coordination. CO has two modes of coordination, terminal and bridging. In either mode, the CO bond order is reduced resulting in lower energy absorption. In the bridging modes it can coordinate to two or three metal atoms. The CO stretching frequencies generally follow the order

CO	>	MCO	>	M_2CO	>	M_3CO
$2141 \ cm^{-1}$		$2100\text{-}1900 \ cm^{-1}$		$1900\text{–}1700 \ cm^{-1}$		$1700\text{–}1550 \ cm^{-1}$
(stretching frequency)		(stretching frequency)		(stretching frequency)		(stretching frequency)

For a tetrahedral molecule $M(CO)_4$ only one IR active stretching C—O vibration (T_2 symmetry) and two Raman active (A_1 and T_2) C—O stretching vibrations are expected from group theory. For $Ni(CO)_4$ compound stretching frequencies are observed at $2057 \ cm^{-1}$, $2121 \ cm^{-1}$ and $2039 \ cm^{-1}$ respectively along with two new Ni—C vibrational bands at $421 \ cm^{-1}$ and $80 \ cm^{-1}$. The position of absorption bands in the triple bond region of CO(—C≡O) is an evidence that in $Ni(CO)_4$, the CO exists as Ni—C≡O.

It has been observed that for the isoelectronic carbonyl complexes, i.e. $Ni(CO)_4$, $[Co(CO)_4]^-$ and $[Fe(CO)_4]^{2-}$, the force constant of M—C bond increases as the force constant of C—O bond decreases. This can be explained on the basis that there is increasing d_π–p_π back bonding as the oxidation state of the metal becomes more negative.

Considering the character table of O_h point group, group theory predicts that there should be one IR active (T_{1u}) and two Raman active (A_{1g} and E_g) fundamentals for C—O and M—C stretching vibrational modes for octahedral metal carbonyl compounds, $M(CO)_6$. $[Cr(CO)_6]$ shows two Cr—C vibrational bonds at $440 \ cm^{-1}$ and $67.9 \ cm^{-1}$ as expected for octahedral structure.

The probable structure of $Fe(CO)_5$ may be either trigonal bipyramidal or square pyramidal and group theory predicts that if trigonal bipyramidal is the chosen structrue of $Fe(CO)_5$ (D_{3h}) then (from character table of D_{3h} point group) there should be two IR active (A_2'', E′) and three Raman active ($2A_1'$, E′) normal modes of vibration for both C—O and Fe—C stretching. But if $Fe(CO)_5$ has square pyramidal structrue (C_{4v}) then there should be three IR active ($2A_1$, E) and four Raman active ($2A_1$, B_1, E) C—O stretching vibrations.

Trigonal bipyramidal
(D_{3h})

Square pyramidal
(C_{4v})

The complete IR and Raman spectra of $Fe(CO)_5$ shows close agreement with predicted bands by group theory for trigonal bipyramidal structrue and it was further confirmed by X-ray studies.

Metal carbonyl compounds of the type $ML_2(CO)_2$ exist in *cis-* and *trans-* isomers.

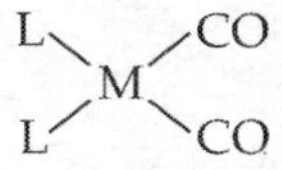

cis-$ML_2(CO)_2$
(C_{3v})

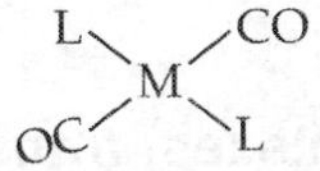

trans-$ML_2(CO)_2$
(D_{2h})

cis-dicarbonyl Complex, *cis*-ML$_2$ (CO)$_2$

The principal axis (C_2) is the z-axis, with the xz plane assigned as the plane of the molecule. The possible C—O stretching motions are shown by arrows (Fig. 11.14).

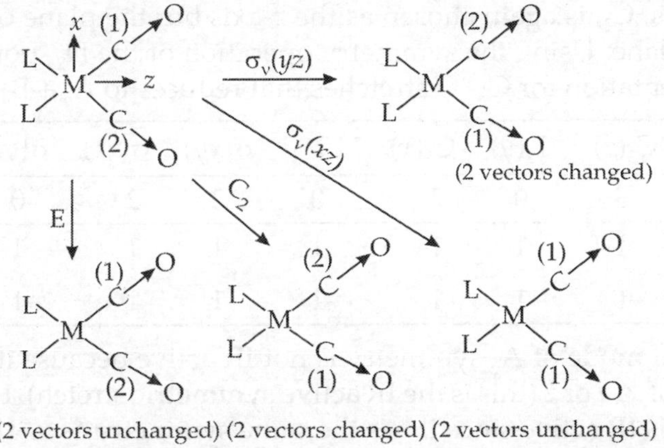

Fig. 11.14 Carbonyl stretching vibrations of *cis*-ML$_2$ (CO)$_2$ square planar complex

The vectors are used to create the reducible representation below using symmetry operations of the C_{2v} point group. A C—O bond will transform with a character of one (1) if it remains unchanged by the symmetry operations, and with a character of zero (0) if it is changed. These operations and their characters are shown below.

	E	C_2	$\sigma_v(xz)$	$\sigma_v(yz)$
Γ	2	0	2	0

As the stretches are unchanged in the identity operation (E) and in the reflection through the plane of the molecule $\sigma_v(xz)$, so each contributes 1 to the character, for a total of 2 for each operation. But both the vectors move to new locations on rotation (C_2 operation or reflection perpendicular to the plane of the molecule ($\sigma_v(yz)$). So these two characters are zero. Thus reducible representation (Γ) reduces to $A_1 + B_1$.

C_{2v}	E	C_2	$\sigma_v(xz)$	$\sigma_v(yz)$		
Γ	2	0	2	0		
A_1	1	1	1	1	z	x^2, y^2, z^2
B_1	1	−1	1	−1	x, R_y	xz

A_1 is IR active band since it transforms as the Cartesian coordinate z. Again, the vibrational mode corresponding to B_1 should be IR active, since it transform as the Cartesian coordinate x.

In C_{2v} the quadratic forms x^2, y^2, z^2, xz transform as A_1, B_1 and they are both Raman active also.

trans-dicarbonyl Complexes, *trans*-ML$_2$ (CO)$_2$

trans-dicarbonyl complex

The principal axis, C_2, is again chosen as the z-axis but the plane of the molecule this time will be xy plane. Using the symmetry operation of the D_{2h} point group, we get a reducible representation for C—O stretches that reduces to $A_g + B_u$.

D_{2h}	E	$C_2(z)$	$C_2(y)$	$C_2(x)$	i	$\sigma(xy)$	$\sigma(xz)$	$\sigma(yz)$		
Γ	2	0	0	2	0	2	2	0		
A_g	1	1	1	1	1	1	1	1		x^2, y^2, z^2
B_{3u}	1	−1	−1	1	−1	1	1	−1	x	

The vibrational mode of A_g symmetry is not IR active because it does not have the same symmetry of x, y or z (this is the IR active symmetric stretch). However, the mode of symmetry B_{3u} is IR active since it has the same symmetry as x. Therefore we expect only one C—O stretch in the IR.

In D_{2h} point group only A_g is Raman active and B_{3u} is Raman inactive. It is thus possible to distinguish *cis*- and *trans*-ML$_2$(CO)$_2$ by taking IR and Raman spectra. If one C—O stretching band appears, the molecule is *trans*-, if two bands appear, the molecule is *cis*-.

Bibliography

1. DS Schonland, *Molecular Symmetry, an Introduction to Group Theory and its Uses in Chemistry*, Van Nostrand Rainhol Company Ltd, New York (1965).

2. BE Douglas and CA Hollingsworth, *Symmetry in Bonding and Spectra - An Introduction*, Academic Press, Inc, New York (1965).

3. LH Hall, *Group Theory and Symmetry in Chemistry*, McGraw-Hill Book Company, New York (1969).

4. JN Murrell, SFA Kettle and JM Tedder, *Valency Theory*, 2nd ed, ELBS & John Wiley & Sons Ltd, London (1970).

5. FA Cotton, *Chemical Application of Group Theory*, 3rd ed, Wiley (2013).

6. ABP Lever, *Inorganic Electronic Spectroscopy*, Elsevier, Amsterdam (1968–84).

7. KV Reddy, *Symmetry and Spectroscopy of Molecules*, New Age International Publishers (2009).

8. SC Rakshit, *Atomic & Molecular Symmetry Group and Chemistry*, Levant Books, Kolkata.

9. PK Bhattacharya, *Group Theory and Its Chemical Applications*, Himalaya Publishing House (1999).

10. R Ameta, *Symmetry and Group Theory in Chemistry*, New Age International (P) Ltd, New Delhi (2013).

11. RS Drago, *Physical Methods in Chemistry*, 2nd ed, WB Sounders, Philadelphia (1977).

12. M Tinkham, *Group Theory and Quantum Mechanics*, McGraw-Hill, New York (1964).

13. M Orchin and H H Jaffe, *J Chem Educ*, 47, 246, 372, 510 (1970); *Symmetry, Orbitals, and Spectra*, Wiley-Interscience (1971); *Symmetry in Chemistry*, 1st ed, Wiley Eastern Pvt Ltd, New Delhi (1971).

14. GL Miessler, DA Tarr, *Inorganic Chemistry*, 3rd ed, Pearson, Delhi (2009).

15. JE Huheey, EA Keiter, RL Keiter, *Inorganic Chemistry*, Harper Collins College Publishers, 4th ed, (1993).

16. BE Douglas, DH Mc Daniel, JJ Alexander, *Concepts and Models of Inorganic Chemistry*, 3rd ed, John Wiley & Sons, New York (1994).

17. SFA Kettle, *Symmetry and Structure (Readable Group Theory for Chemists)*, 2nd ed, John Wiley & Sons, New York (1965); *Physical Inorganic Chemistry, A Coordination Chemistry Approach*, Spectrum University Science Book (1996).

18. KV Raman, *Group Theory and Its Applications to Chemistry*, Tata McGraw-Hill Publishing Company Ltd., New Delhi (1990).

19. KL Kapoor, *A Textbook of Physical Chemistry*, Vol. 4, 4th ed, McMillan, New Delhi (2011).

20. WK Li, GD Zhou, TCW Mak, *Advanced Structural Inorganic Chemistry*, Oxford, University Press, (2008).

21. RK Ray, *Electronic Spectra of Transition Metal Complexes*, New Central Book Agency (P) Ltd., Kolkata (2011).

22. K Nakamoto, *Infrared and Raman Spectra of Inorganic and Coordination Compounds*, 4th ed, John Wiley & Sons, New York (1986).

23. K Nakamoto, *Infrared Spectra of Inorganic and Coordination Compounds*, 2nd ed, John Wiley & Sons Inc, New York (1963)

24. K Nakamoto and PJ Mc Carthy, *Spectroscopy and Structure of Metal Chelate Compounds*, John Wiley and Sons, (1968).

25. M Hammermesh, *Group Theory and Its Applications to Physical Problems*, Pergamon Press, London (1962).

26. M Hammermesh, *Group Theory and Its Applications to Physical Problems*, Pergamon Press, London (1962).

27. M Ladd, *Symmetry and Group Theory in Chemistry*, Horwood Publishing Ltd, Chichester, England (1998).

28. LH Hall, *Group Theory and Symmetry in Chemistry*, McGraw-Hill, New York (1969).

29. A Streitwieser, Jr, *Molecular Orbital Theory for Organic Chemistry*, John Wiley & Sons, Inc, New York (1961).

30. DC Harris and MD Bertolucci, *Symmetry and Spectroscopy : An Introduction to Vibration and Electronic Spectra*, Oxford University Press, (1978).

31. RB Woodword and R Hoffman, *The Conversation of Orbital Symmetry*, Verlag Chemic GmbH, Weinheim/Academic Press Inc (1971).

32. RB Woodward and R Hoffman, *J Am Chem Soc*, (1965).

33. REK Winter, *Tetrahedron Letters*, (1965).

Appendix I

Character Tables of Molecular Symmetry Groups

1. *The nonaxial groups (molecules of low symmetry)*

C_1	E
A	1

C_3	E	σ_h	Basis vectors and components	Basis functions
A'	1	1	$x, y. R_x, T_x, T_y'$	x^2, y^2, z^2, xy
A''	1	−1	z, R_x, R_y, T_z	yz, xz

C_i	E	i	Basis vectors and components	Basis functions
A_g	1	1	R_x, R_y, R_z	$x^2, y^2, z^2, xy, xz, yz$
A_u	1	−1	x, y, z, T_x, T_y, T_z	

2. *The C_n groups (molecules of high symmetry)*

C_2	E	C_2	Basis vectors and components	Basis functions
A	1	1	z, R_z, T_z	$x^2, y^2, z^2, xy, yz, xz$
B	1	−1	x, y, R_z, R_y, T_x, T_y	

C_3	E	C_3	C_3^2	Basis vectors and components	$\epsilon = \exp(2\pi i/3)$ basis functions
A	1	1	1	z, R_z, T_z	$x^2 + y^2, z^2$
E	1	$\begin{cases} \epsilon \\ \epsilon^* \end{cases}$	$\begin{cases} \epsilon^* \\ \epsilon \end{cases}$	$(x, y)(R_x, R_y)$ (T_x, T_y)	$(x^2 - y^2, xy)(yz, xz)$

C_4	E	C_4	C_2	C_4^3	Basis vectors and components	Basis functions
A	1	1	1	1	z, R_z, T_z	$x^2 + y^2, z^2$
B	1	−1	1	−1		$x^2 - y^2, xy$
E	$\begin{cases} 1 \\ 1 \end{cases}$	$\begin{cases} i \\ -i \end{cases}$	$\begin{cases} -1 \\ -1 \end{cases}$	$\begin{cases} -i \\ i \end{cases}$	$(x, y)(R_x, R_y)$ (T_x, T_y)	(yz, xz)

C_5	E	C_5	C_5^2	C_5^3	C_5^4	Basis vectors and components	$\epsilon = \exp(2\pi i/5)$ basis functions
A	1	1	1	1	1	$z, R_z (T_z)$	$x^2 + y^2, z^2$
E_1	$\begin{cases} 1 \\ 1 \end{cases}$	$\begin{cases} \epsilon \\ \epsilon^* \end{cases}$	$\begin{cases} \epsilon^2 \\ \epsilon^{2*} \end{cases}$	$\begin{cases} \epsilon^{2*} \\ \epsilon^2 \end{cases}$	$\begin{cases} \epsilon^* \\ \epsilon \end{cases}$	$(x, y)(R_x, R_y)$	(yz, xz)
E_2	$\begin{cases} 1 \\ 1 \end{cases}$	$\begin{cases} \epsilon^2 \\ \epsilon^{2*} \end{cases}$	$\begin{cases} \epsilon^* \\ \epsilon \end{cases}$	$\begin{cases} \epsilon \\ \epsilon^* \end{cases}$	$\begin{cases} \epsilon^{2*} \\ \epsilon^2 \end{cases}$		$(x^2 - y^2, xy)$

C_6	E	C_6	C_3	C_2	C_3^2	C_5^5	Basis vectors and components	$\epsilon = \exp(2\pi i/6)$ basis functions
A	1	1	1	1	1	1	z, R_z, T_z	x^2+y^2, z^2
B	1	-1	1	-1	1	-1		
E_1	1	ϵ	$-\epsilon^*$	-1	$-\epsilon$	ϵ^*	$(x,y)\,(T_x, T_y)$	
	1	ϵ^*	$-\epsilon^{2*}$	-1	$-\epsilon^*$	ϵ	(R_x, R_y)	(xz, yz)
E_2	1	$-\epsilon^2$	$-\epsilon$	1	$-\epsilon^*$	$-\epsilon$		(x^2-y^2, xy)
	1	$-\epsilon$	$-\epsilon^*$	1	$-\epsilon$	$-\epsilon^*$		

C_7	E	C_7	C_7^2	C_7^3	C_7^4	C_7^5	C_7^6	Basis vectors and components	$\epsilon = \exp(2\pi i/7)$ basis functions
A	1	1	1	1	1	1	1	z, R_z, T_z	x^2+y^2, z^2
E_1	1	ϵ	ϵ^2	ϵ^3	ϵ^{3*}	ϵ^{2*}	ϵ^*		
	1	ϵ^*	ϵ^{2*}	ϵ^{3*}	ϵ^3	ϵ^2	ϵ	$(x,y)\,(T_x, T_y)$	
E_2	1	ϵ^2	ϵ^{3*}	ϵ^*	ϵ	ϵ^3	ϵ^{2*}	(R_x, R_y)	(xz, yz)
	1	ϵ^{2*}	ϵ^3	ϵ	ϵ^*	ϵ^{3*}	ϵ^2		
E_3	1	ϵ^3	ϵ^*	ϵ^2	ϵ^{2*}	ϵ	ϵ^{3*}		(x^2-y^2, xy)
	1	ϵ^{3*}	ϵ	ϵ^{2*}	ϵ^2	ϵ^*	ϵ^3		

C_8	E	C_8	C_4	C_2	C_4^3	C_8^3	C_8^5	C_8^7	Basis vectors and components	$\epsilon = \exp(2\pi i/8)$ basis functions
A	1	1	1	1	1	1	1	1	z, R_z, T_z	x^2+y^2, z^2
B	1	-1	1	1	1	-1	-1	-1		
E_1	1	ϵ	i	-1	$-i$	$-\epsilon^*$	$-\epsilon$	ϵ^*		
	1	ϵ^*	$-i$	-1	i	$-\epsilon$	$-\epsilon^*$	ϵ	$(x,y)\,(T_x, T_y)$	
E_2	1	i	-1	1	-1	$-i$	i	$-i$	(R_x, R_y)	(xz, yz)
	1	$-i$	-1	1	-1	i	$-i$	i		
E_3	1	$-\epsilon$	i	-1	$-i$	ϵ^*	ϵ	$-\epsilon^*$		(x^2-y^2, xy)
	1	$-\epsilon^*$	$-i$	-1	i	ϵ	ϵ^*	$-\epsilon$		

3. *The D_n groups*

D_2	E	$C_2(z)$	$C_2(y)$	$C_2(x)$	Basis vectors and components	Basis functions
A	1	1	1	1		x^2, y^2, z^2
B_1	1	1	-1	-1	z, T_z, R_z	xy
B_2	1	-1	1	-1	y, T_y, R_y	xz
B_3	1	-1	-1	1	x, T_x, R_x	yz

D_3	E	$2C_3$	$3C_2$	Basis vectors and components	Basis functions
A_1	1	1	1		x^2+y^2, z^2
A_2	1	1	-1	z, R_z, T_z	
E	2	-1	0	$(x,y)\,(T_x, T_y)$ (R_x, R_y)	$(x^2-y^2, xy)\,(xz, yz)$

D_4	E	$2C_4$	$C_2(=C_4^2)$	$2C_2'$	$2C_2''$	Basis vectors and components	Basis functions
A_1	1	1	1	1	1		x^2+y^2, z^2
A_2	1	1	1	-1	-1	z, R_z, T_z	
B_1	1	-1	1	1	-1		x^2-y^2
B_2	1	-1	1	-1	1		xy
E	2	0	-2	0	0	$(x,y)\,(R_x, R_y)$ (T_x, T_y)	(xz, yz)

D_5	E	$2C_5$	$2C_5^2$	$5C_2$	Basis vectors and components	Basis functions
A_1	1	1	1	1		$(x^2+y^2), z^2$
A_2	1	1	1	-1	z, R_z, T_z	
E_1	2	$2\cos$ $72°$	$2\cos$ $144°$	0	$(x,y)\,(R_x, R_y)$ (T_x, T_y)	(xz, yz)
E_2	2	$2\cos$ $144°$	$2\cos$ $72°$	0		(x^2-y^2, xy)

D_6	E	$2C_6$	$2C_3$	C_2	$3C_2'$	$3C_2''$	Basis vectors and components	Basis functions
A_1	1	1	1	1	1	1		$(x^2+y^2), z^2$
A_2	1	1	1	1	-1	-1	z, R_z, T_z	
B_1	1	-1	1	-1	1	-1		
B_2	1	-1	1	-1	-1	1		
E_1	2	1	-1	-2	0	0	$(x,y)\,(R_x, R_y)$	(xz, yz)
E_2	2	-1	-1	2	0	0		$(x^2-y^2)(xy)$

4. *The C_{nv} groups*

C_{2v}	E	C_2	$\sigma_v(xz)$	$\sigma_v(yz)$	Basis vectors and components	Basis functions
A_1	1	1	1	1	z, T_z	x^2, y^2, z^2
A_2	1	1	-1	-1	R_z	xy
B_1	1	-1	1	-1	x, R_y, T_x	xz
B_2	1	-1	-1	1	y, R_x, T_y	yz

C_{3v}	E	$2C_3$	$3\sigma_v$	Basis vectors and components	Basis functions
A_1	1	1	1	z, T_z	x^2+y^2, z^2
A_2	1	1	-1	R_z	
E	2	-1	0	$(x,y)\,(R_z, R_y)$ (T_x, T_y)	$((x^2-y^2)\,xy)\,(xz, yz)$

C_{4v}	E	$2C_4$	C_2	$2\sigma_v$	$2\sigma_d$	Basis vectors and components	Basis functions
A_1	1	1	1	1	1	z, T_z	$x^2 + y^2, z^2$
A_2	1	1	1	-1	-1	R_z	
B_1	1	-1	1	1	-1		$x^2 - y^2$
B_2	1	-1	1	-1	1		xy
E	2	0	-2	0	0	$(x, y) (R_x, R_y)$ (T_x, T_y)	(xz, yz)

C_{5v}	E	$2C_5$	$2C_5^2$	$5\sigma_v$	Basis vectors and components	Basis functions
A_1	1	1	1	1	z, T_z	$x^2 + y^2, z^2$
A_2	1	1	1	-1	R_z	
E_1	2	2cos 72°	2cos 144°	0	$(x, y) (R_x, R_y)$ (T_x, T_y)	(xz, yz)
E_2	2	2cos 144°	2cos 72°	0		$(x^2 - y^2, z^2)$

C_{6v}	E	$2C_6$	$2C_3$	C_2	$3\sigma_v$	$3\sigma_d$	Basis vectors and components	Basis functions
A_1	1	1	1	1	1	1	z, T_z	$x^2 + y^2, z^2$
A_2	1	1	1	1	-1	-1	R_z	
B_1	1	-1	1	-1	1	-1		
B_2	1	-1	1	-1	-1	1		
E_1	2	1	-1	-2	0	0	$(x, y) (R_x, R_y)$	(xz, yz)
E_2	2	-1	-1	2	0	0		$(x^2 - y^2), (xy)$

5. *The C_{nh} groups*

C_{2h}	E	C_2	i	σ_h	Basis vectors and components	Basis functions
A_g	1	1	1	1	R_z	x^2, y^2, z^2, xy
A_g	1	-1	1	-1	R_x, R_y	xz, yz
B_u	1	1	-1	-1	$z, T_z,$	
B_u	1	-1	-1	1	x, y, T_x, T_y	

C_{3h}	E	C_3	C_3^2	σ_h	S_3	S_3^5	Basis vectors and components	$\epsilon = \exp(2\pi i/3)$ basis functions
A'	1	1	1	1	1	1	R_z	$x^2 + y^2, z^2$
E'	$\left\{\begin{matrix} 1 \\ 1 \end{matrix}\right.$	$\begin{matrix} \epsilon \\ \epsilon^* \end{matrix}$	$\begin{matrix} \epsilon^* \\ \epsilon \end{matrix}$	$\begin{matrix} 1 \\ 1 \end{matrix}$	$\begin{matrix} \epsilon \\ \epsilon^* \end{matrix}$	$\left.\begin{matrix} \epsilon^* \\ \epsilon \end{matrix}\right\}$	$(x, y) (T_x, T_y)$	$(x^2 - y^2, xy)$
A''	1	1	1	-1	-1	-1	z, T_z	
E''	$\left\{\begin{matrix} 1 \\ 1 \end{matrix}\right.$	$\begin{matrix} \epsilon \\ \epsilon^* \end{matrix}$	$\begin{matrix} \epsilon^* \\ \epsilon \end{matrix}$	$\begin{matrix} -1 \\ -1 \end{matrix}$	$\begin{matrix} -\epsilon \\ -\epsilon^* \end{matrix}$	$\left.\begin{matrix} -\epsilon^* \\ -\epsilon \end{matrix}\right\}$	(R_x, R_y)	(xz, yz)

C_{4h}	E	C_4	C_2	C_4^3	i	S_4^3	σ_h	S_4	Basis vectors and components	Basis functions
A_g	1	1	1	1	1	1	1	1	R_z	$x^2+y^2,\ z^2$
b_g	1	-1	1	-1	1	-1	1	-1		$x^2-y^2,\ xy$
E_g	1	i	-1	$-i$	1	i	-1	$-i$	(R_x, R_y)	(xz, yz)
	1	$-i$	-1	i	1	$-i$	-1	i		
A_u	1	1	1	1	-1	-1	-1	-1	$z,\ T_z$	
B_u	1	-1	1	-1	-1	1	-1	1		
E_u	1	i	-1	$-i$	-1	$-i$	1	i	$(x, y)\ (T_x, T_y)$	
	1	$-i$	-1	i	-1	i	1	$-i$		

C_{5h}	E	C_5	C_5^2	C_5^3	C_5^4	σ_h	S_5	S_5^7	S_5^3	S_5^9	Basis vectors and components	$\epsilon=\exp(2\pi i/5)$ basis functions
A'	1	1	1	1	1	1	1	1	1	1	R_z	$x^2+y^2,\ z^2$
E_1'	1	ϵ	ϵ^2	ϵ^{2*}	ϵ^*	1	ϵ	ϵ^2	ϵ^{2*}	ϵ^*	$(x, y)\ (T_x, T_y)$	
	1	ϵ^*	ϵ^{2*}	ϵ^2	ϵ	1	ϵ^*	ϵ^{2*}	ϵ^2	ϵ		
E_2'	1	ϵ^2	ϵ^*	ϵ	ϵ^{2*}	1	ϵ^2	ϵ^*	ϵ	ϵ^{2*}		$(x^2-y^2,\ xy)$
	1	ϵ^{2*}	ϵ	ϵ^*	ϵ^2	1	ϵ^{2*}	ϵ	ϵ^*	ϵ^2		
A''	1	1	1	1	1	-1	-1	-1	-1	-1	$z,\ T_z$	
E_1''	1	ϵ	ϵ^2	ϵ^{2*}	ϵ^*	-1	$-\epsilon$	$-\epsilon^2$	$-\epsilon^{2*}$	$-\epsilon^*$	(R_x, R_y)	(xz, yz)
	1	ϵ^*	ϵ^{2*}	ϵ^2	ϵ	-1	$-\epsilon^*$	$-\epsilon^{2*}$	$-\epsilon^2$	$-\epsilon$		
E_2''	1	ϵ^2	ϵ^*	ϵ	ϵ^{2*}	-1	$-\epsilon^2$	$-\epsilon^*$	$-\epsilon$	$-\epsilon^{2*}$		
	1	ϵ^{2*}	ϵ	ϵ^*	ϵ^2	-1	$-\epsilon^{2*}$	$-\epsilon$	$-\epsilon^*$	$-\epsilon^2$		

C_{6h}	E	C_6	C_3	C_2	C_3^2	C_6^5	i	S_3^5	S_6^5	σ_h	S_6	S_3	Basis vectors and components	$\epsilon=\exp(2\pi i/6)$ basis functions
A_g	1	1	1	1	1	1	1	1	1	1	1	1	R_z	$x^2+y^2,\ z^2$
B_g	1	-1	1	-1	1	-1	1	-1	1	-1	1	-1		
E_{1g}	1	ϵ	$-\epsilon^*$	-1	$-\epsilon$	ϵ^*	1	ϵ	$-\epsilon^*$	-1	$-\epsilon$	ϵ^*	(R_x, R_y)	(xz, yz)
	1	ϵ^*	$-\epsilon$	-1	$-\epsilon^*$	ϵ	1	ϵ^*	$-\epsilon$	-1	$-\epsilon^*$	ϵ		
E_{2g}	1	$-\epsilon^*$	$-\epsilon$	1	$-\epsilon^*$	$-\epsilon$	1	$-\epsilon^*$	$-\epsilon$	1	$-\epsilon^*$	$-\epsilon$		$(x^2-y^2,\ xy)$
	1	$-\epsilon$	$-\epsilon^*$	1	$-\epsilon$	$-\epsilon^*$	1	$-\epsilon$	$-\epsilon^*$	1	$-\epsilon$	$-\epsilon^*$		
A_u	1	1	1	1	1	1	-1	-1	-1	-1	-1	-1	$z,\ T_z$	
B_u	1	-1	1	-1	1	-1	-1	1	-1	1	-1	1		
E_{1u}	1	ϵ	$-\epsilon^*$	-1	$-\epsilon$	ϵ^*	-1	$-\epsilon$	ϵ^*	1	ϵ	$-\epsilon^*$	$(x, y)\ (T_x, T_y)$	
	1	ϵ^*	$-\epsilon$	-1	$-\epsilon^*$	ϵ	-1	$-\epsilon^*$	ϵ	1	ϵ^*	$-\epsilon$		
E_{2u}	1	$-\epsilon^*$	$-\epsilon$	1	$-\epsilon^*$	$-\epsilon$	-1	ϵ^*	ϵ	-1	ϵ^*	ϵ		
	1	$-\epsilon$	$-\epsilon^*$	1	$-\epsilon$	$-\epsilon^*$	-1	ϵ	ϵ^*	-1	ϵ	ϵ^*		

6. *The D_{nh} Groups*

D_{2h}	E	$C_2(z)$	$C_2(y)$	$C_2(x)$	i	$\sigma(xy)$	$\sigma(xz)$	$\sigma(yz)$	Basis vectors and components	Basis functions
A_g	1	1	1	1	1	1	1	1		x^2+y^2, z^2
B_{1g}	1	1	-1	-1	1	1	-1	-1	R_z	xy
B_{2g}	1	-1	1	-1	1	-1	1	-1	R_y	xz
B_{3g}	1	-1	-1	1	1	-1	-1	1	R_x	yz
A_u	1	1	1	1	-1	-1	-1	-1		
B_{1u}	1	1	-1	-1	-1	-1	1	1	z, T_z	
B_{2u}	1	-1	1	-1	-1	1	-1	1	y, T_y	
B_{3u}	1	-1	-1	1	-1	1	1	-1	x, T_x	

D_{3h}	E	$2C_3$	$3C_2$	σ_h	$2S_3$	$3\sigma_v$	Basis vectors and components	Basis functions
A_1'	1	1	1	1	1	1		x^2+y^2, z^2
A_2'	1	1	-1	1	1	-1	R_z	
E'	2	-1	0	2	-1	-0	$(x, y)(T_x, T_y)$	(x^2-y^2, xy)
A_1''	1	1	1	-1	-1	-1		
A_2''	1	1	-1	-1	-1	1	z, T_z	
E''	2	-1	0	-2	1	0	(R_x, R_y)	(xz, yz)

D_{4h}	E	$2C_4$	C_2	$2C_2'$	$2C_2''$	i	$2S_4$	σ_h	$2\sigma_v$	$2\sigma_d$	Basis vectors and components	Basis functions
A_{1g}	1	1	1	1	1	1	1	1	1	1		x^2+y^2, z^2
A_{2g}	1	1	1	-1	-1	1	1	1	-1	-1	R_z	
B_{1g}	1	-1	1	1	-1	1	-1	1	1	-1		x^2-y^2
B_{2g}	1	-1	1	-1	1	1	-1	1	-1	1		xy
E_g	2	0	-2	0	0	2	0	-2	0	0	(R_x, R_y)	(xz, yz)
A_{1u}	1	1	1	1	1	-1	-1	-1	-1	-1		
A_{2u}	1	1	1	-1	-1	-1	-1	-1	1	1	z, T_z	
B_{1u}	1	-1	1	1	-1	-1	1	-1	-1	1		
B_{2u}	1	-1	1	-1	1	-1	1	-1	1	-1		
E_u	2	0	-2	0	0	-2	0	2	0	0	$(x, y)(T_x, T_y)$	

D_{5h}	E	$2C_5$	$2C_5^2$	$5C_2$	σ_h	$2S_5$	$2S_5^3$	$5\sigma_v$	Basis vectors and components	Basis functions
A_1'	1	1	1	1	1	1	1	1		x^2+y^2, z^2
A_2'	1	1	1	-1	1	1	1	-1	R_z	
E_1'	2	$2\cos 72°$	$2\cos 144°$	0	2	$2\cos 72°$	$2\cos 144°$	0	(R_x, R_y)	
E_2'	2	$2\cos 144°$	$2\cos 72°$	0	2	$2\cos 144°$	$2\cos 72°$	0		(x^2-y^2, xy)
A_1''	1	1	1	1	-1	-1	-1	-1		
A_2''	1	1	1	-1	-1	-1	-1	1	z, T_z	
E_1''	2	$2\cos 72°$	$2\cos 144°$	0	-2	$-2\cos 72°$	$-2\cos 144°$	0	$(x, y)(T_x, T_y)$	(xy, yz)
E_2''	2	$2\cos 144°$	$2\cos 72°$	0	-2	$-2\cos 144°$	$-2\cos 72°$	0		

D_{6h}	E	$2C_6$	$2C_3$	C_2	$3C_2'$	$3C_2''$	i	$2S_3$	$2S_6$	σ_h	$3\sigma_d$	$3\sigma_v$	Basis vectors and components	Basis functions
A_{1g}	1	1	1	1	1	1	1	1	1	1	1	1	R_z	x^2+y^2, z^2
A_{2g}	1	1	1	1	-1	-1	1	1	1	1	-1	-1		
B_{1g}	1	-1	1	-1	1	-1	1	-1	1	-1	1	-1		
B_{2g}	1	-1	1	-1	-1	1	1	-1	1	-1	-1	1		
E_{1g}	2	1	-1	-2	0	0	2	1	-1	-2	0	0	(R_x, R_y)	(xy, yz)
E_{2g}	2	-1	-1	2	0	0	2	-1	-1	2	0	0		(x^2-y^2, xy)
A_{1u}	1	1	1	1	1	1	-1	-1	-1	-1	-1	-1		
A_{2u}	1	1	1	1	-1	-1	-1	-1	-1	-1	1	1	z, T_z	
B_{1u}	1	-1	1	-1	1	-1	-1	1	-1	1	-1	1		
B_{2u}	1	-1	1	-1	-1	1	-1	1	-1	1	1	-1		
E_{1u}	2	1	-1	-2	0	0	-2	-1	1	2	0	0	(x, y) (T_x, T_y)	
E_{2u}	2	-1	-1	2	0	0	-2	1	1	-2	0	0		

7. *The D_{nd} groups*

D_{2d}	E	$2S_4$	C_2	$2C_2'$	$2\sigma_d$	Basis vectors and components	Basis functions
A_1	1	1	1	1	1		x^2+y^2, z^2
A_2	1	1	1	-1	-1	R_z	
B_1	1	-1	1	1	-1		x^2-y^2
B_2	1	-1	1	-1	1	z, T_z	xy
E	2	0	-2	0	0	$(x, y)(T_x, T_y)$ (R_x, R_y)	(xz, yz)

D_{3d}	E	$2C_3$	$3C_2$	i	$2S_6$	$3\sigma_d$	Basis vectors and components	Basis functions
A_{1g}	1	1	1	1	1	1		x^2+y^2, z^2
A_{2g}	1	1	-1	1	1	-1	R_z	
E_g	2	-1	0	2	-1	0	(R_x, R_y)	$(x^2-y^2, xy), (xz, yz)$
A_{1u}	1	1	1	-1	-1	-1		
A_{2u}	1	1	-1	-1	-1	1	z, T_z	
E_u	2	-1	0	-2	1	0	$(x, y)(T_x, T_y)$	

D_{4d}	E	$2S_8$	$2C_4$	$2S_8^3$	C_2	$4C_2'$	$4\sigma_d$	Basis vectors and components	Basis functions
A_1	1	1	1	1	1	1	1		x^2+y^2, z^2
A_2	1	1	1	1	1	-1	-1	R_z	
B_1	1	-1	1	-1	1	1	-1		
B_2	1	-1	1	-1	1	-1	1	z, T_z	
E_1	2	$\sqrt{2}$	0	$-\sqrt{2}$	-2	0	0	$(x, y)(T_x, T_y)$	
E_2	2	0	-2	0	2	0	0		(x^2-y^2, xy)
E_3	2	$-\sqrt{2}$	0	$\sqrt{2}$	-2	0	0	(R_x, R_y)	(xz, yz)

D_{5d}	E	$2C_5$	$2C_5^2$	$5C_2$	i	$2S_{10}^3$	$2S_{10}$	$5\sigma_d$	Basis vectors and components	Basis functions
A_{1g}	1	1	1	1	1	1	1	1		x^2+y^2, z^2
A_{2g}	1	1	1	−1	1	1	1	−1	R_z	
E_{1g}	2	2cos 72°	2cos 144°	0	2	2cos 72°	2cos 144°	0	(R_x, R_y)	(xy, yz)
E_{2g}	2	2cos 144°	2cos 72°	0	2	2cos 144°	2cos 72°	0		(x^2-y^2, xy)
A_{1u}	1	1	1	1	−1	−1	−1	−1		
A_{2u}	1	1	1	−1	−1	−1	−1	1	z, T_z	
E_{1u}	2	2cos 72°	2cos 144°	0	−2	−2cos 72°	−2cos 144°	0	$(x, y)(T_x, T_y)$	
E_{2u}	2	2cos 144°	2cos 72°	0	−2	−2cos 144°	−2cos 72°	0		

8. *The S_n groups*

D_{6d}	E	$2S_{12}$	$2C_6$	$2S_4$	$2C_3$	$2S_{12}^5$	C_2	$6C_2'$	$6\sigma_d$	Basis vectors and components	Basis functions
A_1	1	1	1	1	1	1	1	1	1		x^2+y^2, z^2
A_2	1	1	1	1	1	1	1	−1	−1	R_z	
B_1	1	−1	1	−1	1	−1	1	1	−1		
B_2	1	−1	1	−1	1	−1	1	−1	1	z, T_z	
E_1	2	$\sqrt{3}$	1	0	−1	$-\sqrt{3}$	−2	0	0	$(x, y)(T_x, T_y)$	
E_2	2	1	−1	−2	−1	1	2	0	0		(x^2-y^2, xy)
E_3	2	0	−2	0	2	0	−2	0	0		
E_4	2	−1	−1	2	−1	−1	2	0	0		
E_5	2	$-\sqrt{3}$	1	0	−1	$\sqrt{3}$	−2	0	0	(R_x, R_y)	(xz, yz)

S_4	E	S_4	C_2	S_4^3	Basis vectors and components	Basis functions
A	1	1	1	1	R_z	x^2+y^2, z^2
B	1	−1	1	−1	z, T_z	x^2-y^2, xy
E	$\begin{cases}1 \\ 1\end{cases}$	$\begin{matrix}i \\ -i\end{matrix}$	$\begin{matrix}-1 \\ -1\end{matrix}$	$\begin{matrix}-i \\ i\end{matrix}$	$(x, y), (T_x, T_y)$ (R_x, R_y)	(xz, yz)

S_6	E	C_3	C_3^2	i	S_6^5	S_6	Basis vectors and components	$\epsilon = \exp(2\pi i/3)$ basis functions
A_g	1	1	1	1	1	1	R_z	x^2+y^2, z^2
E_g	$\begin{cases}1 \\ 1\end{cases}$	$\begin{matrix}\epsilon \\ \epsilon^*\end{matrix}$	$\begin{matrix}\epsilon^* \\ \epsilon\end{matrix}$	$\begin{matrix}1 \\ 1\end{matrix}$	$\begin{matrix}\epsilon \\ \epsilon^*\end{matrix}$	$\begin{matrix}\epsilon^* \\ \epsilon\end{matrix}$	(R_x, R_y)	(x^2-y^2, xy) (xz, yz)
A_u	1	1	1	−1	−1	−1	z, T_z	
E_u	$\begin{cases}1 \\ 1\end{cases}$	$\begin{matrix}\epsilon \\ \epsilon^*\end{matrix}$	$\begin{matrix}\epsilon^* \\ \epsilon\end{matrix}$	$\begin{matrix}-1 \\ -1\end{matrix}$	$\begin{matrix}-\epsilon \\ -\epsilon^*\end{matrix}$	$\begin{matrix}-\epsilon^* \\ -\epsilon\end{matrix}$	$(x, y), (T_x, T_y)$	

9. *The cubic groups*

S_8	E	S_8	C_4	S_8^3	C_2	S_8^5	C_4^3	S_8^7	Basis vectors and components	$\epsilon = \exp(2\pi i/8)$ basis functions
A	1	1	1	1	1	1	1	1	R_z	x^2+y^2, z^2

S_8	E	S_8	C_4	S_8^3	C_2	S_8^5	C_4^3	S_8^7	Basis vectors and components	$\epsilon = \exp(2\pi i/8)$ basis functions
B	1	−1	1	−1	1	−1	1	−1	z, T_z	
E_1	1	ϵ	i	$-\epsilon^*$	−1	$-\epsilon$	$-i$	ϵ^*	$(x, y)(T_x, T_y);$	
	1	ϵ^*	$-i$	$-\epsilon$	−1	$-\epsilon^*$	i	ϵ	(R_x, R_y)	
E_2	1	i	−1	$-i$	1	i	−1	$-i$		$(x^2 - y^2, xy)$
	1	$-i$	−1	i	1	$-i$	−1	i		
E_3	1	$-\epsilon^*$	$-i$	ϵ	−1	ϵ^*	i	$-\epsilon$		(xz, yz)
	1	$-\epsilon$	i	ϵ^*	−1	ϵ	$-i$	$-\epsilon^*$		

Cubic groups

T	E	$4C_3$	$4C_3^2$	$3C_2$		$\epsilon = \exp(2\pi i/3)$
A	1	1	1	1		$x^2 + y^2 + z^2$
E	1	ϵ	ϵ^*	1		$(2z^2 - x^2 - y^2,$
	1	ϵ^*	ϵ	1		$x^2 - y^2)$
T	3	0	0	−1	$(R_x, R_y, R_z); (x, y, z)$	(xy, xz, yz)

T_h	E	$4C_3$	$4C_3^2$	$3C_4$	i	$4S_6$	$4S_6^5$	$3\sigma_h$		$(\epsilon = \exp(2\pi i/3))$
A_g	1	1	1	1	1	1	1	1		$x^2 + y^2 + z^2$
A_u	1	1	1	1	−1	−1	−1	−1		
E_g	1	ϵ	ϵ^*	1	1	ϵ	ϵ^*	1		$(2z^2 - z^2 - y^2,$
	1	ϵ^*	ϵ	1	1	ϵ^*	ϵ	1		$x^2 - y^2)$
E_u	1	ϵ	ϵ^*	1	$-\epsilon^*$	−1	$-\epsilon^*$	−1		
	1	ϵ^*	ϵ	1	−1	$-\epsilon^*$	$-\epsilon$	−1		
T_g	3	0	0	−1	−3	0	0	−1	(R_z, R_y, R_z)	(xz, yz, xy)
T_u	3	0	0	−1	−3	0	0	1	(x, y, z)	

T_d	E	$8C_3$	$3C_2$	$6S_4$	$3\sigma_h$	Basis vectors and components	Basis function
A_1	1	1	1	1	1		$x^2 + y^2 + z^2$
A_2	1	1	1	−1	−1		
E	2	−1	2	0	0		$(2z^2 - x^2 - y^2, x^2 - y^2)$
T_1	3	0	−1	1	−1	$(R_z, R_g, R_g); (x, y, z) (T_x, T_y, T_z)$	
T_2	3	0	−1	−1	1		(xy, xy, yz)

O	E	$8C_3$	$3C_2 (= C_4^2)$	$6C_4$	$6C_2$		
A_1	1	1	1	1	1		$x^2 + y^2 + z^2$
A_2	1	1	1	−1	−1		
E	2	−1	2	0	0		$(2z^2 - x^2 - y^2, x^2 - y^2)$
T_1	3	0	−1	1	−1	$(R_z, R_y, R_z); (x, y, z)$	
T_2	3	0	−1	−1	1		(xy, xy, yz)

O_h	E	$8C_2$	$6C_2$	$6C_4$	$3C_2$ $(=C_4^2)$	i	$6C_4$	$8C_6$	$3\sigma_h$	$6\sigma_d$	Basis vectors and components	Basis function
A_{1g}	1	1	1	1	1	1	1	1	1	1		$(z^2+y^2+z^2)$
A_{2g}	1	1	-1	-1	1	1	-1	1	1	-1		
E_g	2	-1	0	0	2	2	0	-1	2	0		$(2x^2-x^2-y^2, x^2-y^2)$
T_{1g}	3	0	-1	1	-1	3	1	0	-1	-1		
T_{2g}	3	0	1	-1	-1	3	-1	0	-1	1	(R_z, R_y, R_z)	(xz, yz, xy)
A_{1u}	1	1	1	1	1	-1	-1	-1	-1	-1		
A_{2u}	1	1	-1	-1	1	-1	1	-1	-1	1		
E_u	2	-1	0	0	2	-2	0	1	-2	0		
T_{1u}	3	0	-1	1	-1	-3	-1	0	1	1	(x, y, z) (T_z, T_y, T_z)	
T_{2u}	3	0	1	-1	-1	-3	1	0	1	-1		

$C_{\alpha v}$	E	$2C_x(\phi)$	...	$\alpha\sigma_v$	Basis vectors and components	Basis functions
$A_1 \equiv \Sigma^+$	1	1	...	1	z, T_z	x^2+y^2, z^2
$E_1 \equiv \Sigma^-$	1	1	...	-1	R_z	
$E_2 \equiv \Pi$	2	$2\cos\theta$	...	0	$(x, y)\,(T_x, T_y)\,(R_z, R_y)$	(xz, yz)
$E_2 \equiv \Delta$	2	$2\cos 2\theta$	...	0		(x^2-y^2, xy)
$E_2 \equiv \phi$	2	$2\cos 3\theta$	...	0		
...	...	...	...	...		

$D_{\alpha h}$	E	$2C_\alpha(\phi)$	...	$\alpha\sigma_v$	i	$2S_\alpha(\phi)$	...	αC_2	Basis vectors and components	Basis functions
Σ_h^+	1	1	...	1	1	1	...	1	R_z	x^2+y^2, z^2
Σ_h^-	1	1	...	-1	1	1	...	-1	(R_z, R_y)	
Π_g	2	$2\cos\phi$	...	0	2	$-2\cos\phi$	...	0		(xz, yz)
Δ_g	2	$2\cos 2\phi$	...	0	2	$2\cos 2\phi$	...	0		(x^2-y^2, xy)
...	...	...	...	...	...	...	...	...	...	
Σ_h^+	1	1	...	1	-1	-1	...	-1	z, T_z	
Σ_h^-	1	1	...	-1	-1	-1	...	1		
Π_u	2	$2\cos\phi$	...	0	-2	$2\cos 2\phi$	...	0	$(x, y)\,(T_x, T_y)$	
Δ_u	2	$2\cos 2\phi$	...	0	-2	$-2\cos 2\phi$	...	0		
...	...	...	...	...	...	...	...	...		

Correlation Tables—O_h and T_d Groups

These tables show correlations of the octahedral and tetrahedral groups with their subgroups which occur commonly in the study of transition metal complexes.

O_h	O	T_d	D_{4h}	D_{2d}	C_{4v}	C_{2v}	D_{3d}	D_3	C_{2y}
A_{1g}	A_1	A_1	A_{1g}	A_1	A_1	A_1	A_{1g}	A_1	A_g
A_{2g}	A_2	A_2	B_{1g}	B_1	B_1	A_2	A_{2g}	A_2	B_g
E_g	E	E	$A_{1g}+B_{1g}$	A_1+B_1	A_1+B_2	A_1+A_2	E_g	E	A_g+B_g
T_{1g}	T_1	T_1	$A_{2g}+E_g$	A_2+E	A_2+E	$A_2+B_1+B_2$	$A_{2g}+E_g$	A_2+E	A_g+2B_g
T_{2g}	T_2	T_2	$B_{2g}+E_g$	B_2+E	B_2+E	$A_1+B_1+B_2$	$A_{1g}+E_g$	A_1+E	$2A_g+B_g$
A_{1u}	A_1	A_2	A_{1u}	B_1	A_2	A_2	A_{1u}	A_1	A_u
A_{2u}	A_2	A_1	B_{1u}	A_1	B_2	A_1	A_{2u}	A_2	B_u
E_u	E	E	$A_{1u}+B_{1u}$	A_1+B_1	A_2+B_2	A_1+A_2	E_u	E	A_u+B_u
T_{1u}	T_1	T_2	$A_{2u}+E_u$	B_2+E	A_1+E	$A_1+B_1+B_2$	$A_{1u}+E_u$	A_2+E	A_u+2B_u
T_{2u}	T_2	T_1	$B_{2u}+E_u$	A_2+E	B_1+E	$A_2+B_1+B_2$	$A_{1u}+E_u$	A_1+E	$2A_u+B_u$

T_d	T	C_{3v}	C_{2v}	D_{2d}
A_1	A	A_1	A_1	A_1
A_2	A	A_2	A_2	B_1
E	E	E	A_1+A_2	A_1+B_1
T_1	T	A_2+E	$A_2+B_1+B_2$	A_2+E
T_2	T	A_1+E	$A_1+B_1+B_2$	B_2+E

K	O	D_4	D_3
S	A_1	A_1	A_1
P	T_1	A_2+E	A_2+E
D	$E+T_2$	$A_1+B_1+B_2+E$	A_1+2E
E	$A_2+T_1+T_2$	$A_2+B_1+B_2+2E$	A_1+2A_2+2E
G	$A_1+E+T_1+T_2$	$2A_1+A_2+B_1+2E$	$2A_1+A_2+3E$
H	$E+2T_1+T_2$	$A_1+2A_2+B_1+B_2+3E$	A_1+2A_2+4E

Correlation Table for O_h Groups

This table shows how the representations of group O_h are changed or decomposed into those of its subgroups when the symmetry is altered or lowered. This table covers only representations of use, in dealing with the more common symmetries of complexes.

O_h	O	T_d	D_{4h}	D_{2d}	C_{4v}	C_{2y}	D_{3d}	D_3	C_{2y}
A_{1g}	A_1	A_1	A_{1g}	A_1	A_1	A_1	A_{1g}	A_1	A_g
A_{2g}	A_2	A_2	B_{1g}	B_1	B_1	A_2	A_{2g}	A_2	B_g
E_g	E	E	$A_{1g} + B_{1g}$	$A_1 + B_1$	$A_1 + B_2$	$A_1 + A_2$	E_g	E	$A_g + B_g$
T_{1g}	T_1	T_1	$A_{2g} + E_g$	$A_2 + E$	$A_2 + E$	$A_2 + B_1 + B_2$	$A_{2g} + E_g$	$A_2 + E$	$A_g + 2B_g$
T_{2g}	T_2	T_2	$B_{2g} + E_y$	$B_2 + E$	$B_2 + E$	$A_1 + B_1 + B_2$	$A_{1g} + E_g$	$A_1 + E$	$2A_g + B_g$
A_{1u}	A_1	A_2	A_{1u}	B_1	A_2	A_2	A_{1u}	A_1	A_u
A_{2u}	A_2	A_1	B_{1u}	A_1	B_2	A_1	A_{2u}	A_2	B_u
E_u	E	E	$A_{1u} + B_{1u}$	$A_1 + B_1$	$A_2 + B_2$	$A_1 + A_2$	E_u	E	$A_u + B_u$
T_{1u}	T_1	T_2	$A_{2u} + E_{1u}$	$B_2 + E$	$A_1 + E$	$A_1 + B_1 + B_2$	$A_{2u} + E_u$	$A_2 + E$	$A_u + 2B_u$
T_{2u}	T_2	T_1	$B_{2u} + E_u$	$A_2 + E$	$B_1 + E$	$A_2 + B_1 + B_2$	$A_{1u} + E_u$	$A_1 + E$	$2A_u + B_u$

Correlation Tables for Some Selected Point Groups and their Important Subgroups

For some selected point groups and their important subgroups.

C_{4v}	C_4	$C_{2v}(\sigma_v)$	$C_{2v}(\sigma'_v)$	$C_s(\sigma_v)$	$C_s(\sigma'_v)$
A_1	A	A_1	A_1	A	A'
A_2	A	A_2	A_2	A''	A''
B_1	B	A_1	A_2	A'	A''
B_2	B	A_2	A_1	A'	A'
E	E	$B_1 + B_2$	$B_1 + B_2$	$A' + A'$	$A' + A'$

D_{2v}	D_2	$C_{2v}(\sigma_{(z)})$	$C_{2h}(C_{2(z)})$	$C_z(\sigma_{xy})$
A_g	A	A_1	A_g	A'
B_{1g}	B_1	A_2	A_g	A'
B_{2g}	B_2	B_1	B_g	A''
B_{3g}	B_3	B_2	B_g	A''
A_u	A	A_2	A_u	A''
B_{1u}	B_1	A_1	A_u	A''
B_{2u}	B_2	B_2	B_u	A'
B_{3u}	B_3	B_1	B_u	A'

D_{3h}	D_3	D_{3v}	$C_{3h}(\sigma_{zy})$	C_{2v}	C_s
A'_1	A_1	A_1	A'	A_1	A'
A'_2	A_2	A_2	A'	B_2	A'
E'	E	E	E'	$A_1 + B_2$	$2A'$
A''_1	A_1	A_2	A''	A_2	A''
A''_2	A_2	A_1	A''	B_1	A''
E''	E	E	E''	$A_2 + B_1$	$2A''$

D_{4h}	D_4	D_{2d}	$D_{2h}(C'_2)$	C_{4v}	C_{4h}	$C_{2v}(C_2, \sigma_v)$	$C_2(C_2)$
A_{1g}	A_1	A_1	A_g	A_1	A_g	A_1	A_g
A_{2g}	A_2	A_2	B_{1g}	A_2	A_g	A_2	A_g
B_{1g}	B_1	B_1	A_g	B_1	B_g	A_1	A_g
B_{2g}	B_2	B_2	B_{1g}	B_2	B_g	A_2	A_g
E_g	E	E	$B_{2g} + B_{3g}$	E	E_g	$B_1 + B_2$	$2B_g$
A_{1u}	A_1	B_1	A_u	A_2	A_u	A_2	A_u
A_{2u}	A_2	B_2	B_{1u}	A_1	A_u	A_1	A_u
B_{1u}	B_1	A_1	A_u	B_2	B_u	A_2	A_u
B_{2u}	B_2	A_2	B_{1u}	B_1	B_u	A_1	A_u
B_u	E	E	$B_{2v} + B_{3v}$	E	E_u	$B_1 + B_2$	$2B_v$

D_{2d}	D_2	C_{2v}	C_S	C_2
A_1	A	A_1	A'	A
A_2	B_1	A_2	A''	A
B_1	A	A_2	A''	A
B_2	B_1	A_1	A'	A
E	$B_2 + B_3$	$B_1 + B_2$	$A' + A''$	$2B$

T_d	D_{2d}	D_2	C_{3v}	C_{2v}
A_1	A_1	A	A_1	A_1
A_2	B_1	A	A_2	A_2
E	$A_1 + B_1$	$2A$	E	$A_1 + A_2$
T_1	$A_2 + E$	$B_1 + B_2 + B_1$	$A_2 + E$	$A_2 + B_1 + B_2$
T_2	$B_2 + E$	$B_1 + B_2 + B_3$	$A_1 + E$	$A_1 + B_1 + B_2$

Appendix V

Character Tables of Two-Valued Representation of Point Groups

D_2'		E	R	$C_2(z)$ $RC_2(z)$	$C_2(x)$ $RC_2(x)$	$C_2(x)$ $RC_2(x)$
Γ'_1	A'_1	1	1	1	1	1
Γ'_2	B'_1	1	1	1	-1	-1
Γ'_3	B'_2	1	1	-1	1	-1
Γ'_4	B'_3	1	1	-1	1	1
Γ'_5	E'	2	-2	0	0	0

D_3'		E	R	C_2^3R C_3	C_3R RC_2^3	$3C_2$	$3C_2R$
Γ'_1	A'_1	1	1	1	1	1	1
Γ'_2	A'_2	1	1	1	-1	-1	-1
Γ'_3	E'_1	2	1	-1	1	-1	0
Γ'_4	E'_2	$\{$ 1	1	-1	1	1	$-i$
		1	-1	-1	1	-1	i
Γ'_5	B'_3	2	-2	1	-1	0	0

D_4'		E	R	C_4^2R C_4	C_4R C_4^3	C_2R C_2	$2C_2'R$ $2C_1$	$2C_2''R$ $2C_2''$
Γ'_1	A'_1	1	1	1	1	1	1	1
Γ'_2	A'_2	1	1	1	1	1	-1	-1
Γ'_3	B'_1	1	1	-1	-1	1	1	-1
Γ'_4	B'_2	1	1	-1	-1	1	-1	1
Γ'_5	E'_1	2	2	0	0	-2	0	0
Γ'_6	E'_2	2	-2	$\sqrt{2}$	$-\sqrt{2}$	0	0	0
Γ'_7	E'_3	2	-2	$-\sqrt{2}$	$\sqrt{2}$	0	0	0

	O'	E	R	4RC$_3^2$ 4C$_3$	4RC$_3$ 4C$_3^2$	3RC$_2$ 3C$_2$	3RC$_4^3$ 3C$_4$	3RC$_4$ 3C$_4^3$	6RC$_2'$ 6C$_2'$
Γ'$_1$	A'$_1$	1	1	1	1	1	1	1	1
Γ'$_2$	A'$_2$	1	1	1	1	1	−1	−1	−1
Γ'$_3$	E'$_1$	2	2	−1	−1	2	0	0	0
Γ'$_4$	T'$_1$	3	3	0	0	−1	1	1	−1
Γ'$_5$	T'$_3$	3	3	0	0	−1	−1	−1	1
Γ'$_6$	E'$_2$	2	−2	1	−1	0	$\sqrt{2}$	0	0
Γ'$_7$	E'$_3$	2	−2	1	−1	0	$\sqrt{2}$	0	0
Γ'$_8$	G'	4	−4	−1	1	0	0	0	0

Appendix VI

Stereographic Projections for the 32 Crystallographic Point Groups*

	Triclinic	Monoclinic (Ist setting)		Tetragonal	
X	1		2		4
$\bar{X}$ (even)	—		m (= $\bar{2}$)		$\bar{4}$
X (even) plus center and $\bar{X}$ (odd)	$\bar{1}$		2/m		4/m
	Monoclinic (2nd setting)		**Orthorhombic**		
X^2		2		222	422
Xm		m		mm^2	4 mm
$\bar{X}^2$ (even) or $\bar{X}m$ (even)	—		—		$\bar{4}2$ m
X^2 or Xm plus center and $\bar{X}$ (odd)		2/m		mmm	4/mmm

* (Reproduced, by permission, from the International Tables for X-Ray Crystallography, Kynock Press, Birmingham)

371

Trigonal	Hexagonal	Cubic	
 3	 6	 23	X
—	 $\bar{6}$	—	$\bar{X}$ (even)
 $\bar{3}$	 6/m	 m^3	X (even) plus center and $\bar{X}$ (odd)
 32	 622	 432	X^2
 3 m	 6 mm	—	Xm
—	 $\bar{6}m2$	 $\bar{4}3m$	$\bar{X}2$ (even) or $\bar{X}m$ (even)
 $\bar{3}m$	 6/mmm	 m^3m	X^2 or Xm plus center and Xm (odd)

Symmetry in Crystals

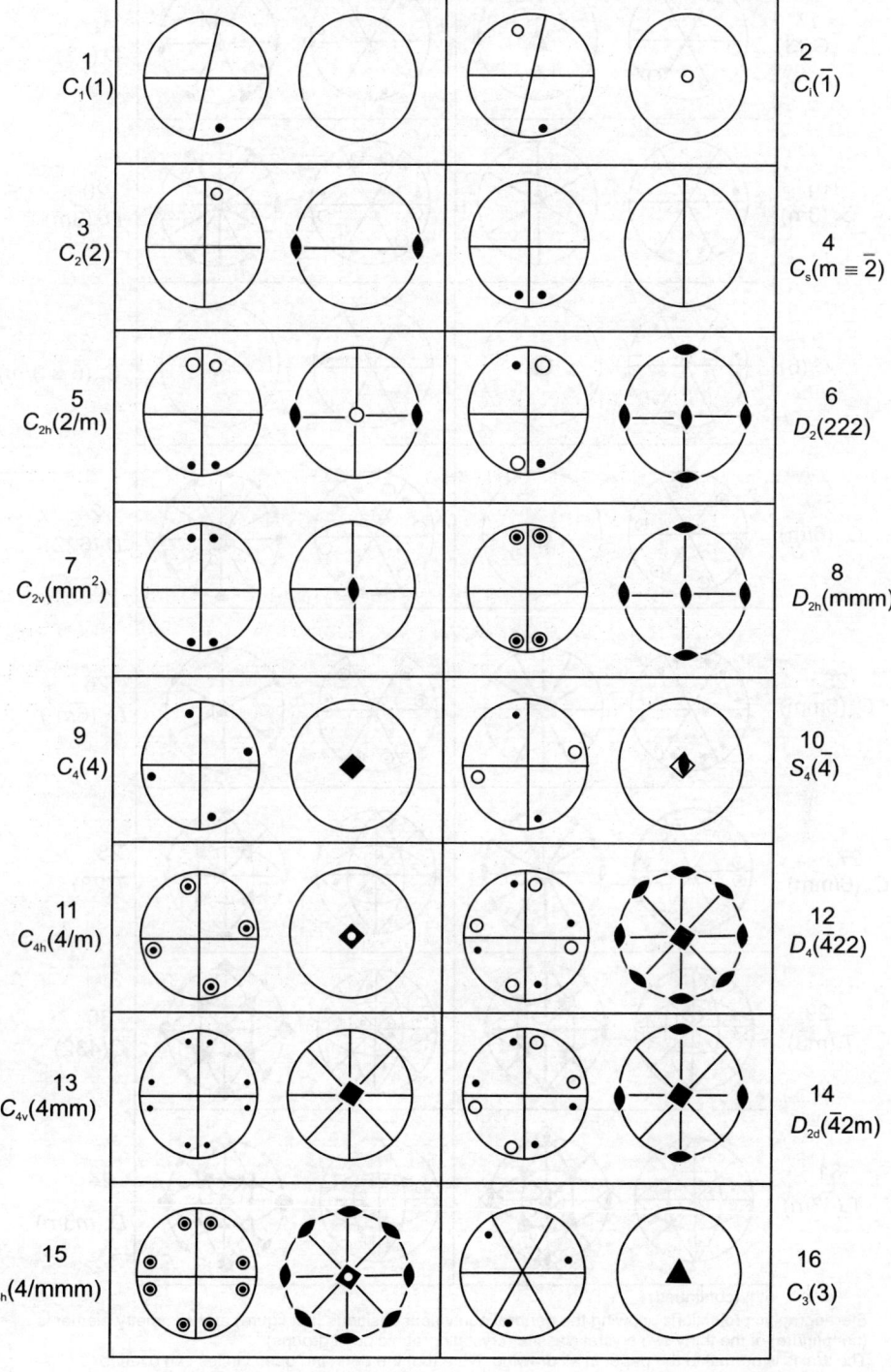

Symmetry in Chemistry

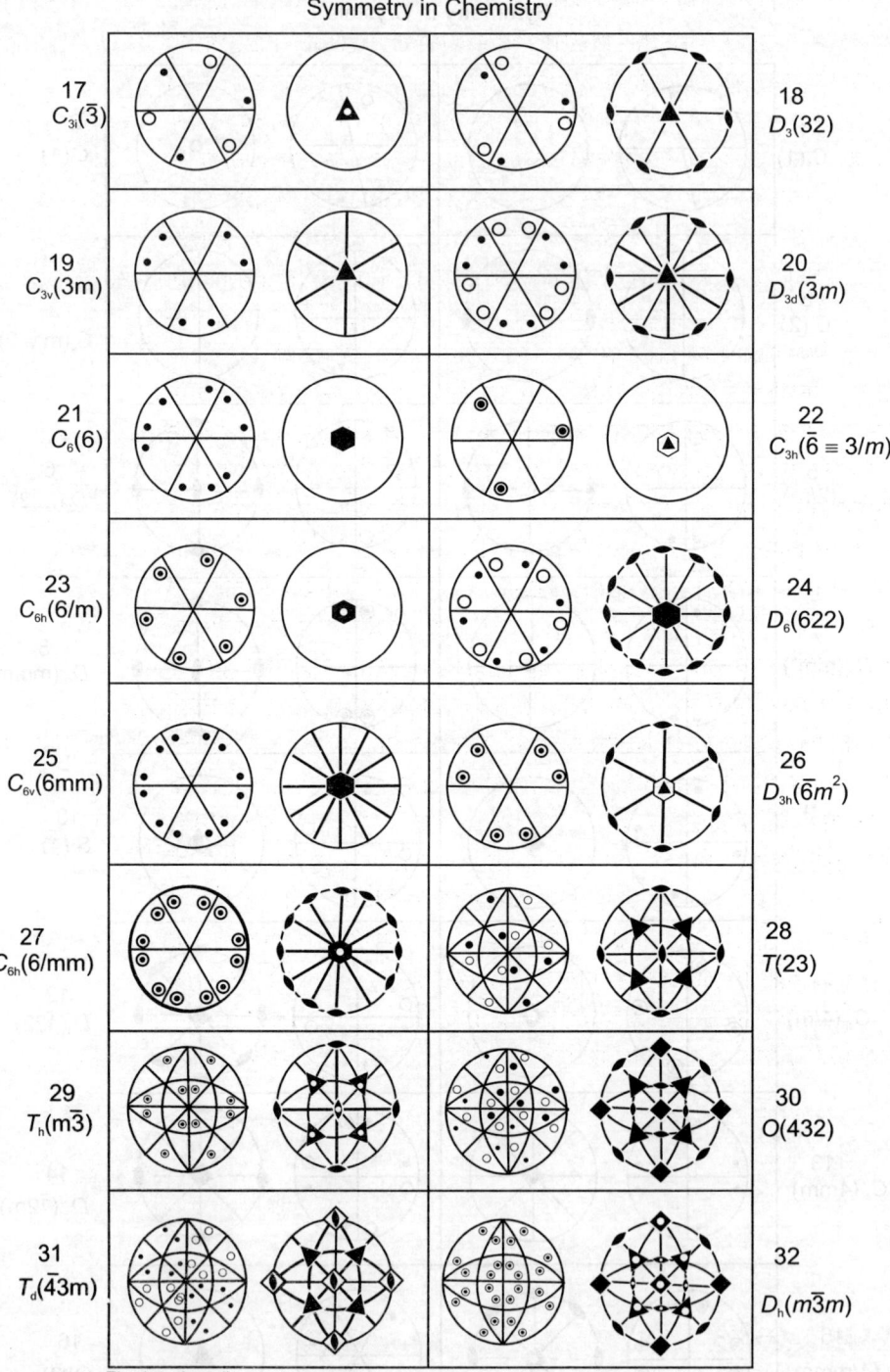

Fig. (continued)

Stereographic projections showing the general equivalent positions (left figure) and symmetry elements (right figure) of the thirty-two crystal classes (crystallographic point groups).

The z-axis is normal to the paper in all drawing. Note that the dots and open circles can overlap, and the thick lines represent mirror planes. For each point group, the Hermann–Mauguin symbol is given in parentheses after the Schönflies symbol.

Infrared and Raman Activity of Overtone and Combination Bands in Selected Point Groups

The infrared (IR) and Raman (R) spectral activity of a fundamental can be inferred from the character table of a molecule. The activity of the combination bands (sum, difference or overtone bands) can be found out from the symmetry of the symmetry of the direct product of the individual components.

I. C_{2v} MOLECULES

(a)   Fundamentals

$$A_1 \text{ (IR, R)}, A_2, \text{ (R)}, B_1 \text{ (IP, R)}, B_2 \text{ (IR, R)}$$

(b)   Combination bands (binary)

$A_1 \times A_1 = A_1$	$A_2 \times A_2 = A_1$	$B_1 \times B_1 = A_1$
$A_1 \times A_2 = A_2 \text{ (R)}$	$A_2 \times B_1 = B_2$	$B_1 \times B_2 = A_2 \text{ (R)}$
$A_1 \times B_1 = B_1$	$A_2 \times B_2 = B_1$	$B_2 \times B_2 = A_1$
$A_1 \times B_2 = B_2$		

All are active in both the IR and R except the A_2 modes which are active only in Raman.

(c)   Overtones

The activity of the product is inferred from that of the fundamental as in (a)

$$A_1^n = A_1$$
$$A_2^n = A_1 \, (n = \text{even}) \text{ or } A_2, \, (n = \text{odd})$$
$$B_2^n = A_1 \, (n = \text{even}) \text{ or } B_1, \, (n = \text{odd})$$
$$B_2^n = A_1 \, (n = \text{even}) \text{ or } B_2, \, (n = \text{odd})$$

II. C_{3v} MOLECULES

(a)   Fundamentals

$$A_1 \text{ (IR, R)}, E \text{ (IR, R)}$$

A_2 is inactive in both the IR and R.

(b)   Combination (binary)

$A_1 \times A_1 = A_1$	$A_2 \times A_2 = A_1$
$*A_1 \times A_2 = A_2$	$A_2 \times E = E$
$A_1 \times E = E$	$E \times E = A_1 + A_2 + E$

---

*inactive.

(c) Overtones

$$A_1^n = A_1; A_2^n = A_1 \text{ or } A_2 \ (n = \text{even or odd}); E_2 = A_1 + E$$

III. D_{3h} MOLECULES

(a) Fundamentals

$$A'_1(R), A''_2(IR), E'(IR, R), E'(R)$$

(b) Allowed binary combinations

$$A_1' \times A_1' = A_1'(R) \quad A_1'' \times A_1' = A_1'(R) \quad A_2' \times A_2' = A_1'(R) \quad A'_2 \times E'' = E'$$

$$A_1' \times A_2'' = A_2''(IR) \ A_1' \times A_2' = A_2'(IR) \quad A_2' \times E' = E' \quad\quad E' \times E' = A_1' A_2' + E'$$

$$A_1' \times E' = E' \quad\quad A_1' \times E' = E''(R) \quad A_2' \times E'' = E'' (R) \quad E' \times E'' = A_1'' + A_2'' + E'$$

$$A_1' \times E'' = E''(R) \quad A_1' \times E' = E' \quad\quad A_2'' \times A_2'' = A_1' (R) \ E'' \times E'' = A_1' + A_2'' + E'$$

$$A''_2 \times E' = E''(R)$$

Unmarked direct products are both IR active.

(c) Overtones

$$(A_1')^n = A_1 (R) = A_1' \text{ or } A_1'' \ (n = \text{even or odd}).$$

$$(A_2')^n = A_1' \text{ or } A_2' \ (n = \text{even or odd}), (A_2'')^n = A_1' \text{ or } A_1'' \ (n = \text{even or odd}).$$

$$(E')^2 = A_1' + E' \ (E'')^2 = A' + E'$$

Activity is same as that of the fundamental. For all nondegenerate IRs similar procedures are used as those of combination bands.

IV. T_d MOLECULES

(a) Fundamentals

$$A_1(R), E(R), T_2(IR, R).$$

(b) Allowed binary combinations

$$A_1 \times A_1 = A_2 \times A_2 = A_1(R) \quad\quad T_1 \times E = T_2 \times E = T_1 \times T_1 \ (R, IR)$$

$$A_1 \times E = A_2 \times E = E(R) \quad\quad T_1 \times T_2 = A_2 \times E + T_2 \times T_1 \times (R, IR)$$

$$A_1 \times T_2 = A_2 \times T_1 = T_2(R, IR) \quad T_1 \times T_1 = T_2 \times T_2 = A_1 + E + T_1 + T_2 \ (IR, IR)$$

$$E \times E = A_1 \times A_2 + E(R)$$

Through A_2 and T_1 are not allowed, some of their combinations are certainly allowed.

(c) Overtones (allowed)

$$A_1^n = A_2^n \ (n = \text{even}) = A_1(R) \quad E^3 = A_1 + A_2 + E(R)$$

$$E^2 = A_1 + E \ (R) \quad\quad T_1^3 = A_2 + 2T_1 \times T_2(IR, R)$$

$$T_1^2 = T_2^2 = A_1 + E + T_2 \ (IR, R) \quad T_2^3 = A_1 + T_1 + 2T_2 \ (IR, R)$$

For degenerate IRs, the symmetries of overtones are calculated by a different procedure since the direct product gives incorrect results. The formula

$$\chi_n (R) = 1/2 \ [\chi(R)\chi_n (R) + \chi(R^n)]$$

can be used where $\chi_n (R)$ is the character for any operation R, $\chi(R^n)$ is the character for the operation R carried out n times, here n is the quantum number ($n = 1$ for fundamental, $n = 2$ for the first overtone, etc.). The resulting representation can be then reduced by the standard reduction formula.

The 230 Space Groups

Crystal System	Crystal Class	Space Groups				
Triclinic	1	P1				
	$\bar{1}$	P$\bar{1}$				
Monoclinic	2	P2	P2$_1$	A2(C2)		
	m	Pm	Pa(Pc)	Am(Cm)	Aa(Cc)	
	2/m	P2/m	P2$_1$/m	A2/m(C2/m)	P2/a(P2/c)	P21/a(P2$_1$/c)	A2/a(C2/c)
orthorhombic	222	P222	P222$_1$	P2$_1$2$_1$2$_1$	P2$_1$2$_1$2$_1$	C222$_1$	C222
		F222	I222	I$_1$2$_1$2$_1$2			
	mm2	Pmm2	Pmc2$_1$	Pcc2	Pma2	Pca2$_1$	Pnc2
		Pmn2$_1$	Pba2	Pna2$_1$	Pnn2	Cmm2	Cmc2$_1$
		Ccc2	Amm2	Abm2	Ama2	Aba2	Fmm2
		Fdd2	Imm2	Iba2	Ima2		
	mmm	Pmmm	Pnnn	Pccm	Pban	Pmma	Pnna
		Pmna	Pcca	Pbam	Pccn	Pbcm	Pnnm
		Pmmn	Pbcn	Pbca	Pnma	Cmcm	Cmca
		Cmmm	Cccm	Cmma	Ccca	Fmmm	Fddd
		Immm	Ibam	Ibca	Imma		
Tetragonal	4	P4	P4$_1$	P4$_2$	P4$_3$	I4	I4$_1$
	$\bar{4}$	P$\bar{4}$	I$\bar{4}$				
	4/m	P4/m	P4$_2$/m	P4/n	P4$_2$/n	I4/m	I4$_1$/a
	422	P422	P42$_1$2	P4$_1$22	P4$_2$2$_1$2	P42$_2$22	P4$_2$2$_1$2
		P4$_3$22	P4$_3$2$_1$2	I422	I4$_1$22		
	4mm	P4mm	P4bm	P4$_2$cm	P4$_2$nm	P4cc	P4nc
		P42mc	P4$_2$bc	I4mm	I4cm	I4$_1$md	I4$_1$cd
	$\bar{4}$2m	P$\bar{4}$2m	P$\bar{4}$2c	P$\bar{4}$2$_1$m	P$\bar{4}$2$_1$c	P$\bar{4}$m2	P$\bar{4}$c2
		P$\bar{4}$b2	P$\bar{4}$n2	I$\bar{4}$m2	I$\bar{4}$c2	I$\bar{4}$2m	I$\bar{4}$2d
	4/mmm	P4/mmm	P4/mcc	P4/nbm	P4/nnc	P4/mbm	P4/mnc
		P4/nmm	P4/ncc	P4$_2$/mmc	P4$_2$mcm	P4$_2$/nbc	P4$_2$/nnm

		P4$_2$/mbc I4$_1$/amd	P4$_2$/mn I4$_1$/acd	P4$_2$/nmc	P4$_2$/ncm	I4/mmm	I4/mcm
	3	P3	P3$_1$	P3$_2$	R3		
	$\bar{3}$	P$\bar{3}$	R$\bar{3}$				
	32	P312 R32	P321	P3$_1$12	P3$_1$21	P3$_2$12	P3$_2$21
	3m	P3m1	P31m	P3c1	P31c	R3m	R3c
	$\bar{3}$m	P$\bar{3}$1m	P$\bar{3}$1c	P$\bar{3}$m1	P$\bar{3}$c1	R$\bar{3}$m	R$\bar{3}$c
Trigonal hexagonal	6	P6	P6$_1$	P6$_2$	P6$_2$	P6$_4$	P6$_3$
	$\bar{6}$	P$\bar{6}$					
	6/m	P6/m	P6$_3$/m				
	622	P622	P6$_1$22	P6$_2$22	P6$_4$22	P6$_3$22	
	6mm	P6mm	P6cc	P6$_3$cm	P6$_2$mc		
	$\bar{6}$m2	P$\bar{6}$m2	P$\bar{6}$c2	P$\bar{6}$2m	P$\bar{6}$2c		
	6mmm	P6/mmm	P6/mcc	P6$_3$/mcc	P6$_3$/mcc		
	23	P23	F23	I23	F2,3	I2$_1$3	
	m3	Pm3 Ia3	Pn3	Fm3	Fd3	Im3	Pa3
	432	P432 P4$_1$32	P4$_2$32 I4$_1$32	F432	F4$_1$32	I432	P4$_3$32
	$\bar{4}$3m	P$\bar{4}$3m	F$\bar{4}$3m	I$\bar{4}$3m	P$\bar{4}$3n	P$\bar{4}$3c	I$\bar{4}$3d
	m3m	Pm3m FI3m	Pn3n Fd3c	Pm3n Im3m	Pn3m Ia3d	Fm3m	Fm3c

The Shapes of *f*-Orbital

When the wave equation for a hydrogen-like atom is solved in the most direct way for orbitals with the angular momentum quantum number $l = 3$, the following results are obtained for the purely angular parts (i.e. omitting all numerical factors):

ψ_0: $(5 \cos^3 \theta - 3 \cos \theta)$ $\qquad m_l = 0$

$\psi_{\pm1}$: $\sin \theta (5 \cos^2 \theta - 1)^{e \pm i\phi}$ $\qquad m_l = \pm 1$

$\psi_{\pm2}$: $(\sin^2 \theta \cos^2 \theta)^{e \pm 2i\phi}$ $\qquad m_l = \pm 2$

$\psi_{\pm2}$: $(\sin^3 \theta)^{e \pm 3i\phi}$ $\qquad m_l = \pm 3$

The seven functions are grouped into sets having projections of the orbital angular momentum on the z-axis of $0, \pm 1, \pm 2$, and ± 3. Each of the functions in the pairs with m_l equal to $\pm1, \pm 2$, and ± 3 is complex as written above, but by taking linear combinations of each pair, for example

$$\frac{1}{\sqrt{2}} (\psi_{+3} + \psi_{-3}) \text{ and } \frac{1}{i\sqrt{2}} (\psi_{+3} - \psi_{-3})$$

the imaginary parts are eliminated. In this way, the seven real, normalized (to unity) orbitals listed in Table A IX.1 are obtained. As shown in the table, the *f* orbitals are in a convenient form for problems involving only a single high-order symmetry axis. For instance, in treating bis (cyclo-octatetraene) metal compounds, where the point group is D_{8h}, we note that the orbitals are already grouped into sets belonging to the representations A_{2u}, E_{1u}, E_{2u}, and E_{3u}.

Table A IX.1

True polynomial	Simplified polynomial	Normalizing factor	Angular function
$z(5z^2 - 3r^2)$	z^3	$\dfrac{\sqrt{7/\pi}}{4}$	$5 \cos^3 \theta - 3 \cos \theta$
$x(5z^2 - r^2)$	xz^3	$\dfrac{\sqrt{42/\pi}}{8}$	$\sin \theta (5 \cos^2 \theta - 1) \cos \phi$
$y(5z^2 - r^2)$	yz^3	$\dfrac{\sqrt{42/\pi}}{8}$	$\sin \theta (5 \cos^2 \theta - 1) \sin \phi$
$z(xy)$		$\dfrac{\sqrt{105/\pi}}{4}$	$\sin^2 \theta, \cos \theta, \sin 2\phi$

$z(x^2 - y^2)$	$\dfrac{\sqrt{105/\pi}}{4}$	$\sin^2 \theta \cos \theta \cos 2\phi$
$x(x^2 - 3y^2)$	$\dfrac{\sqrt{70/\pi}}{8}$	$\sin^3 \theta \cos 3\phi$
$y(3x^2 - y^2)$	$\dfrac{\sqrt{70/\pi}}{8}$	$\sin^3 \theta \sin 3\phi$

However, for problems involving cubic symmetry, the functions given in Table AIX. 1 are awkward to use since they do not directly form triply degenerate sets, despite the fact that the entire set of f functions spans the representations. A_{2u}, T_{1u}, and T_{2u} in the group O_h.

A set of functions directly useful in problems with cubic symmetry can be obtained by taking the following linear combinations of those in Table AIX.1.

A_{2u}: $f_{xyz} = f_{xyz}$ (as before)

$$T_{1u}: \begin{cases} f_{x^3} = -\dfrac{1}{4}\left[\sqrt{6}\, f_{xz^2} - \sqrt{10}\, f_{x(x^2 - 3y^2)} \right] \\[2mm] f_{y^3} = -\dfrac{1}{4}\left[\sqrt{6}\, f_{yz^2} + \sqrt{10}\, f_{y(3x^2 - y^2)} \right] \\[2mm] f_{z^3} = f_{z^3} \text{ (as before)} \end{cases}$$

$$T_{2u}: \begin{cases} f_{z(x^2 - y^2)} = f_{z(x^2 - y^2)} \text{ (as before)} \\[2mm] f_{x(z^2 - y^2)} = \dfrac{1}{4}\left[\sqrt{10}\, f_{xz^2} + \sqrt{6}\, f_{x(x^2 - 3y^2)} \right] \\[2mm] f_{y(z^2 - x^2)} = \dfrac{1}{4}\left[\sqrt{10}\, f_{yz^2} - \sqrt{6}\, f_{y(3x^2 - y^2)} \right] \end{cases}$$

Symmetry Elements and Symmetry Operations Associated with a Square Planer Complex

This, the axis of highest symmetry, is rather complicated because it is the axis of three distinct symmetry operations. The first is the operation of rotation by $90_{\prime\prime}$ either clockwise or anticlockwise, each denoted by C_4. The second is the operation of rotation by $180_{\prime\prime}$ denoted C_2. The third is a composite operation: rotate by $90_{\prime}$ and then reflect in the σ_h mirror plane (see below). Denoted S_4, this operation takes the top of one ligand into the bottom of the next (cf. C_4, which sends top into top).

A pair of equivalent mirror planes, σ_d

A pair of equivalent mirror plane, σ'_d

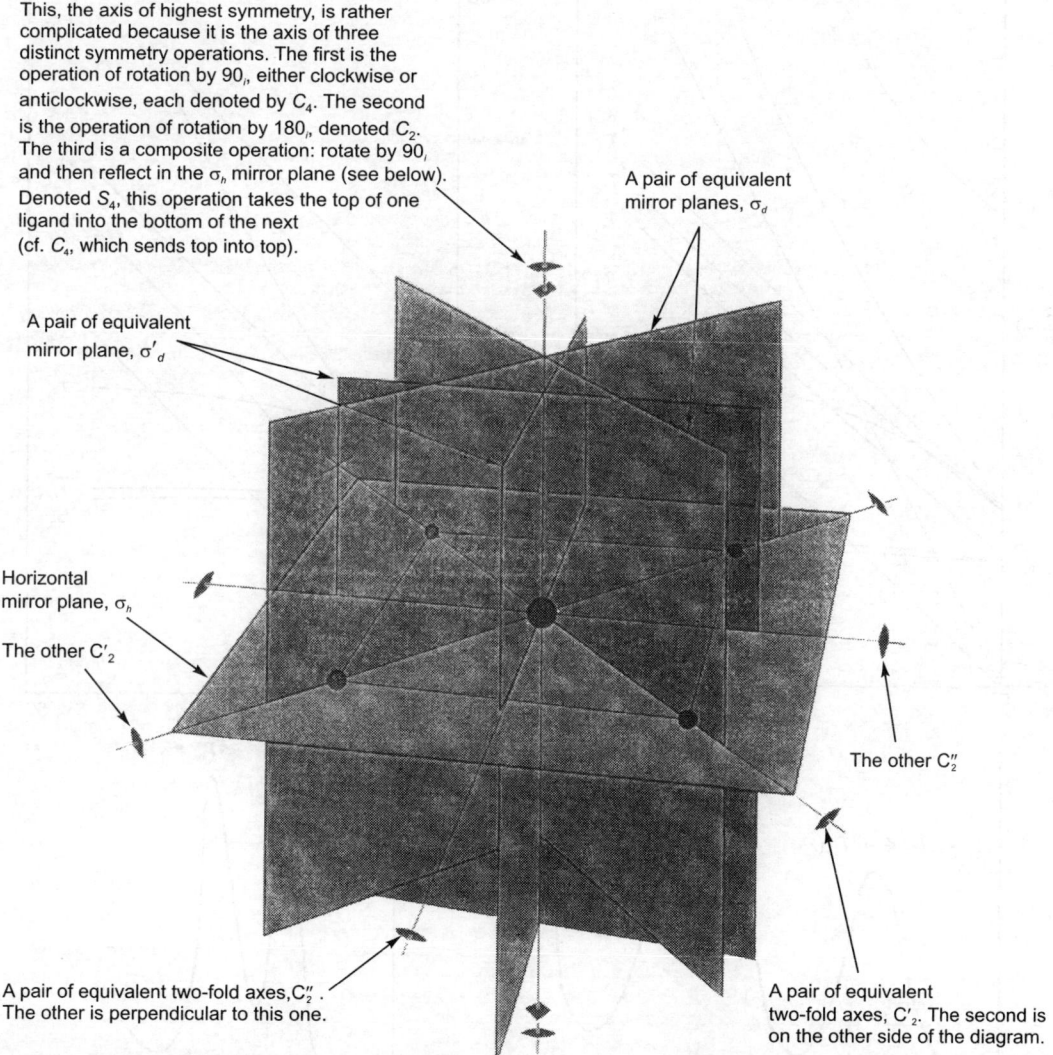

Horizontal mirror plane, σ_h

The other C'_2

The other C''_2

A pair of equivalent two-fold axes,C''_2 . The other is perpendicular to this one.

A pair of equivalent two-fold axes, C'_2. The second is on the other side of the diagram.

Fig. AX.1 Symmetry elements and operations associated with a square planar complex. There are two other symmetry operations, not shown on the diagram. The first, apparently initial operation, is 'leave everything alone' and is denoted by E. The other operation of inversion i in center of symmetry (action is at the central atom), denoted by i

Tanabe–Sugano Diagrams and Some Illustrative Spectra

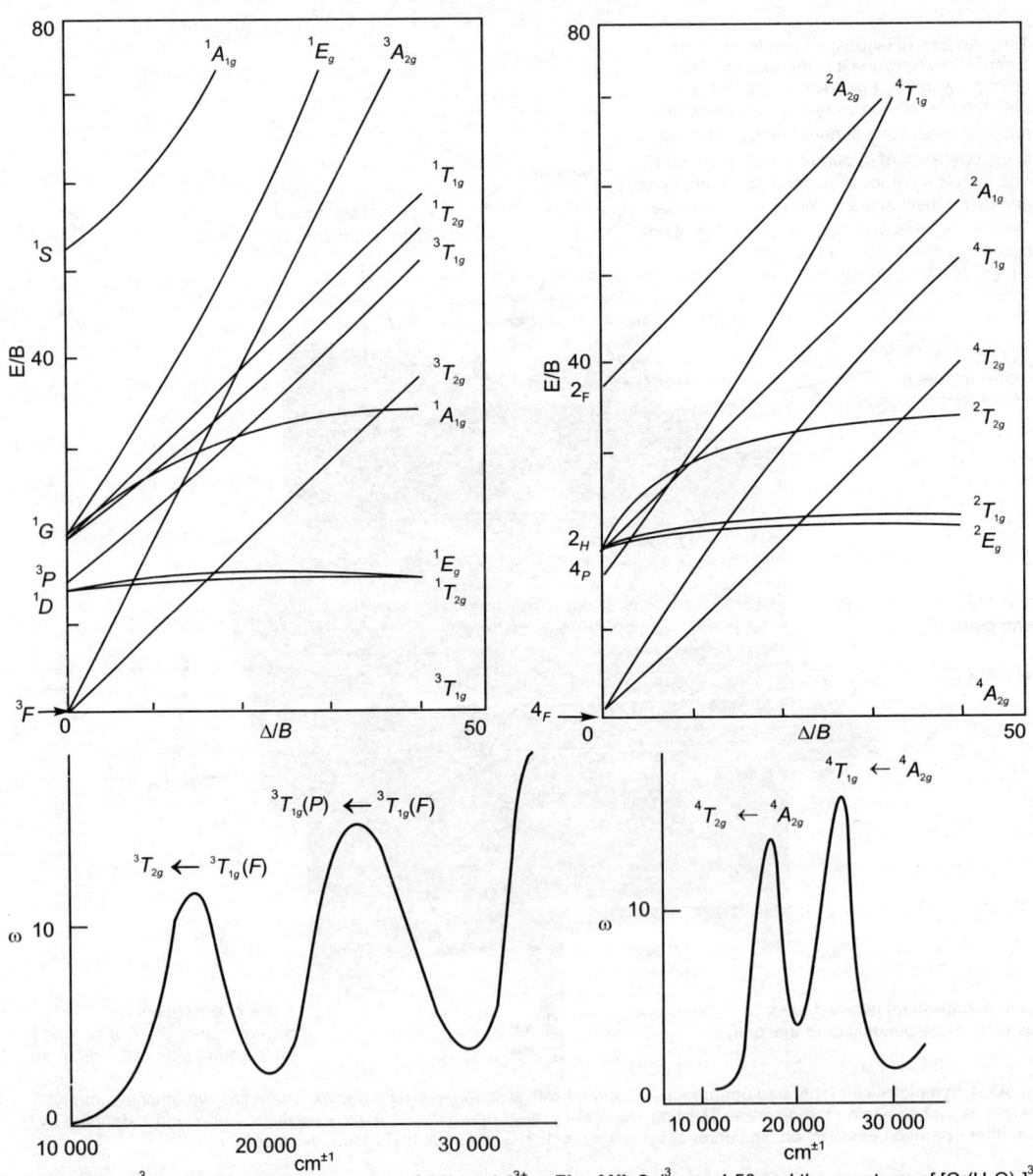

Fig. AXI. 1 d^3; $\gamma = 4.42$ and the spectrum of $[V(urea)_6]^{3+}$ **Fig. AXI. 2** d^3; $\gamma = 4.50$ and the spectrum of $[Cr(H_2O)_6]^{3+}$

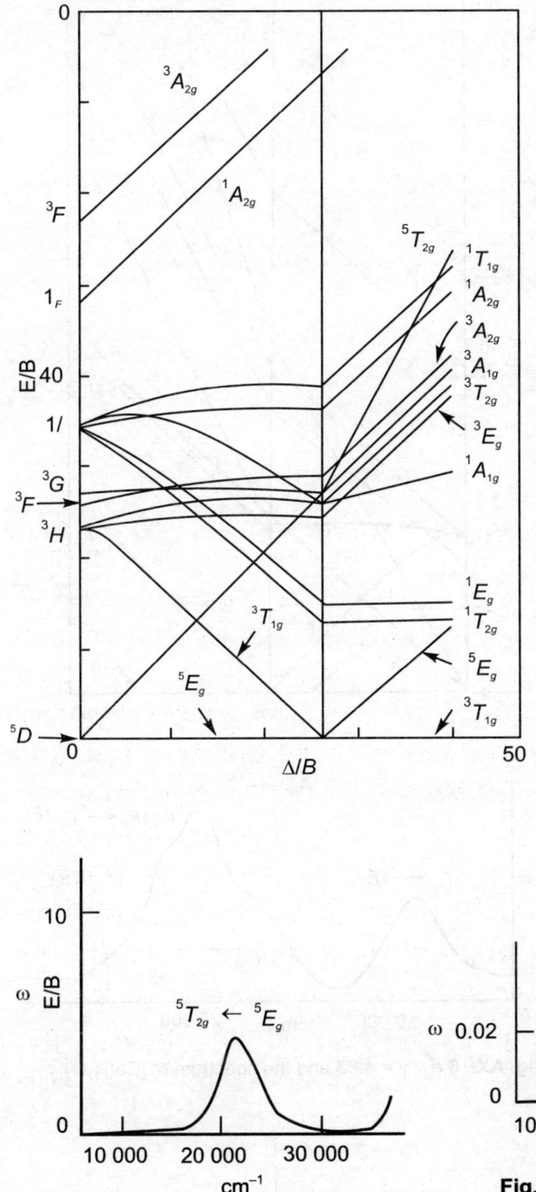

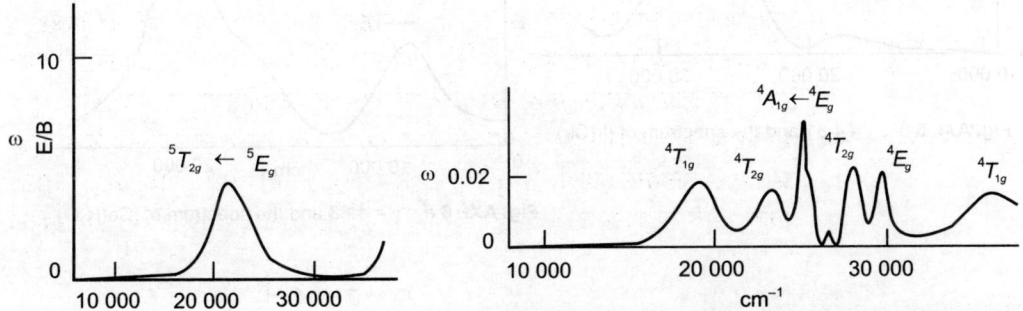

Fig. AXI. 3 d^4 ; γ = 4.61 and the spectrum of [Mn(H$_2$O)$_6$]$^{3+}$

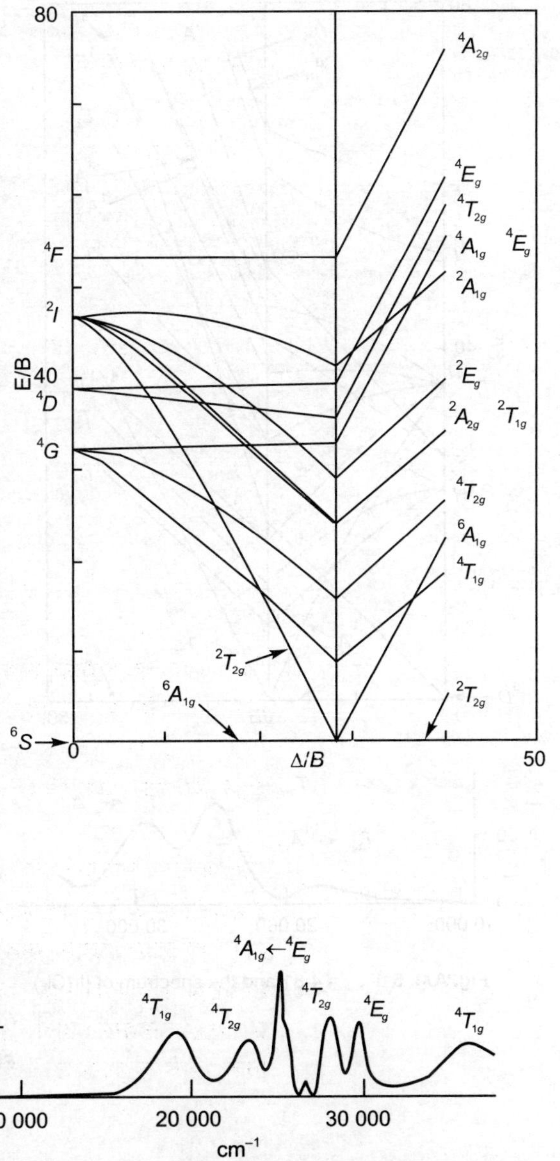

Fig. AXI. 4 d^5 ; γ = 4.48 and the spectrum of [Mn(H$_2$O)$_6$]$^{2+}$ (the ground state for all the transitions in the spectrum is $^6A_{1g}$)

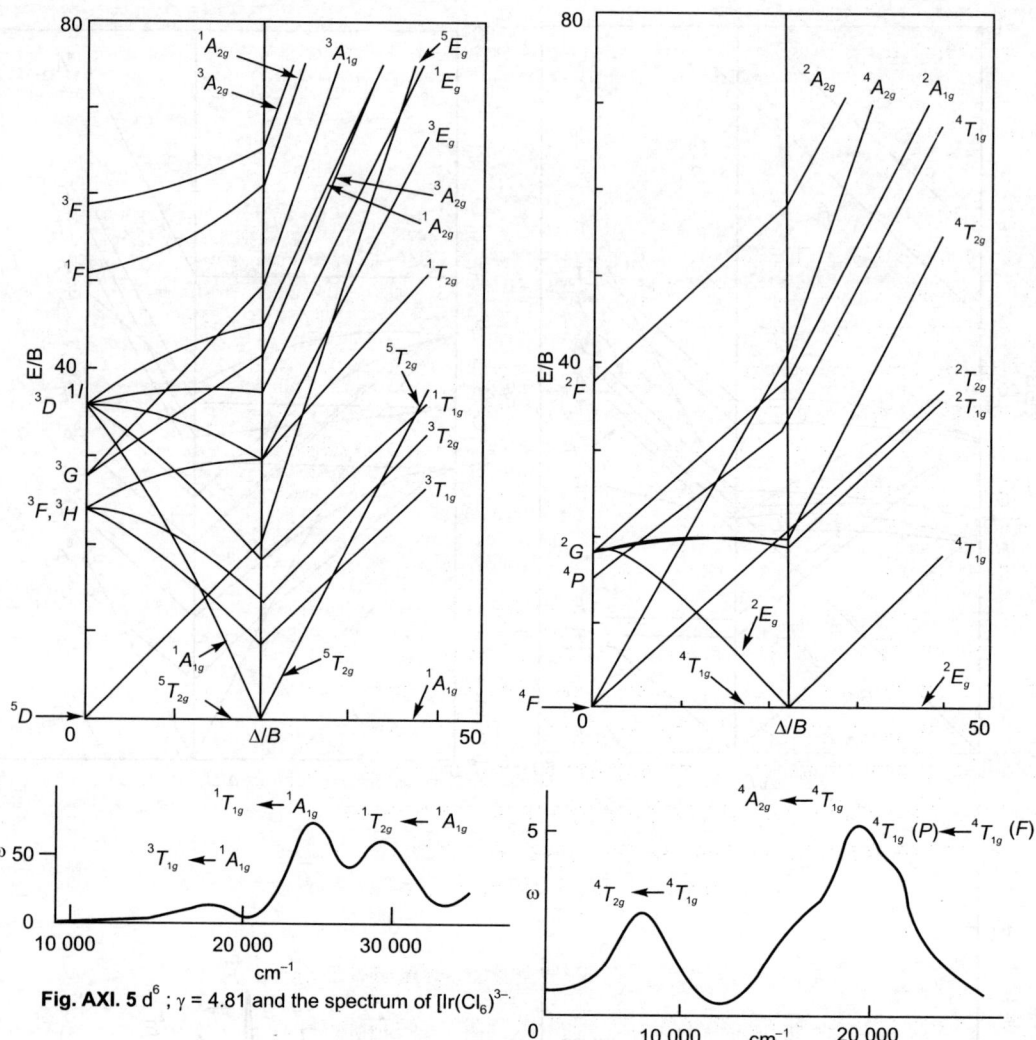

Fig. AXI. 5 d^6 ; $\gamma = 4.81$ and the spectrum of [Ir(Cl$_6$)]$^{3-}$

Fig. AXI. 6 d^7 ; $\gamma = 4.63$ and the spectrum of [Co(H$_2$O)]$^{2+}$

Index

385

READERS GUIDE:

HARRY POTTER AND THE CURSED CHILD –

PARTS I & II

|

Context and Critical Analysis

Authored By

Slim Reads

FREE GIFT SPECIAL REPORT

10 Little-Known Facts Even Potterheads Don't Know

Pop quiz hot shot! You think you know EVERYTHING about the Harry Potter series and its amazing rise in worldwide fandom? THINK AGAIN! I'm sure you know your Dumbledores from your Longbottoms but it is time to push your fandom to the next level (that's right, level 9 and 3/4)!

As our **free gift** for being a **SLIM READS enthusiast** we are happy to give you a special report about the **Little-Known Facts Even Potterheads Don't Know**.

Don't let Voldemort keep you from getting this awesome report!

Get your **free copy** at:

http://sixfigureteen.com/potter